Netty 源码全解与架构思维

黄 俊 著

清华大学出版社
北 京

内 容 简 介

Netty 是目前市面上使用频率较高的网络编程库。它的架构设计非常明确并且层次分明，源码较为易懂，其中包含了 Java NIO（New IO，新 IO）的三大组件：Selector（选择器）、Channel（通道）、ByteBuffer（缓冲区），提供了简单、易用、高效的网络通信库，还实现了自己的内存池管理。它的思想基于 Jemalloc 内存管理库来设计，也提供了很多开封即用的应用层协议编码与解码器。同时，笔者在调研市场后发现，市场上需要一本 Netty 的书，将 Netty 的核心骨架源码和架构思想进行统一的描述，帮助读者通过该书掌握所有难点、重点的代码。本书剥离其他诸如 UDP 协议处理等不常用的源码，在减少篇幅的同时，通过常用的 TCP 协议完整诠释了 Netty 的架构设计与思想。由于本书内容必不可少地涉及 NIO、线程模型、网络编程的知识，本书并没有介绍 Java 语言层面的一些基础知识，例如变量、面向对象、泛型等 Java SE 的基础，读者只要拥有 Java 语言的基础，阅读本书并没有太大的难度。

本书适用于以下读者：希望进入互联网公司工作的读者，研究 Netty 底层知识的读者，在工作中遇见瓶颈、希望通过学习 Netty 提升底层知识的读者，从事开发高并发支撑中间件的读者，从事互联网高并发业务支撑的读者，对多线程编程感兴趣的读者，以及希望通过 Netty 源码找到调优点的读者。

本书封面贴有清华大学出版社防伪标签，无标签者不得销售。
版权所有，侵权必究。举报：010-62782989，beiqinquan@tup.tsinghua.edu.cn。

图书在版编目（CIP）数据

Netty 源码全解与架构思维 / 黄俊著. —北京：清华大学出版社，2023.1
ISBN 978-7-302-62498-1

Ⅰ．①N… Ⅱ．①黄… Ⅲ．①JAVA 语言—程序设计 Ⅳ．①TP312.8

中国国家版本馆 CIP 数据核字（2023）第 005518 号

责任编辑：贾旭龙
封面设计：姜　龙
版式设计：文森时代
责任校对：马军令
责任印制：朱雨萌

出版发行：清华大学出版社
　　　　　网　　址：http://www.tup.com.cn，http://www.wqbook.com
　　　　　地　　址：北京清华大学学研大厦 A 座　　　　　**邮　编**：100084
　　　　　社 总 机：010-83470000　　　　　**邮　购**：010-62786544
　　　　　投稿与读者服务：010-62776969，c-service@tup.tsinghua.edu.cn
　　　　　质量反馈：010-62772015，zhiliang@tup.tsinghua.edu.cn
印 装 者：三河市科茂嘉荣印务有限公司
经　　销：全国新华书店
开　　本：203mm×260mm　　　　　**印　张**：32.25　　　　　**字　数**：861 千字
版　　次：2023 年 1 月第 1 版　　　　　**印　次**：2023 年 1 月第 1 次印刷
定　　价：128.00 元

产品编号：095328-01

前 言
Preface

为什么要写这本书

Netty 是目前市面上使用率较高的网络编程库。它的架构设计非常明确且层次分明，源码简洁易懂，包含了 Java NIO（New IO）的三大组件：Selector（选择器）、Channel（通道）、ByteBuffer（缓冲区）。Netty 提供了简单、易用、高效的网络通信库，实现了内存池管理。它的思想基于 Jemalloc 内存管理库设计，还提供了很多开箱即用的应用层协议编码与解码器。笔者调研市场后发现，市场上需要一本 Netty 的书，来统一阐述 Netty 的核心源码和架构思想，帮助读者掌握所有代码的难点和重点。本书剥离了 UDP 协议处理等不常用的源码，不仅减少了篇幅，而且通过常用的 TCP 协议完整地诠释了 Netty 的架构设计与思想。

本书适合以下读者阅读。
- 希望进入互联网公司工作的读者。
- 研究 Netty 底层知识的读者。
- 在工作中遇到瓶颈，希望通过学习 Netty 提升底层知识的读者。
- 从事开发高并发支撑中间件的读者。
- 从事互联网高并发业务支撑的读者。
- 对多线程编程感兴趣的读者。
- 希望通过 Netty 源码找到调优点的读者。

背景知识

本书没有介绍 Java 语言层面的基础知识，例如变量、面向对象、泛型等 Java SE 的基础知识。所以读者需要首先理解掌握 Java SE，才能更好阅读本书。由于本书致力于研究 Netty 架构和源码层面的知识，必不可少地涉及 NIO、线程模型、网络编程等方面的知识，读者只要拥有 Java 语言基础，阅读本书并没有太大的难度。

本书详细讲解了 Netty 使用的设计模式，注释了源码中的重点和难点，并在每部分内容前都提供了流程和总结。读者可以按照书中的流程来阅读源码，加深理解与记忆。

如何阅读这本书

本书按照 Netty 的模块组织架构进行设计。
- 第 1 篇，讲解 Java 网络编程的基本原理。

➢ 第 2 篇，讲解 Netty 的事件循环组的设计。
➢ 第 3 篇，讲解 Netty 基于 Jemalloc 内存库实现的内存管理模块。
➢ 第 4 篇，讲解 Netty 对原生 NIO 处理的包装和通道处理器的设计。

读者可以根据实际需求，查阅对应的内容进行学习。在第 1 篇中，笔者详细解释了 Java Socket 编程的基本原理，同时展示了 Linux 2.6 的 Epoll 处理源码分析。Epoll 是目前 Selector 选择器底层实现的常用选择。它足够高效，采用了 RB 树和 Callback 进行处理。由于 Linux 2.6 内核实现的 Epoll 比较简单，所以笔者将其源码进行关键解释，为需要了解底层的读者解惑。

对于第 4 篇的通道处理器，笔者没有阐述 UDP（User Datagram Protocol，用户数据报协议）的 Socket（套接字），因为它非常简单。在此，笔者以 TCP（Transmission Control Protocol，传输控制协议）为例，诠释 Netty 的网络通道框架。读者在了解框架实现原理后，查看 UDP 的实现也非常简单。同时，本书以 HTTP（HyperText Transfer Protocol，超文本传输协议）为例介绍应用层协议，但笔者没有详细解释 HTTP 的处理细节，因为这部分内容并不属于 Netty，而属于协议处理。

勘误和支持

由于笔者水平有限，加之编写的时间也很仓促，书中难免会出现一些不准确的地方，恳请读者批评指正。读者可扫描下方二维码，观看视频讲解和资源分享，进行学习交流，期待广大读者提出宝贵的意见。

致谢

首先，我要感谢 Netty 的开发人员。他们在 Netty 架构设计上展现的功力令人折服。正是因为他们的工作成果，本书才有了诞生的意义。

这半年时间里，笔者一边工作一边写作，这给我带来了极大的压力。所以首先我要感谢我的父母，感谢他们在生活上对我无微不至的照顾，使我可以全身心地投入到写作工作中。其次，繁忙的工作之余，写作又占用了绝大部分休息时间，在此感谢我的太太罗亚萍对我的体谅和鼓励，让我始终能以高昂的斗志投入到本书的写作工作中。

再次，感谢我工作中的同事们。正是与他们一起战斗在一线的日子，我才不断地对技术有着日新月异的感悟和理解。正是那些充满激情的岁月，才使我越来越热衷于底层知识的学习和研究。

谨以此书，献给一直鼓励我前进的伙伴们，以及众多热爱底层技术的朋友们。

目 录
Contents

第 1 篇　Netty 世界漫游与 Java 网络编程回顾

第 1 章　Java 网络编程 2
- 1.1　Socket 介绍 2
- 1.2　Socket 编程 3
  - 1.2.1　Java 客户端编程 3
  - 1.2.2　Java 服务端编程 4
  - 1.2.3　C 语言服务端编程 4
- 1.3　BIO 编程 5
- 1.4　NIO 编程 6
  - 1.4.1　NIO 模型原理 6
  - 1.4.2　Buffer 原理 13
  - 1.4.3　Channel 原理 19
  - 1.4.4　Selector 原理 36
- 1.5　AIO 编程 38
  - 1.5.1　Java AIO 描述 39
  - 1.5.2　Java AIO 实例 39
  - 1.5.3　AsynchronousServerSocketChannel.open() 原理 42
  - 1.5.4　AsynchronousServerSocketChannel.bind() 原理 52
  - 1.5.5　ServerSocketChannel.accept(null, new AcceptHandler())原理 52
  - 1.5.6　socketChannel.read(byteBuffer, byteBuffer, new ReadHandler())原理 55
  - 1.5.7　Linux 服务端网络编程 59
  - 1.5.8　Linux 客户端网络编程 60
  - 1.5.9　Java AIO 涉及的 Native JNI 实现 61

第 2 章　mmap 网络编程与 sendfile 原理 64
- 2.1　transferTo 方法 64
  - 2.1.1　transferTo 方法定义 64
  - 2.1.2　transferTo 方法实现原理 64
  - 2.1.3　transferToDirectly 方法 65
  - 2.1.4　transferToTrustedChannel 方法 66
  - 2.1.5　transferToArbitraryChannel 方法 ... 67
- 2.2　JVM 层面零复制原理 68
  - 2.2.1　transferTo0 方法 68
  - 2.2.2　map0 方法 69
- 2.3　sendfile64 方法 69

第 3 章　Linux epoll 实现原理 74
- 3.1　三大函数原型 74
- 3.2　epoll_event 与 epoll_data 对象 75
- 3.3　边缘触发与水平触发 75
- 3.4　使用示例 76
- 3.5　三大函数内核原理 77

第 4 章　Netty 架构与源码组成 92
- 4.1　Netty 是什么 92
- 4.2　Netty 架构组成 92
- 4.3　Netty 三大基础模块 93
  - 4.3.1　事件循环模块 93
  - 4.3.2　内存池模块 94
  - 4.3.3　通道处理器模块 94
- 4.4　Netty 源码组成 95

第 2 篇　事件驱动层

第 5 章　JDK Executor 原理 98
5.1　Executor 接口 98
5.2　ExecutorService 接口 98
5.3　AbstractExecutorService 抽象类 99
5.4　ScheduledExecutorService 接口 104

第 6 章　EventExecutor 与 EventExecutorGroup 原理 105
6.1　EventExecutorGroup 类 105
6.2　EventExecutor 接口 105
6.3　AbstractEventExecutorGroup 方法 106
6.4　MultithreadEventExecutorGroup 类 108
 6.4.1　核心变量与构造器 109
 6.4.2　EventExecutorChooserFactory 与 DefaultEventExecutorChooserFactory 111
 6.4.3　ThreadPerTaskExecutor 类 112
 6.4.4　DefaultThreadFactory 类 113
 6.4.5　FastThreadLocalThread 类 115
 6.4.6　FastThreadLocal 类 115
 6.4.7　shutdownGracefully 方法 121
 6.4.8　awaitTermination 方法 121
6.5　DefaultEventExecutorGroup 类 122
6.6　AbstractEventExecutor 类 123
6.7　AbstractScheduledEventExecutor 方法 125
6.8　SingleThreadEventExecutor 类 129
 6.8.1　核心变量与构造器 129
 6.8.2　execute 核心方法实现 131
 6.8.3　addTask 核心方法 132
 6.8.4　startThread 核心方法 133
 6.8.5　confirmShutdown 核心方法 135
 6.8.6　runAllTasks 核心方法 136
 6.8.7　runShutdownHooks 核心方法 ..138
 6.8.8　awaitTermination 核心方法 138
 6.8.9　takeTask 核心方法 138
 6.8.10　shutdownGracefully 核心方法 140
6.9　DefaultEventExecutor 142

第 7 章　EventLoop 与 EventLoopGroup 原理 143
7.1　EventLoopGroup 接口与 EventLoop 接口 143
 7.1.1　EventLoopGroup 接口 143
 7.1.2　EventLoop 接口 143
7.2　MultithreadEventLoopGroup 原理 144
7.3　DefaultEventLoopGroup 原理 145
7.4　NioEventLoopGroup 类 145
7.5　ThreadPerChannelEventLoopGroup 原理 147
 7.5.1　核心变量与构造器 147
 7.5.2　newChild 核心方法 148
 7.5.3　next 核心方法 148
 7.5.4　shutdownGracefully 核心方法 149
 7.5.5　awaitTermination 核心方法 149
 7.5.6　register 核心方法 150
 7.5.7　nextChild 核心方法 150
7.6　OioEventLoopGroup 类 151
7.7　SingleThreadEventLoop 原理 151
 7.7.1　核心变量与构造器 152
 7.7.2　next 核心方法 152
 7.7.3　executeAfterEventLoopIteration 核心方法 152
 7.7.4　afterRunningAllTasks 核心方法 ..153
 7.7.5　register 核心方法 153
 7.7.6　hasTasks 核心方法 153
 7.7.7　pendingTasks 核心方法 153
7.8　NioEventLoop 154
 7.8.1　核心变量与构造器 154

7.8.2	run 核心方法 156		8.8.1	核心变量与构造器 173
7.8.3	select 核心方法 158		8.8.2	await 核心方法 174
7.8.4	rebuildSelector 核心方法 ...160		8.8.3	awaitUninterruptibly 核心方法 ...175
7.8.5	processSelectedKeys 核心方法 ...162		8.8.4	cancel 核心方法 175
7.8.6	processSelectedKey 核心频道方法 ...163		8.8.5	sync 核心方法 176
7.8.7	processSelectedKey 核心任务方法 ...165		8.8.6	syncUninterruptibly 核心方法 176
7.9	ThreadPerChannelEventLoop 类 165		8.8.7	setSuccess 核心方法 176
			8.8.8	setFailure 核心方法 177

第 8 章 Future 与 Promise 原理 167

8.1	Future 接口 167		8.8.9	trySuccess 核心方法 177
8.2	GenericFutureListener 与 FutureListener 接口 169		8.8.10	tryFailure 核心方法 178
			8.8.11	addListener 核心方法 178
8.2.1	GenericFutureListener 接口 169		8.8.12	notifyListeners 核心方法 179
8.2.2	FutureListener 接口 169		8.8.13	notifyListenersNow 核心方法 ...180
8.3	AbstractFuture 接口 169		8.8.14	notifyListeners0 核心方法 181
8.4	ChannelGroupFuture 接口 170		8.8.15	notifyProgressiveListeners 核心方法 ...181
8.5	GenericProgressiveFutureListener 监听器 171		8.8.16	progressiveListeners 核心方法 ...182
			8.8.17	notifyProgressiveListener 核心方法 ...183
8.6	ChannelFuture 接口 171		8.9	ChannelPromise 接口 184
8.7	Promise 接口 172		8.9.1	DefaultChannelPromise 类 184
8.8	DefaultPromise 接口 172		8.9.2	DefaultChannelGroupFuture 类 ...188

第 3 篇　内存管理层

第 9 章 ByteBuf 与衍生类原理 192

9.1	ByteBuf 原理 192		9.4	UnpooledHeapByteBuf 类 200
9.1.1	构造器与核心变量 193		9.4.1	核心变量与构造器 200
9.1.2	ReferenceCounted 接口 193		9.4.2	getByte 核心方法 201
9.2	AbstractByteBuf 原理 194		9.4.3	getBytes 核心方法 201
9.2.1	核心变量与构造器 194		9.4.4	setByte 核心方法 201
9.2.2	writeByte 核心方法 194		9.4.5	setBytes 核心方法 202
9.2.3	writeBytes 核心方法 195		9.4.6	setShort 核心方法 202
9.2.4	readByte 核心方法 195		9.4.7	capacity 核心方法 202
9.2.5	readBytes 核心方法 196		9.4.8	nioBuffer 核心方法 203
9.2.6	writeZero 核心方法 196		9.5	UnpooledUnsafeHeapByteBuf 类 ...203
9.2.7	discardReadBytes 核心方法 197		9.6	UnpooledDirectByteBuf 原理 205
9.3	AbstractReferenceCountedByteBuf 类 ...197		9.6.1	核心变量与构造器 205
			9.6.2	setByte 核心方法 206

9.6.3　setBytes 核心方法 206
9.6.4　getByte 核心方法 206
9.6.5　getBytes 核心方法 206
9.6.6　capacity 核心方法 207
9.6.7　freeDirect 核心方法 208
9.7　UnpooledUnsafeDirectByteBuf 方法 209
9.8　UnpooledUnsafeNoCleanerDirectByteBuf 类 210
9.9　PooledByteBuf 原理 211
9.9.1　init 核心方法 212
9.9.2　initUnpooled 核心方法 213
9.9.3　reuse 核心方法 213
9.9.4　capacity 核心方法 213
9.9.5　deallocate 核心方法 214
9.9.6　recycle 核心方法 214

第 10 章　Netty 对象池原理 215

10.1　Recycler 原理 215
 10.1.1　核心构造器与变量 215
 10.1.2　Handle 核心接口 217
 10.1.3　Stack 核心类 218
 10.1.4　WeakOrderQueue 核心类 221
10.2　PooledHeapByteBuf 原理 226
10.3　PooledUnsafeHeapByteBuf 类 227
10.4　PooledDirectByteBuf 类 228
10.5　CompositeByteBuf 原理 229
 10.5.1　核心变量与构造器 229
 10.5.2　addComponents 核心方法 230
 10.5.3　addComponent 核心方法 231
 10.5.4　removeComponent 核心方法 232
 10.5.5　removeComponents 核心方法 233
 10.5.6　capacity 核心方法 233
 10.5.7　decompose 核心方法 233
 10.5.8　findComponent 核心方法 235
 10.5.9　getByte 核心方法 235
 10.5.10　setByte 核心方法 235
10.6　WrappedByteBuf 类 236
10.7　SimpleLeakAwareByteBuf 类 236
10.8　AdvancedLeakAwareByteBuf 类 238
10.9　WrappedCompositeByteBuf 类 240
10.10　SimpleLeakAwareCompositeByteBuf 类 240
10.11　AdvancedLeakAwareCompositeByteBuf 类 241
10.12　ResourceLeak 接口 242
10.13　ResourceLeakDetector 类 242
 10.13.1　核心变量与构造器 243
 10.13.2　newRecord 核心方法 247
 10.13.3　open 核心方法 248
 10.13.4　reportLeak 核心方法 248
 10.13.5　reportXLeak 核心方法 249

第 11 章　Netty 内存池原理 251

11.1　ByteBufAllocator 251
11.2　AbstractByteBufAllocator 类 253
 11.2.1　核心变量与构造器 253
 11.2.2　buffer 核心方法 253
 11.2.3　ioBuffer 核心方法 255
 11.2.4　compositeBuffer 核心方法 255
 11.2.5　toLeakAwareBuffer 核心方法 256
 11.2.6　calculateNewCapacity 核心方法 257
11.3　UnpooledByteBufAllocator 类 258
11.4　PooledByteBufAllocator 类 259
 11.4.1　核心变量与构造器 259
 11.4.2　newHeapBuffer 核心方法 263
 11.4.3　newDirectBuffer 核心方法 263
 11.4.4　PoolThreadLocalCache 核心内部类 264
11.5　PoolThreadCache 类 265
 11.5.1　核心变量和构造器 265
 11.5.2　createSubPageCaches 核心方法 267
 11.5.3　createNormalCaches 核心方法 267
 11.5.4　MemoryRegionCache 核心内部类 268
 11.5.5　SubPageMemoryRegionCache 核心内部类 270
 11.5.6　NormalMemoryRegionCache 核心内部类 270

11.5.7	allocateTiny 核心方法	270
11.5.8	cacheForSmall 核心方法	271
11.5.9	cacheForNormal 核心方法	272
11.5.10	allocate 核心方法	272
11.5.11	trim 核心方法	273
11.5.12	free 核心方法	273
11.6	PoolArena 类	274
11.6.1	核心变量与构造器	274
11.6.2	HeapArena 核心内部类	278
11.6.3	DirectArena 核心内部类	279
11.6.4	allocate 核心方法	280
11.6.5	normalizeCapacity 核心方法	282
11.6.6	allocateNormal 核心方法	283
11.6.7	allocateHuge 核心方法	284
11.6.8	findSubpagePoolHead 核心方法	284
11.6.9	free 核心方法	284
11.6.10	freeChunk 核心方法	285
11.7	PoolChunkList 类	286
11.8	PoolSubpage 类	288
11.9	PoolChunk 原理	292
11.9.1	核心变量与构造器	294
11.9.2	allocate 核心方法	296
11.9.3	allocateRun 核心方法	296
11.9.4	allocateNode 核心方法	296
11.9.5	updateParentsAlloc 核心方法	297
11.9.6	allocateSubpage 核心方法	298
11.9.7	free 核心方法	299
11.9.8	updateParentsFree 核心方法	299
11.9.9	initBuf 核心方法	300
11.9.10	initBufWithSubpage 核心方法	300

第 4 篇　通道管理层

第 12 章	Netty Channel 层原理	304
12.1	ChannelOutboundInvoker 接口	304
12.2	ChannelInboundInvoker 接口	305
12.3	Channel 通道接口	306
12.4	ServerChannel 服务端	309
12.5	ChannelMetadata 类	309
12.6	ChannelOutboundBuffer 通道输出缓冲区	310
12.6.1	核心变量与构造器	310
12.6.2	Entry 核心内部类	311
12.6.3	addMessage 核心方法	313
12.6.4	addFlush 核心方法	315
12.6.5	current 核心方法	315
12.6.6	progress 核心方法	316
12.6.7	remove 核心方法	316
12.6.8	removeBytes 核心方法	317
12.6.9	nioBuffers 核心方法	318
12.6.10	Writable 可写标志位原理	320
12.6.11	close 核心方法	322
12.6.12	bytesBeforeUnwritable 核心方法	323
12.6.13	bytesBeforeWritable 核心方法	323
12.7	RecvByteBufAllocator 接收缓冲区分配器	323
12.7.1	RecvByteBufAllocator 接口	324
12.7.2	MaxMessagesRecvByteBufAllocator 接口	325
12.7.3	DefaultMaxMessagesRecvByteBufAllocator 类	325
12.7.4	AdaptiveRecvByteBufAllocator 分配器	327
12.7.5	FixedRecvByteBufAllocator 分配器	330
12.8	AbstractChannel 类	331
12.8.1	核心变量与构造器	331
12.8.2	实现方法原理	332
12.8.3	AbstractUnsafe 核心内部类	336
12.8.4	核心变量与构造器	336
12.8.5	recvBufAllocHandle 核心方法	336
12.8.6	localAddress 与 remoteAddress 核心方法	336

12.8.7	register 核心方法 337	12.19.3	filterOutboundMessage 核心方法............388
12.8.8	register0 核心方法..........................338	12.19.4	NioByteUnsafe 核心内部类...............388
12.8.9	beginRead 核心方法.........................339	12.20	NioSocketChannel NIO TCP 客户端
12.8.10	bind 核心方法..................................339		实现类 390
12.8.11	disconnect 核心方法......................340	12.20.1	核心变量与构造器.........................390
12.8.12	close 核心方法................................341	12.20.2	isActive 核心方法..........................391
12.8.13	closeForcibly 核心方法.................343	12.20.3	shutdownInput 核心方法...............391
12.8.14	deregister 核心方法.......................344	12.20.4	shutdownOutput 核心方法............392
12.8.15	write 核心方法................................345	12.20.5	shutdown 核心方法.......................393
12.8.16	flush 核心方法................................345	12.20.6	doBind 核心方法...........................394
12.8.17	CloseFuture 核心内部类...............346	12.20.7	doConnect 核心方法....................394
12.9	AbstractServerChannel 类 347	12.20.8	doClose 核心方法.........................395
12.10	LocalChannel 类 348	12.20.9	doReadBytes 核心方法................395
12.10.1	核心变量与构造器.........................349	12.20.10	doWriteBytes 核心方法..............395
12.10.2	doRegister 核心方法.....................350	12.20.11	doWriteFileRegion 核心方法.....396
12.10.3	doBind 核心方法...........................351	12.20.12	doWrite 核心方法.......................396
12.10.4	doDisconnect 核心方法...............352	12.20.13	NioSocketChannelUnsafe 核心
12.10.5	doBeginRead 核心方法................354		内部类..398
12.10.6	doWrite 核心方法.........................355	12.21	AbstractNioMessageChannel NIO 服务端
12.10.7	LocalUnsafe 核心内部类.............356		模板类原理 398
12.11	LocalServerChannel 类 357	12.22	NioServerSocketChannel NIO TCP 服务
12.12	AbstractOioChannel 类 360		端实现类原理 402
12.13	AbstractOioByteChannel 类 362		
12.14	OioByteStreamChannel 类 366	**第 13 章**	**Netty 通道流水线与通道处理器**
12.15	OioSocketChannel 类 368		**原理**..**405**
12.16	AbstractOioMessageChannel 原理.....373	13.1	ChannelPipeline 接口ˎ 405
12.17	OioServerSocketChannel 类 374	13.2	ChannelHandlerContext 接口 409
12.18	AbstractNioChannel NIO 模板类 377	13.3	ChannelHandler 接口 410
12.18.1	核心变量与构造器.........................377	13.4	ChannelInboundHandler 接口 411
12.18.2	doRegister 核心方法.....................378	13.5	ChannelOutboundHandler 接口 412
12.18.3	doDeregister 核心方法................379	13.6	AbstractChannelHandlerContext
12.18.4	doBeginRead 核心方法................379		模板类 412
12.18.5	newDirectBuffer 核心方法...........379	13.6.1	核心变量与构造器.........................413
12.18.6	doClose 核心方法.........................381	13.6.2	fireChannelRegistered 核心方法 414
12.18.7	AbstractNioUnsafe 核心内部类...381	13.6.3	fireChannelUnregistered 核心方法............416
12.19	AbstractNioByteChannel NIO 客户端	13.6.4	fireChannelActive 核心方法ˎ 417
	模板类 385	13.6.5	invokeChannelInactive 核心方法 417
12.19.1	核心变量与构造器.........................385	13.6.6	fireExceptionCaught 核心方法ˎ 418
12.19.2	doWrite 核心方法.........................385		

- 13.6.7 fireUserEventTriggered 核心方法 419
- 13.6.8 fireChannelRead 核心方法 420
- 13.6.9 fireChannelReadComplete 核心方法 420
- 13.6.10 fireChannelWritabilityChanged 核心方法 421
- 13.6.11 bind 核心方法 422
- 13.6.12 connect 核心方法 423
- 13.6.13 disconnect 核心方法 424
- 13.6.14 close 核心方法 425
- 13.6.15 deregister 核心方法 425
- 13.6.16 read 核心方法 426
- 13.6.17 write 核心方法 427
- 13.6.18 flush 核心方法 429
- 13.6.19 writeAndFlush 核心方法 429
- 13.6.20 AbstractWriteTask 核心内部类 430
- 13.6.21 WriteTask 核心内部类 431
- 13.6.22 WriteAndFlushTask 核心内部类 432
- 13.7 ChannelHandlerAdapter 抽象类 433
- 13.8 DefaultChannelPipeline 类 433
  - 13.8.1 核心变量与构造器 434
  - 13.8.2 newContext 核心方法 435
  - 13.8.3 addFirst 核心方法 436
  - 13.8.4 addLast 核心方法 439
  - 13.8.5 addBefore 核心方法 440
  - 13.8.6 addAfter 核心方法 441
  - 13.8.7 remove 核心方法 443
  - 13.8.8 removeFirst 核心方法 444
  - 13.8.9 removeLast 核心方法 444
  - 13.8.10 replace 核心方法 444
  - 13.8.11 destroy 核心方法 446
  - 13.8.12 TailContext 核心内部类 448
  - 13.8.13 HeadContext 核心内部类 449
  - 13.8.14 write 核心方法 451
  - 13.8.15 fireChannelRead 核心方法 451
- 13.9 ChannelInboundHandlerAdapter 抽象类 451
- 13.10 SimpleChannelInboundHandler 处理器类 452
- 13.11 ChannelInitializer 类 454
- 13.12 ChannelOutboundHandlerAdapter 抽象类 455
- 13.13 ChannelDuplexHandler 处理器 456
- 13.14 通道与通道处理器小结 456

第 14 章 Netty 解码器与编码器 458
- 14.1 ByteToMessageDecoder 解码器 458
  - 14.1.1 核心变量与构造器 458
  - 14.1.2 channelRead 核心方法 460
  - 14.1.3 handlerRemoved 核心方法 462
  - 14.1.4 channelReadComplete 核心方法 463
  - 14.1.5 channelInactive 核心方法 463
  - 14.1.6 userEventTriggered 核心方法 464
- 14.2 MessageToByteEncoder 编码器 465
- 14.3 MessageToMessageDecoder 解码器 ... 467
- 14.4 MessageToMessageEncoder 编码器 ... 468
- 14.5 ByteToMessageCodec 双端编码器原理 470
- 14.6 MessageToMessageCodec 双端编码器原理 472
- 14.7 编码器小结 474
- 14.8 HttpObjectDecoder 解码器和 HttpRequestDecoder 解码器 475
- 14.9 HttpObjectEncoder 编码器和 HttpContentEncoder 编码器 476
- 14.10 MessageAggregator 消息对象聚合器 478
- 14.11 HttpObjectAggregator 聚合器 479
- 14.12 TCP 粘包原理 480
- 14.13 DelimiterBasedFrameDecoder 解码器 481
- 14.14 FixedLengthFrameDecoder 类 486
- 14.15 LengthFieldBasedFrameDecoder 解码器 487
- 14.16 LineBasedFrameDecoder 解码器 495
- 14.17 ReplayingDecoder 解码器 497

第 1 篇

Netty 世界漫游与 Java 网络编程回顾

本篇概述了 Netty 框架的基本原理。在研究具体模块的源码前，首先帮助读者从总体上把握全局。这样读者在研究源码时，不会陷入迷茫。笔者认为，框架和人体组成一样，由头、五官、四肢、骨架、血肉等结构组成，缺一不可。

开源项目为了保证项目并行开发，通常会使用模块化的方式进行开发。读者在研究某个单独模块时，往往会涉及另一个模块的知识，所以从一开始就把握全局是非常重要的，本篇出发点便是如此。在讲解 Netty 的整体架构之前，笔者也对 Java 网络编程进行了深入回顾和分析，这将有助于读者对 Netty 的理解和记忆。

第 1 章

Java 网络编程

在开始学习 Netty 前，不妨先回顾或学习 Java 的网络编程和基本网络知识。万丈高楼平地起。Netty 作为基于 Java 语言开发的网络编程框架，底层包装的是 Java 网络编程。所以对 Netty 的掌握程度，取决于对 Java BIO（Blocking IO，阻塞式 IO）、NIO（New IO，新式 IO）和 AIO（Asynchronous IO，异步 IO）编程的掌握程度。本章主要介绍 Java 网络编程的基础知识。

1.1　Socket 介绍

首先，对网络通信进行推理。

（1）两台计算机之间需要通信，双方需要一个 IP 地址标识。

（2）两台计算机中的两个进程需要通信，双方需要一个端口标识。

（3）数据需要通过网线传递到对方，而网线中传输的是二进制数据，需要双方定义数据交互格式。这时就引入了 TCP/IP 协议栈，它将网络协议进行分层。每层使用自身的数据格式，并执行操作，互不影响。

① 应用层：与进程自身处理相关，例如 HTTP 协议。

② 传输层：与操作系统和上层编程语言相关，例如 TCP 协议（面向连接）、UDP 协议（面向数据报）。PORT 便在这一层。

③ 网络层：与操作系统相关，例如 IP 协议。IP 地址便在这一层。

④ 数据链路层：与操作系统与网卡硬件设备相关，例如 PPP 帧、以太网帧。

对于两台计算机的硬件，它们只知道网卡间的彼此通信，所以需要 MAC 地址作为标识，用于表示两个网卡的唯一地址，在硬件实现传输时有选择地获取传递的数据。

Socket 是什么？它是 IP 地址和端口的结合，即唯一标识了互联网中不同计算机的不同进程。其中，IP 标识计算机，PORT（端口）表示计算机中的进程。所以在输入浏览器地址时，使用 IP : PORT 访问服务器。

C/S 架构需要在编程语言上实现。

（1）客户端的 Socket 表示客户端进程与计算机。

（2）服务端的 ServerSocket 表示服务端的进程与计算机。

这时，可以得出如下结论。

（1）服务端的 ServerSocket 需要处理来自不同客户端的请求。

（2）客户端的 Socket 只需连接到 ServerSocket 便可进行交互。

（3）服务端与客户端的展现形式是：一个 ServerSocket 对应多个 Socket，即一对多的关系。

（4）ServerSocket 只需负责接收请求，然后在服务端 ServerSocket 分配一个 Socket 与客户端的 Socket 进行交互即可。

（5）对于服务端 ServerSocket，它接收到客户端请求而创建的 Socket，称为客户端 Socket。

1.2　Socket 编程

了解 Socket 后，我们接下来再来看如何使用 Java 语言进行 Socket 编程。从上述 Socket 定义，可把编程分为如下两种。

（1）客户端编程：连接到服务端。

（2）服务端编程：接收客户端请求。

1.2.1　Java 客户端编程

举例说明。创建一个 Socket 实例，将需要连接的服务端信息放入 InetSocketAddress 对象中，调用 socket.connect(inetSocketAddress)连接服务端。连接成功后，可以获取到 Socket 的输入流和输出流，与服务端进行通信。由源码可知如下信息。

（1）InetSocketAddress 对象用于封装 Socket 的信息：IP:PORT。

（2）Socket 对象在 Java 层面表示客户端对象。

（3）socket.connect(inetSocketAddress)方法通过封装服务端信息的 InetSocketAddress 对象，完成对服务端的连接。

（4）socket.connect(inetSocketAddress)没有抛出异常并正常返回时，对于客户端和服务端两台计算机中的两个进程而言，彼此互相知道了对象，只需获取到输入输出流即可完成通信。

注意，Java 网络编程只有原生的流对象 InputStream 和 OutputStream，均由运输层的 TCP 或 UDP 表示。对于应用层而言，需要定义数据交互格式，即数据以什么样的格式编码后在 OutputStream 传递到对方。而对方在接收到这些格式的数据后，需要 Java 开发人员解码为需要的数据。Java 原生只支持 TCP 与 UDP 的操作。

```java
public class ClientDemo {
    public static void main(String[] args) throws Exception{
        int port= 80;
        Socket socket = new Socket();
        InetSocketAddress inetSocketAddress = new InetSocketAddress("remote address",port);
        socket.connect(inetSocketAddress);         // 连接到服务端
        // 使用流对象与服务端通信
        InputStream inputStream = socket.getInputStream();
        OutputStream outputStream = socket.getOutputStream();
        // doSomething
        // 完成处理后，关闭流与 Socket
        inputStream.close();
        outputStream.close();
        socket.close();
    }
}
```

1.2.2　Java 服务端编程

前文介绍过，服务端 Socket 与客户端 Socket 为一对多的关系。服务端 Socket（ServerSocket）等同于用于建立服务器端的 Socket（注意：这里描述的都是与客户端对等的 Socket，而不是 ServerSocket）的中间产物。以下例子中由代码可知，ServerSocket 表示服务端用来接收客户端连接请求并创建与之对等的 Socket socket。

```java
public class ServerDemo {
    public static void main(String[] args) throws Exception {
        int serverPort = 80;
        ServerSocket serverSocket = new ServerSocket(serverPort);
        Socket socket = serverSocket.accept();     // 发生线程阻塞，接收客户端请求
        // 使用流对象与服务端通信
        InputStream inputStream = socket.getInputStream();
        OutputStream outputStream = socket.getOutputStream();
        // doSomething
        // 完成处理后关闭流与 Socket
        inputStream.close();
        outputStream.close();
        socket.close();
        serverSocket.close();
    }
}
```

1.2.3　C 语言服务端编程

为了方便读者类比学习，笔者这里也给出了 C 语言的服务端编程。因为 Java 的 Socket 编程最终都会通过 JNI（Java Native Interface，Java 本地接口）调用 C 处理。通过 C 的服务端编程，读者应该清楚 Java ServerSocket 类完成了以下操作。

- ☑ 创建 Socket。
- ☑ 绑定端口。
- ☑ 接收连接。
- ☑ 分配缓冲区。
- ☑ 读取数据。
- ☑ 关闭连接。

不过 C 语言较为复杂，而 Java 通过 JVM 和 JNI 封装了 C 语言的操作。源码如下。

```c
int main(void)                                      // 主函数从这里开始运行
{
    int sk,csk;                                     // 服务端 sk 和客户端 csk fd（文件描述符）
    char rbuf[51];                                  // 接收缓冲区
    struct sockaddr_in addr;                        // socket 地址
    sk = socket(AF_INET,SOCK_STREAM,0);             // 创建 socket
    bzero(&addr,sizeof(struct sockaddr));           // 清空内存

    // 设置属性
    addr.sin_family = AF_INET;
```

```c
addr.sin_addr.s_addr = htonl(INADDR_ANY);
addr.sin_port = htons(5000);                    // 设置端口

// 绑定地址
if(bind(sk,(struct sockaddr *)&svraddr,sizeof(struct sockaddr_in))== -1){
    fprintf(stderr,"Bind error:%s\n",strerror(errno));
    exit(1);
}

if(listen(sk,1024) == -1){                      // 开始监听来自客户端的连接
    fprintf(stderr,"Listen error:%s\n",strerror(errno));
    exit(1);
}

// 从完成 TCP 三次握手的队列中获取 client 连接
if((csk = accept(sk,(struct sockaddr *)NULL,NULL)) == -1){
    fprintf(stderr,"accept error:%s\n",strerror(errno));
    exit(1);
}
memset(rbuf,0,51);                              // 重置缓冲区
recv(csk,rbuf,50,0);                            // 从 socket 中读取数据放入缓冲区
printf("%s\n",rbuf);                            // 打印接收到的数据
// 关闭客户端和服务端
close(csk);
close(sk);
}
```

1.3 BIO 编程

学习 ServerSocket 和 Socket 后，研究 BIO 编程就相对容易。BIO 是 Blocking IO 的缩写，表示阻塞式 IO 编程。什么是阻塞？可从以下代码进行推理。

（1）主线程为执行 main 方法的线程。

（2）创建 ServerSocket 后，调用 serverSocket.accept()接收请求。如果请求没有到达，则发生阻塞。

（3）返回 serverSocket.accept()后，建立与客户端对等的 Socket 对象。此时调用 inputStream.read(buffer)读取客户端数据。如果底层 Socket 读缓冲区中不存在数据，则发生阻塞。

（4）调用 bufferedWriter.write(buildHttpResp())将数据写入客户端。如果底层 Socket 写缓冲区充满，则发生阻塞。

可得出如下阻塞点。

（1）serverSocket.accept()接收操作。

（2）inputStream.read(buffer)读操作。

（3）bufferedWriter.write(buildHttpResp())写操作。

这样就可以定义什么是 BIO 了，即所有涉及 Socket 的 IO 操作都可能阻塞当前线程，称之为阻塞

式 IO。

```java
public class BIODemo {
    public static void main(String[] args) throws Exception {
        int serverPort = 80;
        ServerSocket serverSocket = new ServerSocket(serverPort);
        System.out.println("启动服务器");
        while (true) {                                      // 循环接收每一个来自客户端的连接
            Socket socket = serverSocket.accept();          // 线程阻塞在这里，接收客户端请求
            byte[] buffer=new byte[1024];
            InputStream inputStream = socket.getInputStream();
            inputStream.read(buffer);                       // 读取客户端数据
            BufferedWriter bufferedWriter = buildBufferedWriter(socket.getOutputStream());
            // 模拟业务操作
            doSomeWork();
            bufferedWriter.write(buildHttpResp());          // 将数据写入客户端数据
            bufferedWriter.flush();
        }
    }
}
```

1.4　NIO 编程

与 BIO 相对应的是 NIO(Non-Blocking IO，非阻塞 IO)。但对于 Java 来说，Java NIO 不是 Non-Blocking IO，而是 New IO。因为它定义了一个新模型，即缓冲区模型，并基于该模型实现了 Non-Blocking IO。本节详细介绍 Java NIO 的原理。

1.4.1　NIO 模型原理

为了展示 NIO 模型的推理，笔者在这里以文件 IO 形式展示给读者。因为相对于 Socket 的网络 IO 来说，本地磁盘上的文件分析最为简单，所以通过该方法来类比学习 Socket 的 IO。这里先看看如果使用传统的 FileInputStream 读取文件时会发生的情况（注意：这里省略了对于 FileInputStream 类和 FileOutStream 类的使用，因为这属于 Java SE 的基础知识）。由源码可知，调用 FileInputStream.read 方法时，将通过 JNI 调用 C 函数库的 read 函数完成读取。源码如下。

```java
public int read() throws IOException {
    return read0();
}

private native int read0() throws IOException;
```

```
// JNI 实现
JNIEXPORT jint JNICALL Java_java_io_FileInputStream_read(JNIEnv *env, jobject this) {
    return readSingle(env, this, fis_fd);
}

jint readSingle(JNIEnv *env, jobject this, jfieldID fid) {
```

```c
    jint nread;
    char ret;
    FD fd = GET_FD(this, fid);
    if (fd == -1) {
        JNU_ThrowIOException(env, "Stream Closed");
        return -1;
    }
    nread = IO_Read(fd, &ret, 1);    // 完成实际读取操作,指定缓冲区只有一个 byte(C 语言中一个 char 等于 1B)
    if (nread == 0) {                // 到达文件末尾: EOF
        return -1;
    } else if (nread == -1) {        // 发生异常
        JNU_ThrowIOExceptionWithLastError(env, "Read error");
    }
    return ret & 0xFF;               // 由于只读一个 byte,所以取低 8 位即可
}

ssize_t handleRead(FD fd, void *buf, jint len)
{
    ssize_t result;
    RESTARTABLE(read(fd, buf, len), result);    // 最终调用 C 函数库的 read 函数完成读取
    return result;
}
```

以下是 FileOutStream 类的写入过程。由源码可知,调用 FileOutStream.write 方法时,将通过 JNI 调用 C 函数库的 write 函数完成读取。源码如下。

```c
public void write(int b) throws IOException {
    write(b, append);      // append 表示当前写入文件是否为追加操作
}

private native void write(int b, boolean append) throws IOException;

JNIEXPORT void JNICALL Java_java_io_FileOutputStream_write(JNIEnv *env, jobject this, jint byte, jboolean append) {
    writeSingle(env, this, byte, append, fos_fd);
}

void writeSingle(JNIEnv *env, jobject this, jint byte, jboolean append, jfieldID fid) {
    char c = (char) byte;
    jint n;
    FD fd = GET_FD(this, fid);
    if (fd == -1) { // 代表文件的描述符必须存在(描述符的含义接下来要在内核的 VFS(Virtual File System,虚拟文件系统)中介绍,这里就只需要知道它代表文件就可以了)
        JNU_ThrowIOException(env, "Stream Closed");
        return;
    }
    if (append == JNI_TRUE) {        // 文件追加操作
        n = IO_Append(fd, &c, 1);
    } else {                         // 文件写入操作
        n = IO_Write(fd, &c, 1);
    }
    if (n == -1) {                   // 发生写错误
        JNU_ThrowIOExceptionWithLastError(env, "Write error");
    }
}
```

```c
// 实际调用 C 函数库的 write 函数完成写入
ssize_t handleWrite(FD fd, const void *buf, jint len)
{
    ssize_t result;
    RESTARTABLE(write(fd, buf, len), result);
    return result;
}
```

以下是 Java NIO 的文件读入和写出的原理。可能有些读者不知道如何使用，笔者这里给出了用例。源码如下。

```java
public class FileNIO {
    public static void main(String[] args) throws Exception {
        File file = new File("file path");
        FileInputStream fileInputStream = new FileInputStream(file);
        FileChannel channel = fileInputStream.getChannel();           // 打开文件通道
        ByteBuffer readBuffer = ByteBuffer.allocate(1);               // 读缓冲区
        ByteBuffer writeBuffer = ByteBuffer.wrap("1".getBytes());     // 读缓冲区
        channel.read(readBuffer);                                     // 读取数据放入缓冲区中
        channel.write(writeBuffer);                                   // 将缓冲区中数据写入通道
    }
}
```

从上述代码中，可发现如下核心操作。

（1）fileInputStream.getChannel()基于 FileInputStream 类打开文件通道。

（2）channel.read(readBuffer)读入通道数据放入缓冲区。

（3）channel.write(writeBuffer)将缓冲区数据写入通道。

接下来逐一分析这三个方法的源码，再进行总结。首先由源码可知 fileInputStream.getChannel()的原理，打开通道的实现类为 FileChannelImpl。源码如下。

```java
public FileChannel getChannel() {
    synchronized (this) { // 上锁以保证线程安全（这里可以优化为 DCL。笔者当前版本为 JDK 8，高版本可能会进行优化）
        if (channel == null) {
            channel = FileChannelImpl.open(fd, path, true, false, this);
        }
        return channel;
    }
}

public static FileChannel open(FileDescriptor fd, boolean readable, boolean writable,Object parent){
    return new FileChannelImpl(fd, readable, writable, false, parent);
}
```

下面是 channel.read(readBuffer)的源码。由源码可知。

（1）read(ByteBuffer dst)方法最终调用 IOUtil.read(fd, dst, -1, nd)完成读取操作。

（2）在 IOUtil.read(fd, dst, -1, nd)中，对缓冲区中的 ByteBuffer 类型进行判断。Java 内存分为：堆内内存（例如，byte[]数组）、堆外内存（JVM 进程的堆内存，笔者会在后面介绍 ByteBuffer 时详细描述）。而 C 语言操作的内存为堆外内存。所以，如果是堆内内存，需要执行复制操作，将读取到的数据读入堆内内存中，这就是所谓的多一次的复制。

第 1 章　Java 网络编程

（3）最终读取操作和 FileInputStream read 方法一样，调用 C 函数库的 read 函数。

```java
public int read(ByteBuffer dst) throws IOException {
    ensureOpen();
    if (!readable)
        throw new NonReadableChannelException();
    synchronized (positionLock) {         // 上锁保证文件位点写入下标线程安全
        int n = 0;
        int ti = -1;
        try {
            begin();
            ti = threads.add();
            if (!isOpen())
                return 0;
            do {                  // 打开通道时，将缓冲区中的数据写入文件
                n = IOUtil.read(fd, dst, -1, nd);
            } while ((n == IOStatus.INTERRUPTED) && isOpen());
            return IOStatus.normalize(n);
        } finally {
            threads.remove(ti);
            end(n > 0);
            assert IOStatus.check(n);
        }
    }
}

static int read(FileDescriptor fd, ByteBuffer dst, long position, NativeDispatcher nd) throws IOException
{
    if (dst.isReadOnly())        // 缓冲区必须可写
        throw new IllegalArgumentException("Read-only buffer");
    if (dst instanceof DirectBuffer)   // 堆外内存缓冲区，直接读取文件然后写入
        return readIntoNativeBuffer(fd, dst, position, nd);

    // 堆内缓冲区，首先获取一个堆外的临时缓冲区
    ByteBuffer bb = Util.getTemporaryDirectBuffer(dst.remaining());
    try {                        // 然后指定读取操作，读取完毕后将临时缓冲区中的数据复制到堆内缓冲区中
        int n = readIntoNativeBuffer(fd, bb, position, nd);
        bb.flip();
        if (n > 0)
            dst.put(bb);
        return n;
    } finally {                  // 完毕后归还临时缓冲区
        Util.offerFirstTemporaryDirectBuffer(bb);
    }
}

private static int readIntoNativeBuffer(FileDescriptor fd, ByteBuffer bb, long position, NativeDispatcher nd)
    throws IOException
{
    // 判断缓冲区写入 position，了解即可，后面在描述缓冲区时会详细讲解
    int pos = bb.position();
    int lim = bb.limit();
    assert (pos <= lim);
    int rem = (pos <= lim ? lim - pos : 0);
```

```
    if (rem == 0)
        return 0;
    int n = 0;
    // 注意，这里操作的都是堆外内存地址
    if (position != -1) {                           // 从文件中间读
        n = nd.pread(fd, ((DirectBuffer)bb).address() + pos,
                    rem, position);
    } else {                                         // 从文件头部开始读
        n = nd.read(fd, ((DirectBuffer)bb).address() + pos, rem);
    }
    if (n > 0)
        bb.position(pos + n);
    return n;
}

// 通过 JNI 读取
static native int read(int fildes, long buf, int nbyte) throws UnixException;

JNIEXPORT jint JNICALL Java_sun_nio_fs_UnixNativeDispatcher_read(JNIEnv* env, jclass this, jint fd,jlong address, jint nbytes){
    ssize_t n;
    void* bufp = jlong_to_ptr(address);
    RESTARTABLE(read((int)fd, bufp, (size_t)nbytes), n);  // 和普通 FileInputStream 一样，调用 C 函数库的 read 函数
    if (n == -1) {
        throwUnixException(env, errno);
    }
    return (jint)n;
}
```

下面是 channel.write(writeBuffer)源码。由源码可知。

（1）write(ByteBuffer src)方法最终调用 IOUtil.write(fd, src, -1, nd)完成写入操作。

（2）在 IOUtil.write(fd, src, -1, nd)中，对缓冲区中的 ByteBuffer 类型进行判断。如果是堆内内存，则需要进行复制操作，将需写入的数据复制到堆外内存 ByteBuffer 中。这就是所谓的多一次的复制。

（3）最终，写入操作同 FileInputStream write 方法一样，要调用 C 函数库的 write 函数。

```
public int write(ByteBuffer src) throws IOException {
    ensureOpen();
    if (!writable)                                   // 通道必须可写
        throw new NonWritableChannelException();
    synchronized (positionLock) {                    // 上锁保证文件位点线程安全
        int n = 0;
        int ti = -1;
        try {
            begin();
            ti = threads.add();
            if (!isOpen())
                return 0;
            do {
                n = IOUtil.write(fd, src, -1, nd);    // 完成实际写入操作
            } while ((n == IOStatus.INTERRUPTED) && isOpen());
            return IOStatus.normalize(n);
        } finally {
```

```java
                threads.remove(ti);
                end(n > 0);
                assert IOStatus.check(n);
            }
        }
    }

    static int write(FileDescriptor fd, ByteBuffer src, long position, NativeDispatcher nd) throws IOException{
        if (src instanceof DirectBuffer)                    // 直接写入堆外缓冲区
            return writeFromNativeBuffer(fd, src, position, nd);
        // 否则分配临时堆外缓冲区
        int pos = src.position();
        int lim = src.limit();
        assert (pos <= lim);
        int rem = (pos <= lim ? lim - pos : 0);
        ByteBuffer bb = Util.getTemporaryDirectBuffer(rem);
        try {
            // 将数据复制到临时堆外缓冲区后，执行写入操作
            bb.put(src);
            bb.flip();
            src.position(pos);
            int n = writeFromNativeBuffer(fd, bb, position, nd);
            if (n > 0) {
                src.position(pos + n);
            }
            return n;
        } finally {                                         // 使用完毕后归还临时缓冲区
            Util.offerFirstTemporaryDirectBuffer(bb);
        }
    }

    private static int writeFromNativeBuffer(FileDescriptor fd, ByteBuffer bb, long position, NativeDispatcher nd)
        throws IOException
    {
        int pos = bb.position();
        int lim = bb.limit();
        assert (pos <= lim);
        int rem = (pos <= lim ? lim - pos : 0);
        int written = 0;
        if (rem == 0)
            return 0;
        if (position != -1) {                               // 从文件中间写入
            written = nd.pwrite(fd,
                                ((DirectBuffer)bb).address() + pos,
                                rem, position);
        } else {                                            // 从文件头部写入
            written = nd.write(fd, ((DirectBuffer)bb).address() + pos, rem);
        }
        if (written > 0)
            bb.position(pos + written);
        return written;
    }

    // JNI 完成写入
```

```
static native int write(int fildes, long buf, int nbyte) throws UnixException;

JNIEXPORT jint JNICALL Java_sun_nio_fs_UnixNativeDispatcher_write(JNIEnv* env, jclass this, jint fd, jlong address, jint nbytes){
    ssize_t n;
    void* bufp = jlong_to_ptr(address);    // 将 Java 的 jlong 类型转为 C 的指针，即 jlong_to_ptr(a) ((void*)(a))
    RESTARTABLE(write((int)fd, bufp, (size_t)nbytes), n);    // 调用 C 函数库的 write 方法完成实际写入
    if (n == -1) {
        throwUnixException(env, errno);
    }
    return (jint)n;
}
```

接下来，根据上述内容总结 Java NIO 编程模型。

（1）将原来传输双方数据的 OutputStream、InputStream 抽象为 Channel 通道，上述例子实现类为 FileChannelImpl。

（2）将原来传输双方数据的字节流，抽象为 ByteBuffer。可以支持堆内内存与堆外内存。

可以说，Java NIO 编程模型是对传统编程模型的封装，原生 Java IO 是面向字节流编程，Java NIO 是面向缓冲区编程。下面来看一组图。

（1）图 1-1 是 Java IO 的模型图。

（2）图 1-2 是使用 BufferedStream 带缓冲区的 IO 模型。

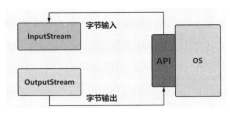
图 1-1　Java 原生 IO 模型

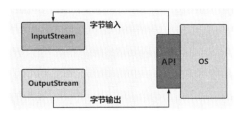
图 1-2　Java 带缓冲区 IO 模型

（3）图 1-3 和图 1-4 是 Java NIO 模型。

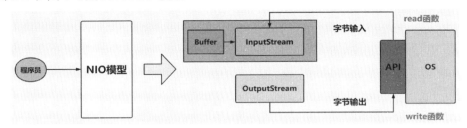
图 1-3　Java NIO 模型与 IO 模型

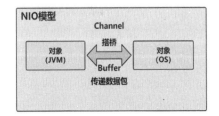
图 1-4　Java NIO 模型

1.4.2 Buffer 原理

Buffer 封装了需要在通道中传输的数据。它既然要存放数据，必然需要容器，而 Java NIO 编程模型是对传统编程模型的封装。从上面的源码分析中可知，ByteBuffer 分为 DirectBuffer（堆外内存）、HeapBuffer（堆内内存）。有些读者可能并不完全理解什么是堆外内存，什么是堆内内存。

可以这样理解，堆外内存为 JVM 进程的内存，堆内内存为堆外内存中分割出来的一部分空间，用于存放对象信息。这就好比一套房子，其中一间是主卧室，房子称为堆外内存，主卧室为堆内内存。如果要走出大门（数据传输），首先要从主卧室走到房子内部的客厅（将堆内内存的数据复制到堆外内存），再走出大门（将堆外内存的数据传递给 OS），如图 1-5 所示。

图 1-5 堆外内存与堆内内存

Buffer 是容器，存放容器的类型有 byte、char、short、int、long、float、double。如下为 Buffer 的继承体系，将它们的共用特性放入 Buffer 抽象类中，然后根据不同的容器类型定义不同的容器存放数据。由于在网络传输中通常将传输的数据编码为字节数据，将传递的字节数据解码为对象，所以本节中的 Buffer 实际类型为 ByteBuffer。同时，ByteBuffer 还有以下两个继承类。

（1）HeapByteBuffer：表示堆内内存的字节缓冲区。
（2）MappedByteBuffer：表示堆外内存的字节缓冲区。

```
Buffer (java.nio)
    IntBuffer (java.nio)                // 定义容器为 int[] hb 数组
    FloatBuffer (java.nio)              // 定义容器为 float[] hb 数组
    CharBuffer (java.nio)               // 定义容器为 char[] hb 数组
    DoubleBuffer (java.nio)             // 定义容器为 double[] hb 数组
    ShortBuffer (java.nio)              // 定义容器为 short[] hb 数组
    LongBuffer (java.nio)               // 定义容器为 long[] hb 数组
    ByteBuffer (java.nio)               // 定义容器为 byte[] hb 数组
        HeapByteBuffer (java.nio)       // 定义堆内内存的字节缓冲区
        HeapByteBufferR (java.nio)      // 定义堆内内存的只读字节缓冲区
        MappedByteBuffer (java.nio)     // 定义堆外内存的字节缓冲区（包括映射方法）
            DirectByteBuffer (java.nio) // 完整实现堆外内存的相关操作
            DirectByteBufferR (java.nio)// 定义堆外内存只读字节缓冲区
```

对于 Buffer 抽象的核心定义，由源码可知。

（1）在堆内内存中，可以直接操作 Java byte[]数组。这时 byte 为数组对象的首地址，可以直接使用数组操作符 byte[n]访问。而对于堆外内存，并没有实际的引用。这时定义 long address 表示指向堆外内存的数组的首地址。

（2）定义四个变量来操作缓冲区。

① position：表示当前正在操作的数组下标。
② limit：表示当前正在操作的数组下标限制，读写限制。
③ capacity：表示当前正在操作的数组的容量。
④ mark：表示保存当前正在操作的数组下标，可以在需要时将 position 还原为 mark 值。

假设有一个数组 byte[]。要知道当前数组的大小，则使用 capacity 变量表示。数组需要读写操作，则需要一个下标保存当前读写的位点，使用 position 变量表示。而这时对于 limit 而言就很好理解了，在进行读写操作时,需要一个下标限制读写的长度,它最大为 capacity,最小为 position,而在读写时,只读写 position<=limit 之间的下标即可。对于 mark 变量而言,有时候在读写时,需要一个变量表示当前 position 的位置,方便回退到该位置重新读写。

定义多个方法,方便子类调整 position、limit、capacity 和 mark 这四个变量。

Buffer 类的核心就是定义四个变量以及操作它们的方法,而这些方法和变量又正好是 Buffer 类操作子类的数据容器——数组的下标值。下面来看一组图,该组图描述了如下流程时的变量位置。

(1)初始 ByteBuffer,如图 1-6 所示。
(2)写入 2B 数据,如图 1-7 所示。
(3)调用 flip 方法转为读操作,如图 1-8 所示。
(4)读出 2B 数据,如图 1-9 所示。
(5)还原缓冲区,如图 1-10 所示。

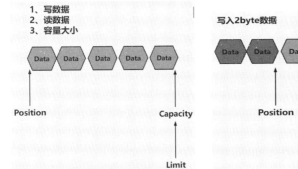

图 1-6　初始 ByteBuffer　　　　图 1-7　写入 2B 数据　　　　图 1-8　调用 flip 方法转为读操作

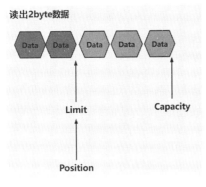
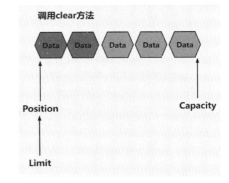

图 1-9　读出 2B 数据　　　　图 1-10　调用 clear 方法还原缓冲区

源码如下。

```
public abstract class Buffer {
    // 约束: mark <= position <= limit <= capacity
    private int mark = -1;
    private int position = 0;
    private int limit;
    private int capacity;

    // 只能被 DirectBuffer 堆内缓冲区对象使用,保存堆外内存的数组地址
```

```java
long address;

// 用于转换读写。将 limit 设置为当前 position，然后将 position 设置为 0，这时就可以读取或写入 position <= limit 中间的数组项
public final Buffer flip() {
    limit = position;
    position = 0;
    mark = -1;
    return this;
}

// 用于还原读取或写入 position 下标
public final Buffer rewind() {
    position = 0;
    mark = -1;
    return this;
}

// 将变量还原为默认值
public final Buffer clear() {
    position = 0;
    limit = capacity;
    mark = -1;
    return this;
}

// 记录当前 position 到 mark 变量中
public final Buffer mark() {
    mark = position;
    return this;
}

// 将保存的 mark 变量还原到 position 中
public final Buffer reset() {
    int m = mark;
    if (m < 0)
        throw new InvalidMarkException();
    position = m;
    return this;
}
}
```

然后是定义 ByteBuffer。由源码可知，该类提供了工具方法，方便外部创建缓冲区对象，同时声明了保存堆内内存数据的 byte[] hb 数组。源码如下。

```java
public abstract class ByteBuffer extends Buffer implements Comparable<ByteBuffer>{
    final byte[] hb;        // 堆内内存的数组
    final int offset;       // 当前操作该数组的起始下标，默认为 0
    // 构造器初始化变量
    ByteBuffer(int mark, int pos, int lim, int cap, byte[] hb, int offset){
        super(mark, pos, lim, cap);
        this.hb = hb;
```

```java
        this.offset = offset;
    }

    // 工具方法：创建堆外缓冲区
    public static ByteBuffer allocateDirect(int capacity) {
        return new DirectByteBuffer(capacity);
    }

    // 工具方法：创建堆内缓冲区
    public static ByteBuffer allocate(int capacity) {
        if (capacity < 0)
            throw new IllegalArgumentException();
        return new HeapByteBuffer(capacity, capacity);
    }

    // 工具方法：创建堆内缓冲区，并将 array 中 offset-length 之间的数据放入 HeapByteBuffer 中
    public static ByteBuffer wrap(byte[] array,
                                  int offset, int length)
    {
        try {
            return new HeapByteBuffer(array, offset, length);
        } catch (IllegalArgumentException x) {
            throw new IndexOutOfBoundsException();
        }
    }

    /*其他工具方法：分片、复制等*/
}
```

HeapByteBuffer 类主要实现了 Buffer 中定义的工具方法，对于下标的操作已经在 Buffer 中实现。源码如下。

```java
class HeapByteBuffer extends ByteBuffer{
    // 根据 buf 数组、偏移量、长度来初始化堆内内存
    HeapByteBuffer(byte[] buf, int off, int len) {
        super(-1, off, off + len, buf.length, buf, 0);
    }

    // 指定容量和初始 limit 来初始化堆内内存，来创建 byte 数组
    HeapByteBuffer(int cap, int lim) {
        super(-1, 0, lim, cap, new byte[cap], 0);
    }

    /*实现 Buffer 中定义的工具方法：分片、复制等*/
}
```

MappedByteBuffer 类提供了对文件映射 mmap 的支持。如图 1-11 所示，可以将磁盘上的文件通过系统调用 mmap 函数直接映射到用户空间中，读取和写入文件数据时不需要再经过以下流程：复制数据到用户空间缓冲区、复制数据到内核缓冲区和写入磁盘。源码中提供了三个本地方法和一个 FileDescriptor fd 变量，其中 FileDescriptor fd 变量用于表示磁盘中的文件，三个本地方法用于操作 mmap 函数。

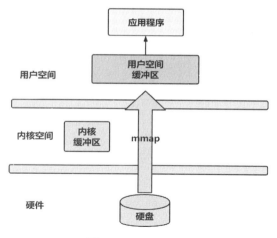

图 1-11 mmap 原理

由源码可知，只使用堆外内存时，不会涉及 FileDescriptor fd。源码如下。

```java
public abstract class MappedByteBuffer extends ByteBuffer{
    private final FileDescriptor fd;                              // 用于文件映射使用的 fd

    // 指定 fd 对文件进行映射
    MappedByteBuffer(int mark, int pos, int lim, int cap, FileDescriptor fd)
    {
        super(mark, pos, lim, cap);
        this.fd = fd;
    }

    // 不使用文件映射，仅使用堆外内存
    MappedByteBuffer(int mark, int pos, int lim, int cap) {
        super(mark, pos, lim, cap);
        this.fd = null;
    }

    private native boolean isLoaded0(long address, long length, int pageCount); // 判断当前文件中的内容是否
    已经映射到 JVM 内存中
    private native void load0(long address, long length);                       // 将文件映射到 JVM 内存中
    private native void force0(FileDescriptor fd, long address, long length);   // 将对于缓冲区的修改强制刷新
    到映射的文件中
}
```

最后是 DirectByteBuffer 类。由源码可知。

（1）内部使用 Unsafe unsafe 来分配和释放堆外内存。

（2）在 Cleaner.create(this, new Deallocator(base, size, cap))方法中传入 Deallocator 内部类。该类在回收 DirectByteBuffer 对象时释放占用的堆外内存，调用 unsafe.freeMemory(address)完成释放。只要 DirectByteBuffer 对象被回收，就不会发生堆外内存泄漏（关于这是如何做到的呢？这是一个虚引用的基础问题：只需要让 Cleaner 类继承虚引用，当 DirectByteBuffer 对象不被垃圾回收（Garbage Collection）时，由于 DirectByteBuffer 持有该对象，所以不会回收 Cleaner 对象。但是如果 DirectByteBuffer 对象被垃圾回收，不存在唯一的强引用对象了，就会把 Cleaner 对象放入 ReferenceQueue 队列，用 Reference

线程调用 Cleaner 的 clean 方法完成回调方法。此时就可以释放该对象中保存的 long address 堆外内存)。

（3）同时将用 Cleaner.create(this, new Deallocator(base, size, cap))方法创建的 Cleaner 对象保存到实例变量。这时就可以在外部通过反射调用该方法来手动释放堆外内存。Netty 也是这样实现的。

（4）DirectByteBuffer 没有使用文件映射，所以并没有初始化 FileDescriptor fd（关于如何使用映射，详见 FileChannel.map 方法）。

```java
// 堆外内存接口
public interface DirectBuffer {
    // 获取堆外内存的地址
    long address();

    // 获取与 DirectBuffer 绑定的 attach 对象
    Object attachment();

    // 获取用于释放堆外内存的 cleaner
    Cleaner cleaner();
}

class DirectByteBuffer extends MappedByteBuffer implements DirectBuffer{
    protected static final Unsafe unsafe = Bits.unsafe();  // 用于操作堆外内存
    private final Object att;        // 保存当前缓冲区绑定的对象，但通常不用，也可以使用该属性保存关联对象

    // 用于在 DirectByteBuffer 对象被回收后，回调释放堆外内存
    private static class Deallocator implements Runnable{

        private static Unsafe unsafe = Unsafe.getUnsafe();
        private long address;               // 保存堆外内存地址
        private long size;                  // 大小
        private int capacity;               // 容量

        private Deallocator(long address, long size, int capacity) {
            assert (address != 0);
            this.address = address;
            this.size = size;
            this.capacity = capacity;
        }

        public void run() {
            if (address == 0) {
                return;
            }
            unsafe.freeMemory(address);             // 直接使用 unsafe 释放内存
            address = 0;
            Bits.unreserveMemory(size, capacity);   // 减少 Bits 类中记录的已使用的堆外内存空间
        }
    }

    private final Cleaner cleaner;          // 清理缓冲区的对象

    // 构造器，使用容量来初始化成员变量
    DirectByteBuffer(int cap) {
```

```
        super(-1, 0, cap, cap);                              // 初始化父类
        boolean pa = VM.isDirectMemoryPageAligned();         // 获取页对齐大小,通常为 4KB
        int ps = Bits.pageSize();
        long size = Math.max(1L, (long)cap + (pa ? ps : 0));
        Bits.reserveMemory(size, cap);                       // 增加堆外内存使用计数
        long base = 0;
        try {
            base = unsafe.allocateMemory(size);              // 分配内存
        } catch (OutOfMemoryError x) {
            Bits.unreserveMemory(size, cap);
            throw x;
        }
        unsafe.setMemory(base, size, (byte) 0);              // 将内存中的值清零
        if (pa && (base % ps != 0)) {                        // 进行页对齐
            address = base + ps - (base & (ps - 1));
        } else {
            address = base;
        }
        cleaner = Cleaner.create(this, new Deallocator(base, size, cap));  // 创建 Cleaner 对象,同时将
Deallocator 类的实例挂入 Cleaner 列表
        att = null;                                          // 不使用 attach 对象
    }
}
```

1.4.3 Channel 原理

FileChannel 的读写操作都是调用同 FileInputStream 和 FileOutputStream 类一样的函数,即 read 和 write 函数。唯一不同的是,使用 FileChannel 传递的数据为 Buffer 缓冲区,可以说是 Channel 包装了 FileInputStream 和 FileOutputStream 类,实现了基于缓冲区通信的 Channel 通道。本节分析了 Channel 的整体结构,由于本书不是专门介绍 NIO 的 Java 原生类库方面的书,所以本节将向读者展示整个 Channel 的常用继承体系。对于 FileChannel 而言,笔者在描述 NIO 编程一节时,已经向读者展示了源码。对于 SocketChannel 而言,由于其类相对复杂且会占用大量篇幅,所以将其省略,读者只需要了解这些类拥有什么样的功能即可。如果读者想要研究接口的实现,则可以根据前面介绍的 FileChannelImpl 的流程来学习。

1. Channel 接口

首先是最底层的 Channel 接口定义。Channel 接口通道表示硬件设备、文件、网络套接字或程序组件等对象的连接。这些对象能够执行一个或多个不同的 I/O 操作,例如读或写。Channel 通道处于打开或者关闭状态。在创建通道时默认保持打开状态,一旦执行 Close 方法将其关闭,它将一直保持关闭状态。一旦通道关闭,任何尝试调用 I/O 的操作都将抛出 ClosedChannelException。通道是否打开可以通过调用 isOpen 方法来判断。通常,通道的实现是多线程安全的,具体将在扩展和实现该接口的接口和类的规范中有所描述。源码如下:

```
public interface Channel extends Closeable {

    // 判断当前通道是否处于打开状态
    public boolean isOpen();
```

```
// 关闭该通道。当通道关闭后，尝试调用通道的 IO 操作会抛出 ClosedChannelException 异常。如果通道已
经关闭，则调用该接口不会执行
public void close() throws IOException;
}
```

2．ReadableByteChannel 接口

ReadableByteChannel 通道表示可以读取字节数据的通道。同一时间段内，只能有一个线程在可读通道上执行读操作。一个线程在该通道上执行读操作，其他任何尝试读操作的线程都会阻塞，直到其他线程读取完成。其他类型的 I/O 操作是否可以与读操作同时进行取决于该通道的子类实现。源码如下。

```
public interface ReadableByteChannel extends Channel {
    // 从这个通道读取字节放入给定的字节缓冲区。假如此时调用此方法，从通道读取 r 字节，其中 r 是缓冲区中
剩余的字节数，即 dst.remaining 方法的返回值，也就是 limit - position 的长度。假设此时读取长度为 n 的字节，
其中 0<=n<=r，这些字节将被存放到缓冲区的 position - position + n - 1 的区间中。返回时，缓冲区的 limit 不会改
变。一个读操作可能不会将整个缓冲区填满，实际上它也有可能根本不会存放任何字节信息，这取决于通道当前
的状态。例如，在非阻塞模式下，Socket 通道不能读取比 Socket 输入缓冲区多的字节，同样，文件通道只能读
取文件中剩余的字节。但是可以保证的是，如果一个通道处于阻塞模式，并且缓冲区中至少还剩下一个字节，则
这个方法将阻塞，直到至少读取一个字节为止。此方法可以在任何时候调用。但是，如果另一个线程已经在这个
通道上执行读操作，则这个方法的调用会阻塞，直到该线程操作完成
    public int read(ByteBuffer dst) throws IOException;
}
```

3．WritableByteChannel 接口

该接口描述同 ReadableByteChannel 接口，只不过这里定义的是可写操作的方法。源码如下。

```
public interface WritableByteChannel extends Channel
{
    // 将给定的缓冲区中的字节数据写入通道。假如此时需要尝试向通道写入 r 个字节，其中 r 是缓冲区中剩余的
字节数，即 src.remaining 方法返回值(limit - position 的长度)。假设写入了一个长度为 n 的字节数据，其中 0<=n<=r。
这些字节信息将从 position 下标处开始的缓冲区传输，最后写入的字节的下标是 position + n - 1。返回时，缓冲区
的 limit 不会改变。写操作只有在写完所有 r 个字节后才会返回，除非另有说明。在某些特定类型的通道中，可以
根据它们的状态，只写部分字节，或者根本不写。例如，Socket 通道在非阻塞模式下，不能写入超过 Socket 输
出缓冲区大小的字节。此方法可以在任何时候调用。但是，如果另一个线程已经在这个通道上执行写操作，则这
个方法的调用会阻塞，直到该操作完成
    public int write(ByteBuffer src) throws IOException;
}
```

4．ByteChannel 接口

ByteChannel 接口很简单，它组合了 ReadableByteChannel 和 WritableByteChannel 接口，本身没有声明任何新方法，表示一个可读且可写的通道。源码如下。

```
public interface ByteChannel extends ReadableByteChannel, WritableByteChannel {}
```

5．SeekableByteChannel 接口

SeekableByteChannel 接口定义了一个可以保存当前操作位置并允许改变该位置的通道，该通道通常代表一个文件对象。可以通过 position 方法获取当前操作下标，通过 position(long)方法修改操作下标。该接口还定义了对通道所表示的对象大小的访问，例如文件的大小。源码如下。

```
public interface SeekableByteChannel extends ByteChannel {
```

```java
    // 获取当前操作下标
    long position() throws IOException;

    // 修改操作下标
    SeekableByteChannel position(long newPosition) throws IOException;

    // 获取当前通道代表的对象大小
    long size() throws IOException;

    // 从当前通道中截取长度为 size 的通道
    SeekableByteChannel truncate(long size) throws IOException;
}
```

6．GatheringByteChannel 接口

GatheringByteChannel 接口表示一个可以从缓冲区数组中批量写入字节数据的通道。即在一次写操作调用中，从一个或多个给定的缓冲区数组中写入一个或多个字节。当实现网络协议或文件格式时，聚集写操作非常有用。例如，将数据分组到由一个或多个固定长度的头和可变长度的内容体组成的 Buffer 缓冲区中批量写入。源码如下。

```java
public interface GatheringByteChannel extends WritableByteChannel{

    // 将给定缓冲区数组中的内容写入该通道。写入该通道的数据长度最多为 r 个字节，其中 r 是给定缓冲区数组中
    // 剩余可读的总字节数，即 srcs[offset].remaining()+srcs[offset+1].remaining()+...+srcs[offset+length-1].remaining()。
    // 假设写了一个长度为 n 的字节序列，其中 0<=n<=r。此时，第一个缓冲区 srcs[offset].remaining()字节写入通道；
    // 然后是第二个缓冲区 srcs[offset+1].remaining()字节写入通道。以此类推，直到满足 n 个字节，数据写入停止。默
    // 认情况下，写操作只有在写完所有指定的 r 长度字节后才会返回。当然在某些特定类型的通道实现中，会根据通道的
    // 状态，来选择只写入通道部分字节，也可能根本不写入通道。例如，Socket 通道在非阻塞模式下，不能写入超过 Socket
    // 输出缓冲区大小的字节数。如果另一个线程已经在这个通道上执行写操作，则调用这个方法的其他线程会阻塞，直
    // 到另一个线程操作完成。其中，参数 offset 表示缓冲区数组中的第一个需要写入的缓冲区的下标值，必须为非负且不
    // 大于 src.length；length 表示从 offset 处写入多少个 ByteBuffer 数据，必须为非负且不大于 src.length
    public long write(ByteBuffer[] srcs, int offset, int length)
        throws IOException;

    // 同 write(ByteBuffer[] srcs, int offset, int length)方法一样。默认从 0 下标处写入 ByteBuffer[] srcs 数组中所
    // 有的缓冲区
    public long write(ByteBuffer[] srcs) throws IOException;
}
```

7．ScatteringByteChannel 接口

ScatteringByteChannel 接口表示一种可以将多个字节批量读入缓冲区数组的通道。Scattering 读操作可以在一次调用中将多个字节数据读入一个或多个给定的缓冲区数组中。当实现网络协议数据传输或文件数据传输时，Scattering 读操作作用处非常多。例如，可以将通道数据拆分为由一个或多个固定长度的头和可变长度的数据主体，然后这些分组后的数据放入缓冲区数组。源码如下。

```java
public interface ScatteringByteChannel extends ReadableByteChannel {

    // 从通道中读取多个字节数据到给定缓冲区的数组中。尝试从该通道读取最多 r 个字节。其中 r 是给定缓冲区数组
```

中剩余的可写总字节数,即 dsts[offset].remaining()+dsts[offset+1].remaining()+...+dsts[offset+length-1].remaining()。假设此时通道中有长度为 n 的数据可以读取,其中 0<=n<=r,则此时将从缓冲区数组中的第一个 ByteBuffer 开始的可读下标 readIndex 处开始放入数据,直到整个 n 个数据被传输到缓冲区数组中。如果另一个线程已经在这个通道上执行了一个读操作,则该方法的调用会阻塞,直到占用线程操作完成

```
public long read(ByteBuffer[] dsts, int offset, int length)
    throws IOException;

// 同 read(ByteBuffer[] dsts, int offset, int length)方法一样。默认从 0 下标处,读取 ByteBuffer[] dsts 数组中所有的缓冲区
public long read(ByteBuffer[] dsts) throws IOException;
}
```

8. InterruptibleChannel 接口

InterruptibleChannel 为标识性接口,表示可以异步关闭和中断的通道。实现此接口的通道可以异步关闭。如果一个线程在可中断通道的 I/O 操作中被阻塞,则另一个线程可能会调用该通道的 close 方法,这将导致阻塞线程接收一个 AsynchronousCloseException 异常。实现这个接口的通道也是可中断的。如果一个线程在可中断通道上的 I/O 操作被阻塞,则另一个线程可能会调用被阻塞线程的 Thread.interrupt() 中断方法,这将导致通道被关闭。阻塞线程接收到一个 ClosedByInterruptException 异常,并且阻塞线程的中断状态位被设置。如果一个线程的中断状态位已经被设置,并且它在一个通道上执行了一个阻塞 I/O 操作,则这个通道将被关闭。并且线程将立即收到一个 ClosedByInterruptException 异常,且它的中断状态将保持不变。只有当通道实现了此接口时,它才支持异步关闭和中断。源码如下。

```
public interface InterruptibleChannel extends Channel {}
```

9. NetworkChannel 接口

NetworkChannel 接口表示一种描述网络套接字的通道。bind(SocketAddress)方法用于将套接字绑定到本地 SocketAddress 地址,getLocalAddress 方法返回套接字绑定到的地址,setOption 和 getOption 方法用于设置和查询套接字选项。源码如下。

```
public interface NetworkChannel extends Channel
{
    // 将通道表示的套接字绑定到本地地址,即在套接字和本地地址之间建立关联。一旦建立了关联,套接字将保持绑定状态,直到通道关闭。如果 local 参数的值为 null,则套接字将绑定到一个自动分配的地址上
    NetworkChannel bind(SocketAddress local) throws IOException;

    // 获取通道表示的套接字绑定的本地地址
    SocketAddress getLocalAddress() throws IOException;

    // 设置套接字选项
    <T> NetworkChannel setOption(SocketOption<T> name, T value) throws IOException;

    // 获取名为 name 的套接字选项
    <T> T getOption(SocketOption<T> name) throws IOException;

    // 列出通道表示的套接字支持的选项
    Set<SocketOption<?>> supportedOptions();
}
```

10. AsynchronousChannel 接口

AsynchronousChannel 接口为标识性接口,用于表示一种支持异步 I/O 操作的通道。异步 I/O 操作

通常有两种类型。

（1）Future operation(…)。

（2）void operation(… A attachment, CompletionHandler <V,? super A> handler)。

其中 operation 表示 I/O 操作的名称（例如读或写）。V 表示 I/O 操作的结果类型。A 表示关联到 I/O 操作的对象类型，主要用于在操作结果时提供上下文信息。CompletionHandler 表示异步执行完毕后回调的对象。A 关联对象对于使用 CompletionHandler 异步回调来说非常重要，因为需要在回调时获取到上文传递过来的元数据完成进一步的操作。

对于第一种类型，可以使用 Future 接口的方法判断操作是否已完成，或等待操作完成并获取结果。

对于第二种类型，当异步 I/O 操作完成或失败时，会调用 CompletionHandler 处理 IO 操作的结果。

实现该接口的通道是可进行异步关闭。如果未完成通道上的 I/O 操作，并调用通道的 close 方法，则 I/O 操作会失败，抛出 AsynchronousCloseException 异常。异步通道是线程安全的，一些通道实现可能支持并发读和写。

Future 接口定义的 Future.cancel 方法可用于取消异步任务。如果此时对异步执行的 IO 通道操作执行该方法，会导致所有等待 I/O 操作结果的线程抛出 java.util.concurrent.CancellationException 异常。但是底层 I/O 操作是否可以取消取决于具体实现。如果取消操作导致了通道和它所连接的对端处于数据不一致的状态，该通道将被置于特定的错误状态，以防止进一步尝试发起类似于被取消的 I/O 操作。例如，如果取消读操作，但实现不能保证是否继续能从通道中读取字节，它会将通道置于错误状态，再次尝试发起读操作会抛出异常。同样，如果取消写操作，但实现不能保证没有字节被写入通道，随后发起写操作的尝试将失败，并抛出异常。

注意，调用 Future.cancel 方法时，如果指定 mayInterruptIfRunning 参数设置为 true，I/O 操作可能会关闭通道而中断响应。在这种情况下，所有等待 I/O 操作结果的线程都会抛出 CancellationException，并且通道上任何其他未完成的 I/O 操作也都会抛出 AsynchronousCloseException 异常。源码如下。

```
public interface AsynchronousChannel extends Channel{}
```

11. AbstractInterruptibleChannel 抽象类

AbstractInterruptibleChannel 抽象类为可中断通道的基本实现类。该类实现了通道的异步关闭和中断所需的底层机制。具体的通道类必须在调用可能永久阻塞的 I/O 操作前后分别调用 begin 和 end 方法，所以要在 try… finally 代码块中使用这些方法，如下所示。

```
boolean completed = false;
try {
    begin();
    completed = ...;     // 执行 IO 阻塞操作
    return ...;          // 返回结果
} finally {
    end(completed);
}
```

上述例子中，end 方法的 completed 参数用于表示 I/O 操作是否完成。例如，在读取字节的操作中，当且仅当某些字节实际上被复制到调用者的传入的目标缓冲区中时，该参数为 true。源码如下。

```
public abstract class AbstractInterruptibleChannel implements Channel, InterruptibleChannel {
```

```java
private final Object closeLock = new Object();        // 关闭锁对象
private volatile boolean open = true;                  // 表示当前通道是否处于打开状态

// 关闭该通道
public final void close() throws IOException {
    synchronized (closeLock) {                         // 获取锁，以保证线程安全
        if (!open)
            return;
        open = false;
        implCloseChannel();
    }
}

// 子类完成实际关闭方法
protected abstract void implCloseChannel() throws IOException;

// 判断当前通道是否已经关闭
public final boolean isOpen() {
    return open;
}

private Interruptible interruptor;                     // 线程中断时回调该对象的 interrupt 方法
private volatile Thread interrupted;

// 标记当前 IO 操作可能一直阻塞
protected final void begin() {
    if (interruptor == null) {
        interruptor = new Interruptible() {            // 初始化中断时回调对象
            public void interrupt(Thread target) {
                synchronized (closeLock) {             // 获取对象锁关闭当前通道
                    if (!open)
                        return;
                    open = false;
                    interrupted = target;
                    try {
                        AbstractInterruptibleChannel.this.implCloseChannel();
                    } catch (IOException x) { }
                }
            }
        };
    }
    blockedOn(interruptor);                            // 设置当前通道对象阻塞在 interruptor 对象上，用于性能分析
    Thread me = Thread.currentThread();
    if (me.isInterrupted())                            // 如果当前线程已经被中断，则回调 interrupt 方法
        interruptor.interrupt(me);
}

// 标识异步结果结束，completed 标识异步执行是否完成，线程被中断返回时抛出 ClosedByInterruptException
// 异常。否则抛出 AsynchronousCloseException 异常，标识通道被其他线程异步关闭
protected final void end(boolean completed) throws AsynchronousCloseException{
    blockedOn(null);                                   // 清除标识对象
    Thread interrupted = this.interrupted;
    if (interrupted != null && interrupted == Thread.currentThread()) { // 线程被中断
        interrupted = null;
```

```java
            throw new ClosedByInterruptException();
        }
        if (!completed && !open)                              // 通道关闭
            throw new AsynchronousCloseException();
    }

    // -- sun.misc.SharedSecrets --
    static void blockedOn(Interruptible intr) {
        sun.misc.SharedSecrets.getJavaLangAccess().blockedOn(Thread.currentThread(), intr);
    }
}
```

12. SelectableChannel 抽象类

SelectableChannel 抽象类表示一种可通过选择器（Selector）进行多路复用的通道。它继承自 AbstractInterruptibleChannel 抽象类，也具备中断通道的功能。为了与选择器一起使用，这个类的实例对象必须首先通过 register 方法将通道注册到选择器中。当注册成功后，该方法将返回一个新的 SelectionKey 对象。该对象表示注册到选择器中的通道，一旦注册到选择器中，通道将保持注册状态，直到取消注册。注册到选择器中的通道不能直接解除注册，而是需要先将 SelectionKey 对象从选择器中取消。调用 SelectionKey.cancel() 方法执行解除注册操作时，SelectionKey 对象可以在选择器的下一个选择操作期间取消通道的注册。通道关闭时（无论是通过调用 Channel.close 方法关闭，还是通过中断在通道上的 I/O 操作中阻塞的线程关闭），通道的所有 SelectionKey 对象都将被取消。如果关闭选择器，则通道也将随之注销。表示其注册的键将立即失效，而不是延迟到下一次 Selector.select 选择期间移除。一个通道只能被同一个选择器注册一次。通道是否注册到一个或多个选择器中，可以使用 isRegistered 方法来判断。可选择的通道是多线程安全的。

SelectableChannel 通道实例或者处于阻塞模式，或者处于非阻塞模式。在阻塞模式下，在通道上调用的每个 I/O 操作都将阻塞，直到它完成。在非阻塞模式下，I/O 操作永远不会阻塞，可能实际传输的字节比要求处理的字节少，或根本没有传输任何字节，这取决于通道当前的状态（例如，当通道底层的缓冲区不可写时，要求写入 10B，则不会写入任何字节。这是因为缓冲区已经满了，此时会立即返回）。可选择通道的阻塞模式可以通过调用它的 isBlocking 方法确定，新创建的可选择通道实例总是处于阻塞模式。非阻塞模式与选择器（Selector）多路复用器结合在一起能够显著提高性能。在向选择器注册之前，通道必须置于非阻塞模式，并且在取消注册之前不能返回到阻塞模式。源码如下。

```java
public abstract class SelectableChannel extends AbstractInterruptibleChannel implements Channel
{
    // 返回创建该通道实例的 SelectorProvider 对象
    public abstract SelectorProvider provider();

    // 返回标识此通道支持的操作集，在这个整数值中设置的位精确地表示对这个通道有效的操作
    // (SelectionKey.OP_READ、SelectionKey.OP_WRITE、SelectionKey.OP_CONNECT、SelectionKey.OP_ACCEPT)
    public abstract int validOps();

    // 判断此通道当前是否注册到选择器中。因为新创建的通道实例未注册，所以返回 false。由于 SelectionKey
    // 的取消和通道注销之间存在一定的延迟（取消 SelectionKey 将在下一次执行 select 操作时执行），取消所有
    // SelecttionKey 后，通道可能在一段时间内仍保持注册状态，通道关闭后由于同样原因也将在一段时间内保持注册
    // 状态
```

```
public abstract boolean isRegistered();

// 获取表示通道与给定选择器的注册的键（SelectionKey）
public abstract SelectionKey keyFor(Selector sel);

// 将该通道注册到给定的选择器中，并返回一个选择键。如果该通道当前已注册到给定的选择器中，则返回代
表该注册的选择键，键的感兴趣集将被更改为参数 ops，就像直接调用 SelectionKey.interestOps(int)方法一样。如
果 att 参数不为空，则键的 attach 对象将被设置为该值。如果键已经被取消，则会抛出 CancelledKeyException。如
果该通道还没有在给定的选择器中注册，就会被注册并返回生成的新键 SelectionKey，SelectionKey 对象的初始
兴趣集将是 ops，它的 attach 对象是 att。如果该方法被调用时，另一个线程调用 configureBlocking(boolean)方法
正在配置通道，则它将首先阻塞，直到其他操作完成。该方法随后将获取选择器的 SelectionKey 键集锁保证线程
安全。因此，涉及相同选择器的键集锁的其他操作会被阻塞。例如，另一个注册或选择操作同时调用该方法
public abstract SelectionKey register(Selector sel, int ops, Object att)
    throws ClosedChannelException;

// 同 register(Selector sel, int ops, Object att)方法一样，只是不指定 Object att 对象
public final SelectionKey register(Selector sel, int ops) throws ClosedChannelException{
    return register(sel, ops, null);
}

// 将通道配置为非阻塞模式
public abstract SelectableChannel configureBlocking(boolean block)
    throws IOException;

// 判断通道是否阻塞
public abstract boolean isBlocking();

// 获取对象锁，用于 configureBlocking 方法与 register 方法之间的同步，保证两个方法的执行顺序
public abstract Object blockingLock();
}
...
```

13．FileChannel 抽象类

FileChannel 抽象类是用于读取、写入、映射和操作文件的通道。FileChannel 是连接到磁盘文件的 SeekableByteChannel 通道的实例。它在其连接到的文件中有一个当前操作的位置（读或写），可以通过 position()方法查询，也可以通过 position(long)方法修改该位置。磁盘文件本身包含多个数据字节信息，可以读取和写入。可以使用 size 方法查询当前文件大小：当字节写入超过当前大小时，文件的大小会相应增加；当文件被截断时，文件的大小会相应减小。文件还可能有一些相关的元数据，如访问权限、内容类型和最后修改时间。该类没有定义元数据的访问方法。除了简单的通道的读、写和关闭操作外，这个类还定义了以下文件特定的操作。

（1）可以调用 read(ByteBuffer, long)或 write(ByteBuffer, long)，在指定的文件的偏移量处操作文件，不会影响通道当前操作的 position 位置。

（2）文件的一块数据区域可以是 map 方法直接映射到当前 JVM 进程中，对于大文件，这通常比调用读或写方法要高效得多（在 Linux 平台其实就是 mmap 函数原理）。

（3）修改映射到 JVM 进程中的文件，可以通过 force 方法将修改内容写入磁盘文件中，以确保在系统崩溃时数据不会丢失。

（4）字节数据可以从一个文件通道通过 transferTo/transferFrom 方法传输到其他的通道中或从其他

通道传输到当前通道中。这两个方法的实现将使用一种可以被许多操作系统优化的方式，直接在文件系统缓存中进行非常快速地传输（在 Linux 平台其实就是 sendfile 函数原理）。

（5）文件的一个数据区域可以通过 lock 方法禁止其他线程访问。

文件通道是并发安全的。在任何时候，只能有一个修改通道位置或可以改变其文件大小的操作执行，因为对于通道，操作文件的位置和文件本身而言是线程互斥的。执行这些存在互斥的方法时，线程将获取锁执行，此时会串行化修改操作。对于其他没有明确规定的操作，当它们不存在多线程互斥时可以并行执行，而是否真的并行执行取决于底层的实现。

由该类的实例提供的文件信息应该与同一程序中其他实例提供的同一文件的信息保持一致，即同一文件不同对象间看到的文件信息应该是一致的。但是，由于底层操作系统存在文件缓存和网络文件系统协议导致的数据更新延迟，所以最终此类实例当中，在同一文件上所提供的信息可能与其他并发运行的程序所看到的信息一致，也可能不一致。不管这些程序是用什么语言编写的，也不管它们是在同一台机器上，还是在其他机器上运行的，都可能出现不一致的现象。

文件通道实例是通过调用该类定义的 open 方法创建的。文件通道也可以从现有的 java.io.FileInputStream 的 getChannel 方法，java.io.FileOutputStream 的 getChannel 方法或 java.io.RandomAccessFile 的 getChannel 方法创建。对于同一个文件而言，上述方法都返回连接到相同文件的文件通道对象。如果文件通道是从现有的输入输出流或 RandomAccessFile 中获取的，文件通道的状态与 getChannel 方法返回该通道的对象的状态密切相关。通过通道读写文件的数据会改变原始对象的文件操作位置，通过文件通道改变文件长度也会改变原始 File 文件对象的长度。反之亦然，即通过通道做的一切都会被原始的 File 文件对象所感知。源码及相关描述如下。

```
public abstract class FileChannel extends AbstractInterruptibleChannel implements
SeekableByteChannel, GatheringByteChannel, ScatteringByteChannel {

    // 打开或创建一个文件，并返回一个文件通道访问该文件。path 参数表示文件的路径信息，options 参数决定
    如何打开文件。StandardOpenOption.READ 和 StandardOpenOption.WRITE 选项决定是否打开文件进行读取或
    写入。如果数组中没有包含任何选项，默认文件以 StandardOpenOption.READ 打开。默认情况下，读或写将从
    文件的 0 下标处开始执行。除了 READ 和 WRITE，还可能存在以下选项。
    // StandardOpenOption.APPEND 把写入的数据追加在文件的数据末尾
    // StandardOpenOption.TRUNCATE_EXISTING 把已经存在的文件大小设置为 0，即清空文件
    // StandardOpenOption.CREATE_NEW 创建一个新文件，如果文件已存在，则抛出异常
    // StandardOpenOption.CREATE 创建一个新文件，如果文件已存在，就用新文件覆盖旧文件
    // StandardOpenOption.DELETE_ON_CLOSE 当调用 close 方法关闭文件时，将尝试将磁盘文件删除
    // StandardOpenOption.SPARSE 给操作系统提示当前操作的文件为稀疏存储的文件（稀疏存储的文件可以
    在文件内容中间进行插入，而不是覆盖）
    // StandardOpenOption.SYNC 每次更新文件时，将文件的内容和元数据信息同步到磁盘上
    // StandardOpenOption.DSYNC 每次更新文件时，将文件的内容同步到磁盘上
    public static FileChannel open(Path path,
                        Set<? extends OpenOption> options,
                        FileAttribute<?>... attrs)
        throws IOException
    {
        FileSystemProvider provider = path.getFileSystem().provider();
        return provider.newFileChannel(path, options, attrs);
    }
```

```java
    // 文件属性和值数组
    private static final FileAttribute<?>[] NO_ATTRIBUTES = new FileAttribute[0];

    // 同 open(Path path,Set<? extends OpenOption> options,FileAttribute<?>... attrs)方法，只不过这里不提供
文件属性列表
    public static FileChannel open(Path path, OpenOption... options)
        throws IOException
    {
        Set<OpenOption> set = new HashSet<OpenOption>(options.length);
        Collections.addAll(set, options);
        return open(path, set, NO_ATTRIBUTES);
    }

    // -- 通道操作 --

    // 默认从 0 下标处开始放入通道数据
    public final long read(ByteBuffer[] dsts) throws IOException {
        return read(dsts, 0, dsts.length);
    }
    // 默认从 0 下标处开始写入通道数据
    public final long write(ByteBuffer[] srcs) throws IOException {
        return write(srcs, 0, srcs.length);
    }
    // 子类完成实际方法实现
    public abstract int read(ByteBuffer dst) throws IOException;
    public abstract long read(ByteBuffer[] dsts, int offset, int length) throws IOException;
    public abstract int write(ByteBuffer src) throws IOException;
    public abstract long write(ByteBuffer[] srcs, int offset, int length) throws IOException;
    public abstract long position() throws IOException;
    public abstract FileChannel position(long newPosition) throws IOException;
    public abstract long size() throws IOException;
    public abstract FileChannel truncate(long size) throws IOException;
    public abstract void force(boolean metaData) throws IOException;
    // 将当前通道的数据传输到 WritableByteChannel target 通道中
    public abstract long transferTo(long position, long count, WritableByteChannel target)throws IOException;
    // 将 ReadableByteChannel src 通道的数据传输到当前通道中
    public abstract long transferFrom(ReadableByteChannel src, long position, long count)throws IOException;
    public abstract int read(ByteBuffer dst, long position) throws IOException;
    public abstract int write(ByteBuffer src, long position) throws IOException;

    // -- 内存映射操作 --

    // 当进行文件映射时使用的映射模式
    public static class MapMode {

        // 只读模式
        public static final MapMode READ_ONLY
            = new MapMode("READ_ONLY");

        // 可读可写模式
        public static final MapMode READ_WRITE
            = new MapMode("READ_WRITE");
```

```java
    // 当前线程私有映射（即用于代替 malloc 分配内存，指定该选项不会对磁盘文件进行映射）
    public static final MapMode PRIVATE
        = new MapMode("PRIVATE");

    private final String name;

    private MapMode(String name) {
        this.name = name;
    }
}
```

// 将该通道文件的一片区域（position -- position + size）直接映射到内存中。文件区域可以通过以下三种方式映射到内存中
// 只读模式：任何修改缓冲区内容的操作都将抛出 java.nio.ReadOnlyBufferException。（映射模式为 READ_ONLY）
// 读/写模式：对缓冲区所做的更改最终将写入文件中。对于已经映射了相同文件的其他程序来说，它们可能可见，也可能不可见。因为写入的内容可能还在当前进程的页中，还没有同步到磁盘中。（映射模式为 READ_WRITE）
// 私有模式：对缓冲区所做的更改不会写入文件中，并且不会对已经映射了相同文件的其他程序可见。如果当前应用程序对缓冲区修改，将导致操作系统创建已修改部分的私有副本（即 Copy On Write）。（映射模式为 PRIVATE）
// 对于只读映射，通道必须以读模式创建，以供应用程序读取数据。对于读/写或私有映射，通道必须同时以读和写模式创建。这个方法返回的 MappedByteBuffer 对象映射的字节缓冲区的 position 为 0，并且 limit 和 capacity 将由参数 size 决定，它的 mark 将初始化为-1，这在前面已经介绍过。缓冲区及其映射的文件将一直存在，直到缓冲区本身被垃圾回收。文件映射一旦建立，就不独立于创建它的文件通道对象，所以关闭通道不会影响映射的有效性。内存映射文件的许多细节依赖于底层操作系统的实现，因此对于不同系统会存在差异，所以没有具体说明底层的实现原理，不过本节的后面会介绍在 Linux 下的文件映射函数原理。需要映射的区域大于在此通道连接的文件大小时，该方法的行为是未指定的。有可能会映射失败抛出异常，也有可能会映射成功，但是在访问到不属于文件的部分时会抛出异常。该程序或其他程序对底层文件的内容或大小所做的更改是否会影响映射的缓冲区，将由操作系统实现，并且应用程序对缓冲区的更改写入文件的速率也由操作系统控制。对于大多数操作系统来说，将一个文件映射到内存中比通过普通的读写方法读写几十 KB 的数据要昂贵得多（因为映射建立时，存在 page fault 缺页异常，导致 OS 去处理该异常而不是执行读写操作）。从性能角度看，通常文件映射适用于相对较大的文件映射到内存，同时已经建立了真实映射，即进程虚拟内存与物理页帧的实际映射。如果没有建立该映射，随着文件的大小增加，page fault 缺页中断会更多

```java
    public abstract MappedByteBuffer map(MapMode mode, long position, long size)
        throws IOException;
```

// -- 锁实现 --

// 对文件 position - position + size 区域上锁，参数 shared 表明是否为共享锁
// 获取该通道表示的文件在区域 position - position + size 上的锁。此方法的调用将被阻塞，直到锁定该区域，或关闭通道，或中断调用线程。如果该通道在调用此方法期间被另一个线程关闭，则将抛出 AsynchronousCloseException 异常。如果调用线程在等待获取锁时被中断，则将设置中断状态位，并抛出 FileLockInterruptionException 异常；如果调用该方法时调用线程已经处于中断状态，则会立即抛出异常，线程的中断状态位不会被改变
// 由 position 和 size 参数指定的区域不需要包含在实际的底层文件中，即可以不在文件的大小区域内。被锁定的区域大小是固定的。如果锁定区域最初包含了文件的结尾，但随着数据写入，该文件的大小超出了该区域，文件的新增部分不会被锁覆盖。如果预计文件的大小会增长，并且需要对整个文件加锁，则应该锁定一个从 0 开始的区域，并且该区域不小于文件大小预估增长的最大大小。不带参数的 lock 方法会锁定一个大小为 Long.MAX_VALUE 的区域，这已经完全足够覆盖整个文件
// shared 参数用于指明是否使用共享锁，即读读共享，读写互斥。一些操作系统不支持共享锁，在这种情况下，对共享锁的请求会自动转换为对排他锁的请求。通过调用锁对象的 FileLock.isShared 方法可以判断新获得的锁是共享的，还是排他的
// 注意，整个 JVM 进程持有文件锁，所以文件锁不适合用于控制同一虚拟机进程中的多个线程对文件的访问互斥

```
public abstract FileLock lock(long position, long size, boolean shared)
    throws IOException;

// 同 lock(long position, long size, boolean shared)，表示锁定整个文件
public final FileLock lock() throws IOException {
    return lock(0L, Long.MAX_VALUE, false);
}

// 尝试获取锁，不进行阻塞
public abstract FileLock tryLock(long position, long size, boolean shared)
    throws IOException;

// 同 tryLock(long position, long size, boolean shared)方法一样，表示锁定整个文件
public final FileLock tryLock() throws IOException {
    return tryLock(0L, Long.MAX_VALUE, false);
}
}
```

14．SocketChannel 抽象类

SocketChannel 抽象类表示一个可选择的面向流连接套接字的通道，即 TCP 客户端通道。通过调用该类的 open 方法可以创建套接字通道。新创建的套接字通道处于已打开状态，但尚未连接。尝试在未连接的通道上调用 I/O 操作将导致抛出 NotYetConnectedException 异常。套接字通道可以通过调用它的 connect 方法连接对端 Socket，一旦连接成功，套接字通道将保持连接状态，直到关闭。通过调用 socket 通道的 isConnected 方法可以确定它是否被连接。

套接字通道继承自 AbstractSelectableChannel 抽象类，所以它支持非阻塞连接。非阻塞模式的套接字通道在创建后，建立到远程套接字的连接过程可以通过 connect 方法启动，此时该方法不会被阻塞。当 Socket 可以连接时，随后可以通过 finishConnect 方法完成实际连接。通过调用 isConnectionPending 方法可以确定连接操作是否正在进行。

套接字通道支持异步关闭，这类似于 Channel 类中指定的异步关闭的操作。如果套接字的输入端被一个线程关闭，而另一个线程在套接字的通道上的读操作被阻塞，则阻塞线程中的读操作不会读取到任何字节，并且会返回-1。如果套接字的输出端被一个线程关闭，而另一个线程在套接字的通道上的写操作中被阻塞，则被阻塞的线程将会收到 AsynchronousCloseException 异常。

Socket 选项使用 setOption(SocketOption,Object)方法来配置。套接字通道支持以下选项。

（1）java.net.StandardSocketOptions.SO_SNDBUF Socket，发送缓冲区大小。

（2）java.net.StandardSocketOptions.SO_RCVBUF Socket，接收缓冲区大小。

（3）java.net.StandardSocketOptions.SO_KEEPALIVE，是否对连接保活（保持连接并判断对方主机是否宕机，避免当前 Socket 永远阻塞并等待另一端发送数据。设置该选项后，如果 2 小时内此 Socket 连接与对端 Socket 没有发生数据交换，TCP 就自动给对方发一个保持存活探测分节(keepalive probe)）。

（4）java.net.StandardSocketOptions.SO_REUSEADDR 复用地址，存在五种情况。

① 同一个 IP：Socket 相同的 Socket-A 处于 TIME_WAIT 状态时（什么是 TIME_WAIT？参考下 TCP 的四次挥手（Four-Way Wavehand）中，接收到对端的 FIN 时，发送完 ACK 需要等待 2MSL 的时间），其他程序进程 B 创建的 Socket-B 要占用该地址和端口，此时可以设置 SO_REUSEADDR 使进程 B 可以绑定成功。

② 在同一端口上启动多个进程，并且这些进程同时绑定同一个端口，就需要设置 SO_REUSEADDR 选项。但要注意，每个进程绑定的 IP 地址不能相同，此时由操作系统对多个进程的输入负载均衡。

③ 需要在完全相同的地址和端口的重复绑定时，可以使用 SO_REUSEADDR，但只能用于 UDP。

④ java.net.StandardSocketOptions.SO_LINGER，如果缓冲区中存在数据，则在调用 close 方法关闭 socket 时，等待缓冲区中的数据写入对端（仅适用于阻塞模式）。

⑤ java.net.StandardSocketOptions.TCP_NODELAY 关闭 TCP Nagle 算法。

套接字通道是线程安全的。它们支持并发读和并发写，具体实现将保证在同一时间段内最多只能有一个线程读/写数据。connect 和 finishConnect 方法和其他方法都是互斥的，调用其中一个方法时，尝试进行读或写操作将阻塞，直到该调用完成。源码及其相关描述如下。

```java
public abstract class SocketChannel extends AbstractSelectableChannel implements ByteChannel,
ScatteringByteChannel, GatheringByteChannel, NetworkChannel {

    // 初始化 SocketChannel，参数 SelectorProvider 表示创建该通道的 SelectorProvider 对象
    protected SocketChannel(SelectorProvider provider) {
        super(provider);
    }

    // 通过 SelectorProvider 打开通道对象
    public static SocketChannel open() throws IOException {
        return SelectorProvider.provider().openSocketChannel();
    }

    // 打开指定连接地址，先使用 open 方法获取到该实例，然后调用 connect 方法连接到对端
    public static SocketChannel open(SocketAddress remote)
        throws IOException
    {
        SocketChannel sc = open();
        try {
            sc.connect(remote);
        } catch (Throwable x) {
            try {
                sc.close();
            } catch (Throwable suppressed) {
                x.addSuppressed(suppressed);
            }
            throw x;
        }
        assert sc.isConnected();
        return sc;
    }

    // 返回当前通道支持的在 Selector 选择器中的感兴趣事件集，支持读、写、连接
    public final int validOps() {
        return (SelectionKey.OP_READ
                | SelectionKey.OP_WRITE
                | SelectionKey.OP_CONNECT);
    }
```

```java
// 绑定本地地址
public abstract SocketChannel bind(SocketAddress local)
    throws IOException;

// 设置 TCP 参数
public abstract <T> SocketChannel setOption(SocketOption<T> name, T value)
    throws IOException;

// 关闭输入端
public abstract SocketChannel shutdownInput() throws IOException;

// 关闭输出端
public abstract SocketChannel shutdownOutput() throws IOException;

// 获取当前通道表示的 Socket
public abstract Socket socket();

// 判断是否连接成功
public abstract boolean isConnected();

// 判断是否仍处于连接处理中
public abstract boolean isConnectionPending();

// 处理实际连接过程
public abstract boolean connect(SocketAddress remote) throws IOException;

// 完成连接套接字通道的过程。非阻塞连接操作是通过将套接字通道置于非阻塞模式，然后调用其 connect
方法启动。一旦连接建立或失败，套接字通道将变为可连接的，并且可以调用此方法完成连接后续操作。如果连
接操作失败，调用这个方法将抛出适当的 java.io.IOException。如果已经连接通道，这个方法不会阻塞，立即返
回 true。如果通道处于非阻塞模式，且连接进程尚未完成，则该方法返回 false；如果通道处于阻塞模式，则这个
方法将阻塞直到连接完成或失败，并且总是返回 true 或抛出异常
public abstract boolean finishConnect() throws IOException;

// 获取连接的远程地址
public abstract SocketAddress getRemoteAddress() throws IOException;

// 读写缓冲区操作

public abstract int read(ByteBuffer dst) throws IOException;
public abstract long read(ByteBuffer[] dsts, int offset, int length)throws IOException;
public final long read(ByteBuffer[] dsts) throws IOException {
    return read(dsts, 0, dsts.length);
}
public abstract int write(ByteBuffer src) throws IOException;
public abstract long write(ByteBuffer[] srcs, int offset, int length)
    throws IOException;
public final long write(ByteBuffer[] srcs) throws IOException {
    return write(srcs, 0, srcs.length);
}

// 获取本地地址
public abstract SocketAddress getLocalAddress() throws IOException;
}
...
```

15. ServerSocketChannel 抽象类

ServerSocketChannel 抽象类表示一个可选择的面向流监听 Socket 的通道，即 TCP 服务端 Socket。服务器套接字通道是通过调用该类的 open 方法创建的。新创建的服务器套接字通道处于打开状态，但尚未绑定地址。尝试调用未绑定地址的服务器套接字通道的 accept 方法将抛出 NotYetBoundException。服务器套接字通道可以通过调用 bind 方法绑定本地地址，以接收来自客户端的连接。套接字选项使用 setOption(SocketOption,Object)方法配置。服务器套接字通道支持以下选项。

（1）java.net.StandardSocketOptions.SO_RCVBUF，接收缓冲区大小。

（2）java.net.StandardSocketOptions.SO_REUSEADDR，复用地址。

相关源码与描述如下。

```java
public abstract class ServerSocketChannel extends AbstractSelectableChannel implements NetworkChannel {
    // provider 表示创建该服务端通道的 SelectorProvider 对象
    protected ServerSocketChannel(SelectorProvider provider) {
        super(provider);
    }

    // 通过 SelectorProvider 打开通道
    public static ServerSocketChannel open() throws IOException {
        return SelectorProvider.provider().openServerSocketChannel();
    }

    // 返回服务端通道所支持的在 Selector 选择器中的感兴趣时间集，这里只是 OP_ACCEPT
    public final int validOps() {
        return SelectionKey.OP_ACCEPT;
    }

    // -- ServerSocket 特定操作 --

    // 绑定地址
    public final ServerSocketChannel bind(SocketAddress local)
        throws IOException
    {
        return bind(local, 0);
    }

    // backlog 参数表示 TCP 完成三次握手的 accept 队列
    public abstract ServerSocketChannel bind(SocketAddress local, int backlog)
        throws IOException;

    // 设置 socket 选项
    public abstract <T> ServerSocketChannel setOption(SocketOption<T> name, T value)
        throws IOException;

    // 所关联的 socket 对象
    public abstract ServerSocket socket();

    // 用于接收来自客户端的连接。如果处于非阻塞模式，则立即返回；如果处于阻塞模式，将等待阻塞直到接收到来自客户端的连接
```

```
    public abstract SocketChannel accept() throws IOException;

    // 获取本地地址
    public abstract SocketAddress getLocalAddress() throws IOException;
}
```

16．AsynchronousFileChannel 抽象类

AsynchronousFileChannel 抽象类表示读取和写入文件均为异步操作的通道。异步文件通道对象通过调用该类定义的 open 方法打开文件并创建。该文件包含一个可变长度的字节数据，可以读取和写入。文件当前大小可以通过 size 方法查询。字节写入超过其当前大小时，文件的大小增加；文件被截断时，文件的大小减小。异步文件通道对象不保存当前操作文件中数据的位置信息，而是在执行异步读和写方法参数中传入。

CompletionHandler 对象将在异步读写中作为参数传入，并被调用来使用 I/O 操作的结果。这个类还定义了启动异步操作的读和写方法。返回 Future 表示异步正在执行的操作，Future 对象可用于判断操作是否已完成，等待操作完成并获取操作结果。除了读写操作，这个类还定义了以下操作。

（1）force 方法可以将文件修改的内容写入底层存储设备，以确保在系统崩溃时数据不会丢失。

（2）lock 方法可以锁定文件的一个区域，禁止被其他程序访问。

AsynchronousFileChannel 底层实现与一个线程池相关联，将任务提交到该线程池，以处理 I/O 事件，并将 IO 操作的结果传递给设置的 CompletionHandler 处理器处理该结果。

在通道的读写参数中传入的 CompletionHandler 对象的回调将由线程池中的一个线程调用。如果 I/O 操作立即完成，而调用读写方法的线程属于线程池中的一个线程，则 CompletionHandler 对象可以由当前调用线程直接调用。在未指定线程池的情况下创建 AsynchronousFileChannel 时，该通道将与一个系统相关的默认线程池相关联，该默认线程池可以与其他通道共享。默认线程池可以通过 AsynchronousChannelGroup 类定义的系统属性来配置。

该通道是线程安全的。Channel 接口定义的 Channel.close 方法可以在任何时候调用，这将导致通道上所有未完成的异步操作变为 AsynchronousCloseException 异常完成。多个读写操作未完成时，是不指定 I/O 操作的顺序以及 CompletionHandler 对象被调用的顺序的，即不能保证按照操作提交顺序完成执行。因为异步 IO 具体完成时间是不确定的。执行异步读或写时使用的 ByteBuffer 缓冲区对于多个并发 I/O 操作来说是线程不安全的，因为有可能多个操作同时完成，而回调时将在线程池的不同线程中回调。

与 FileChannel 一样，该类的实例提供的文件数据保证与同一程序中其他实例提供的同一文件的数据保持一致。但是由于底层操作系统在执行 IO 操作时存在缓存和网络文件系统协议导致的延迟，该类实例所提供的文件数据可能与其他并发运行的程序所看到的文件数据一致，也可能不一致。不管这些程序是用什么语言编写的，也不管它们是在同一台机器上还是在其他机器上运行的，都会存在上述问题。具体源码以及描述如下。

```
// 完成异步任务时回调的处理器
public interface CompletionHandler<V,A> {

    // 操作完成时回调。result 表示处理结果，attachment 表示 attach 对象
    void completed(V result, A attachment);

    // 操作失败时回调
    void failed(Throwable exc, A attachment);
```

```java
}

public abstract class AsynchronousFileChannel implements AsynchronousChannel{

    protected AsynchronousFileChannel() {
    }

    // 打开异步通道，相关描述同 FileChannel
    public static AsynchronousFileChannel open(Path file,
                                    Set<? extends OpenOption> options,
                                    ExecutorService executor,
                                    FileAttribute<?>... attrs)
        throws IOException
    {
        FileSystemProvider provider = file.getFileSystem().provider();
        return provider.newAsynchronousFileChannel(file, options, executor, attrs);
    }

    // 文件属性
    private static final FileAttribute<?>[] NO_ATTRIBUTES = new FileAttribute[0];

    // 打开异步通道，相关描述同 FileChannel
    public static AsynchronousFileChannel open(Path file, OpenOption... options)
        throws IOException
    {
        Set<OpenOption> set = new HashSet<OpenOption>(options.length);
        Collections.addAll(set, options);
        return open(file, set, null, NO_ATTRIBUTES);
    }

    // 获取文件大小
    public abstract long size() throws IOException;

    // 截断文件并返回新的通道对象
    public abstract AsynchronousFileChannel truncate(long size) throws IOException;

    // 强制将修改的文件数据写入磁盘。metaData 表示在写入磁盘时是否将文件的元数据也写入磁盘
    public abstract void force(boolean metaData) throws IOException;

    // 同 FileChannel，只不过这里为异步操作，完成时调用 handler

    public abstract <A> void lock(long position,
                                    long size,
                                    boolean shared,
                                    A attachment,
                                    CompletionHandler<FileLock,? super A> handler);

    public final <A> void lock(A attachment,
                                    CompletionHandler<FileLock,? super A> handler)
    {
        lock(0L, Long.MAX_VALUE, false, attachment, handler);
    }
    public abstract Future<FileLock> lock(long position, long size, boolean shared);
    public final Future<FileLock> lock() {
```

```java
        return lock(0L, Long.MAX_VALUE, false);
    }
    public abstract FileLock tryLock(long position, long size, boolean shared)
        throws IOException;
    public final FileLock tryLock() throws IOException {
        return tryLock(0L, Long.MAX_VALUE, false);
    }

    // 异步读取数据放入 ByteBuffer 中，完成时回调 CompletionHandler
    public abstract <A> void read(ByteBuffer dst,
                                  long position,
                                  A attachment,
                                  CompletionHandler<Integer,? super A> handler);

    // 不指定回调操作
    public abstract Future<Integer> read(ByteBuffer dst, long position);

    // 将 ByteBuffer 的数据异步写入 Channel 中，完成时回调 CompletionHandler
    public abstract <A> void write(ByteBuffer src,
                                   long position,
                                   A attachment,
                                   CompletionHandler<Integer,? super A> handler);

    // 不指定回调操作
    public abstract Future<Integer> write(ByteBuffer src, long position);
}
```

17．AsynchronousSocketChannel 与 AsynchronousServerSocketChannel 抽象类

这两个抽象类和 AsynchronousFileChannel 抽象类相似，只不过读写操作是异步的，相关描述也与 AsynchronousFileChannel 一模一样，即线程池和 NIO 操作，所以在此不再赘述。读者可以自己查看源码，相信在了解 SocketChannel、ServerSocketChannel、AsynchronousFileChannel 后，就理解了。笔者将在 AIO 编程中，介绍内部实现细节。

1.4.4　Selector 原理

Selector 接口表示应用于可选择通道对象的多路复用器。可以通过调用该类的 open 方法创建选择器对象，该方法使用系统默认的 java.nio.channals.spi.SelectorProvider 选择器创建新的选择器。选择器也可以通过调用自定义选择器提供程序的 java.nio.channels.spi.SelectorProvider.openSelector 方法创建。选择器创建后，将一直保持打开状态，直到调用 close 方法将其关闭。

可选择的通道注册到选择器中后，将返回一个 SelectionKey 对象，用于表示选择器中的可选择的通道。选择器维护以下三组选择键集合。

（1）key set（所有键的键集）表示所有注册到该选择器中的通道 SelectionKey 对象集，可通过 keys 方法获取。

（2）selected-key-set（被选择键的键集）表示当前注册到该选择器中的通道 SelectionKey 对象，在该对象中指定的感兴趣事件集已经产生了对应的事件，即 READ、WRITE、CONNECT、ACCEPT，将这些 SelectionKey 对象放入该键集，可以通过 selectedKeys 方法获取。selected-key-set 是 key set 的子集。

（3）cancelled-key（被取消键的键集）表示已经被取消但其通道尚未从选择器中解除注册的键集，该集合不能直接访问。cancelled-key 是 key set 的子集。

这三个键集在新创建的选择器中都是空的。通道可以调用 SelectableChannel.register(Selector,int)方法将自身添加到选择器中，此时会返回一个 SelectionKey 对象，将通道从选择器中解除注册时，可以调用 SelectionKey.cancel 方法，该方法将 SelectionKey 对象放入 cancelled-key 键集中，在下一次执行 select 方法时进行移除。

通过执行 select 方法，SelectionKey 将被添加到 selected-key-set 键集中（在注册的感兴趣事件到达时才会放入）。通过调用集合的 java.util.Set remove(java.lang.Object)方法，或调用从集合中获得的 java.util.Iterator 的 remove 方法，直接从 selected-key-set 键集中删除指定键。

每次调用 select 方法时，键可以被添加到选择器的 selected-key-set 键集或从 cancelled-key 键集中删除。select 方法有三种类型，即 select 方法、select(long)方法、selectNow()方法，涉及以下三个步骤。

（1）cancelled-key 键集中的每一个 SelectionKey 都将被移除，同时解除注册。

（2）调用底层操作系统的系统调用，以了解每个注册通道的事件准备情况。对于准备好至少一个感兴趣事件的通道，将执行以下两种操作中的一种。

① 如果通道的 SelectionKey 未添加到 selected-key-set 键集中，则将其添加到该集中，并修改就绪事件集为当前准备好的事件类型。

② 如果通道 SelectionKey 已经添加到 selected-key-set 键集中，只需要把就绪事件集修改为当前准备好的事件类型即可。任何先前已经记录在就绪集中的就绪事件信息都会保留。换句话说，底层系统返回的就绪事件按位分离到 SelectionKey 对象的当前就绪集中，即一个整型变量中，按位表示当前类型。

如果在此步骤开始时，键集中的所有 SelectionKey 都没有注册感兴趣事件集，则任何从操作系统获取到的事件都不会放入 selected-key-set 键集中。

（3）如果在步骤（2）进行时，其他线程将 SelectionKey 添加到 cancelled-key 键集中，则它们将按步骤（1）进行处理。

select 方法、select(long)方法、selectNow()方法的区别在于，选择操作是否阻塞等待一个或多个通道准备好。如果阻塞，则设定阻塞多长时间。选择器本身是线程安全的，但是它们的键集不是。如果存在多个线程直接修改其中一个集合，则应通过同步操作来控制并发访问。源码如下。

```java
public abstract class Selector implements Closeable {

    // 打开选择器
    public static Selector open() throws IOException {
        return SelectorProvider.provider().openSelector();
    }

    // 判断当前选择器是否处于打开状态
    public abstract boolean isOpen();

    // 获取创建当前选择器的 SelectorProvider
    public abstract SelectorProvider provider();

    // 获取所有注册到选择器中的 SelectionKey，包括 selectedKeys 和 cancelKeys
    public abstract Set<SelectionKey> keys();
```

```
// 获取当前被选择的 SelectionKey 集合
public abstract Set<SelectionKey> selectedKeys();

// 立即查询底层 OS 中是否存在准备好的事件。如果没有，不阻塞等待
public abstract int selectNow() throws IOException;

// 立即查询底层 OS 中是否存在准备好的事件。如果没有，阻塞 timeout 毫秒数
public abstract int select(long timeout)
    throws IOException;

// 立即查询底层 OS 中是否存在准备好的事件。如果没有，一直阻塞，直到产生准备好的事件
public abstract int select() throws IOException;

// 唤醒阻塞在 select 方法上的线程
public abstract Selector wakeup();

// 关闭当前选择器
public abstract void close() throws IOException;
}
```

1.5 AIO 编程

本节详细描述了 Java AIO 的实现原理，逐行解释了 AsynchronousServerSocketChannel 的 open、bind、accept 方法的原理和 AsynchronousSocketChannel 的 read 方法实现原理。这里由于篇幅有限，没有介绍 write 方法，但相信读者看完了解之后也能推测出 write 的作用。读者看完之后就会明白，AIO 就是选择器和线程池的结合，即 Epoll 和 ThreadPoolExecutor。读者可根据实际需要阅读不同的小节。同时，为了满足不懂 C 语言但想了解 JNI 的底层原理的朋友，笔者这里给出了 Linux 下的 C 网络服务端、客户端编程实例和使用到的 JNI 网络编程实例，读者可以先看实例再看描述，也可以先看 Java 的 AIO 再看实例，取决于读者自身的实际情况。因为笔者这里在编写时将两者进行分离，并做出了解释，互不影响。同时，读者可以在 JNI 的描述中看到和 C 网络编程一模一样的函数的影子。整体的 AIO 框架模型如图 1-12 所示。

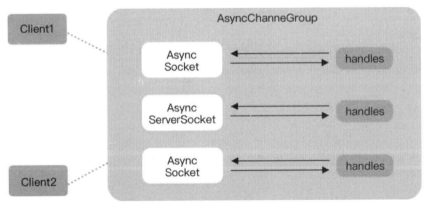

图 1-12　AIO 架构

1.5.1　Java AIO 描述

　　JDK 1.7 推出了两个类支持网络的异步编程 AsynchronousServerSocketChannel 和 AsynchronousSocketChannel，本节从最底层开始介绍 JVM 如何实现异步编程。相信读者肯定知道 NIO 的 selector 选择器编程，它可以通过底层调用 select、poll、epoll 函数让 OS 完成监听端口的连接、读、写等操作。满足条件后，通过 epoll_wait，即 java 应用层的 Selector.select()获取准备好的 channel，对这些端口进行操作，通过选择器将网络事件和应用层进行分离，应用一个单线程就可以完成全部网络事件的处理，这让应用层的线程可以不用关注网络事件，从而实现非阻塞式操作。没有调用 Selector.select 方法时，应用线程可以自由活动。但这根本不是 AIO 异步编程，真正的异步编程是可以设置回调函数，由其他线程处理完毕后自动继续处理后面的步骤，不需要或可以根据业务通知调用线程，不过真正的 AIO 根本不需要通知调用线程。底层知识牢固的读者可以想象一下 CPU 的流水线，也可以想象一下车间的流水线，一个产品的经过一次流水线就完成生产了，不需要在流水线中重复。所以 JDK 1.7 为了满足真正的异步 IO，这里称之为 AIO，推出了 AsynchronousServerSocketChannel 和 AsynchronousSocketChannel 这两个类。

　　但是，读者真的相信存在这样的 IO 吗？如果了解 Linux 内核底层就可以发现，AIO 的支持并不完善，不管是 Linux 的 AIO 原生库、Glibc 的 AIO 库、最新版的内核的 IO_URING 等，都不可能是完全的 AIO，即自己异步回调逐步完成，还是需要应用层创建一个线程轮询完成队列，当然也可以设置好 Callback 的地址，由这个线程轮询到完成后自动回调。这些 AIO 库中最像 AIO 的必然是 IO_URING，毕竟通过测试表明确实很强大，因为它将应用层当作生产者，内核当作消费者，只管提交 IO 任务，由内核完成后放入到完成队列，由应用层来取，仅此而已，不过已经是很大的进步了。

1.5.2　Java AIO 实例

　　有些读者可能不是非常理解上文的内容。那么读者的知识或许有欠缺，需要针对补充相关知识。笔者这里也会在后文逐行解释 Linux 的 AIO 原生库、Glibc 的 AIO 库及最新版内核 IO_URING 的实现原理和细节。这里只介绍 AsynchronousServerSocketChannel、AsynchronousSocketChannel 在 Linux 上的实现原理如何。首先，请看如下实例。

```java
/**
 * @author hj
 * @version 1.0
 * @description: 异步网络 IO 例子
 * @date 2021/5/15 9:53
 */
public class AsyncChannelDemo {
    // main thread wait until complete
    public static final CountDownLatch EXIT_LATCH = new CountDownLatch(1);
    // current server's SocketChannel
    public static AsynchronousServerSocketChannel serverSocketChannel;
    // current connected client socket
    public static AsynchronousSocketChannel curClientSocket;
    // create our SocketChannel via port
```

```java
public static void createAsynchronousServerSocketChannel(int port) {
    try {
        serverSocketChannel = AsynchronousServerSocketChannel.open();
        serverSocketChannel.bind(new InetSocketAddress(port));
    } catch (IOException e) {
        throw new RuntimeException(e);
    }
}
// when we complete,we must close server and make main thread end
public static void closeServer() {
    System.out.println("close aio server");
    try {
        if (curClientSocket != null) {
            curClientSocket.close();
        }
        if (serverSocketChannel != null) {
            serverSocketChannel.close();
        }
    } catch (IOException e) {
        e.printStackTrace();
    }
    // notify main thread
    EXIT_LATCH.countDown();
}
// handle client data
public static class ReadHandler implements CompletionHandler<Integer, ByteBuffer> {

    @Override
    public void completed(Integer result, ByteBuffer buffer) {
        // now buffer is our need data from client
        System.out.println(new String(buffer.array()));
        // close server
        closeServer();
    }

    @Override
    public void failed(Throwable exc, ByteBuffer attachment) {
        exc.printStackTrace();
        closeServer();
    }
}
// handle server accept client request
public static class AcceptHandler implements CompletionHandler<AsynchronousSocketChannel, Object>{

    @Override
    public void completed(AsynchronousSocketChannel socketChannel, Object attachment) {
        System.out.println("receive result:" + socketChannel.toString());
        // now result is socket from client that's your browser
        // we can read some data using handler
        ByteBuffer byteBuffer = ByteBuffer.allocate(2048);
        socketChannel.read(byteBuffer, byteBuffer, new ReadHandler());
        curClientSocket = socketChannel;
```

```java
        }
        @Override
        public void failed(Throwable exc, Object attachment) {
            exc.printStackTrace();
            closeServer();
        }
    }
    public static void main(String[] args) {
        System.out.println("start aio server");
        createAsynchronousServerSocketChannel(8888);
        serverSocketChannel.accept(null, new AcceptHandler());
        // main thread wait exit
        try {
            System.out.println("wait server close");
            EXIT_LATCH.await();
        } catch (InterruptedException e) {
            e.printStackTrace();
        }
        System.out.println("end aio server");
    }
}
```

这里创建了 AsynchronousServerSocketChannel 实例用于接收客户端请求，通过 CompletionHandler 的两个实例 AcceptHandler 和 ReadHandler 分别用于接收客户端的连接请求和读取客户端的数据。CompletionHandler 就是上面介绍的回调函数。线程设置好回调函数后，就不需要再处理其他关于 channel 的事情了，只需要结束方法即可，所有业务在 CompletionHandler 回调函数中编写。读者一定要理解上面的异步回调编写方式，使用过 Netty 或前端 Promise 的读者应该对此并不陌生，其中的原理是相通的。图 1-13 是运行的效果图。

图 1-13　运行效果图 1

刚启动时如图 1-14 所示，在浏览器访问 8888 端口。

读取到了客户端发送的 HTTP 请求，并在读取后关闭了连接，这并不影响研究 AIO。接下来逐步分析它的实现原理。注意在此使用 Linux 作为分析对象。

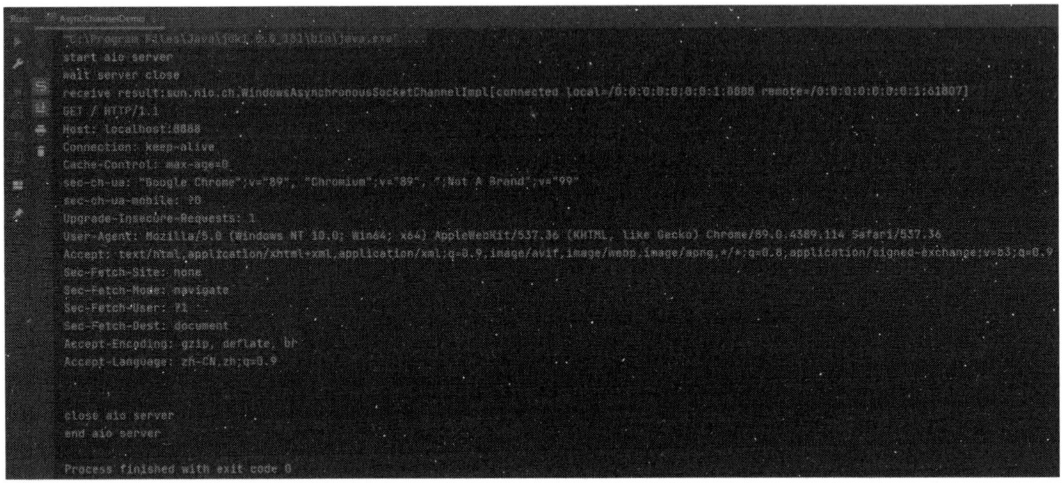

图 1-14 运行效果图 2

1.5.3 AsynchronousServerSocketChannel.open()原理

首先观察 open()方法的源码，如下所示。

```
public static AsynchronousServerSocketChannel open() throws IOException{
    // 不指定异步通道组
    return open(null);
}
public static AsynchronousServerSocketChannel open(AsynchronousChannelGroup group)throws
IOException{
    // 如果是同一个组的，则 AsynchronousServerSocketChannel 的提供者 AsynchronousChannelProvider 要
    保持一致，不过这里不是
    AsynchronousChannelProvider provider = (group == null) ?
        AsynchronousChannelProvider.provider() : group.provider();
    return provider.openAsynchronousServerSocketChannel(group);
}
```

最后，需要提供一个 AsynchronousServerSocketChannel 的 AsynchronousChannelProvider，这是因为 Java 跨平台，有 Windows 和 Linux 等版本。在此选取 Linux 的提供者。下面的代码介绍 AsynchronousChannelProvider.provider()是如何找到提供者的。

```
public static AsynchronousChannelProvider provider() {
    return ProviderHolder.provider;
}
private static class ProviderHolder {
    // 类加载时才初始化 provider
    static final AsynchronousChannelProvider provider = load();

    private static AsynchronousChannelProvider load() {
        return AccessController
            .doPrivileged(new PrivilegedAction<AsynchronousChannelProvider>() {
                public AsynchronousChannelProvider run() {
                    AsynchronousChannelProvider p;
                    // 先通过 Property 获取
```

```java
                    p = loadProviderFromProperty();
                    if (p != null)
                        return p;
                    // 否则通过 ServiceLoader 获取
                    p = loadProviderAsService();
                    if (p != null)
                        return p;
                    // 如果都没有指定，则使用 DefaultAsynchronousChannelProvider 来创建默认的提供者
                    return sun.nio.ch.DefaultAsynchronousChannelProvider.create();
                }});
}

private static AsynchronousChannelProvider loadProviderFromProperty() {
    // 没有设置，则为 null
    String cn = System.getProperty("java.nio.channels.spi.AsynchronousChannelProvider");
    if (cn == null)
        return null;
    Class<?> c = Class.forName(cn, true,
                            ClassLoader.getSystemClassLoader());
    return (AsynchronousChannelProvider)c.newInstance();
}

private static AsynchronousChannelProvider loadProviderAsService() {
    // 没有使用 Java 的 SPI 接口，即没有设置 META-INF/services，所以也没有
    ServiceLoader<AsynchronousChannelProvider> sl =
            ServiceLoader.load(AsynchronousChannelProvider.class,
                            ClassLoader.getSystemClassLoader());
    Iterator<AsynchronousChannelProvider> i = sl.iterator();
    for (;;) {
        try {
            return (i.hasNext()) ? i.next() : null;
        } catch (ServiceConfigurationError sce) {
            if (sce.getCause() instanceof SecurityException) {
                // Ignore the security exception, try the next provider
                continue;
            }
            throw sce;
        }
    }
}
}
```

没有通过系统变量设置，也没有使用 SPI 接口，所以只能创建默认的提供者。源码如下。

```java
public static AsynchronousChannelProvider create() {
    String osname = AccessController
            .doPrivileged(new GetPropertyAction("os.name"));
    if (osname.equals("SunOS"))
        return createProvider("sun.nio.ch.SolarisAsynchronousChannelProvider");
    // 注意这里，研究对象是 Linux
    if (osname.equals("Linux"))
        return createProvider("sun.nio.ch.LinuxAsynchronousChannelProvider");
    if (osname.contains("OS X"))
        return createProvider("sun.nio.ch.BsdAsynchronousChannelProvider");
```

```java
        throw new InternalError("platform not recognized");
}
```

所以创建的 Provider 是 LinuxAsynchronousChannelProvider。

```java
private static AsynchronousChannelProvider createProvider(String cn) {
    Class<AsynchronousChannelProvider> c;
    try {
        c = (Class<AsynchronousChannelProvider>)Class.forName(cn);
    } catch (ClassNotFoundException x) {
        throw new AssertionError(x);
    }
    try {
        return c.newInstance();      // 直接调用默认构造器初始化
    } catch (IllegalAccessException | InstantiationException x) {
        throw new AssertionError(x);
    }
}
```

LinuxAsynchronousChannelProvider 的构造函数如下所示。

```java
// 未执行任何操作
public LinuxAsynchronousChannelProvider() {
}
```

provider.openAsynchronousServerSocketChannel(group)如何创建 AsynchronousServerSocketChannel。代码如下。

```java
public AsynchronousServerSocketChannel
openAsynchronousServerSocketChannel(AsynchronousChannelGroup group) throws IOException{
    return new UnixAsynchronousServerSocketChannelImpl(toPort(group));
}
private Port toPort(AsynchronousChannelGroup group) throws IOException {
    if (group == null) {
        // group 为 null，创建默认的 EventPort
        return defaultEventPort();
    } else {
        if (!(group instanceof EPollPort))
            throw new IllegalChannelGroupException();
        return (Port)group;
    }
}
// 创建默认的 EventPort
private EPollPort defaultEventPort() throws IOException {
    if (defaultPort == null) {
        synchronized (LinuxAsynchronousChannelProvider.class) {
            if (defaultPort == null) {
                // 创建 EPollPort。注意这里的 ThreadPool.getDefault 表示默认线程池，然后还启动了线程
                defaultPort = new EPollPort(this, ThreadPool.getDefault()).start();
            }
        }
    }
    return defaultPort;
}
// 默认线程池
```

```java
static ThreadPool getDefault() {
    return DefaultThreadPoolHolder.defaultThreadPool;
}
// 返回全局单例默认线程池
private static class DefaultThreadPoolHolder {
    final static ThreadPool defaultThreadPool = createDefault();
}
// 创建默认线程池
static ThreadPool createDefault() {
    // 通过系统变量设置。因为没有设置，所以返回-1
    int initialSize = getDefaultThreadPoolInitialSize();
    if (initialSize < 0) // 默认是 CPU 的核心数
        initialSize = Runtime.getRuntime().availableProcessors();
    ThreadFactory threadFactory = getDefaultThreadPoolThreadFactory();
    if (threadFactory == null)
        threadFactory = defaultThreadFactory;
    // 通过 Executors 创建了 Cached 线程池。读者可能会有疑问，这里不是 CachedThreadPool 线程无限大吗？
别着急，下面的 ThreadPool 对它进行了包装
    ExecutorService executor = Executors.newCachedThreadPool(threadFactory);
    return new ThreadPool(executor, false, initialSize);
}
// ThreadPool 构造器，保存真正执行器 executor 和最大线程数。有这个约束，就不担心 CachedThreadPool 线程
超过 poolSize 了
private ThreadPool(ExecutorService executor,boolean isFixed,int poolSize){
    this.executor = executor;
    this.isFixed = isFixed;
    this.poolSize = poolSize;
}
```

下面的源码介绍 new EPollPort(this, ThreadPool.getDefault()).start()的内容。

```java
// EPollPort 构造器。初始化 epfd、socketpair，添加 sv[0]的 epoll 监听数据读事件，最后分配事件队列
EPollPort(AsynchronousChannelProvider provider, ThreadPool pool)throws IOException{
    super(provider, pool);
    // 创建 epoll 的 fd，这里就到 JVM 了，最后又会回到 Glibc 的 epoll_create，了解即可
    this.epfd = epollCreate();
    // 创建套接字对，创建一对 CP 用于唤醒操作
    int[] sv = new int[2];
    try {
        // 这里就又进入到了 JVM 底层，其实是通过 socketpair 函数创建套接字对，用于进程或线程间的通信操
作。也可以在同一个进程中使用这个套接字对进行写入或读取。读者可以与 Java ByteArrayInputStream 和
ByteArrayOutStream 进行类比
        socketpair(sv);
        // 通过 epoll 的 ctl 控制操作添加 sv[0]的 fd，监听的操作为 POLLIN，表明读操作。通过 sv[1]向 sv[0]发
送事件，此时 epoll 会捕捉到这个读事件
        epollCtl(epfd, EPOLL_CTL_ADD, sv[0], POLLIN);
    } catch (IOException x) {
        close0(epfd);
        throw x;
    }
    // 保存进程通信的套接字
    this.sp = sv;
    // 分配轮询事件的数组，这里 MAX_EPOLL_EVENTS 为 512
```

```java
        this.address = allocatePollArray(MAX_EPOLL_EVENTS);
        // 创建 Java 的 ArrayBlockingQueue，用于保存 Java 层面的事件
        this.queue = new ArrayBlockingQueue<Event>(MAX_EPOLL_EVENTS);
        // 向队列中放入事件对象 NEED_TO_POLL，表明当前队列需要线程进行轮询处理事件
        this.queue.offer(NEED_TO_POLL);
}
// 启动线程处理队列事件
EPollPort start() {
    startThreads(new EventHandlerTask());
    return this;
}
// 在线程池中处理 EventHandlerTask 对象
protected final void startThreads(Runnable task) {
    // 由于传入的是 ThreadPool 对象，里面持有了 CachedThreadPool，并且不是 fixed 线程池。所以这时 isFixedThreadPool()返回 false
    if (!isFixedThreadPool()) {
        // 启动内部线程处理 IO 事件，注意，这里的 internalThreadCount 默认为 1
        for (int i=0; i<internalThreadCount; i++) {
            startInternalThread(task);
            threadCount.incrementAndGet();
        }
    }
    // 判断线程池的大小是否大于 0，这里设置的是等于 CPU 核心数，所以满足条件
    if (pool.poolSize() > 0) {
        // 包装外部的 EventHandlerTask 任务对象，处理线程绑定当前 AsynchronousChannelGroupImpl 对象。内部通过 ThreadLocal 来传递
        task = bindToGroup(task);
        try {
            // 向线程池中提交任务执行，并处理 IO 事件
            for (int i=0; i<pool.poolSize(); i++) {
                pool.executor().execute(task);
                threadCount.incrementAndGet();
            }
        } catch (RejectedExecutionException x) {
        }
    }
}
```

EPollPort start 方法中启动了内部线程和线程池，向其中执行了传入的 EventHandlerTask 对象。并且根据设置的线程池大小，放入对应大小的任务对象，所以是多个线程同时处理 EventHandlerTask 对象。下面介绍 EventHandlerTask 的 Runnable 方法。

```java
private class EventHandlerTask implements Runnable {
    // 对 epoll 进行轮询，获取事件准备好的 channel，即网络事件
    private Event poll() throws IOException {
        try {
            for (;;) {
                // 读者还记得调用 allocatePollArray 方法开辟的 C 层面用于服务 epoll 的队列吗？这个队列就用到了。调用 epollWait 时，内核会将准备好的事件放入 address 的队列中返回。返回值 n 为准备好的事件数
                int n = epollWait(epfd, address, MAX_EPOLL_EVENTS);
                fdToChannelLock.readLock().lock(); // 获取通道读锁
                try {
                    // 处理网络事件
```

```java
                    while (n-- > 0) {
                        // 获取已经准备好事件的 fd
                        long eventAddress = getEvent(address, n);
                        int fd = getDescriptor(eventAddress);
                        // 如果 fd 是设置的 socketpair，则唤醒等待数据的线程
                        if (fd == sp[0]) {
                            // 等待数据的线程等待唤醒计数为0，表明没有线程需要当前套接字对的数据，所
// 以调用 drain1 函数，清除队列里的全部数据，避免内存浪费
                            if (wakeupCount.decrementAndGet() == 0) {
                                drain1(sp[0]);
                            }
                            // 通过 n--对 n 进行操作，通过查看减 1 后的 n 是否大于 0，判断队列里是否还有
// 更多准备好的事件。如果有，则向队列中放入 EXECUTE_TASK_OR_SHUTDOWN 事件
                            if (n > 0) {
                                queue.offer(EXECUTE_TASK_OR_SHUTDOWN);
                                continue;
                            }
                            return EXECUTE_TASK_OR_SHUTDOWN;
                        }
                        // fd 是正常网络事件，则从之前保存 fd 到 channel 的映射表中获取到准备好的网络事件
                        PollableChannel channel = fdToChannel.get(fd);
                        if (channel != null) {
                            // 封装 Event 对象并返回
                            int events = getEvents(eventAddress);
                            Event ev = new Event(channel, events);
                            // 如果队列里还需要处理别的事件，则将当前 ev 放入队列，由其他线程获取执行。
// 如果是最后一个事件对象，则直接返回，由当前线程完成
                            if (n > 0) {
                                queue.offer(ev);
                            } else {
                                return ev;
                            }
                        }
                    }
                } finally {
                    fdToChannelLock.readLock().unlock();
                }
            }
        } finally {
            // 向队列中放入 NEED_TO_POLL 事件。注意，这个队列是公用的，有多个线程同时指向前任务，
// 当前线程已经处理了自己获取的事件，设置 NEED_TO_POLL 事件，告诉其他线程可以进行轮询 epoll 事件
            queue.offer(NEED_TO_POLL);
        }
    }
    // 线程启动后直接执行
    public void run() {
        // 从 ThreadLocal 中获取到之前通过 task = bindToGroup(task)放入的 GroupAndInvokeCount 对象
        Invoker.GroupAndInvokeCount myGroupAndInvokeCount =
            Invoker.getGroupAndInvokeCount();
        // 通过该标识判断当前任务是否在 ThreadPool 中运行。通过 startInternalThread 方法开启一个内部线程
// 处理当前任务，剩余的任务对象将在 ThreadPool 中执行
        final boolean isPooledThread = (myGroupAndInvokeCount != null);
        boolean replaceMe = false;
```

```java
Event ev;
try {
    // 循环处理事件
    for (;;) {
        // 如果在 ThreadPool 中执行，则重置 InvokeCount 调用计数
        if (isPooledThread)
            myGroupAndInvokeCount.resetInvokeCount();
        try {
            replaceMe = false;
            // 从队列中获取事件
            ev = queue.take();
            // 注意，在 EPollPort 构造器的最后传入了该事件对象 NEED_TO_POLL，该对象也指示当前工作线程应该调用 poll 方法获取网络事件
            if (ev == NEED_TO_POLL) {
                try {
                    ev = poll();
                } catch (IOException x) {
                    x.printStackTrace();
                    return;
                }
            }
        } catch (InterruptedException x) { // 忽略中断异常，会继续执行事件
            continue;
        }
        // 如果事件设置了 EXECUTE_TASK_OR_SHUTDOWN，则从全局 taskQueue 中获取任务执行。如果任务为空，则执行 shutdown 操作。在 AsynchronousChannelGroupImpl 的构造方法中对 taskQueue 进行赋值，不过这个 taskQueue 是针对 Fixed 固定线程池操作的，所以当前传入的 ThreadPool 不是固定线程池，taskQueue 为 null，而 pollTask 中如果 pollTask 为 null，则返回的 task 也是 null，所以只要遇到 EXECUTE_TASK_OR_SHUTDOWN 事件，则必然是 shutdown 结束当前线程
        if (ev == EXECUTE_TASK_OR_SHUTDOWN) {
            Runnable task = pollTask();
            if (task == null) {
                return;
            }
            replaceMe = true;
            task.run();
            continue;
        }
        // 开始处理事件
        try {
            // 通过调用 channel 的 onEvent 执行事件
            ev.channel().onEvent(ev.events(), isPooledThread);
        } catch (Error x) {
            replaceMe = true; throw x;
        } catch (RuntimeException x) {
            replaceMe = true; throw x;
        }
    }
} finally {
    // 线程退出后，释放资源
    int remaining = threadExit(this, replaceMe);
    if (remaining == 0 && isShutdown()) {
        implClose();
```

```
            }
        }
    }
}
```

读者一定要注意这里的 EventHandlerTask 是多个线程同时处理，如果还需要处理别的事件，则优先将事件放入队列中，由其他线程获取并执行，并且由于没有使用固定线程池，这里的 taskQueue 为空。只要出现 EXECUTE_TASK_OR_SHUTDOWN，就一定会结束线程，并且通过 sv[0]和 sv[1]进行通信。接收到 sv[0]读事件时，即其他线程向 sv[1]中写入数据，就会将 EXECUTE_TASK_OR_SHUTDOWN 放入队列中，或返回结束线程。接下来介绍 ev.channel().onEvent(ev.events()和 isPooledThread)如何执行事件，这里的 channel 对象就是 UnixAsynchronousServerSocketChannelImpl 对象。注意，这里的事件一定是网络事件，因为内部 Java 的特殊事件，只是用于指导线程的动作。

```
public void onEvent(int events, boolean mayInvokeDirect) {
    // 获取更新锁，判断 acceptPending 标志位
    synchronized (updateLock) {
        if (!acceptPending)
            return;
        acceptPending = false;
    }
    FileDescriptor newfd = new FileDescriptor();
    InetSocketAddress[] isaa = new InetSocketAddress[1];    // 用于保存客户端的地址信息
    Throwable exc = null;
    try {
        begin();                                             // 获取读锁
        // 获取准备好的请求事件
        int n = accept0(this.fd, newfd, isaa);
        // 如果返回 UNAVAILABLE，表明当前线程被虚假唤醒，因为通过 epoll wait 拿到的网络事件，并且判断了返回的事件数，所以不可能出现这种情况
        if (n == IOStatus.UNAVAILABLE) {
            // 如果真的出现，则复位 acceptPending，并重新轮询当前 fd
            synchronized (updateLock) {
                acceptPending = true;
            }
            port.startPoll(fdVal, Port.POLLIN);
            return;
        }
    } catch (Throwable x) {
        if (x instanceof ClosedChannelException)
            x = new AsynchronousCloseException();
        exc = x;
    } finally {
        end();                                               // 释放读锁
    }
    // 接收到了客户端的连接且没有发生异常，则调用 finishAccept 完成接收动作，并返回 client 端的 AsynchronousSocketChannel
    AsynchronousSocketChannel child = null;
    if (exc == null) {
        try {
            child = finishAccept(newfd, isaa[0], acceptAcc);
        } catch (Throwable x) {
```

```
            if (!(x instanceof IOException) && !(x instanceof SecurityException))
                x = new IOException(x);
            exc = x;
        }
    }
    // 获取当前设置的 acceptHandler 和 acceptAttachment 对象，并调用复写的 complete 函数
    CompletionHandler<AsynchronousSocketChannel,Object> handler = acceptHandler;
    Object att = acceptAttachment;
    PendingFuture<AsynchronousSocketChannel,Object> future = acceptFuture;
    enableAccept();
    if (handler == null) {
        future.setResult(child, exc);
        if (child != null && future.isCancelled()) {
            try {
                child.close();
            } catch (IOException ignore) { }
        }
    } else {
        Invoker.invoke(this, handler, att, child, exc);
    }
}
```

获取到事件，并调用设置的 handler 的 complete 方法。底层的 AIO 的操作是通过线程池调用 epoll 模拟的。接下来介绍 finishAccept 是怎样创建客户端对象 AsynchronousSocketChannel 的。

```
private AsynchronousSocketChannel finishAccept(FileDescriptor newfd,
                                               final InetSocketAddress remote,
                                               AccessControlContext acc) throws IOException,
                                               SecurityException
{
    AsynchronousSocketChannel ch = null;
    try {
        // 创建客户端 channel, 即 UnixAsynchronousSocketChannelImpl。注意客户端 channel 和服务端 channel
        // 使用同样的 port
        ch = new UnixAsynchronousSocketChannelImpl(port, newfd, remote);
    } catch (IOException x) {
        nd.close(newfd);
        throw x;
    }
    ...
    return ch;
}
// 客户端 socket channel 构造器
UnixAsynchronousSocketChannelImpl(Port port,
                                  FileDescriptor fd,
                                  InetSocketAddress remote) throws IOException
{
    super(port, fd, remote);
    this.fdVal = IOUtil.fdVal(fd);
    IOUtil.configureBlocking(fd, false);    // 同样将 client channel 设置为非阻塞
    try {
        port.register(fdVal, this);         // 保存 socket fd 和 socket channel 的映射
    } catch (ShutdownChannelGroupException x) {
        throw new IOException(x);
```

```
        }
        this.port = port;
}
```

  EPollPort 通过 socketpair 函数进行线程通信。注意，这里的队列分为两个，JVM 底层关于 OS 的事件和 Java 层面处理的事件，所以这里设置了两个队列，即 allocatePollArray（用于放置 epoll 准备好的网络事件）和 new ArrayBlockingQueue（用于存放 Java 层面指示线程执行响应动作的事件），且默认大小都是 512。注意，这里已经持有了等于 CPU 核心数的线程池对象 ThreadPool。还要注意，上面的分析仅是线程启动后且有网络事件之后的一系列步骤。由于还没有绑定端口，所以 epoll_Wait 不会返回任何事件。接下来介绍 new UnixAsynchronousServerSocketChannelImpl(toPort(group)) 使用这个端口执行了哪些操作。

```
abstract class AsynchronousServerSocketChannelImpl extends AsynchronousServerSocketChannel
implements Cancellable, Groupable{
    // 保存 C 调用的 OS 创建的 socket fd
    protected final FileDescriptor fd;
    AsynchronousServerSocketChannelImpl(AsynchronousChannelGroupImpl group) {
        super(group.provider());
        // 这里其实就是调用了函数 socket(domain, type, 0)，创建了 Linux 网络编程的套接字
        this.fd = Net.serverSocket(true);
    }
}
class UnixAsynchronousServerSocketChannelImpl extends AsynchronousServerSocketChannelImpl
implements Port.PollableChannel{
    UnixAsynchronousServerSocketChannelImpl(Port port)throws IOException{
        super(port);         // 初始化父类，这里是 AsynchronousServerSocketChannelImpl
        try {
            // 设置 socket 为非阻塞
            IOUtil.configureBlocking(fd, false);
        } catch (IOException x) {
            nd.close(fd);         // prevent leak
            throw x;
        }
        this.port = port;
        // 上面创建的 FD 最终为 Java 层面的类 FileDescriptor，这里需要直接获取到调用 socket 函数返回的具体值，所以通过 fdVal 拿到 socket 函数返回的具体值
        this.fdVal = IOUtil.fdVal(fd);
        // 将这个 fdVal 和当前的 UnixAsynchronousServerSocketChannelImpl 绑定在一起
        port.register(fdVal, this);
    }
    // PORT 用于绑定 fd 和 channel，就是将其保存在一个 Map 集合中
    protected final Map<Integer,PollableChannel> fdToChannel =
        new HashMap<Integer,PollableChannel>();
    final void register(int fd, PollableChannel ch) {
        fdToChannelLock.writeLock().lock();
        try {
            if (isShutdown())
                throw new ShutdownChannelGroupException();
            fdToChannel.put(Integer.valueOf(fd), ch);
        } finally {
            fdToChannelLock.writeLock().unlock();
        }
    }
```

从上面的源码可以总结出，EPollPort 本身就是 AsynchronousChannelGroup 的实例，其中保存了线程池和 Epoll 的 epfd，以及一个用于通信的 socketpair。UnixAsynchronousServerSocketChannelImpl 是 AsynchronousServerSocketChannel 的实例，其中保存了原生的套接字 socket 的 fd，并且将其在 EPollPort 中进行了绑定。

1.5.4 AsynchronousServerSocketChannel.bind()原理

下面的源码是 serverSocketChannel.bind(new InetSocketAddress(port))绑定地址的过程，上面的源码只创建了结构，但底层的 socket 还没有绑定地址。

```
public final AsynchronousServerSocketChannel bind(SocketAddress local)throws IOException{
    return bind(local, 0);
}
// 接下来的操作由子类完成，对应这里就是 UnixAsynchronousServerSocketChannelImpl
public abstract AsynchronousServerSocketChannel bind(SocketAddress local, int backlog)
    throws IOException;
```

下面是 UnixAsynchronousServerSocketChannelImpl 的 bind 函数。

```
public final AsynchronousServerSocketChannel bind(SocketAddress local, int backlog)throws IOException
{
    InetSocketAddress isa = (local == null) ? new InetSocketAddress(0) :
    try {
        begin(); // 加载读锁
        // 持有修改状态的对象锁
        synchronized (stateLock) {
            if (localAddress != null)
                throw new AlreadyBoundException();
            // 在绑定前调用钩子函数
            NetHooks.beforeTcpBind(fd, isa.getAddress(), isa.getPort());
            // 开始绑定，通过 bind 函数让 fd 对应的 socket 绑定相应的端口
            Net.bind(fd, isa.getAddress(), isa.getPort());
            // 绑定后对端口进行监听。backlog 就是 accept queue，即完成 TCP 三次握手的队列长度。如果不
            // 指定，则默认为 50。这里也是调用 Linux 网络编程的 listen 函数，将当前 socket 变为服务端的 socket，即将其变
            // 为 ServerSocket。因为在网络编程中并没有区分 ServerSocket 和 Socket，如果是服务端 Socket，则需要使用 listen
            // 函数将其设置为被动连接的一端，即服务端，让其开始监听连接，进行 TCP 三次握手，然后再调用 accept 开始
            // 接收请求
            Net.listen(fd, backlog < 1 ? 50 : backlog);
            // 保存 fd 中的本地地址
            localAddress = Net.localAddress(fd);
        }
    } finally {
        end(); // 释放读锁
    }
    return this;
}
```

1.5.5 ServerSocketChannel.accept(null, new AcceptHandler())原理

这一小节分析 ServerSocketChannel.accept(null, new AcceptHandler())的执行原理。注意上面的 bind,

只是设置了套接字，还没有开始接收请求。

```java
// 直接由子类来进行实现
public abstract <A> void accept(A attachment,CompletionHandler<AsynchronousSocketChannel,? super A> handler);
```

下面的源码是 UnixAsynchronousServerSocketChannelImpl 的实现。

```java
public final <A> void accept(A attachment,CompletionHandler<AsynchronousSocketChannel,? super A> handler){
    if (handler == null)
        throw new NullPointerException("'handler' is null");
    // 通过调用 implAccept 完成实际的 accept 流程
    implAccept(attachment, (CompletionHandler<AsynchronousSocketChannel,Object>)handler);
}
// 实际进行 accept 方法
Future<AsynchronousSocketChannel> implAccept(Object att, CompletionHandler<AsynchronousSocketChannel,Object> handler){
    // 如果通道已经被关闭，则直接回调 handler 的 fail 方法处理异常
    if (!isOpen()) {
        Throwable e = new ClosedChannelException();
        if (handler == null) {
            return CompletedFuture.withFailure(e);
        } else {
            Invoker.invoke(this, handler, att, null, e);
            return null;
        }
    }
    ...
    // 设置标志位，表明准备开始接收请求了，确保不会有多个线程同时进行 accept
    if (!accepting.compareAndSet(false, true))
        throw new AcceptPendingException();
    FileDescriptor newfd = new FileDescriptor();
    InetSocketAddress[] isaa = new InetSocketAddress[1];
    Throwable exc = null;
    try {
        // 获取读锁
        begin();
        int n = accept0(this.fd, newfd, isaa); // 尝试开始接收连接，由于通过 IOUtils 设置了 NON_BLOCKING，所以这里会立即返回
        if (n == IOStatus.UNAVAILABLE) {
            // 当前设置了非阻塞，内核返回 IO 状态，表明没有任何客户端连接可用
            PendingFuture<AsynchronousSocketChannel,Object> result = null;
            // 获取更新锁
            synchronized (updateLock) {
                // handler 为空，则构建一个 PendingFuture，表示异步调用结果
                if (handler == null) {
                    this.acceptHandler = null;
                    result = new PendingFuture<AsynchronousSocketChannel,Object>(this);
                    this.acceptFuture = result;
                } else {
                    // 这里不为空，所以保存传递进来的参数
                    this.acceptHandler = handler;
                    this.acceptAttachment = att;
                }
```

```
            // 设置标志位，开始接收请求
            this.acceptPending = true;
        }
        // 通知 port 可以开始轮询获取网络事件了
        port.startPoll(fdVal, Port.POLLIN);
        return result;
    }
} catch (Throwable x) {
    // 发生接收异常
    if (x instanceof ClosedChannelException)
        x = new AsynchronousCloseException();
    exc = x;
} finally {
    // 释放读锁
    end();
}
// 如果正好在 accept 时返回了连接，并且没有发生任何异常，则在当前线程中开始接收当前返回的请求。这
种情况很少发生，除非有客户端不断重试后连接成功
AsynchronousSocketChannel child = null;
if (exc == null) {
    try {
        // 处理 accept
        child = finishAccept(newfd, isaa[0], null);
    } catch (Throwable x) {
        exc = x;
    }
}
// 之前通过 accepting.compareAndSet(false, true)操作，保证只有一个线程能够处理 accept。在 accept 连接
后，调用 handler 处理结果之前，需要释放这里持有的 accepting 锁
enableAccept();
// 如果设置了 handler，则调用 completed 函数
if (handler == null) {
    return CompletedFuture.withResult(child, exc);
} else {
    Invoker.invokeIndirectly(this, handler, att, child, exc);
    return null;
}
}
```

这里通过 accept 接收连接，由于设置了非阻塞，所以大多数情况下会返回 UNAVAILABLE 状态，这表明没有可用连接，即 backlog 队列为空。由于是 AIO，所以不需要阻塞当前线程，将请求通过 port.startPoll(fdVal, Port.POLLIN)在另外一个线程里监听连接。以下是 EPollPort.startPoll 的源码。

```
void startPoll(int fd, int events) {
    // 通过使用 epollCtl 将套接字 fd 添加到 epoll 中，让内核对其进行监听。首先通过 EPOLL_CTL_MOD 尝试
进行修改操作。如果是第一次调用，应该用 EPOLL_CTL_ADD，所以如果返回 ENOENT 状态，则尝试通过
EPOLL_CTL_ADD 添加事件
    int err = epollCtl(epfd, EPOLL_CTL_MOD, fd, (events | EPOLLONESHOT));
    if (err == ENOENT)
        err = epollCtl(epfd, EPOLL_CTL_ADD, fd, (events | EPOLLONESHOT));
    if (err != 0)
        throw new AssertionError();          // should not happen
}
```

以上就是对于 AsynchronousServerSocketChannel 的 open、bind、accept 方法的详细说明。其实读者可以思考，底层就是 epoll 结合线程。事件分为两类：Java 层用于指示线程动作事件和底层 epoll 连接事件，并且在其中也监听了 sv[0]这个套接字。

1.5.6　socketChannel.read(byteBuffer,byteBuffer,new ReadHandler())原理

虽然上面只介绍了 AsynchronousServerSocketChannel，没有涉及客户端通道对象 AsynchronousSocketChannel，但这并不影响理解。举一反三，AsynchronousSocketChannel 也使用了相同的线程池，只不过使用 epoll 监听了读写事件。接下来研究 socketChannel.read(byteBuffer, byteBuffer, new ReadHandler())原理。注意，在 UnixAsynchronousSocketChannelImpl 构造器中，已经将 socket 设置为了非阻塞，且持有外部的 port 对象。

```java
public final <A> void read(ByteBuffer dst,A attachment,CompletionHandler<Integer,? super A> handler)
{
    read(dst, 0L, TimeUnit.MILLISECONDS, attachment, handler);
}
public abstract <A> void read(ByteBuffer dst,
                              long timeout,
                              TimeUnit unit,
                              A attachment,
                              CompletionHandler<Integer,? super A> handler);
```

由子类完成读取并调用 CompletionHandler 回调函数。所以可以直接介绍 UnixAsynchronousSocketChannelImpl。

```java
public final Future<Integer> read(ByteBuffer dst) {
    if (dst.isReadOnly())
        throw new IllegalArgumentException("Read-only buffer");
    // 注意这里的 false，表明 isScatteringRead 为 false
    return read(false, dst, null, 0L, TimeUnit.MILLISECONDS, null, null);
}
private <V extends Number,A> Future<V> read(boolean isScatteringRead,
                                            ByteBuffer dst,
                                            ByteBuffer[] dsts,
                                            long timeout,
                                            TimeUnit unit,
                                            A att,
                                            CompletionHandler<V,? super A> handler)
{
    // socket 已经关闭了直接调用 handler 的 fail 方法
    if (!isOpen()) {
        Throwable e = new ClosedChannelException();
        if (handler == null)
            return CompletedFuture.withFailure(e);
        Invoker.invoke(this, handler, att, null, e);
        return null;
    }
    // 客户端地址信息为空，表明还没有进行连接就直接调用 read
    if (remoteAddress == null)
        throw new NotYetConnectedException();
    boolean hasSpaceToRead = isScatteringRead || dst.hasRemaining();
    boolean shutdown = false;
```

```java
        ...
        return implRead(isScatteringRead, dst, dsts, timeout, unit, att, handler);   // 实际读取动作
}
<V extends Number,A> Future<V> implRead(boolean isScatteringRead,
                                        ByteBuffer dst,
                                        ByteBuffer[] dsts,
                                        long timeout,
                                        TimeUnit unit,
                                        A attachment,
                                        CompletionHandler<V,? super A> handler)
{
    ...
    int n = IOStatus.UNAVAILABLE;
    Throwable exc = null;
    boolean pending = false;
    try {
        begin();
        // 先尝试直接读取数据。由于设置了非阻塞，如果客户端事件没有准备好，则立即返回 UNAVAILABLE
        if (attemptRead) {
            if (isScatteringRead) {
                n = (int)IOUtil.read(fd, dsts, nd);
            } else {
                n = IOUtil.read(fd, dst, -1, nd);
            }
        }
        // 通道没有准备好可读数据
        if (n == IOStatus.UNAVAILABLE) {
            PendingFuture<V,A> result = null;
            // 获取更新状态锁，设置 handler 和 attachment，并调用 updateEvents 更新事件，将当前 client 的
fd 放入 epoll 中监听事件
            synchronized (updateLock) {
                ...
                this.readHandler = (CompletionHandler<Number,Object>)handler;
                this.readAttachment = attachment;
                this.readFuture = null;
                this.readPending = true;
                updateEvents();
            }
            // 设置好当前 channel 等待事件并直接返回
            pending = true;
            return result;
        }
    } catch (Throwable x) {
        if (x instanceof ClosedChannelException)
            x = new AsynchronousCloseException();
        exc = x;
    } finally {
        if (!pending)
            enableReading();
        end();
    }
    // 开始处理读到的数据并调用 handler 的 complete 方法
    Number result = (exc != null) ? null : (isScatteringRead) ?
        (Number)Long.valueOf(n) : (Number)Integer.valueOf(n);
```

```
        if (handler != null) {
            if (invokeDirect) {
                Invoker.invokeDirect(myGroupAndInvokeCount, handler, attachment, (V)result, exc);
            } else {
                Invoker.invokeIndirectly(this, handler, attachment, (V)result, exc);
            }
            return null;
        } else {
            return CompletedFuture.withResult((V)result, exc);
        }
}
// 将 socket fd 添加到 port 的 epoll 中进行监听读事件
private void updateEvents() {
    int events = 0;
    if (readPending)                          // 设置了 readPending,则表明监听读事件
        events |= Port.POLLIN;
    if (connectPending || writePending)       // 如果设置了 connectPending 或 writePending,就是写事件
        events |= Port.POLLOUT;
    if (events != 0)                          // 如果设置了 events,则添加到 Epoll 中监听
        port.startPoll(fdVal, events);
}
```

有数据就读,没数据就设置好 handler,然后将其放入 epoll 中监听。在 epoll 中,事件返回后会由内部线程或用户线程调用 channel 的 onEvent 方法,上面介绍了 AsynchronousServerSocketChannel,接下来介绍 AsynchronousSocketChannel 的 onEvent 方法。

```
// 事件准备好后,由监听 epoll_wait 的线程调用
public void onEvent(int events, boolean mayInvokeDirect) {
    boolean readable = (events & Port.POLLIN) > 0;
    boolean writable = (events & Port.POLLOUT) > 0;
    // 判断读写事件
    if ((events & (Port.POLLERR | Port.POLLHUP)) > 0) {
        readable = true;
        writable = true;
    }
    finish(mayInvokeDirect, readable, writable);
}
// 实际完成读写操作
private void finish(boolean mayInvokeDirect,boolean readable,boolean writable){
    boolean finishRead = false;
    boolean finishWrite = false;
    boolean finishConnect = false;
    ...
    // 根据读写事件完成读操作或写操作,这里只关注读操作
    if (finishRead) {
        if (finishWrite)
            finishWrite(false);
        finishRead(mayInvokeDirect);
        return;
    }
    if (finishWrite) {
        finishWrite(mayInvokeDirect);
    }
```

```java
    if (finishConnect) {
        finishConnect(mayInvokeDirect);
    }
}
// 完成读取操作
private void finishRead(boolean mayInvokeDirect) {
    int n = -1;
    Throwable exc = null;
    // 获取设置的 handler 和 attachment 对象
    boolean scattering = isScatteringRead;
    CompletionHandler<Number,Object> handler = readHandler;
    Object att = readAttachment;
    PendingFuture<Number,Object> future = readFuture;
    Future<?> timeout = readTimer;
    try {
        begin();
        // 开始读取数据
        if (scattering) {
            n = (int)IOUtil.read(fd, readBuffers, nd);
        } else {
            n = IOUtil.read(fd, readBuffer, -1, nd);
        }
        ...
    } catch (Throwable x) {
        enableReading();
        if (x instanceof ClosedChannelException)
            x = new AsynchronousCloseException();
        exc = x;
    } finally {
        // 如果没有发生异常，则将其重新放入 epoll 监听写事件中
        if (!(exc instanceof AsynchronousCloseException))
            lockAndUpdateEvents();
        end();
    }
    ...
    // 调用 handler 的 complete 方法
    Number result = (exc != null) ? null : (scattering) ?
        (Number)Long.valueOf(n) : (Number)Integer.valueOf(n);
    if (handler == null) {
        future.setResult(result, exc);
    } else {
        if (mayInvokeDirect) {
            Invoker.invokeUnchecked(handler, att, result, exc);
        } else {
            Invoker.invokeIndirectly(this, handler, att, result, exc);
        }
    }
}
```

以上就是客户端读取操作的完整分析。复用 PORT 的 ThreadPool 和 epollFd 来操作多线程 AIO 和 Epoll。这就是完整的 Java AIO 在 Linux 上的支持。当前的分析版本是 JDK1.8，未来可能会将 Epoll 替换为 IO_URING，或两者联合一起使用。

1.5.7 Linux 服务端网络编程

```c
#include <stdlib.h>
#include <stdio.h>
#include <errno.h>
#include <string.h>
#include <netdb.h>
#include <sys/types.h>
#include <netinet/in.h>
#include <sys/socket.h>
#include <arpa/inet.h>
#define PORT_NUM 8888
int main(int argc, char *argv[])
{
    int sockfd,new_fd;
    struct sockaddr_in server_addr;              // 服务器地址
    struct sockaddr_in client_addr;              // 客户端地址
    int sin_size;
    char hello[]="Hello\n";
    /* 服务器端开始建立 sockfd 描述符 */
    if((sockfd=socket(AF_INET,SOCK_STREAM,0))==-1)// 使用 AF_INET 协议栈代表 TCP/IP, SOCK_STREAM 表明使用协议栈中的 TCP 协议
    {
        fprintf(stderr,"Socket error:%s\n\a",strerror(errno));
        exit(1);
    }
    /* 构建服务端的 sockaddr 结构 */
    bzero(&server_addr,sizeof(struct sockaddr_in));   // 初始化 server_addr 结构体
    server_addr.sin_family=AF_INET;                   // 设置协议栈为 TCP/IP
    server_addr.sin_addr.s_addr=htonl(INADDR_ANY);    // 网络上传输数据使用大端序进行描述,通过 htonl 函数将字节序转为大端序,INADDR_ANY 表示可以接收任意 IP 地址的数据,即绑定到所有的 IP
    server_addr.sin_addr.s_addr=inet_addr("192.168.1.1"); // 用于绑定到固定 IP, inet_addr 用于转化为整型
    server_addr.sin_port=htons(PORT_NUM);             // 设置端口号
    /* 绑定端口信息 */
    if(bind(sockfd,(struct sockaddr *)(&server_addr),sizeof(struct sockaddr))==-1)
    {
        fprintf(stderr,"Bind error:%s\n\a",strerror(errno));
        exit(1);
    }
    /* 设置允许连接的最大客户端数 */
    if(listen(sockfd,5)==-1)
    {
        fprintf(stderr,"Listen error:%s\n\a",strerror(errno));
        exit(1);
    }
    // 循环处理接收到的客户端请求
    while(1)
    {
        /* 服务器阻塞,直到客户程序建立连接 */
        sin_size=sizeof(struct sockaddr_in);
        if((new_fd=accept(sockfd,(struct sockaddr *)(&client_addr),&sin_size))==-1)
        {
```

```
            fprintf(stderr,"Accept error:%s\n\a",strerror(errno));
            exit(1);
        }
        fprintf(stderr,"Server get connection from %s\n",inet_ntoa(client_addr.sin_addr));  // 将网络地址转换成字符串，并打印到输出终端

        //向客户端程序写入 hello 数组里的字符
        if(write(new_fd,hello,strlen(hello))==-1)
        {
            fprintf(stderr,"Write Error:%s\n",strerror(errno));
            exit(1);
        }
        /* 通信已经结束 */
        close(new_fd);
        /* 循环下一个 */
    }
    /* 结束通信 */
    close(sockfd);
    exit(0);
}
```

1.5.8 Linux 客户端网络编程

```
#include <stdlib.h>
#include <stdio.h>
#include <errno.h>
#include <string.h>
#include <netdb.h>
#include <sys/types.h>
#include <netinet/in.h>
#include <sys/socket.h>
#include <arpa/inet.h>
#define PORT_NUM 8888                     // 服务端的端口号
int main(int argc, char *argv[])
{
    int sockfd;
    char buffer[1024];
    struct sockaddr_in server_addr;       // 描述服务器的地址
    struct hostent *host;
    int nbytes;
    /* 使用 hostname 查询 host 名字 */
    if(argc!=2)
    {
        fprintf(stderr,"Usage:%s hostname \a\n",argv[0]);
        exit(1);
    }
    if((host=gethostbyname(argv[1]))==NULL)
    {
        fprintf(stderr,"Gethostname error\n");
        exit(1);
    }
    /* 客户程序开始建立 sockfd 描述符 */
    if((sockfd=socket(AF_INET,SOCK_STREAM,0))==-1) // AF_INET:Internet;SOCK_STREAM:TCP
```

```c
{
    fprintf(stderr,"Socket Error:%s\a\n",strerror(errno));
    exit(1);
}
/* 初始化并设置服务端连接地址 */
bzero(&server_addr,sizeof(server_addr));           // 初始化，置 0
server_addr.sin_family=AF_INET;                     // 设置协议栈
server_addr.sin_port=htons(PORT_NUM);               // 端口号
server_addr.sin_addr=*((struct in_addr *)host->h_addr);  // 服务端 IP
/* 连接服务端 */
if(connect(sockfd,(struct sockaddr *)(&server_addr),sizeof(struct sockaddr))==-1)
{
    fprintf(stderr,"Connect Error:%s\a\n",strerror(errno));
    exit(1);
}
/* 读取服务端响应数据 */
if((nbytes=read(sockfd,buffer,1024))==-1)
{
    fprintf(stderr,"Read Error:%s\n",strerror(errno));
    exit(1);
}
buffer[nbytes]='\0';
printf("client received:%s\n",buffer);
/* 结束通信 */
close(sockfd);
exit(0);
}
```

1.5.9 Java AIO 涉及的 Native JNI 实现

（1）socket0 方法的代码如下。

```c
JNIEXPORT int JNICALL
    Java_sun_nio_ch_Net_socket0(JNIEnv *env, jclass cl, jboolean preferIPv6,
                                jboolean stream, jboolean reuse)
{
    int fd;
    int type = (stream ? SOCK_STREAM : SOCK_DGRAM);
    int domain = AF_INET;
    fd = socket(domain, type, 0);
    if (fd < 0) {
        return handleSocketError(env, errno);
    }
    ...
    return fd;
}
```

（2）bind0 方法的代码如下。

```c
JNIEXPORT void JNICALL
    Java_sun_nio_ch_Net_bind0(JNIEnv *env, jclass clazz, jobject fdo, jboolean preferIPv6,
                              jboolean useExclBind, jobject iao, int port)
{
    SOCKADDR sa;
```

```c
    int sa_len = SOCKADDR_LEN;
    int rv = 0;
    if (NET_InetAddressToSockaddr(env, iao, port, (struct sockaddr *)&sa, &sa_len, preferIPv6) != 0) {
        return;
    }
    rv = NET_Bind(fdval(env, fdo), (struct sockaddr *)&sa, sa_len);
    if (rv != 0) {
        handleSocketError(env, errno);
    }
}
JNIEXPORT int JNICALL
    NET_Bind(int s, struct sockaddr *him, int len)
{
    int rv;
    rv = bind(s, him, len);
    if (rv == SOCKET_ERROR) {
            if (WSAGetLastError() == WSAEACCES) {
                WSASetLastError(WSAEADDRINUSE);
            }
    }
    return rv;
}
```

（3）listen 方法的代码如下。

```c
JNIEXPORT void JNICALL
    Java_sun_nio_ch_Net_listen(JNIEnv *env, jclass cl, jobject fdo, jint backlog)
{
    if (listen(fdval(env, fdo), backlog) < 0)
        handleSocketError(env, errno);
}
```

（4）accept0 方法的代码如下。

```c
JNIEXPORT jint JNICALL
    Java_sun_nio_ch_ServerSocketChannelImpl_accept0(JNIEnv *env, jobject this,
                                                   jobject ssfdo, jobject newfdo,
                                                   jobjectArray isaa)
{
    jint ssfd = (*env)->GetIntField(env, ssfdo, fd_fdID);
    jint newfd;
    struct sockaddr *sa;
    int alloc_len;
    jobject remote_ia = 0;
    jobject isa;
    jint remote_port;
    NET_AllocSockaddr(&sa, &alloc_len);        // 初始化 sockaddr
    for (;;) {
        socklen_t sa_len = alloc_len;
        newfd = accept(ssfd, sa, &sa_len);
        if (newfd >= 0) {
            break;
        }
        if (errno != ECONNABORTED) {
            break;
```

```c
        }
    }
    // 没有可用的 TCP 三次握手的 socket 信息
    if (newfd < 0) {
        free((void *)sa);
        if (errno == EAGAIN)
            return IOS_UNAVAILABLE;
        if (errno == EINTR)
            return IOS_INTERRUPTED;
        JNU_ThrowIOExceptionWithLastError(env, "Accept failed");
        return IOS_THROWN;
    }
    // 设置 Java 对象 newfdo 的 fd_fdID
    (*env)->SetIntField(env, newfdo, fd_fdID, newfd);
    // 获取远程 client 信息
    remote_ia = NET_SockaddrToInetAddress(env, sa, (int *)&remote_port);
    free((void *)sa);                    // 释放用于接收的 sockaddr
    // 创建 isa_class 指定的 Java 对象，这里对应 InetSocketAddress，可以参考前面的源码描述
    isa = (*env)->NewObject(env, isa_class, isa_ctorID,
                            remote_ia, remote_port);
    // 设置数组元素
    (*env)->SetObjectArrayElement(env, isaa, 0, isa);
    return 1;
}
```

第 2 章

mmap 网络编程与 sendfile 原理

本章将从 Java 层面到内核层面描述对于文件映射零复制和通道 transfer 的原理。

2.1 transferTo 方法

2.1.1 transferTo 方法定义

FileChannel 通道对象的 transferTo 方法可以用于将文件直接传输给 target 变量所指的可写通道中，通常使用此方法完成对 sendfile 函数的调用。该调用是通过 JNI 调用的，可以将该方法与 SocketChannel 一起使用，避免数据先从磁盘传输到用户空间，然后再写回内核，最后放入 socket 的缓冲区增加性能。首先，介绍 transferTo 方法的定义，参数 position 表示当前 FileChannel 操作文件的位置，count 是传输到 target 中的数量，target 是目标通道对象。详细实现如下。

```
public abstract class FileChannel extends AbstractInterruptibleChannel implements SeekableByteChannel,
GatheringByteChannel, ScatteringByteChannel{
    public abstract long transferTo(long position, long count,
                                    WritableByteChannel target)
        throws IOException;
}
```

2.1.2 transferTo 方法实现原理

FileChannelImpl 类对于 FileChannel 类的 transferTo 方法实现，此处省略了参数和通道的校验，直接关注核心方法。共有三种传输方式，第一种方式需要操作系统接口支持，通过操作系统直接传送数据；第二种方式通过 mmap 的方式共享内存传送数据；第三种方式通过传统方式进行传输，这种方式最慢。接下来逐个尝试这三种方式。详细实现如下。

```
public class FileChannelImpl extends FileChannel{
    public long transferTo(long position, long count,
                           WritableByteChannel target)
        throws IOException
    {
        ... // 参数校验
        long n;
        // 尝试直接传输，需要内核支持
        if ((n = transferToDirectly(position, icount, target)) >= 0)
            return n;
        // 尝试通过 mmap 共享内存的方式进行可信通道的传输
        if ((n = transferToTrustedChannel(position, icount, target)) >= 0)
```

```
            return n;
        // 否则通过传统方式进行传输，这种方式最慢
        return transferToArbitraryChannel(position, icount, target);
    }
}
```

2.1.3 transferToDirectly 方法

transferToDirectly 方法直接通过操作系统传送的原理，这种是最快的零复制方式。注意，这里的零复制指的是传送的文件内容不需要从操作系统复制到用户空间，再写入 socket 缓冲区中。此处省略了校验操作，只关注核心操作。首先，获取当前文件通道和目标通道对象的 fd，然后根据是否对 position 上锁，调用 transferToDirectlyInternal 方法完成数据传送。详细实现如下。

```
private long transferToDirectly(long position, int icount,WritableByteChannel target) throws IOException {
    ...
    // 获取当前文件通道和目标通道对象的 fd
    int thisFDVal = IOUtil.fdVal(fd);
    int targetFDVal = IOUtil.fdVal(targetFD);
    if (thisFDVal == targetFDVal)   // 不允许传送给自己
            return IOStatus.UNSUPPORTED;
    // 如果需要使用 position 锁，则获取该锁调用 transferToDirectlyInternal 方法完成数据传送。通常 transferToDirectlyNeedsPositionLock 方法始终返回 true
    if (nd.transferToDirectlyNeedsPositionLock()) {
        synchronized (positionLock) {
            long pos = position();
            try {
                return transferToDirectlyInternal(position, icount,
                                    target, targetFD);
            } finally {
                position(pos);
            }
        }
    } else {
        return transferToDirectlyInternal(position, icount, target, targetFD);
    }
}
```

下面的源码是 transferToDirectlyInternal 方法的实现，在此通过 JNI 调用本地方法 transferTo0 完成传送。详细实现如下。

```
private long transferToDirectlyInternal(long position, int icount, WritableByteChannel target, FileDescriptor targetFD) throws IOException{
    ...
    do {
        // JNI 调用本地方法 transferTo0 完成传送
        n = transferTo0(fd, position, icount, targetFD);
    } while ((n == IOStatus.INTERRUPTED) && isOpen());
    ...
}

// 如果操作系统不支持，将返回-2
private native long transferTo0(FileDescriptor src, long position,
                        long count, FileDescriptor dst);
```

2.1.4　transferToTrustedChannel 方法

　　transferToTrustedChannel 方法用于在可信通道中，使用 mmap 操作通过共享内存的方式进行数据传送。设置了最大 mmap 的大小 MAPPED_TRANSFER_SIZE 为 8MB。如果需要传送的文件数据大于这个值，则需要分阶段映射。获取 MappedByteBuffer 文件映射缓冲区后，调用目标通道的 write 方法完成数据写入。注意，这里使用的是映射缓冲区，所以不存在两次复制。只存在把数据放入页缓存中，通过 mmap 映射到进程的虚拟地址空间，write 函数会直接使用映射的数据，所以不存在复制到 JVM 的堆内存空间，随后再复制到内核。详细实现如下。

```java
private long transferToTrustedChannel(long position, long count,WritableByteChannel target)throws IOException{
    ...
    long remaining = count;
    // 设置了最大 mmap 的大小 MAPPED_TRANSFER_SIZE 为 8MB，如果传送的文件数据大于这个值，则需
    要分阶段映射
    while (remaining > 0L) {
        long size = Math.min(remaining, MAPPED_TRANSFER_SIZE);
        try {
            // 获取当前文件映射缓冲区
            MappedByteBuffer dbb = map(MapMode.READ_ONLY, position, size);
            try {
                // 调用目标通道写入数据
                int n = target.write(dbb);
                ...
            } finally {
                // 写入完成，结束内存映射
                unmap(dbb);
            }
        } catch (ClosedByInterruptException e) {
            ...
        } catch (IOException ioe) {
            ...
        }
    }
    return count - remaining;
}
```

　　以下源码是 map 的方法原理，直接调用 mapInternal 方法进行映射。详细实现如下。

```java
public MappedByteBuffer map(MapMode mode, long position, long size) throws IOException {
    ...
    // 映射并返回解除映射对象
    Unmapper unmapper = mapInternal(mode, position, size, prot, isSync);
    // 根据模式构建只读 MappedByteBufferR 或读写 MappedByteBuffer
    if (unmapper == null) {
        FileDescriptor dummy = new FileDescriptor();
        if ((!writable) || (prot == MAP_RO))
            return Util.newMappedByteBufferR(0, 0, dummy, null, isSync);
        else
            return Util.newMappedByteBuffer(0, 0, dummy, null, isSync);
    } else if ((!writable) || (prot == MAP_RO)) {
```

```
            return Util.newMappedByteBufferR((int)unmapper.cap,
                                    unmapper.address + unmapper.pagePosition,
                                    unmapper.fd,
                                    unmapper, isSync);
        } else {
            return Util.newMappedByteBuffer((int)unmapper.cap,
                                    unmapper.address + unmapper.pagePosition,
                                    unmapper.fd,
                                    unmapper, isSync);
        }
    }
```

mapInternal 方法通过 JNI 调用本地方法完成映射，同时创建解除映射对象，这里使用 DefaultUnmapper，详细实现如下。

```
private Unmapper mapInternal(MapMode mode, long position, long size, int prot, boolean isSync)
    throws IOException
{
    ...
    long addr = -1;
    int ti = -1;
    try {
        ...
        synchronized (positionLock) {
            ...
            try {
                // JNI 调用本地方法完成映射
                addr = map0(prot, mapPosition, mapSize, isSync);
            } catch (OutOfMemoryError x) {
                ...
            }
        }
        ...
        // 构建解除映射对象，这里使用 DefaultUnmapper
        Unmapper um = (isSync
                        ? new SyncUnmapper(addr, mapSize, size, mfd, pagePosition)
                        : new DefaultUnmapper(addr, mapSize, size, mfd, pagePosition));
        return um;
    } finally {
        ...
    }
}

private native long map0(int prot, long position, long length, boolean isSync)
    throws IOException;
```

2.1.5 transferToArbitraryChannel 方法

transferToArbitraryChannel 方法用于完成最慢的传统传输方式。首先，创建默认为 TRANSFER_SIZE 8KB 的堆内缓冲区，随后调用 read 函数将文件数据读入该缓冲区中，接着调用目标 target.write 方法完成对数据的传送。这里发生了多次复制操作：磁盘文件到页缓存，页缓存到 JVM 内存，JVM 内存到堆内内存多次复制，所以性能最差。详细实现如下。

```java
private long transferToArbitraryChannel(long position, int icount, WritableByteChannel target) throws IOException {
    int c = Math.min(icount, TRANSFER_SIZE);
    // 创建堆内缓冲区
    ByteBuffer bb = ByteBuffer.allocate(c);
    long tw = 0;                              // 总写入数据
    long pos = position;
    try {
        // 循环写入
        while (tw < icount) {
            // 把文件数据读入堆内缓冲区中
            bb.limit(Math.min((int)(icount - tw), TRANSFER_SIZE));
            int nr = read(bb, pos);
            if (nr <= 0)
                break;
            // 反转模式，从读模式切换到写模式
            bb.flip();
            // 把文件数据写入堆内缓冲区中
            int nw = target.write(bb);
            tw += nw;
            if (nw != nr)
                break;
            pos += nw;
            // 清空缓冲区，方便下一次读取
            bb.clear();
        }
        return tw;
    } catch (IOException x) {
        ...
    }
}
```

2.2 JVM层面零复制原理

2.1节介绍了两个核心本地方法，即 transferTo0 和 map0，前者用于在操作系统上直接传输文件到 target 缓冲区中，后者用于数据映射。本节详细说明这两个函数在JNI层面上的实现原理。

2.2.1 transferTo0方法

transferTo0 方法用 C 的宏定义选择不同平台下的函数实现，在此只考虑 Linux 内核的实现，直接调用 Linux 内核的 sendfile64 函数进行实现。详细实现如下。

```c
JNIEXPORT jlong JNICALL Java_sun_nio_ch_FileChannelImpl_transferTo0(JNIEnv *env, jobject this,
                                                                    jint srcFD,
                                                                    jlong position, jlong count,
                                                                    jint dstFD)
{
    off64_t offset = (off64_t)position;
    jlong n = sendfile64(dstFD,              // 目标描述符
                         srcFD,              // 源描述符
```

```
            &offset,              // 源传送偏移量
            (size_t)count);       // 传送大小
    ...
    return n;
}
```

2.2.2 map0 方法

map0 方法首先获取映射文件的 fd，随后设置映射的保护权限 protections 和映射属性 flags，然后调用内核的 mmap64 函数进行映射。详细实现如下。

```
JNIEXPORT jlong JNICALL Java_sun_nio_ch_FileChannelImpl_map0(JNIEnv *env, jobject this,
                                                              jint prot, jlong off, jlong len){
    void *mapAddress = 0;

    jobject fdo = (*env)->GetObjectField(env, this, chan_fd);
    jint fd = fdval(env, fdo);
    int protections = 0;
    int flags = 0;
    // 设置映射标志位
    if (prot == sun_nio_ch_FileChannelImpl_MAP_RO) {
        // 只读映射
        protections = PROT_READ;
        flags = MAP_SHARED;
    } else if (prot == sun_nio_ch_FileChannelImpl_MAP_RW) {
        // 读写映射
        protections = PROT_WRITE | PROT_READ;
        flags = MAP_SHARED;
    } else if (prot == sun_nio_ch_FileChannelImpl_MAP_PV) {
        // 私有映射
        protections =  PROT_WRITE | PROT_READ;
        flags = MAP_PRIVATE;
    }
    // 使用 Linux mmap64 函数进行映射
    mapAddress = mmap64(
        0,              // 传入期望映射地址为 0，表明让内核决定映射起始虚拟地址
        len,            // 映射的长度
        protections,    // 映射地址权限：读、读写
        flags,          // 是否为私有映射
        fd,             // 映射的文件描述符
        off);           // 映射的文件数据偏移量

    ...
    return ((jlong) (unsigned long) mapAddress);
}
```

2.3 sendfile64 方法

Linux 内核原理的 sys_sendfile64 函数是 JNI 调用的系统函数。首先，sendfile64 方法判断是否指定了 offset 文件写入时的偏移量。如果使用了该偏移量，则将该指针的所指地址的偏移量值移动到内核内

存中，随后调用 do_sendfile 函数完成数据传送。详细实现如下。

```
asmlinkage ssize_t sys_sendfile64(int out_fd, int in_fd, loff_t __user *offset, size_t count)
{
    loff_t pos;
    ssize_t ret;
    // 如果指定了偏移量，则将用户空间传递的数据传入 pos 中，然后调用 do_sendfile 函数完成传送
    if (offset) {
        if (unlikely(copy_from_user(&pos, offset, sizeof(loff_t))))
            return -EFAULT;
        ret = do_sendfile(out_fd, in_fd, &pos, count, 0);
        // 将最新的 pos 放入 offset 地址中
        if (unlikely(put_user(pos, offset)))
            return -EFAULT;
        return ret;
    }
    return do_sendfile(out_fd, in_fd, NULL, count, 0);
}
```

do_sendfile 函数首先获取输入、输出的文件对象（in_file、out_file），以及输入、输出的 innode 对象（in_inode、out_inode）。参数校验后，调用输入文件对象 in_file 的操作函数 f_op 结构体的 sendfile 函数完成数据写出。详细实现如下。

```
static ssize_t do_sendfile(int out_fd, int in_fd, loff_t *ppos,
                           size_t count, loff_t max)
{
    ...
    // 获取输入文件对象
    in_file = fget_light(in_fd, &fput_needed_in);
    ...
    // 获取输入文件对象 innode
    in_inode = in_file->f_dentry->d_inode;
    ...
    // 获取输出文件对象
    out_file = fget_light(out_fd, &fput_needed_out);
    ...
    // 获取输出文件对象 innode
    out_inode = out_file->f_dentry->d_inode;
    ...
    // 通过输入文件对象的 sendfile 完成写入，file_send_actor 函数地址为内核读入数据放入 Page 页中回调
    retval = in_file->f_op->sendfile(in_file, ppos, count, file_send_actor, out_file);
    ...
    return retval;
}
```

调用文件对象的操作函数 sendfile 函数写入数据。generic_file_sendfile 函数为诸多文件系统 innode 操作函数。首先创建文件读取描述结构 read_descriptor_t，随后调用方法 do_generic_file_read 完成实际写入，写入的数量保存在 written 变量中。详细实现如下。

```
ssize_t generic_file_sendfile(struct file *in_file, loff_t *ppos,
                              size_t count, read_actor_t actor, void __user *target)
{
    // 创建文件读取描述结构
```

```c
read_descriptor_t desc;
if (!count)
    return 0;
desc.written = 0;
desc.count = count;
desc.buf = target;
desc.error = 0;
// 完成具体数据传送操作
do_generic_file_read(in_file, ppos, &desc, actor);
// 返回写入 target 的数量
if (desc.written)
    return desc.written;
return desc.error;
}
```

do_generic_file_read 方法直接调用 do_generic_mapping_read 方法读取磁盘文件数据，然后回调 read_actor_t 所指向的回调函数 filp->f_dentry->d_inode->i_mapping。在内核中通过 address_space 结构体保存从该 inode 代表的文件中读取数据存放的 page 页，&filp->f_ra 指向的 file_ra_state 用于表示预读数据的状态。详细实现如下。

```c
static inline void do_generic_file_read(struct file * filp, loff_t *ppos,
                read_descriptor_t * desc,
                read_actor_t actor)
{
    do_generic_mapping_read(filp->f_dentry->d_inode->i_mapping,// 文件数据页信息
            &filp->f_ra,                                        // 预读状态信息
            filp,                                               // 当前文件结构
            ppos,                                               // 读取的文件 position 地址
            desc,                                               // 文件读取描述结构
            actor);                                             // 回调函数
}
```

do_generic_mapping_read 方法是完成文件数据读取的核心。首先，尝试从 address_space 结构体中获取缓存的物理页帧，address_space 结构体维护了一个保存当前 inode 所指数据的所有页，其中通过基数维护这些物理页帧。如果没有发现该页存在，则进入 handle_ra_miss 方法记录状态，同时跳转到 no_cached_page 中分配新的物理页帧，随后跳转到 readpage 将磁盘中的数据读取到该物理页帧中，然后调用 read_actor_t 回调函数，该函数用于处理该页数据。

```c
void do_generic_mapping_read(struct address_space *mapping,
                struct file_ra_state *ra,
                struct file * filp,
                loff_t *ppos,
                read_descriptor_t * desc,
                read_actor_t actor)
{
    // 获取 address_space 结构所属的 inode
    struct inode *inode = mapping->host;
    ...
    for (;;) {
        ...
        find_page:
```

```c
            // 尝试从 address_space 地址空间中直接获取物理页帧（为了方便理解原理，这里没有写关于 index 下
标的计算）
            page = find_get_page(mapping, index);
            // 如果页为空，需要调用 handle_ra_miss 获取新的物理页帧，随后跳转到 no_cached_page 地址处执行
            if (unlikely(page == NULL)) {
                handle_ra_miss(mapping, ra, index);
                goto no_cached_page;
            }
            // 验证获取到的物理页帧中的内容是否有效。如果无效，则调用 page_not_up_to_date 方法从磁盘中读
取数据
            if (!PageUptodate(page))
                goto page_not_up_to_date;
page_ok:
            ...
            // 调用回调函数处理该页
            ret = actor(desc, page, offset, nr);
            ...
page_not_up_to_date:
            // 物理页帧的数据无效
            if (PageUptodate(page))
                goto page_ok;
            ...
readpage:
            // 从磁盘中读取数据放入该页
            error = mapping->a_ops->readpage(filp, page);
            ...
no_cached_page:
            // 还未缓存该页，需要获取一个新的物理页帧
            if (!cached_page) {
                // 从内存管理系统中分配一个不在 CPU 高速缓存中的物理页帧放入 address_space 中
                cached_page = page_cache_alloc_cold(mapping);
                if (!cached_page) {
                    desc->error = -ENOMEM;
                    break;
                }
            }
            // 将其添加到页缓存的 lru 队列
            error = add_to_page_cache_lru(cached_page, mapping,
                                    index, GFP_KERNEL);
            ...
            // 获取到新页后，读取磁盘数据放入其中
            goto readpage;
    }
    ...
}
```

file_send_actor 函数在 do_generic_mapping_read 中将数据放入物理页帧 page 后回调。调用输出文件指针 file 的 sendpage 文件操作函数使用该页。详细实现如下。

```c
int file_send_actor(read_descriptor_t * desc, struct page *page, unsigned long offset, unsigned long size)
{
    ...
    written = file->f_op->sendpage(file, page, offset,
                            size, &file->f_pos, size<count);
    ...
```

```
        // 返回实际写入数量
        return written;
}
```

假设使用的输出通道对象是 SocketChannel，即网络端，则会进入 Socket 的 sock_sendpage 方法中，该方法获取到了 file 代表的 socket 结构，随后调用该结构的 sendpage 函数。详细实现如下。

```
ssize_t sock_sendpage(struct file *file, struct page *page,
                      int offset, size_t size, loff_t *ppos, int more)
{
    ...
    // 根据 d_inode 结构获取 socket 的地址
    sock = SOCKET_I(file->f_dentry->d_inode);
    ...
    // 调用 socket 的 sendpage 操作
    return sock->ops->sendpage(sock, page, offset, size, flags);
}
```

由于 TCP 相比于 UDP 较为复杂，这里只关注页的去向，以 UDP 协议为例。在 udp_sendpage 中调用 ip_append_page 将该页放入写缓冲区中，而在 ip_append_page 中将调用 skb_fill_page_desc(skb, i, page, offset, len)方法，向 UDP 发送队列添加一个数据报。同时，ip_append_page 不会复制物理页帧的数据，只是将 skb_frag_t 结构指向该页帧。

```
int udp_sendpage(struct sock *sk, struct page *page, int offset, size_t size, int flags)
{
    ...
    ret = ip_append_page(sk, page, offset, size, flags);
    ...
    return ret;
}
```

小结

可以总结得出，使用 sendfile 系统调用时，不会先将数据传输到用户空间，然后再复制到内核空间进行复制，而是直接在 OS 层面操作页帧直接传输数据，如图 2-1 所示。使用该系统调用后，将由内核直接从 in_fd 读取数据放入 out_fd 网络栈中直接传输，与用户空间无关。

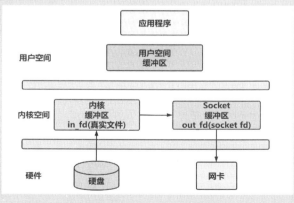

图 2-1 sendfile 原理

第 3 章 Linux epoll 实现原理

select 选择操作的实现在 Linux 上有三种方式。

（1）select：每次调用 select，都要把整个 fd 集合从用户态复制到内核态，同时每次调用 select 都要在内核遍历传递进来的所有 fd，select 支持的文件描述符数量太小了，默认是 1024，这些操作会导致非常大的开销。

（2）poll：poll 的实现和 select 相似，只是 fd 集合存在形式不同。poll 函数使用 pollfd 结构表示键集，而不是 select 的 fd_set 结构，其他的流程一致。poll 函数没有最大文件描述符数量的限制。

（3）epoll：对上述两个函数的改进，只返回准备好的事件，不会受到最大文件描述符数量的限制，内部不使用轮询方式查看准备好的事件，而是通过回调的方式来处理。

本章介绍 Linux 2.6 下的 Epoll 执行原理。

Epoll 函数用于实现 Java 和众多语言底层的 IO 多路复用，创建一个 epoll 对象，将 fd（File Descriptor，文件描述符对象）注册到 epoll 对象中，通过调用 epoll 对象的查询函数获取准备好事件的 fd 对象（Java 开发者请参考 Selector 与 Channel 的使用方式）。本章详细解释了在 Linux 2.6 版本中，Epoll 三大函数的源码与原理分析，详细说明其中的设计技巧。Linux 现在发展到了非常高的版本，本书选取最开始的 Epoll 实现，此版本没有经过任何优化的修改，最为纯粹，方便读者理解。读者可以在此基础上学习高版本对于 Epoll 的优化，包括红黑树的引入等。本章内容稍长，需静下心来阅读。

3.1 三大函数原型

1. epoll_create 函数

`int epoll_create(int __size);`

epoll_create 函数用于创建一个代表 Epoll 对象的 fd。Linux 继承了 UNIX 的优良特性，将一切均视为文件对象处理，这里的文件有两层含义：真实的磁盘上的文件、虚拟文件。而这里的 Epoll 便是一个虚拟文件对象。该函数的参数定义如下。

（1）__size 参数：用于提供给内核一个提示，当前需要监听的 fd 个数，具体怎么做由内核来处理。

（2）返回值：代表当前 Epoll 对象的 fd。

2. epoll_ctl 函数

`int epoll_ctl(int __epfd, int __op, int __fd, struct epoll_event *__event);`

epoll_ctl 函数用于控制由 epoll_create 方法创建的 Epoll 对象。通过该函数操作 Epoll 实现添加、修改、删除监听的 fd 对象。参数定义如下。

（1）__epfd：表示通过 epoll_create 方法创建的 Epoll 对象。
（2）__op：表示操作类型，EPOLL_CTL_ADD（添加监听的 fd）、EPOLL_CTL_DEL（删除监听的 fd）、EPOLL_CTL_MOD（修改监听的 fd）。
（3）__fd：表示需要操作目标的 fd 对象。
（4）__event：表示用于描述需要监听的 fd 对象的感兴趣事件类型。
（5）返回值：0 为成功，-1 为失败。

3. epoll_wait 函数

epoll_wait(int __epfd, **struct** epoll_event *__events, int __maxevents, int __timeout);

epoll_wait 函数用于获取 Epoll 对象监听的 fd 对象列表，监听使用 epoll_ctl 函数添加到 Epoll 对象中准备好事件的 fd 对象。该函数的参数定义如下。
（1）__epfd：表示通过 epoll_create 方法创建的 Epoll 对象。
（2）__events：表示用于接收准备好事件的 fd 对象的缓冲区。
（3）__maxevents：表示这一次调用可以接收多少准备好的 fd 对象，该参数通常设置为 events 参数的长度。
（4）__timeout：表示如果没有准备好事件对象，需要等待多久返回。
（5）返回值：返回 events 缓冲区中有效的 fd 个数，即准备好事件的 fd 个数。

3.2 epoll_event 与 epoll_data 对象

epoll_data 结构用于保存用户数据，epoll_event 结构用于表示监听的 fd 对象的事件和用户数据。读者很容易可以理解，Epoll 要监听 fd，并且要指定监听的事件类型，这时就需要 epoll_event 数据载体，同时可能要设置一些与之关联的数据，如 fd 对象，用户自己的数据 ptr 等（Java 开发者请参考 Selector 和 Channel 绑定时指定的 Attachment）。详细实现如下。

```
typedef union epoll_data{
    void *ptr;              // 保存数据指针
    int fd;                 // 监听的 fd 对象
    uint32_t u32;           // 保存一个 32 位值
    uint64_t u64;           // 保存一个 64 位值
} epoll_data_t;

struct epoll_event{
    uint32_t events;        // 监听的事件类型
    epoll_data_t data;      // 用户数据载体
} __EPOLL_PACKED;
```

3.3 边缘触发与水平触发

Epoll 对象有两种监听的 fd 处理方式：边缘触发（edge-triggered，ET）和水平触发（level-triggered，

LT）。在 ET 模式下，通过 epoll_wait 函数获取到准备好的事件后，如果没有处理完所有事件，再次调用 epoll_wait 函数时不会再次返回该 fd。而 LT 模式下如果 fd 的事件没有处理完成，下一次调用 epoll_wait 函数时会返回该 fd。示例如下。

（1）将一个读取数据的 rfd 注册到 epoll 对象中。
（2）向这个 rfd 中写入 2KB 的数据。
（3）调用 epoll_wait 函数时，由于可以读取 2KB 的数据，会返回已经准备好事件的 rfd。
（4）读取 rfd 中 1KB 的数据，注意，rfd 中还有 1KB 没有处理。
（5）再次调用 epoll_wait 函数。

如果在添加 rfd 时指定 EPOLLET，即 ET 模式，尽管 rfd 中还有 1KB 的数据可以读取，但是因为处于 ET 模式，此时再次调用 epoll_wait 函数会阻塞当前线程，因为 ET 模式需要用户自己处理当前事件的数据。如果使用 EPOLLLT，即 LT 模式，再次调用 epoll_wait 函数，会返回剩余 1KB 读取的 rfd 对象。

3.4 使用示例

这里以一个服务端监听连接事件的 listen_sock fd 为例说明 Epoll 的使用。本例将创建 Epoll 对象，使用三个函数进行 CRUD（Create、Read、Update、Delete）（Java 开发者可以参照 NIO 的 ServerScoketChannel 和 ScoketChannel，操作方式一模一样，只不过屏蔽了 Epoll 的细节）。

```c
#define MAX_EVENTS 10                              // 定义处理的最大事件个数
struct epoll_event ev, events[MAX_EVENTS];         // 初始化 ev 数据载体和接收准备好的 fd 的事件数组
int listen_sock, conn_sock, nfds, epollfd;         // 初始化监听连接的 listen_sock fd、客户端连接 conn_sock
                                                   // fd、接收 epoll_wait 函数返回的准备好事件的个数变量 nfds、Epoll 对象 epollfd fd
epollfd = epoll_create(10);                        // 创建 epoll 对象
if (epollfd == -1) {
    perror("epoll_create");
    exit(EXIT_FAILURE);
}
// 设置 ev 数据载体。设置感兴趣事件为 EPOLLIN 代表读事件
ev.events = EPOLLIN;
ev.data.fd = listen_sock;                                           // 设置用户数据载体中的 fd 为 listen_sock
if (epoll_ctl(epollfd, EPOLL_CTL_ADD, listen_sock, &ev) == -1) {   // 将其添加到 epoll 监听列表中
    perror("epoll_ctl: listen_sock");
    exit(EXIT_FAILURE);
}
// 循环处理所有事件
for (;;) {
    nfds = epoll_wait(epollfd, events, MAX_EVENTS, -1);    // 查询 epoll 函数中是否有准备好的事件，这里使
                                                           // 用 events 数组接收准备好的事件 fd，使用 MAX_EVENTS 指定接收的最大事件数量，使用-1 表明 timeout 为无限
                                                           // 期，即当没有事件时阻塞当前进程
    if (nfds == -1) {
        perror("epoll_pwait");
        exit(EXIT_FAILURE);
    }
    for (n = 0; n < nfds; ++n) {                           // 循环处理已经准备好事件的 fd
        if (events[n].data.fd == listen_sock) {            // 这里以处理监听客户端连接的 listen_sock fd 为例说明
```

```
                conn_sock = accept(listen_sock,(struct sockaddr *) &local, &addrlen); // 接收客户端连接
                if (conn_sock == -1) {
                    perror("accept");
                    exit(EXIT_FAILURE);
                }
                setnonblocking(conn_sock);              // 设置客户端 fd 为非阻塞模式
                ev.events = EPOLLIN | EPOLLET;  // 设置触发模式为 ET 且感兴趣事件类型为 EPOLLIN 读事件
                ev.data.fd = conn_sock;                 // 指定数据载体 fd 为 conn_sock
                if (epoll_ctl(epollfd, EPOLL_CTL_ADD, conn_sock,&ev) == -1) { // 将其添加到 Epoll 监听对象中
                    perror("epoll_ctl: conn_sock");
                    exit(EXIT_FAILURE);
                }
            } else {                                    // 如果是其他事件，还可以继续处理
                do_use_fd(events[n].data.fd);
            }
        }
    }
```

小结

epoll_create 函数用于创建对象，epoll_ctl 函数用于对 Epoll 对象增删改操作，epoll_wait 函数用于对 Epoll 对象实现查询操作。epoll_event 与 epoll_data 对象用于承载与 Epoll 对象交互的数据结构。对于 ET 和 LT 触发而言，ET 模式下，一个事件只会触发一次，如果在该事件中的数据没有处理完毕，下一个事件到来时，不会再次返回监听的 fd；对于 LT 而言，如果数据没有处理完毕，可以再次调用 epoll_wait 函数处理未处理完成数据的 fd。

3.5 三大函数内核原理

了解如何使用 Epoll 后，本节会从三个函数的 Linux 内核源码进行原理讲解。因为 Linux 中一切皆文件，但是由于文件系统是另外一个模块，本书不会占用大量篇幅介绍文件系统相关的概念，但还是有必要介绍一下 VFS 的相关概念。

Linux 中一切皆是文件，但对于文件来说，在磁盘上有真实存在的文件和虚拟文件，例如 Epoll 文件。为了兼容这两者，Linux 提出了 VFS（虚拟文件系统），用其作为访问文件系统的抽象层。

对于应用而言，只要面对 VFS，不需要知道底层是虚拟文件还是真实文件，反正一切皆是文件，按文件形式处理即可。这时就引入了 file 结构体，该结构体代表了一个文件对象；inode 结构体表示文件的元数据信息，也称之为 index node 索引结点。每个进程都需要打开文件，这时就等同于创建了一个 file 对象，不可能把 file 对象暴露给用户空间。读者一定要记住，内核永远不信任用户空间的数据，这就意味着从用户空间传递过来的数据都要复制到内核中才能使用（当然，对于基本数据类型变量，只需要强制转换为不会造成内核瘫痪的数据即可，例如下面的 size 变量）。

对于从内核到用户空间的数据，也需要复制，同时不会将内核的数据结构暴露给用户空间。这时就需要一个映射，将 file 文件对象映射到一个整型变量，将这个整型变量返回给用户空间，而用户空间如果要操作 file 对象，传入该 fd，内核就可以反映射到 file 文件对象操作即可。这个映射信息即保存

在进程的 PCB 控制块中，对于 Linux 而言就是 task_struct（后文中，笔者会基于该版本详细介绍进程管理模块，读者这里了解下即可）。

1. sys_epoll_create 函数

sys_epoll_create 函数用于创建 epoll 对象，同时返回 fd。首先，通过 ep_get_hash_bits 函数计算出 hashbits 变量，随后调用 ep_getfd 函数创建一个 Epoll 对象，并且将其与 Epoll fd 关联。最后调用 ep_file_init 函数初始化 epoll 对象。

Java 开发者可能不太熟悉 goto 语句，但还是需要习惯一下，这里定义了两个退出点：eexit_2 和 eexit_1，分别用于在不同错误下进行返回执行清理工作或打印错误信息。详细描述如下（再次强调一下，Linux 中一切皆是文件，Epoll 对象也是文件）。

```
long sys_epoll_create(int size){
    int error, fd;
    unsigned int hashbits;
    struct inode *inode;
    struct file *file;
    // 根据传入的 hint size 计算出 hash 位数
    hashbits = ep_get_hash_bits((unsigned int) size);
    // 创建一个 Epoll 对象，并将其与 Epoll fd 关联
    error = ep_getfd(&fd, &inode, &file);
    if (error)
        goto eexit_1;
    // 初始化 Epoll 对象
    error = ep_file_init(file, hashbits);
    if (error)
        goto eexit_2;
    return fd; // 返回 fd
eexit_2:
    sys_close(fd);
eexit_1:
    DNPRINTK(3, (KERN_INFO "[%p] eventpoll: sys_epoll_create(%d) = %d\n",
            current, size, error));
    return error;
}
```

2. ep_getfd 函数

ep_getfd 函数用于初始化 efd 指针、einode 指针、efile 指针，分别表示 Epoll 对象的 fd、元数据对象、文件对象，同时将 efd 与 dfile 文件对象进行了关联。由于篇幅有限，并且涉及内存模块相关的知识，所以笔者这里没有展开 ep_eventpoll_inode 与 d_alloc 等与 VFS 相关的内容，这些不属于 Epoll 的研究范畴，只需要注意该函数中 file->f_op = &eventpoll_fops 这行代码即可。

文件是需要操作的，而操作的函数便在 eventpoll_fops 结构中。调用文件对象函数时，由于设置地址为 eventpoll_fops，所以会调用这里面的函数进行操作。对于 Java 开发者而言，这就是 C 语言的接口。定义了一堆接口方法，则需要实现，而 eventpoll_fops 就是实现。C 语言没有接口的概念，不过 C 语言有指针，设置函数指针指向不同的函数便实现了抽象的过程，这也是 VFS 的核心。详细实现如下。

```
static int ep_getfd(int *efd, struct inode **einode, struct file **efile)
{
    struct qstr this;
```

```c
    char name[32];
    struct dentry *dentry;                              // Epoll 文件目录对象
    struct inode *inode;                                // Epoll 索引结点对象
    struct file *file;                                  // Epoll 文件对象
    int error, fd;
    error = -ENFILE;
    file = get_empty_filp(); // 获取一个空的文件对象，判断 max_files 打开的最大文件对象限制，同时分配 Epoll
文件对象 file
    if (!file)
        goto eexit_1;
    inode = ep_eventpoll_inode();   // 分配 Epoll 索引结点
    error = PTR_ERR(inode);
    if (IS_ERR(inode))
        goto eexit_2;
    // 从进程的 files_struct 结构中获取一个空闲的 fd，判断进程打开的最大文件描述符限制，当前版本中的限制
为：#define INR_OPEN 1024
    error = get_unused_fd();
    if (error < 0)
        goto eexit_3;
    fd = error;
    error = -ENOMEM;
    sprintf(name, "[%lu]", inode->i_ino);
    this.name = name;
    this.len = strlen(name);
    this.hash = inode->i_ino;
    dentry = d_alloc(eventpoll_mnt->mnt_sb->s_root, &this);   // 分配目录结点
    if (!dentry)
        goto eexit_4;
    dentry->d_op = &eventpollfs_dentry_operations;            // 设置目录操作
    d_add(dentry, inode);
    // 设置 file 对象文件属性
    file->f_vfsmnt = mntget(eventpoll_mnt);
    file->f_dentry = dget(dentry);
    file->f_pos = 0;
    file->f_flags = O_RDONLY;
    file->f_op = &eventpoll_fops;                             // 初始化文件对象的基础操作回调函数，了解即可
    file->f_mode = FMODE_READ;
    file->f_version = 0;
    file->private_data = NULL;
    // 将 fd 与 file 对象进行关联，读者这里就把 fd 当成数组下标，然后数组中的对象为 file 即可
    fd_install(fd, file);
    // 将 efd、einode、efile 指针指向前面分配的三大结构的地址
    *efd = fd;
    *einode = inode;
    *efile = file;
    return 0;
eexit_4:
    put_unused_fd(fd);
eexit_3:
    iput(inode);
eexit_2:
    put_filp(file);
eexit_1:
```

```
        return error;
}

// Epoll 文件操作的函数结构体
static struct file_operations eventpoll_fops = {
    .release    = ep_eventpoll_close,
    .poll       = ep_eventpoll_poll
};
```

3．ep_file_init 函数

ep_getfd 函数中都是与 VFS 相关的对象，而 Epoll 文件是虚拟文件，虚拟文件需要使用数据载体表示 Epoll 结构。eventpoll 结构便是用来实际操作 Epoll 的核心数据结构。

首先，分配一个 eventpoll 结构的空间，随后调用 ep_init 函数初始化该结构，为了简单明了，这里不展开讲解该方法，读者可以在下一小节中直接查看该结构的数据定义。详细实现如下。

```
static int ep_file_init(struct file *file, unsigned int hashbits)
{
    int error;
    struct eventpoll *ep;
    // 首先，调用内存管理模块的 kmalloc 函数分配一个 eventpoll 结构
    if (!(ep = kmalloc(sizeof(struct eventpoll), GFP_KERNEL)))
        return -ENOMEM;
    memset(ep, 0, sizeof(*ep));          // 对分配的内存进行清零
    error = ep_init(ep, hashbits);       // 初始化 eventpoll 结构
    if (error) {
        kfree(ep);
        return error;
    }
    file->private_data = ep;             // 将文件对象关联到当前 eventpoll 结构
    return 0;
}
```

4．eventpoll 结构

该结构存放在 efile 文件对象的 private_data 中，再次强调，Linux 一切皆是文件，此时需要真实的文件载体，而该结构就是这个载体。可能读者会问如果不介入 VFS，是不是直接使用该结构就可以了？答案是肯定的。详细的参数描述如下。

```
struct eventpoll {
    rwlock_t lock;                          // 保护该结构的读写锁
    struct rw_semaphore sem;                // 用于保护 eventpoll 文件对象的读写信号量
    wait_queue_head_t wq;                   // sys_epoll_wait 函数时，用于保存阻塞进程的等待队列
    wait_queue_head_t poll_wait;            // 用于调用 file->poll 函数时，阻塞进程的等待队列
    struct list_head rdllist;               // 已经准备好事件的 fd 列表
    unsigned int hashbits;                  // 通过传入的 size 计算的 hash 位数
    char *hpages[EP_MAX_HPAGES];            // 用于存放 struct epitem 的数据页
};
```

5．epitem 结构原理

epitem 结构用于表示添加到 Epoll 中的文件信息，每个添加到 Epoll 中监听的 fd 都拥有这样的结构。

```
struct epitem {
```

```
    struct list_head llink;              // 用于将该 epitem 结构关联到对应的 Epoll 对象中
    struct list_head rdllink;  // 用于将该 epitem 结构关联到对应的 Epoll 对象中的 rdllist 准备好事件的 fd 列表中
    int nwait;                            // poll 操作的活动等待队列数
    struct list_head pwqlist;             // 包含轮询等待队列的列表
    struct eventpoll *ep;                 // 所属 epoll 对象
    int fd;                               // 关联的文件 fd
    struct file *file;                    // 关联的文件对象
    struct epoll_event event;             // fd 感兴趣事件集
    atomic_t usecnt;                      // 保存当前结构的引用计数
    struct list_head fllink;              // 将该结构连接到 struct file 文件的 items 列表
    struct list_head txlink;              // 将该 item 连接到 transfer 转移列表
    unsigned int revents;                 // 返回给用户空间的事件集
};
```

6. sys_epoll_ctl 函数

sys_epoll_ctl 函数用于操作 Epoll 对象,对 Epoll 对象执行添加、删除、修改操作。首先,通过 epfd、fd 取出 epoll 文件对象和操作的文件对象,然后进行校验。注意,只有支持 tfile->f_op->poll 操作的对象才可以用于 epoll 监听,随后根据 op 操作调用不同函数操作 epoll 对象。同时读者应该知道 epoll_event 事件结构中的 events 变量是整型变量,所以是按位作为标志位开启不同的感兴趣事件,通过或运算符可以组合这些位。详细实现如下。

```
long sys_epoll_ctl(int epfd, int op, int fd, struct epoll_event __user *event){
    int error;
    struct file *file, *tfile;
    struct eventpoll *ep;
    struct epitem *epi;
    struct epoll_event epds;

    error = -EFAULT;
    if (copy_from_user(&epds, event, sizeof(struct epoll_event))) // 首先将用户空间传递的 epoll_event 复制到
内核空间(内核不应该直接使用用户空间的数据)
        goto eexit_1;
    error = -EBADF;
    file = fget(epfd);                    // 获取 epfd 所代表的 epoll 文件对象(通过 fd->file 的映射)
    if (!file)
        goto eexit_1;
    tfile = fget(fd);                     // 获取要操作的 fd 文件对象
    if (!tfile)
        goto eexit_2;
    error = -EPERM;
    if (!tfile->f_op || !tfile->f_op->poll) // 判断 tfile 文件对象是否支持 poll 操作
        goto eexit_3;

    error = -EINVAL;
    if (file == tfile || !IS_FILE_EPOLL(file)) // 判断 epoll 的 file 对象是否为 epoll 文件
        goto eexit_3;
    ep = file->private_data;              // 获取 epoll 文件对象的 epoll 核心结构体 eventpoll
    down_write(&ep->sem);                 // 获取写信号量
    epi = ep_find(ep, tfile, fd);         // epoll 对象的监听文件中是否存在当前需要操作的 fd 的 epitem 结构
(这里使用链表管理 epitem,因此可通过遍历链表对比 epitem 的 tfile 和 fd 属性)
    error = -EINVAL;
```

```c
switch (op) {
    case EPOLL_CTL_ADD:                                    // 添加操作
        if (!epi) {
            epds.events |= POLLERR | POLLHUP;              // 自动添加 ERR 和 HUP 事件
            error = ep_insert(ep, &epds, tfile, fd);       // 将监听文件对象插入 epoll 对象中监听
        } else
            error = -EEXIST;                               // 只能添加一次
        break;
    case EPOLL_CTL_DEL:                                    // 删除操作
        if (epi)                                           // 如果 epitem 存在,则执行移除
            error = ep_remove(ep, epi);
        else
            error = -ENOENT;
        break;
    case EPOLL_CTL_MOD:                                    // 修改操作
        if (epi) {
            epds.events |= POLLERR | POLLHUP;              // 自动添加 ERR 和 HUP 事件
            error = ep_modify(ep, epi, &epds);
        } else
            error = -ENOENT;
        break;
}
//  ep_find 函数中,如果 epi 存在,则增加 epitem 的计数。现在需要释放这个计数,就是对 usecnt 变量原子性减 1
if (epi)
    ep_release_epitem(epi);
up_write(&ep->sem);                                        // 释放写信号量
eexit_3:
fput(tfile);
eexit_2:
fput(file);
eexit_1:
DNPRINTK(3, (KERN_INFO "[%p] eventpoll: sys_epoll_ctl(%d, %d, %d, %p) = %d\n",
        current, epfd, op, fd, event, error));
return error;
}
```

7. ep_insert 函数

ep_insert 函数用于将监听的 tfile 文件对象放入 epitem 中,然后将其放入对应的链表中。首先,初始化 epi 中的链表结构,接着初始化 epitem 结构体变量,然后初始化 poll table 的回调函数为 ep_ptable_queue_proc(稍后会详细介绍该回调函数的作用及其调用的原理),最后把 epitem 添加到 epoll 的监听链表中,判断当前文件对象在添加到队列后是否马上就发生了感兴趣事件,如果是这样,就将其放入到 epollevent 的 rdllink 准备链表的末尾,并唤醒等待进程。详细实现如下。

```c
static int ep_insert(struct eventpoll *ep, struct epoll_event *event,
        struct file *tfile, int fd){
    int error, revents, pwake = 0;
    unsigned long flags;
    struct epitem *epi;
    struct ep_pqueue epq;
    error = -ENOMEM;
```

```c
    if (!(epi = EPI_MEM_ALLOC()))              // 分配一个新的 epitem 结构
        goto eexit_1;
    // 初始化 epi 中的链表结构
    INIT_LIST_HEAD(&epi->llink);
    INIT_LIST_HEAD(&epi->rdllink);
    INIT_LIST_HEAD(&epi->fllink);
    INIT_LIST_HEAD(&epi->txlink);
    INIT_LIST_HEAD(&epi->pwqlist);
    // 初始化 epitem 结构体变量
    epi->ep = ep;
    epi->file = tfile;
    epi->fd = fd;
    epi->event = *event;
    atomic_set(&epi->usecnt, 1);               // 设置引用计数为 1
    epi->nwait = 0;
    // 初始化 poll table 的回调函数为 ep_ptable_queue_proc
    epq.epi = epi;
    init_poll_funcptr(&epq.pt, ep_ptable_queue_proc);
    // 执行目标文件的 poll 函数，回调 ep_ptable_queue_proc 函数
    revents = tfile->f_op->poll(tfile, &epq.pt);
    if (epi->nwait < 0)
        goto eexit_2;
    // 把 epitem 添加到目标文件对象的 epoll hook 链表
    spin_lock(&tfile->f_ep_lock);
    list_add_tail(&epi->fllink, &tfile->f_ep_links);
    spin_unlock(&tfile->f_ep_lock);
    write_lock_irqsave(&ep->lock, flags);
    // 把 epitem 添加到 epoll 的监听链表中（注意这里使用的是基于 hash 表的链表，即找到索引下标，然后链
接到不同的下标中，见 hashmap 的链地址法）
    list_add(&epi->llink, ep_hash_entry(ep, ep_hash_index(ep, tfile, fd)));
    // 如果文件已经准备好了，且当前 epi 没有被放入到 epollevent 的 rdllink 准备链表的末尾
    if ((revents & event->events) && !EP_IS_LINKED(&epi->rdllink)) {
        list_add_tail(&epi->rdllink, &ep->rdllist);
        // 如果有进程正在阻塞，则唤醒进程处理该准备好的事件
        if (waitqueue_active(&ep->wq))
            wake_up(&ep->wq);
        if (waitqueue_active(&ep->poll_wait))   // 唤醒通过 poll 函数阻塞进程
            pwake++;
    }
    write_unlock_irqrestore(&ep->lock, flags);
    if (pwake)
        ep_poll_safewake(&psw, &ep->poll_wait);
    return 0;
eexit_2:
    ep_unregister_pollwait(ep, epi);
    write_lock_irqsave(&ep->lock, flags);
    if (EP_IS_LINKED(&epi->rdllink))
        EP_LIST_DEL(&epi->rdllink);
    write_unlock_irqrestore(&ep->lock, flags);
    EPI_MEM_FREE(epi);
eexit_1:
    return error;
}
```

8. ep_pqueue 结构体

ep_pqueue 结构体仅作为 poll_table 和 epitem 结构体的包装，而 poll_table_struct 结构体只是一个回调函数 poll_queue_proc，该函数接收一个文件对象和 wait_queue_head_t 等待队列指针。poll_table_struct 结构体，即 poll_table 结构体。

```
struct ep_pqueue {
    poll_table pt;
    struct epitem *epi;
};
typedef void (*poll_queue_proc)(struct file *, wait_queue_head_t *, struct poll_table_struct *);
typedef struct poll_table_struct {
    poll_queue_proc qproc;
} poll_table;
```

init_poll_funcptr(&epq.pt,ep_ptable_queue_proc)的原理，该函数仅是把 poll_table 的函数指针指向 qproc。

```
static inline void init_poll_funcptr(poll_table *pt, poll_queue_proc qproc){
    pt->qproc = qproc;
}
```

传入的 ep_ptable_queue_proc 函数的实现。首先，通过 poll_table 地址取出 epitem（读者可以思考如何获取？前文中，ep_pqueue 结构中包含 poll_table，而这个地址的下面便是 epitem。用 poll_table 的地址转为 ep_pqueue 结构体指针，然后直接取 epitem 即可），分配一个 eppoll_entry 结构体，调用 init_waitqueue_func_entry 初始化 eppoll_entry 的变量和回调函数为 ep_poll_callback。这时相当于把 eppoll_entry 的 wait_queue_t 的变量的回调函数设置为 ep_poll_callback 函数入口，同时将 eppoll_entry 的 wait_queue_t wait 结点添加到 wait_queue_head_t whead 的链表中（该等待结点已经设置了回调函数为 ep_poll_callback，当数据可用时会回调该函数），并且将 eppoll_entry 添加到 epitem 的 pwqlist 链表中。此时读者是否发现，既可以从操作的 fd 的等待连接表中回调 wait 中的函数，又可以从 epitem 结点中通过遍历 pwqlist 链表获取到等待信息结构 eppoll_entry。所以可以提取出共同点，将结构通过 list_head 结构和其变种 wait_queue_head_t 结构关联在一起。关联是为了查询操作。详细实现如下。

```
static void ep_ptable_queue_proc(struct file *file, wait_queue_head_t *whead, poll_table *pt){
    struct epitem *epi = EP_ITEM_FROM_EPQUEUE(pt);    // 根据 poll_table 地址取出 epitem
    struct eppoll_entry *pwq;
    if (epi->nwait >= 0 && (pwq = PWQ_MEM_ALLOC())) {    // 分配 eppoll_entry 结构体
        init_waitqueue_func_entry(&pwq->wait, ep_poll_callback); // 初始化 eppoll_entry 的变量和回调函数为 ep_poll_callback
        pwq->whead = whead;
        pwq->base = epi;
        add_wait_queue(whead, &pwq->wait); // 将 eppoll_entry 中的 wait_queue_t 添加到 wait_queue_head_t 等待结点中
        list_add_tail(&pwq->llink, &epi->pwqlist);    // 将 eppoll_entry 添加到 epitem 的 pwqlist 链表中
        epi->nwait++;                                 // 增加等待计数
    } else {
        epi->nwait = -1;
    }
}

struct eppoll_entry {
```

```
    struct list_head llink;              // 用于将该结构连接到 struct epitem 的 pwqlist 链表结点
    void *base;                          // 指向关联的 epitem 结构指针
    wait_queue_t wait;                   // 用于将该结构添加到目标文件的等待链表结点
    wait_queue_head_t *whead;            // 用于保存当前 eppoll_entry 添加到目标文件的链表结点指针
};
```

这个回调函数就是初始化 eppoll_entry 设置回调函数为 ep_poll_callback，然后将其添加到目标文件对象的等待结点中。如何知道这个链表结点是什么呢？另外读者是否还有一个疑惑，是什么来回调这个 ep_ptable_queue_proc 函数呢？在上文看到最后会调用 tfile->f_op->poll(tfile, &epq.pt)的 poll 函数，这里以网络函数来看下 poll 函数做了什么？

对于网络 Socket 文件来说，将其插入 Epoll 后，会通过上面的 poll 函数调用 sock_poll。该函数进一步调用 tcp_poll（因为假定使用 TCP 协议），而在 tcp_poll 函数中调用了 poll_wait 函数，该函数就是回调 ep_ptable_queue_proc。详细实现如下。

```
static unsigned int sock_poll(struct file *file, poll_table * wait)
{
    struct socket *sock;
    sock = SOCKET_I(file->f_dentry->d_inode);
    return sock->ops->poll(file, sock, wait);      // 通过预设的函数指针调用 tcp_poll
}

unsigned int tcp_poll(struct file *file, struct socket *sock, poll_table *wait)
{
    unsigned int mask;
    struct sock *sk = sock->sk;
    struct tcp_opt *tp = tcp_sk(sk);
    poll_wait(file, sk->sk_sleep, wait);           // 回调 poll_table*wait 的 qproc 函数
    ...
}

static inline void poll_wait(struct file * filp, wait_queue_head_t * wait_address, poll_table *p)
{
    if (p && wait_address)
        p->qproc(filp, wait_address, p);
}
```

笔者一直说：Talk is cheap, show me the code。笔者在给出任何结论时，都会给出相应的论据，所以这里还要给出一个论据。在 Socket fd 中，调用了 tcp_poll 函数，将 eppoll_entry 结构中的 wait_queue_t wait 结构添加到了 sk->sk_sleep 链表中，此时会被回调其中的 func 函数，但这是怎样被回调的呢？请看以下代码。

sock_init_data 函数用于初始化 sock 结构，不清楚 Linux 的网络模块的读者可能会问，为什么会有 socket 和 sock 结构？这里稍微提示一下，Linux 兼容对 BSD 对网络的规范，而这个 socket 结构就是 general BSD socket 结构，而兼容是兼容，毕竟 Linux 有自己的 socket，而这个 sock 就是 Linux 实现网络模块的结构。

这里设置了多个回调函数，当 sock 发生对应事件时回调这些函数。以 sk_state_change 事件来举例，当 sock 的状态改变后，在 sock_def_wakeup 方法中调用 wake_up_interruptible_all，该函数名为唤醒所有的进程，但是由于在 Epoll 添加过程中不存在进程的阻塞，所以设置了 wait_queue_t 的 func 为

ep_poll_callback 函数，所以最后在 __wake_up_common 链表遍历中调用了该回调函数。当然，sock 其他的回调函数也是如此。详细实现如下。

```c
void sock_init_data(struct socket *sock, struct sock *sk){
    ...
    sk->sk_state_change =   sock_def_wakeup;       // sock 状态改变后回调
    sk->sk_data_ready   =   sock_def_readable;     // sock 数据可用时回调
    sk->sk_write_space  =   sock_def_write_space;  // sock 写数据的空间可用时回调
    sk->sk_error_report =   sock_def_error_report; // sock 发生错误时回调
    ...
}

// sock 状态改变后回调
void sock_def_wakeup(struct sock *sk){
    read_lock(&sk->sk_callback_lock);              // 获取读锁
    if (sk->sk_sleep && waitqueue_active(sk->sk_sleep)) // 等待队列不为空，则调用 wake_up_interruptible_all 唤醒 sk_sleep 链表中的等待结点
        wake_up_interruptible_all(sk->sk_sleep);
    read_unlock(&sk->sk_callback_lock);
}

#define wake_up_interruptible_all(x)    __wake_up((x),TASK_INTERRUPTIBLE, 0);  // 0 表示非互斥唤醒，即唤醒全部

void __wake_up(wait_queue_head_t *q, unsigned int mode, int nr_exclusive){
    unsigned long flags;
    spin_lock_irqsave(&q->lock, flags);
    __wake_up_common(q, mode, nr_exclusive, 0);    // 调用该函数完成唤醒过程
    spin_unlock_irqrestore(&q->lock, flags);
}

static void __wake_up_common(wait_queue_head_t *q, unsigned int mode, int nr_exclusive, int sync){
    struct list_head *tmp, *next;
    list_for_each_safe(tmp, next, &q->task_list) { // 遍历等待链表 wait_queue_head_t *q
        wait_queue_t *curr;
        unsigned flags;
        curr = list_entry(tmp, wait_queue_t, task_list); // 从当前 task_list 地址中获取到 wait_queue_t 结构
        flags = curr->flags;
        if (curr->func(curr, mode, sync) &&        // 调用其设置的回调函数，如果是 epoll，就是回调了 ep_poll_callback 函数
            (flags & WQ_FLAG_EXCLUSIVE) &&         // 如果指定了互斥唤醒，则使用 nr_exclusive 决定唤醒多少个进程（注意这里的标志位，因为在 ep_poll 函数中添加等待进程时，没有指定这个标志位，考虑下惊群效应）
            !--nr_exclusive)
            break;
    }
}
```

以下代码是 ep_poll_callback 回调函数的原理。

```c
static int ep_poll_callback(wait_queue_t *wait, unsigned mode, int sync){
    int pwake = 0;
    unsigned long flags;
    struct epitem *epi = EP_ITEM_FROM_WAIT(wait);  // 从 wait 结点中获取到 epitem，eppoll_entry 中包含
```

了 wait_queue_t wait 结构，可以根据 wait 的地址-wait 相对于 eppoll_entry 结构的偏移量即可
```
    struct eventpoll *ep = epi->ep;              // 从 epitem 中获取到关联的 eventpoll 结构
    write_lock_irqsave(&ep->lock, flags);
    // 如果该文件代表的 epitem 已经存在于 eventpoll 中，则直接退出，因为前面直接调用了 poll 函数，考虑下如
果数据在调用 poll 时就已经可用了呢？这时是不是在前面就将其添加到 eventpoll 的 rdllink 准备好事件的链表中？
    if (EP_IS_LINKED(&epi->rdllink))
        goto is_linked;
    list_add_tail(&epi->rdllink, &ep->rdllist);   // 添加到 eventpoll 的 rdllink 准备好事件的链表
is_linked:
    // 唤醒所有等待 eventpoll 准备好事件的进程，这里就是唤醒了通过 epoll_wait 调用而阻塞的进程（注意，这
里导致了惊群效应的发生，由于没有使用互斥唤醒标志 WQ_FLAG_EXCLUSIVE，所以这里的 wake_up 函数会唤
醒所有在 wq 等待链表上的进程）
    if (waitqueue_active(&ep->wq))
        wake_up(&ep->wq);
    if (waitqueue_active(&ep->poll_wait))
        pwake++;
    write_unlock_irqrestore(&ep->lock, flags);
    if (pwake)
        ep_poll_safewake(&psw, &ep->poll_wait);
    return 1;
}
```

9. ep_remove 函数

ep_remove 函数用于从 epoll 监听 fd 链表中移除对应的 fd。其实这只不过是 insert 的逆向过程罢了。做了哪些连接，就需要将其从对应的链表中摘除，然后将其从 epoll 结构中移除，最后释放 epitem 所占用的内存空间。详细实现如下。

```
static int ep_remove(struct eventpoll *ep, struct epitem *epi){
    int error;
    unsigned long flags;
    struct file *file = epi->file;
    ep_unregister_pollwait(ep, epi);              // 移除所有 epitem 结构的 pwqlist 链表中，eppoll_entry 结构添加
的 wait 等待结点，因为移除这个 fd 后不需要再监听回调了
    spin_lock(&file->f_ep_lock);
    if (EP_IS_LINKED(&epi->fllink))               // 从监听文件对象的链表中移除 epitem 结构
        EP_LIST_DEL(&epi->fllink);
    spin_unlock(&file->f_ep_lock);
    write_lock_irqsave(&ep->lock, flags);
    error = ep_unlink(ep, epi);                   // 从 epoll 结构中移除 epitem 结构
    write_unlock_irqrestore(&ep->lock, flags);
    if (error)
        goto eexit_1;
    ep_release_epitem(epi);                       // 释放 epitem 占用的内存空间
    error = 0;
eexit_1:
    DNPRINTK(3, (KERN_INFO "[%p] eventpoll: ep_remove(%p, %p) = %d\n",
                current, ep, file, error));
    return error;
}
```

10. ep_modify 函数

ep_modify 函数用于修改监听文件对象的感兴趣事件集。首先，修改感兴趣事件集合，调用修改 fd

的 poll 函数，该函数会查看当前是否有准备好的事件集。然后进一步判断当前 epitem 是否被移除，如果没有移除，则根据当前事件集和 epitem 的状态选择将其从 eventpoll 准备好的事件链表中放入还是移除。详细实现如下。

```c
static int ep_modify(struct eventpoll *ep, struct epitem *epi, struct epoll_event *event){
    int pwake = 0;
    unsigned int revents;
    unsigned long flags;
    epi->event.events = event->events;                  // 更新感兴趣事件集
    revents = epi->file->f_op->poll(epi->file, NULL);   // 回调修改 fd 的 poll 函数，注意，传递的 poll_table
    *wait 为 NULL，所以并没有修改之前设置的回调函数，只是查看是否有事件发生
    write_lock_irqsave(&ep->lock, flags);
    epi->event.data = event->data;                       // 更新用户数据结构
    // 如果已经删除了当前修改文件的 epitem，则不执行任何操作
    if (EP_IS_LINKED(&epi->llink)) {
        // 判断当前是否有准备好的事件，revents 是通过修改文件的 poll 回调获取的事件集合
        if (revents & event->events) {
            if (!EP_IS_LINKED(&epi->rdllink)) {
                // 将当前 epitem 添加到 epoll 准备好的事件链表
                list_add_tail(&epi->rdllink, &ep->rdllist);
                // 唤醒等待进程
                if (waitqueue_active(&ep->wq))
                    wake_up(&ep->wq);
                if (waitqueue_active(&ep->poll_wait))
                    pwake++;
            }
        } else if (EP_IS_LINKED(&epi->rdllink))          // 如果当前 epitem 已经连接了准备好的事件，但是由
于已经修改了感兴趣事件集，则需要将其从准备好的事件链表中移除
            EP_LIST_DEL(&epi->rdllink);
    }
    write_unlock_irqrestore(&ep->lock, flags);
    if (pwake)
        ep_poll_safewake(&psw, &ep->poll_wait);
    return 0;
}
```

11. sys_epoll_wait 函数

sys_epoll_wait 函数用于实现对 Epoll 对象的查询操作。从 epoll_wait 函数中可以知道，struct epoll_event __user *events 用于存放准备好的事件集，int maxevents 用于指定大小，通常等于 events 的数组大小，int timeout 用于指明超时等待时间。Linux 内核不相信任何用户空间的数据，所以对这些参数进行了详细校验，然后获取到 eventpoll 对象，调用 ep_poll 函数完成事件获取。

ep_poll 函数中，首先计算超时时间，然后判断准备好的事件链表是否为空。如果为空，则创建进程等待节点，并放入 eventpoll 对象的 wq 链表中，同时设置进程状态为 TASK_INTERRUPTIBLE，然后判断当前进程有没有需要处理的信号，使用进程调度器选择其他进程执行（Linux 内核不同于用户空间，需要手动将控制权交由进程管理模块来调度，而这里的 schedule_timeout 函数便是进程调度模块的核心函数，该函数使用进程调度器和调度算法选取下一个进程完成进程切换。笔者会在后文中详细介绍该切换过程）。

由此可以得出结论，该等待进程有三种返回情况，即由 ep_poll_callback()回调函数唤醒、进程传递

信号唤醒、超时时间到唤醒。准备好的事件发生后，会调用 ep_events_transfer 函数构建 txlist，并将其复制到用户指定的内存空间中。

ep_events_transfer 函数中，首先调用 ep_collect_ready_items 函数将 rdllist 中的准备好的事件取出来放入 txlist 中，获取的数量由 maxevents 指定，然后调用 ep_send_events 将 txlist 复制到用户指定的 struct epoll_event __user *events 空间中。

笔者没有展开介绍该函数，因为这里又会涉及 Linux 的内存管理原理，例如，copy_to_user 函数的使用方式，这已经不再属于 Epoll 的范畴，读者也不需要在本文了解。最后调用 ep_reinject_items 函数判断是否需要将 txlist 中的事件重新放入 epoll 的 rdllist 中。

ep_reinject_items 函数中，EP_IS_LINKED(&epi->llink) && !(epi->event.events & EPOLLET) 核心判断逻辑，判断如果 epitem 没有被删除，且设置的事件处理类型不为 EPOLLET（边缘触发），就会将其重新放入 rdllist 中。最后判断如果重复放入了，则唤醒所有等待进程。

那么问题就来了，很多人把这一步当作惊群效应，但其实它并不是。这正是 Epoll 高性能的体现。如果返回的 fd 数量较大，且使用了水平触发，一个进程处理不过来，是否可以使用多个进程唤醒处理，然后由处理函数保证互斥即可，这正是高性能的体现。而需要注意的惊群在这个版本中确实有，存在于之前介绍的 ep_poll_callback() 回调函数和 ep_poll 函数中。正是由于在添加进程等待节点时并没有指定 WQ_FLAG_EXCLUSIVE 标志位，从而在前面介绍的 __wake_up_common 函数中不会响应 nr_exclusive 变量，尽管 wake_up 函数的 nr_exclusive 为 1，但由于没有设置标志位，所以并不响应，前文中已经详细介绍过。函数实现如下。

```c
long sys_epoll_wait(int epfd, struct epoll_event __user *events,int maxevents, int timeout){
    int error;
    struct file *file;
    struct eventpoll *ep;
    // 最大获取事件必须大于 0
    if (maxevents <= 0)
        return -EINVAL;
    // 验证传入的存放准备好事件的内存区域合法性
    if ((error = verify_area(VERIFY_WRITE, events, maxevents * sizeof(struct epoll_event))))
        goto eexit_1;
    error = -EBADF;
    file = fget(epfd);                          // 根据 epoll fd 获取到 epoll 文件对象
    if (!file)
        goto eexit_1;
    error = -EINVAL;
    if (!IS_FILE_EPOLL(file))                   // 判断当前 epoll file 对象是否为 epoll 文件
        goto eexit_2;
    ep = file->private_data;                    // 获取到 epoll 文件对象的实际承载体：eventpoll 结构
    error = ep_poll(ep, events, maxevents, timeout);  // 调用该函数获取事件
eexit_2:
    fput(file);
eexit_1:
    DNPRINTK(3, (KERN_INFO "[%p] eventpoll: sys_epoll_wait(%d, %p, %d, %d) = %d\n",
             current, epfd, events, maxevents, timeout, error));
    return error;
}
```

```c
// 获取准备好的事件集。如果指定了 timeout，则等待超时后返回
static int ep_poll(struct eventpoll *ep, struct epoll_event __user *events,int maxevents, long timeout){
    int res, eavail;
    unsigned long flags;
    long jtimeout;
    wait_queue_t wait;
    // 计算超时时间。如果 timeout 为-1，则超时时间戳为 MAX_SCHEDULE_TIMEOUT，该值为 long 的最大值，
    // 否则根据内核的时钟频率 Hz 来计算，当前内核的频率为 100Hz
    jtimeout = timeout == -1 || timeout > (MAX_SCHEDULE_TIMEOUT - 1000) / HZ ?
        MAX_SCHEDULE_TIMEOUT: (timeout * HZ + 999) / 1000;

retry:
    write_lock_irqsave(&ep->lock, flags);
    res = 0;
    if (list_empty(&ep->rdllist)) { // 如果准备好的事件链表为空，则构建当前进程的等待节点结构 wait_queue_t
    // wait，并将其添加到 eventpoll 的 wq 等待链表
        init_waitqueue_entry(&wait, current);
        add_wait_queue(&ep->wq, &wait);         // 注意，由于这里添加节点时没有指定互斥状态
    // WQ_FLAG_EXCLUSIVE，从而在 callback 唤醒时会唤醒所有等待进程
        // 循环等待直到满足条件
        for (;;) {
            set_current_state(TASK_INTERRUPTIBLE); // 设置当前进程状态为 TASK_INTERRUPTIBLE（可
    // 中断阻塞状态）
            if (!list_empty(&ep->rdllist) || !jtimeout) // 如果准备好事件集此时不为空，或超时，则退出循环
                break;
            if (signal_pending(current)) {          // 当前进程有需要处理的信号，退出并设置返回值为 EINTR
                res = -EINTR;
                break;
            }
            write_unlock_irqrestore(&ep->lock, flags);
            jtimeout = schedule_timeout(jtimeout); // 调用该函数让当前进程阻塞，并且调用 Linux 内核的进程
    // 调度模块来选择其他进程运行
            write_lock_irqsave(&ep->lock, flags);
        }
        // 满足条件后将当前进程从 eventpoll 的 wq 等待链表中移除，并设置 TASK_RUNNING 的标志位表示当
    // 前进程处于运行状态
        remove_wait_queue(&ep->wq, &wait);
        set_current_state(TASK_RUNNING);
    }
    eavail = !list_empty(&ep->rdllist);
    write_unlock_irqrestore(&ep->lock, flags);
    // 如果当前 epoll 中存在准备好的事件集，则调用 ep_events_transfer 函数将其复制到用户指定的 events 内
    // 存中，否则返回 retry 处继续尝试
    if (!res && eavail &&                         // 如果指定了 res，则直接返回 res
        !(res = ep_events_transfer(ep, events, maxevents)) && jtimeout)
        goto retry;

    return res;
}

// 将准备好的事件传输到用户指定的内存空间 events 中
static int ep_events_transfer(struct eventpoll *ep,struct epoll_event __user *events, int maxevents){
    int eventcnt = 0;
```

```c
    struct list_head txlist;
    INIT_LIST_HEAD(&txlist);                            // 初始化传输结点
    down_read(&ep->sem);
    // 将 rdllist 准备好事件集的链表节点转移到 txlist 链表中，最大个数为 maxevents
    if (ep_collect_ready_items(ep, &txlist, maxevents) > 0) {
        eventcnt = ep_send_events(ep, &txlist, events);    // 将 txlist 传输到用户空间中
        ep_reinject_items(ep, &txlist); // 根据 Epoll 指定的 LT 和 ET 模式选择是否重新将 txlist 放入 rdllist 中
    }
    up_read(&ep->sem);
    return eventcnt;
}

// 根据 Epoll 指定的 LT 和 ET 模式选择是否重新将 txlist 放入 rdllist 中
static void ep_reinject_items(struct eventpoll *ep, struct list_head *txlist){
    int ricnt = 0, pwake = 0;
    unsigned long flags;
    struct epitem *epi;
    write_lock_irqsave(&ep->lock, flags);
    while (!list_empty(txlist)) {                       // 循环直到 txlist 链表为空
        epi = list_entry(txlist->next, struct epitem, txlink);  // 根据 txlist 的地址获取到所属 epitem 结构
        EP_LIST_DEL(&epi->txlink);
        if (EP_IS_LINKED(&epi->llink) && !(epi->event.events & EPOLLET) &&    // epitem 没有被删除，且设置的事件处理类型不为 EPOLLET（边缘触发）
            (epi->revents & epi->event.events) && !EP_IS_LINKED(&epi->rdllink)) {
            // 将 tx 中的 epitem 重新放入 rdllist 准备好事件集的链表中
            list_add_tail(&epi->rdllink, &ep->rdllist);
            ricnt++;                                    // 计算重放入的次数
        }
    }
    // 如果重复放入准备好事件集，则唤醒所有等待结点
    if (ricnt) {
        if (waitqueue_active(&ep->wq))
            wake_up(&ep->wq);
        if (waitqueue_active(&ep->poll_wait))
            pwake++;
    }
    ...
}
```

第 4 章 Netty 架构与源码组成

4.1 Netty 是什么

Netty 是 NIO 客户端服务器框架，基于 Netty 可以快速、简单、高效的开发网络应用程序，例如协议服务器和客户端。

Netty 简化了网络编程，例如 TCP 和 UDP 套接字服务器。快速和简单意味着基于 Netty 构建的应用程序不需要考虑可维护性或性能问题。Netty 经过了精心的设计，实现了许多协议（FTP、SMTP、HTTP 和各种基于二进制和文本协议），这些协议直接拆封即用，不需要应用程序自行实现。

下面根据前面的 Java 网络编程基础知识进行推理。

（1）Java 使用 NIO 模型：SocketChannel、ServerChannel 对象表示客户端通道与服务端通道。

（2）Java 使用 Selector 提供了对可选择通道的多路复用。

（3）Java 使用 Buffer 作为通道的数据载体，在互相关联的通道中传递数据。

（4）在网络通信中，需要考虑两大难点。

① IO 模型，即底层通过什么样的方式进行通信。

② 多线程模型，即采用什么样的线程组织方式处理 IO 操作。

这两个难点将决定网络通信框架的性能，对于 Netty 而言，它的核心便是如此。

（5）Netty 是 NIO 的通信框架，利用了 Java 高性能的 NIO 模型进行封装，同时考虑了多线程的模型与实际 IO 模型处理。

（6）从 Socket 读取数据时，必然需要对数据进行编码和解码，再进行业务处理，Netty 在设计时也需要考虑数据处理的模型。

所以，Netty 是什么呢？Netty 就是封装了 Java NIO 细节，并提供了高层次的线程模型与 IO 模型，实现了常用应用层协议的 NIO 类库。

4.2 Netty 架构组成

如图 4-1 所示，Netty 的架构设计分为四个模块。

（1）Transport Service：通信服务，包括对 Java 底层 Channel 的封装，定义了 IO 模型。

（2）Protocol Support：提供丰富的应用层协议，应用程序可以直接拆封即用。

（3）Extensible Event Model：弹性事件模型，定义了多线程模型。

（4）Zero-Copy-Capable Rich Byte Buffer：实现了自己的内存管理机制，对堆内存的 byte[]数组和 NIO 的 ByteBuffer 进行了封装，提供了零复制的实现。

图 4-1 Netty 整体架构

4.3 Netty 三大基础模块

Netty 有三个核心模块，分别是事件循环组、内存池、通道处理器。

本书将这三个模块作为独立篇章进行讲解，本节先初步介绍每个模块，后面的每一篇中将根据整体脉络进行详细介绍。

4.3.1 事件循环模块

ServerSocket 可以接收多个客户端的连接，生成 Socket 对象表示与客户端的连接，并在多线程中处理这些 Socket。如何定义线程、处理 Socket 是非常重要的。因为在多线程环境中，会涉及线程互斥的问题，所以定义 Netty 的线程模型，必须妥善处理这些连接的 Socket。

如图 4-2 所示，Netty 将多个线程组成一组，成为事件循环组。而每个线程中唯一持有一个 Selector 选择器，有 N 个线程时，就会存在 N 个选择器。Netty 为 IO 密集型任务，通常设置的 N 为 CPU 核心数的两倍。为什么要这样设计？这主要是为了确保线程竞争切换的开销与线程等待 IO 事件时 CPU 空闲的平衡。ServerScoketChannel 和 SocketChannel 会唯一绑定其中一个选择器，这时就等同于唯一绑定一个线程，这有何好处？好处是在多线程之间操作不需要上锁了，因为 Netty 规定所有的 Socket 处理都要在它绑定的线程中处理，其他线程需要处理 Channel 时，需要进行如下判断。

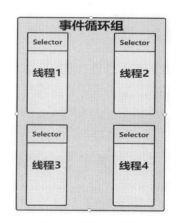

图 4-2 事件循环组

```
if(inEventLoop()){         // 当前线程属于 socket 绑定的事件循环线程，则直接调用
    // 调用 socket
}else{
    execute(()->{// 调用 socket });   // 否则封装调用任务，放到事件循环线程的队列中等待执行
}
```

为什么是事件循环组？因为线程在内部组成一个组，提供了 next 方法获取该组中的下一个事件循环线程，即彼此关联，所以成为循环组，默认循环算法为数组轮询。

4.3.2　内存池模块

频繁申请和释放内存会影响性能，为了解决这个问题，Netty 在内存中实现了一套内存管理器，通过该管理器管理以下两种类型的内存。

（1）堆内内存：new byte[length]。

（2）堆外内存：DirectByteBuffer。

这两种类型为实际存储类型，就像 ByteBuffer 一样，Netty 通过 ByteBuf 类包装实际存储数据的这两种类型的内存，提供了 readIndex 和 writeIndex 作为当前对存储数据内存的读写下标。如图 4-3 所示，Netty 通过以下几个模块管理内存。

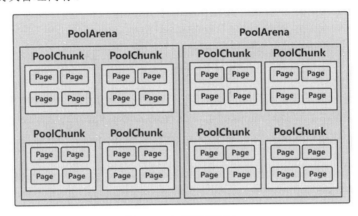

图 4-3　内存管理模块

（1）PoolArena：表示内存池管理区，每个线程通过 ThreadLocal 保存，类型分为 Direct 和 Heap，分别表示堆外和堆内区域。通过该区域减少线程分配时的竞争，默认情况下等于事件循环线程的数量，当线程数量大于该区域时，线程在分配时会上锁。

（2）PoolChunk：基于伙伴算法来管理 Page 页，默认一个 Chunk 的大小为 16MB。

（3）Page：基本管理单元页，默认为 8KB。

4.3.3　通道处理器模块

如图 4-4 所示，Netty 通过以下组件构建通道处理。

图 4-4　通道处理器模块

（1）ChannelPipeline：表示处理器流水线的集合。

（2）ChannelHandler：表示通道处理器。

为什么要这样设计？当事件循环线程从 Socket 中读取到数据时，这些数据通过内存管理模块创建的 ByteBuf 对象包装，其中为字节数据。而这时需要从其中解码为应用层协议（如 HTTP），将其递交给业务处理器处理自身业务。这时很容易联想到责任链模式，将一系列处理方法通过该模式可以很好地满足要求。在 Netty 中通过责任链模式将通道处理器形成一条处理器链，前一个处理器的结果可以作为后一个处理器的输入，以实现业务与实际编码解码器进行解耦，在输出处理器的末端，也会把输出的对象转为字节信息写入底层 Socket。

4.4 Netty 源码组成

Netty 使用 Maven 的聚合工程进行开发，源码组织如下。
（1）Netty-buffer 模块：内存池管理模块。
（2）Netty-codec-x 模块：编码器与解码器模块（基于通道处理器实现的编码和解码器）。
（3）Netty-common 模块：包含各个模块共用的工具类。
（4）Netty-handler 模块：通道处理器模块。
（5）Netty-resolver 模块：DNS 域名解析模块。
（6）Netty-transport 模块：对 Java NIO 通道的包装实现模块。
（7）Netty-example 模块：提供 Netty 的使用示例。

第 2 篇

事件驱动层

本篇详细介绍 Netty 中的线程模型。

（1）事件循环。

（2）事件循环组。

由于事件循环与事件循环组会操作 Channel 通道，笔者没有展开说明通道相关的操作，具体的实现会在第 3 篇介绍，读者只需要了解事件循环组的架构设计，以及如何调用和创建 Selector 选择器配合通道对象完成 IO 操作即可。

注意在文中出现的三个队列 scheduledTaskQueue、taskQueue、tailTasks 的作用以及处理时机。同时，要注意使用事件循环组时，如何通过 Chooser 选择事件循环组中的事件循环线程。

最后，原生 Java 使用 Future 接口表示异步执行任务，可以通过它等待、取消、获取任务执行的结果，但是这通常需要调用方来实际调用处理结果，有时需要执行异步回调处理，即调用方不需要调用 Future 来手动获取结果并执行，而是在处理完成任务后，由异步执行线程回调异步方法。Netty 对 Future 进行了增强，使得观察者模式实现了 Promise 的定义，这是需要特别注意的。

第 5 章 JDK Executor 原理

5.1 Executor 接口

Executor 接口是 JDK 的 JUC（java.util.concurrent）顶层接口，该接口定义了一个用于执行提交的 Runnable 接口实例的执行器对象。该接口主要用于解耦需要执行的 Runnable 任务与调用者之间的关联，子类负责处理 Runnable 的执行方式，即单线程执行、多线程执行、周期性调度执行等待。通常使用该接口的实例执行任务，而不是直接创建线程来处理。

```
new Thread(new(RunnableTask())).start();   // 不推荐

// 推荐方式
Executor executor = anExecutor;
executor.execute(new RunnableTask1());
executor.execute(new RunnableTask2());
```

接口定义源码如下所示。读者可以考虑，为什么不直接使用线程的方式启动？提示：可以从接口的定义出发考虑这个问题。

（1）接口本身是为了解耦，需要把解耦的方法定义在接口中。

（2）执行任务：实现了 Runnable 接口的 RunnableTask，可以开启一个线程执行，也可以同步执行。

（3）如果觉得启动线程执行任务较好，可使用 new Thread(new(RunnableTask())).start()；如果觉得在当前线程执行较好，并且没有抽取该接口，可以执行该方法的逻辑，并耦合在业务代码中，找到该代码进行修改。

（4）这时就定义了一个接口，将封装的变化放到 Executor 接口的实现类中执行，只需要定义多个 Executor 子类，完成自己的逻辑，在需要时可以通过类似 Spring 的 IOC 方式进行修改，而不需要改动源码。

（5）这就是接口解耦合，代码如下。

```
public interface Executor {
    void execute(Runnable command);
}
```

5.2 ExecutorService 接口

ExecutorService 接口是 Executor 接口的次一级接口，在 Executor 接口中定义一个执行方法，而在

ExecutorService 接口中需要定义一些用于服务执行器的方法。这里的思路是，Executor 接口用于执行抽象任务，ExecutorService 接口用于在执行抽象任务时的一些功能函数。源码如下。

```java
public interface ExecutorService extends Executor {
    // 关闭服务
    void shutdown();

    // 立刻关闭服务
    List<Runnable> shutdownNow();

    // 判断服务是否调用了 shutdown
    boolean isShutdown();

    // 判断服务是否完全终止
    boolean isTerminated();

    // 等待当前服务停止
    boolean awaitTermination(long timeout, TimeUnit unit)
        throws InterruptedException;

    // 提交一个任务 Callable task，返回一个存根 stub
    <T> Future<T> submit(Callable<T> task);

    // 提交一个任务 Runnable task，返回一个存根 stub
    Future<?> submit(Runnable task);

    // 提交一组 callable 任务执行，返回所有任务的存根 stub
    <T> List<Future<T>> invokeAll(Collection<? extends Callable<T>> tasks)
        throws InterruptedException;

    // 提交一组任务 Callable task，返回所有任务的存根 stub。注意，这里包含了等待执行的时间
    <T> List<Future<T>> invokeAll(Collection<? extends Callable<T>> tasks,
                    long timeout, TimeUnit unit)
        throws InterruptedException;

    // 提交一组任务，等待其中任何一个任务完成后返回
    <T> T invokeAny(Collection<? extends Callable<T>> tasks)
        throws InterruptedException, ExecutionException;

    // 提交一组任务，等待其中任何一个任务完成后返回。注意，这里包含了等待执行的时间
    <T> T invokeAny(Collection<? extends Callable<T>> tasks, long timeout, TimeUnit unit)
        throws InterruptedException, ExecutionException, TimeoutException;
}
```

5.3 AbstractExecutorService 抽象类

对于面向对象的设计模式来说，面对这种又长又多的接口，怎样能让子类轻松继承呢？可以使用模板方法设计模式。可以通过抽象类来实现部分公用算法，这样子类即线程池的实现就会轻松一些，不用自己再写一遍这些算法了。

AbstractExecutorService 抽象类的作用便是如此。该类实现了大部分 ExecutorService 定义的功能方法，将核心的 execute 方法放到子类实现。源码如下。

```java
public abstract class AbstractExecutorService implements ExecutorService {
    // 将 Runnable 对象封装为 FutureTask
    protected <T> RunnableFuture<T> newTaskFor(Runnable runnable, T value) {
        return new FutureTask<T>(runnable, value);
    }

    // 将 Callable 对象封装为 FutureTask
    protected <T> RunnableFuture<T> newTaskFor(Callable<T> callable) {
        return new FutureTask<T>(callable);
    }
    // 封装 Runnable task, 然后调用 execute 执行
    public Future<?> submit(Runnable task) {
        if (task == null) throw new NullPointerException();
        RunnableFuture<Void> ftask = newTaskFor(task, null);
        execute(ftask);
        return ftask;
    }

    // 封装 Runnable task, 调用 execute 执行。注意这里还封装了结果
    public <T> Future<T> submit(Runnable task, T result) {
        if (task == null) throw new NullPointerException();
        RunnableFuture<T> ftask = newTaskFor(task, result);
        execute(ftask);
        return ftask;
    }

    // 封装 Callable task, 调用 execute 执行
    public <T> Future<T> submit(Callable<T> task) {
        if (task == null) throw new NullPointerException();
        RunnableFuture<T> ftask = newTaskFor(task);
        execute(ftask);
        return ftask;
    }

    /* 执行传入的任务，只要执行任何一个任务完成后就返回，然后取消其他任务，
       这里 Any 指任何一个 tasks 中的任务执行完毕，timed 和 nanos 用来表明执行超时时间 */
    private <T> T doInvokeAny(Collection<? extends Callable<T>> tasks, boolean timed, long nanos)
        throws InterruptedException, ExecutionException, TimeoutException {
        // 任务判空
        if (tasks == null) throw new NullPointerException();
        int ntasks = tasks.size();
        if (ntasks == 0) throw new IllegalArgumentException();
        // 创建用于保存执行 stub 的 future 对象的集合
        ArrayList<Future<T>> futures = new ArrayList<Future<T>>(ntasks);
        // 创建 ExecutorCompletionService 用于执行任务
        ExecutorCompletionService<T> ecs = new ExecutorCompletionService<T>(this);
        try {
            // 记录异常，如果不能获取任何 result，可以抛出最后一个异常
            ExecutionException ee = null;
```

```java
final long deadline = timed ? System.nanoTime() + nanos :   0L;
// 获取集合的迭代器
Iterator<? extends Callable<T>> it = tasks.iterator();
// 先开始一项任务，其余的增量执行
futures.add(ecs.submit(it.next()));
// 任务数减 1
--ntasks;
// 正在执行的任务为 1
int active = 1;
for (;;) {
    /* 从 ExecutorCompletionService 中获取一个执行任务的 future。如果 future 为空，
       判断任务数量是否大于 0，如果还有下一个任务，继续递交执行。如果 future 不为空，
       直接调用 f.get()返回执行结果，然后在 finally 中取消所有任务。读者可能会问，
       ExecutorCompletionService 是什么？其实很简单，看看它的构造器：
       ExecutorCompletionService(Executor executor, BlockingQueue<Future<V>>
       completionQueue)，包括一个执行器，一个阻塞队列，功能是在执行器 executor
       中执行任务，任务完成后将结果放入 completionQueue 队列中，然后生产端从这个
       队列中获取结果。这就是生产者消费者模型 */
    Future<T> f = ecs.poll();
    if (f == null) {
        if (ntasks > 0) {
            --ntasks;
            futures.add(ecs.submit(it.next()));
            ++active;
        }
        else if (active == 0)
            break;
        else if (timed) {
            /* 如果指定了超时时间，等待 nanos 纳秒后如果 future 还是为空，
               则抛出超时异常 */
            f = ecs.poll(nanos, TimeUnit.NANOSECONDS);
            if (f == null)
                throw new TimeoutException();
            nanos = deadline - System.nanoTime();
        }
        else
            /* 如果所有任务都放到线程池里执行了，则直接调用 take 方法
               阻塞当前线程，等待任务执行完毕后返回。poll 方法非阻塞或带超时
               时间的阻塞，take 方法阻塞直到队列里放入结果，这里是 future */
            f = ecs.take();
    }
    if (f != null) {
        --active;
        try {
            return f.get();
        } catch (ExecutionException eex) {
            ee = eex;
        } catch (RuntimeException rex) {
            ee = new ExecutionException(rex);
        }
    }
}
if (ee == null)
```

```java
            ee = new ExecutionException();
        throw ee;

    } finally {
        for (int i = 0, size = futures.size(); i < size; i++)
            // 传入 true 表明中断执行线程
            futures.get(i).cancel(true);
    }
}

// 间接调用 doInvokeAny
public <T> T invokeAny(Collection<? extends Callable<T>> tasks)
    throws InterruptedException, ExecutionException {
    try {
        return doInvokeAny(tasks, false, 0);
    } catch (TimeoutException cannotHappen) {
        return null;
    }
}

// 带超时时间的间接调用 doInvokeAny
public <T> T invokeAny(Collection<? extends Callable<T>> tasks,
                       long timeout, TimeUnit unit)
    throws InterruptedException, ExecutionException, TimeoutException {
    return doInvokeAny(tasks, true, unit.toNanos(timeout));
}

/* 上面的 invokeAny 为其中一个任务执行完毕后返回，这里的 invokeAll 直接翻译成 All 就可以了，
   接着执行所有传入的任务，并等待所有任务执行完毕后返回 */
public <T> List<Future<T>> invokeAll(Collection<? extends Callable<T>> tasks)
    throws InterruptedException {
    if (tasks == null) throw new NullPointerException();
    // 保存 future 对象的 arraylist
    ArrayList<Future<T>> futures = new ArrayList<Future<T>>(tasks.size());
    boolean done = false;
    try {
        // 直接遍历所有 tasks 来调用 execute，将其放入线程池中全部执行，将 future 放入集合中
        for (Callable<T> t : tasks) {
            RunnableFuture<T> f = newTaskFor(t);
            futures.add(f);
            execute(f);
        }
        /* 遍历 future，等待每一个任务执行。注意这里的异常处理，当其中一个 future
           发生执行异常时则直接忽略掉，还是会设置 done 为 true 表明完成执行，
           最后返回 future，所以调用者应自行判断 future 是否发生了异常 */
        for (int i = 0, size = futures.size(); i < size; i++) {
            Future<T> f = futures.get(i);
            if (!f.isDone()) {
                try {
                    f.get();
                } catch (CancellationException ignore) {
                } catch (ExecutionException ignore) {
                }
```

```java
            }
        }
        done = true;
        return futures;
    } finally {
        /* 如果发生异常导致任务执行失败，即不属于 CancellationException 和
           ExecutionException，则取消所有的任务 */
        if (!done)
            for (int i = 0, size = futures.size(); i < size; i++)
                futures.get(i).cancel(true);
    }
}

// 带超时时间的执行所有任务
public <T> List<Future<T>> invokeAll(Collection<? extends Callable<T>> tasks,
                                     long timeout, TimeUnit unit)
    throws InterruptedException {
    if (tasks == null) throw new NullPointerException();
    // 将超时时间转为纳秒
    long nanos = unit.toNanos(timeout);
    ArrayList<Future<T>> futures = new ArrayList<Future<T>>(tasks.size());
    boolean done = false;
    try {
        /* 遍历执行所有任务，将其转变为 FutureTask，放入 future 集合中，
           计算 deadline（超时时间） */
        for (Callable<T> t : tasks)
            futures.add(newTaskFor(t));
        final long deadline = System.nanoTime() + nanos;
        final int size = futures.size();
        /* 遍历所有的任务，开始执行。如果达到超时时间，直接返回 futures。
           若超时时间设置过短，会导致任务未开始执行就返回，而且没有取消任务，
           这时返回的 futures 中可能还在线程池中执行。这点在使用时务必小心 */
        for (int i = 0; i < size; i++) {
            execute((Runnable)futures.get(i));
            nanos = deadline - System.nanoTime();
            if (nanos <= 0L)
                return futures;
        }
        /* 所有任务都放到了线程池中执行。遍历所有的 future，根据超时时间选择返回或等待 */
        for (int i = 0; i < size; i++) {
            Future<T> f = futures.get(i);
            if (!f.isDone()) {
                if (nanos <= 0L)
                    return futures;
                try {
                    f.get(nanos, TimeUnit.NANOSECONDS);
                } catch (CancellationException ignore) {
                } catch (ExecutionException ignore) {
                } catch (TimeoutException toe) {
                    return futures;
                }
                nanos = deadline - System.nanoTime();
            }
```

```
            }
            done = true;
            return futures;
        } finally {
            // 如果出现未知异常,则直接取消所有任务
            if (!done)
                for (int i = 0, size = futures.size(); i < size; i++)
                    futures.get(i).cancel(true);
        }
    }
}
```

5.4 ScheduledExecutorService 接口

ScheduledExecutorService 接口定义了任务周期性调度或延迟调度的方法。可以说,接口的继承等同于功能的扩展,ExecutorService 定义了执行器的服务方法,ScheduledExecutorService 继承自 ExecutorService,等同于扩展了 ExecutorService 的方法。这些方法由源码可知,可以对提交的 Runnable command 执行周期性调度。源码如下。

```
public interface ScheduledExecutorService extends ExecutorService {

    // 在延迟 delay 时间后执行 command。unit 为时间单位。只调度一次
    public ScheduledFuture<?> schedule(Runnable command, long delay, TimeUnit unit);

    // 在延迟 delay 时间后执行 callable。unit 为时间单位。只调度一次
    public <V> ScheduledFuture<V> schedule(Callable<V> callable, long delay, TimeUnit unit);

    // 周期性调度 command,在延迟 initialDelay 时间后执行 command。period 为周期时间,unit 为时间单位。
    // 每次任务基于上一次开始时间决定下次启动的时间,如果一次任务执行时间大于间隔时间,会导致下一次延缓执
    // 行,不会有两次任务并行执行(任务 M 在 A 时间开始执行,执行了 B 时间,这时任务 M 的下一次执行事件为 A
    // 时间加周期时间)
    public ScheduledFuture<?> scheduleAtFixedRate(Runnable command,
                                                   long initialDelay,
                                                   long period,
                                                   TimeUnit unit);

    // 周期性调度 command,在延迟 initialDelay 时间后执行 command。period 为周期时间,unit 为时间单位。
    // 每次任务基于上一次结束时间来延迟固定时间后执行下一次任务(任务 M 在 A 时间开始执行,执行了 B 时间,这
    // 时任务 M 的下一次执行事件为 A 时间、B 时间、周期时间之和)
    public ScheduledFuture<?> scheduleWithFixedDelay(Runnable command,
                                                      long initialDelay,
                                                      long delay,
                                                      TimeUnit unit);
}
```

第 6 章 EventExecutor 与 EventExecutorGroup 原理

6.1 EventExecutorGroup 类

EventExecutorGroup 类继承自 ScheduledExecutorService，在提供 ScheduledExecutorService 接口及其父接口的功能外，增加了一系列接口，用于执行关闭和获取事件执行器组中的下一个执行器 EventExecutor。从源码中可以了解，该接口管理了一组 EventExecutor，并提供管理它们生命周期的方法和获取 EventExecutor 的方法。源码如下。

```
public interface EventExecutorGroup extends ScheduledExecutorService, Iterable<EventExecutor> {
    // 判断当前 EventExecutorGroup 管理的所有 EventExecutor 是否正在关闭
    boolean isShuttingDown();

    // 用于安全关闭执行器组
    Future<?> shutdownGracefully(); // 通过默认值调用下面有参的方法完成关闭
    Future<?> shutdownGracefully(long quietPeriod, long timeout, TimeUnit unit); // 通知执行器组当前需要关闭。一旦调用了该方法，isShuttingDown 方法将返回 true，并且执行器开始执行关闭动作

    // 获取 Future 对象，在当前 EventExecutorGroup 管理的所有 EventExecutor 都终止后，设置该对象
    Future<?> terminationFuture();

    // 获取当前 EventExecutorGroup 管理的 EventExecutor
    EventExecutor next();
}
```

6.2 EventExecutor 接口

这里将直接通过推理说明 EventExecutor 接口的原理。

（1）EventExecutorGroup 接口定义了一个事件执行器组，这个组可以表示 1 个或多个。

（2）EventExecutor 继承自 EventExecutorGroup，从名字可以看出，它只代表了一个执行器。

（3）EventExecutorGroup 接口可以在执行器组中只定义一个执行器。

（4）EventExecutor 继承自 EventExecutorGroup，表示一个特殊的执行器组，里面只有一个执行器，即它本身。

（5）这样，next 方法将永远返回它自己。

该接口对 EventExecutorGroup 扩展，进一步定义了事件执行器的功能函数。这种接口设计有什么

好处？假设定义一个 EventExecutorGroup，即 EEG-A，它包括一组 EventExecutor，EventExecutor 内部又包括一组 EventExecutorGroup，即 EEG-B，同理依次向下包括多个子执行器。源码如下。

```java
public interface EventExecutor extends EventExecutorGroup {
    /**
     * 当前事件执行器组中只有它自己，next 永远返回自身
     */
    @Override
    EventExecutor next();

    /**
     * 返回该事件执行器所属的时间执行器组 EventExecutorGroup
     */
    EventExecutorGroup parent();

    /**
     * 通过 Thread.currentThread()判断当前线程是否为当前 EventExecutor 的绑定执行线程
     */
    boolean inEventLoop();

    /**
     * 指定线程对象判断，同 inEventLoop 方法
     */
    boolean inEventLoop(Thread thread);

    /**
     * 生成新的 Promise 实例
     */
    <V> Promise<V> newPromise();

    /**
     * 生成新的 ProgressivePromise 实例
     */
    <V> ProgressivePromise<V> newProgressivePromise();

    /**
     * 创建新的 Future 实例，但通过 result 参数指定 Future 的结果，并标记已经成功完成
     */
    <V> Future<V> newSucceededFuture(V result);

    /**
     * 创建新的 Future 实例，但通过 cause 参数指定 Future 的异常结果，并标记已经失败
     */
    <V> Future<V> newFailedFuture(Throwable cause);
}
```

6.3　AbstractEventExecutorGroup 方法

AbstractEventExecutorGroup 方法提供了 EventExecutorGroup 的服务方法默认实现。由源码可知，该类为了确保方法都是异步，所以默认实现均是调用 next 方法获取到执行器组中的下一个执行器，然后调用该执行器的同样方法。源码如下。

```java
public abstract class AbstractEventExecutorGroup implements EventExecutorGroup {

    // 向 EventExecutorGroup 执行器组提交 Runnable 任务
    @Override
    public Future<?> submit(Runnable task) {
        return next().submit(task);
    }

    // 向 EventExecutorGroup 执行器组提交 Runnable 任务并指定结果值
    @Override
    public <T> Future<T> submit(Runnable task, T result) {
        return next().submit(task, result);
    }

    // 向 EventExecutorGroup 执行器组提交 Callable 任务并指定结果值
    @Override
    public <T> Future<T> submit(Callable<T> task) {
        return next().submit(task);
    }

    // 向 EventExecutorGroup 执行器组提交延迟执行的 Runnable 任务
    @Override
    public ScheduledFuture<?> schedule(Runnable command, long delay, TimeUnit unit) {
        return next().schedule(command, delay, unit);
    }

    // 向 EventExecutorGroup 执行器组提交延迟执行的 Callable 任务
    @Override
    public <V> ScheduledFuture<V> schedule(Callable<V> callable, long delay, TimeUnit unit) {
        return next().schedule(callable, delay, unit);
    }

    // 向 EventExecutorGroup 执行器组提交周期性执行的 Runnable 任务
    @Override
    public ScheduledFuture<?> scheduleAtFixedRate(Runnable command, long initialDelay, long period, TimeUnit unit) {
        return next().scheduleAtFixedRate(command, initialDelay, period, unit);
    }

    // 向 EventExecutorGroup 执行器组提交周期性执行的 Runnable 任务
    @Override
    public ScheduledFuture<?> scheduleWithFixedDelay(Runnable command, long initialDelay, long delay, TimeUnit unit) {
        return next().scheduleWithFixedDelay(command, initialDelay, delay, unit);
    }

    // 通过指定默认参数调用 shutdownGracefully 方法完成优雅关闭
    // DEFAULT_SHUTDOWN_QUIET_PERIOD = 2、DEFAULT_SHUTDOWN_TIMEOUT = 15
    @Override
    public Future<?> shutdownGracefully() {
        return shutdownGracefully(DEFAULT_SHUTDOWN_QUIET_PERIOD, DEFAULT_SHUTDOWN_TIMEOUT, TimeUnit.SECONDS);
    }

    // 已废弃，推荐使用 shutdownGracefully()
```

```java
@Override
@Deprecated
public abstract void shutdown();

// 已废弃，推荐使用 shutdownGracefully()
@Override
@Deprecated
public List<Runnable> shutdownNow() {
    shutdown();
    return Collections.emptyList();
}

// 执行所有提交的集合中的任务
@Override
public <T> List<java.util.concurrent.Future<T>> invokeAll(Collection<? extends Callable<T>> tasks)
    throws InterruptedException {
    return next().invokeAll(tasks);
}

// 执行所有提交的集合中的任务，可以指定任务超时时间
@Override
public <T> List<java.util.concurrent.Future<T>> invokeAll(
    Collection<? extends Callable<T>> tasks, long timeout, TimeUnit unit) throws InterruptedException {
    return next().invokeAll(tasks, timeout, unit);
}

// 执行集合中的任务，直到有一个任务被执行完毕
@Override
public <T> T invokeAny(Collection<? extends Callable<T>> tasks) throws InterruptedException, ExecutionException {
    return next().invokeAny(tasks);
}

// 执行集合中的任务，直到有一个任务被执行完毕，可以指定等待时间
@Override
public <T> T invokeAny(Collection<? extends Callable<T>> tasks, long timeout, TimeUnit unit)
    throws InterruptedException, ExecutionException, TimeoutException {
    return next().invokeAny(tasks, timeout, unit);
}

// 执行传入的任务对象
@Override
public void execute(Runnable command) {
    next().execute(command);
}
}
```

6.4　MultithreadEventExecutorGroup 类

AbstractEventExecutorGroup 类对上层接口的方法进行实现，每个方法都调用 next 方法找到下一个线程来异步执行，同时返回 Future 表示异步执行的任务。而 MultithreadEventExecutorGroup 则是完成了多线程的事件执行器组的实现。

6.4.1 核心变量与构造器

从成员变量的声明中可知。

（1）EventExecutor[] children 表示子事件执行器数组，具备执行器组的特性，即拥有子执行器。

（2）AtomicInteger terminatedChildren 用于记录已经终止的子事件执行器数量，结合 Promise<?> terminationFuture，可以让需要执行器组终止的线程进行异步等待和事件回调。

（3）EventExecutorChooserFactory.EventExecutorChooser chooser 用于从子执行器数组中根据该接口的实例完成下一个子执行器的获取（负载均衡器），可以在 next 方法中调用该选择器完成对下一个子执行器的选择。

最终的构造函数流程如下。

（1）创建子执行器数组。

（2）调用子类实现的 newChild 方法创建子执行器的方法，使用其返回的实例完成初始化。在创建过程中，只要有一个子执行器创建实例失败，就需要全部关闭。

（3）通过选择器工厂创建选择器实例。

（4）创建终止监听器，并将该监听器放入子执行器中。

（5）把子执行器数组设置为只读执行器，不允许任何修改。

流程代码如下。

```java
public abstract class MultithreadEventExecutorGroup extends AbstractEventExecutorGroup {

    private final EventExecutor[] children;                                    // 子事件执行器数组（执行器组）

    private final Set<EventExecutor> readonlyChildren;                         // 只读子事件执行器

    private final AtomicInteger terminatedChildren = new AtomicInteger();      // 已经终止的子事件执行器数量

    private final Promise<?> terminationFuture = new DefaultPromise(GlobalEventExecutor.INSTANCE);// 子
    事件执行器终止 Promise 对象，可以在关闭执行器组时通过该 Future 接口来异步获取关闭结果

    private final EventExecutorChooserFactory.EventExecutorChooser chooser;    // 用于从子执行器数组中根据
    该接口的实例完成下一个子执行器的获取（负载均衡器）

    // 构造函数，提供了默认的线程工厂，即 ThreadPerTaskExecutor
    protected MultithreadEventExecutorGroup(int nThreads, ThreadFactory threadFactory, Object... args) {
        this(nThreads, threadFactory == null ? null : new ThreadPerTaskExecutor(threadFactory), args);
    }

    // 构造函数提供默认的 EventExecutorChooser 实例，即 DefaultEventExecutorChooserFactory.INSTANCE
    protected MultithreadEventExecutorGroup(int nThreads, Executor executor, Object... args) {
        this(nThreads, executor, DefaultEventExecutorChooserFactory.INSTANCE, args);
    }

    // 完整构造器
    protected MultithreadEventExecutorGroup(int nThreads, Executor executor,
                                EventExecutorChooserFactory chooserFactory, Object... args)
    {
```

```java
        if (nThreads <= 0) {        // 参数校验
            throw new IllegalArgumentException(String.format("nThreads: %d (expected: > 0)", nThreads));
        }

        if (executor == null) { // 未指定 Executor 实例的情况下，创建默认的 ThreadPerTaskExecutor 实例
            executor = new ThreadPerTaskExecutor(newDefaultThreadFactory());
        }

        children = new EventExecutor[nThreads];        // 创建子执行器数组

        for (int i = 0; i < nThreads; i ++) {                    // 初始化子执行器数组
            boolean success = false;
            try {
                children[i] = newChild(executor, args); // 调用子类创建子执行器的方法完成初始化（为什么这
样设计的原因很简单，当前类不知道子执行器的实例是谁，只知道需要创建，这里再一次展现了抽象方法的魅力）
                success = true;
            } catch (Exception e) {
                // 创建出现异常，则包装异常对象
                throw new IllegalStateException("failed to create a child event loop", e);
            } finally {
                if (!success) {                                // 只要任一子执行器创建实例失败，就要全部关闭
                    for (int j = 0; j < i; j ++) {
                        children[j].shutdownGracefully();
                    }
                    for (int j = 0; j < i; j ++) {            // 当前线程等待子执行器全部关闭
                        EventExecutor e = children[j];
                        try {
                            while (!e.isTerminated()) {
                                e.awaitTermination(Integer.MAX_VALUE, TimeUnit.SECONDS);
                            }
                        } catch (InterruptedException interrupted) {   // 等待期间被中断唤醒，重置当前线程
中断标志位（中断异常会消耗标志位），由调用方处理该中断
                            Thread.currentThread().interrupt();
                            break;
                        }
                    }
                }
            }
        }

        chooser = chooserFactory.newChooser(children); // 通过工厂创建选择器

        // 创建终止监听器，并将该监听器放入子执行器中。该监听器在操作完成时，对当前类的
terminatedChildren 终止线程数进行自增，最后一个子执行器执行完毕后，把 terminationFuture 的成功结果设置
为 null，表示终止完成
        final FutureListener<Object> terminationListener = new FutureListener<Object>() {
            @Override
            public void operationComplete(Future<Object> future) throws Exception {
                if (terminatedChildren.incrementAndGet() == children.length) {
                    terminationFuture.setSuccess(null);
                }
            }
        };
```

```
    for (EventExecutor e: children) {
        e.terminationFuture().addListener(terminationListener);
    }

    // 创建完成执行器数组后，就把它设置为只读执行器，不允许任何修改
    Set<EventExecutor> childrenSet = new LinkedHashSet<EventExecutor>(children.length);
    Collections.addAll(childrenSet, children);
    readonlyChildren = Collections.unmodifiableSet(childrenSet);
}
```

6.4.2 EventExecutorChooserFactory 与 DefaultEventExecutorChooserFactory

EventExecutorChooserFactory 接口定义了创建 EventExecutorChooser 实例的方法，由于 EventExecutorChooser 接口属于 EventExecutorChooserFactory，所以在接口内部定义了 EventExecutorChooser 选择器实例的选择功能函数。

在默认的选择器 DefaultEventExecutorChooserFactory 工厂中，定义了一个全局的 INSTANCE 单例对象。同时在创建 EventExecutorChooser 实例时，会根据子执行器数组的长度是否为 2 的倍数来选择合适的取余算法。内部取余的过程是通过原子性的自增实现的，所以也可以理解为默认执行器选择器执行算法为轮询。源码如下：

```
public interface EventExecutorChooserFactory {

    /**
     * 返回新的 EventExecutorChooser 实例
     */
    EventExecutorChooser newChooser(EventExecutor[] executors);

    // 定义选择功能方法
    interface EventExecutorChooser {

        /**
         * 返回下一个子执行器
         */
        EventExecutor next();
    }
}
// 默认选择器工厂
public final class DefaultEventExecutorChooserFactory implements EventExecutorChooserFactory {

    // 默认单例
    public static final DefaultEventExecutorChooserFactory INSTANCE = new DefaultEventExecutorChooserFactory();

    private DefaultEventExecutorChooserFactory() { }

    // 通过 executors 子执行器数组完成创建。通过判断子执行器数组的长度选择对应的 EventExecutorChooser 实例
    @Override
    public EventExecutorChooser newChooser(EventExecutor[] executors) {
```

```
        if (isPowerOfTwo(executors.length)) {
            return new PowerOfTowEventExecutorChooser(executors);
        } else {
            return new GenericEventExecutorChooser(executors);
        }
    }
```

// 二进制的与操作符等于二进制截断，这里相当于对 val 本身和其负数值截断。例如，val 为 8，二进制为 1000（这里为 32 位，省略前面的 28 个 0），-8 为 1000（省略 28 个 1）。执行与操作符，截断前面 29 个 1 的值，这时为原值，表明为 2 的倍数。但是如果 val 为 7，二进制为 111，-7 为 001（省略 29 个 1），此时执行与运算，将截断为 1

```
    private static boolean isPowerOfTwo(int val) {
        return (val & -val) == val;
    }
```

// 数组长度是 2 的倍数时，可以用二进制截断实现方式 idx.getAndIncrement() & executors.length - 1，通过与运算加速取余数操作。还是以 8 为例，8-1=7，二进制为 0111，此时截断 idx.getAndIncrement()的第三位，正好落在数组 0 - 7 的 index 上

```
    private static final class PowerOfTowEventExecutorChooser implements EventExecutorChooser {
        private final AtomicInteger idx = new AtomicInteger(); // 原子性自增（即使溢出为负数，也不影响）
        private final EventExecutor[] executors;

        PowerOfTowEventExecutorChooser(EventExecutor[] executors) {
            this.executors = executors;
        }

        @Override
        public EventExecutor next() {
            return executors[idx.getAndIncrement() & executors.length - 1];
        }
    }
```

// 数组长度不为 2 的倍数，只能用相对较慢的 abs 取绝对值，%用来计算长度。请思考：为什么要取绝对值？AtomicInteger 溢出为负数怎么办？

```
    private static final class GenericEventExecutorChooser implements EventExecutorChooser {
        private final AtomicInteger idx = new AtomicInteger();
        private final EventExecutor[] executors;

        GenericEventExecutorChooser(EventExecutor[] executors) {
            this.executors = executors;
        }

        @Override
        public EventExecutor next() {
            return executors[Math.abs(idx.getAndIncrement() % executors.length)];
        }
    }
}
```

6.4.3　ThreadPerTaskExecutor 类

ThreadPerTaskExecutor 是 MultithreadEventExecutorGroup 默认的 Executor 执行器，该执行器在

newChild(executor, args)方法中传递给子执行器完成对子执行器 EventExecutor 对象的创建，同时用于启动线程执行子执行器任务。该执行器同它的名字一样，在 execute 方法中通过 ThreadFactory 线程工厂创建新线程处理传递过来的任务 command。

由于 EventExecutor 子执行器对象只需要一个线程处理它的任务（只有一个执行器的组），所以很容易推理得出有多少个子执行器，就会产生出多少个线程。

为什么不用线程池？线程池用于缓存线程，减少每次执行任务时对线程的创建和销毁时间，这里需要线程池吗？答案是否定的，因为只会提交与 EventExecutor 数组长度相同的子执行器任务。源码如下。

```java
public final class ThreadPerTaskExecutor implements Executor {
    private final ThreadFactory threadFactory;

    public ThreadPerTaskExecutor(ThreadFactory threadFactory) {
        if (threadFactory == null) {
            throw new NullPointerException("threadFactory");
        }
        this.threadFactory = threadFactory;
    }

    @Override
    public void execute(Runnable command) {
        threadFactory.newThread(command).start();
    }
}
```

6.4.4 DefaultThreadFactory 类

默认的线程对象通过 poolId 和 nextId 生成线程名前缀，根据构造函数设置线程的属性，即守护线程、线程优先级、线程组。注意以下两点。

（1）对传入的 Runnable 执行对象进行了包装，将其包装为 DefaultRunnableDecorator，目的是为了在任务执行完毕后对 ThreadLocal 进行清理。

（2）创建的线程对象是 FastThreadLocalThread。

```java
public class DefaultThreadFactory implements ThreadFactory {
    private static final AtomicInteger poolId = new AtomicInteger();     // 线程池 ID，用于生成线程名
    private final AtomicInteger nextId = new AtomicInteger();            // 线程 ID，用于生成线程名
    private final String prefix;                                          // 生成的线程名前缀
    private final boolean daemon;                                         // 是否为守护线程
    private final int priority;                                           // 线程优先级
    protected final ThreadGroup threadGroup;                              // 线程组

    // 完整构造器。校验参数后，保存生成线程前缀、是否为 daemon 守护线程、优先级、所属线程组
    public DefaultThreadFactory(String poolName, boolean daemon, int priority, ThreadGroup threadGroup) {
        if (poolName == null) {
            throw new NullPointerException("poolName");
        }
        if (priority < Thread.MIN_PRIORITY || priority > Thread.MAX_PRIORITY) {
            throw new IllegalArgumentException(
                "priority: " + priority + " (expected: Thread.MIN_PRIORITY <= priority <=
```

```java
                Thread.MAX_PRIORITY)");
        }

        prefix = poolName + '-' + poolId.incrementAndGet() + '-';
        this.daemon = daemon;
        this.priority = priority;
        this.threadGroup = threadGroup;
    }

    // 生成新的线程对象
    @Override
    public Thread newThread(Runnable r) {
        Thread t = newThread(new DefaultRunnableDecorator(r), prefix + nextId.incrementAndGet());
        // 设置线程属性
        try {
            if (t.isDaemon()) {
                if (!daemon) {
                    t.setDaemon(false);
                }
            } else {
                if (daemon) {
                    t.setDaemon(true);
                }
            }

            if (t.getPriority() != priority) {
                t.setPriority(priority);
            }
        } catch (Exception ignored) {
            // Doesn't matter even if failed to set.
        }
        return t;
    }

    // 创建线程对象
    protected Thread newThread(Runnable r, String name) {
        return new FastThreadLocalThread(threadGroup, r, name);
    }

    // 装饰需要执行的 Runnable 任务,标准的装饰者模式。主要对任务进行封装,在任务执行完毕后,移除
ThreadLocal 的存储,避免内存泄漏
    private static final class DefaultRunnableDecorator implements Runnable {

        private final Runnable r;

        DefaultRunnableDecorator(Runnable r) {
            this.r = r;
        }

        @Override
        public void run() {
            try {
                r.run();
            } finally {
```

```
            FastThreadLocal.removeAll();
        }
    }
}
```

6.4.5 FastThreadLocalThread 类

　　线程工厂创建的默认线程对象是 FastThreadLocalThread，该线程工厂是在 ThreadPerTaskExecutor 执行器中调用，而 ThreadPerTaskExecutor 线程工厂又在 MultithreadEventExecutorGroup 中传递给了子执行器。这很明显，子执行器在操作任务时，创建的线程对象就是 FastThreadLocalThread。注意，Netty 中的执行线程对象就是该类的实例。

　　FastThreadLocalThread 类继承自 Thread 类，因为 Thread 类定义了一个 Java 线程对象，只能通过继承该类对其进行扩展。这里面主要扩展了对 InternalThreadLocalMap 的操作，那么 InternalThreadLocalMap 又是什么？它和线程自带的 ThreadLocal.ThreadLocalMap 有何区别？本节将首先了解这个类，下一节笔者会详细介绍，只需要知道这里扩展了这个 InternalThreadLocalMap 类即可。源码如下。

```java
public class FastThreadLocalThread extends Thread {
    private InternalThreadLocalMap threadLocalMap;

    public final InternalThreadLocalMap threadLocalMap() {
        return threadLocalMap;
    }

    public final void setThreadLocalMap(InternalThreadLocalMap threadLocalMap) {
        this.threadLocalMap = threadLocalMap;
    }
}
```

6.4.6 FastThreadLocal 类

　　FastThreadLocalThread 又叫 FastThreadLocal，顾名思义，Fast 就是快。到底哪里快呢？还得从 JDK 自带的 ThreadLocal.ThreadLocalMap 说起。ThreadLocalMap 是 JDK 的 ThreadLocal 类中的一个内部类，可以创建 ThreadLocal 对象，并在线程中保存线程本地变量。以下是放入的过程。

　　（1）根据 ThreadLocalMap 是否存在，决定是否需要创建 ThreadLocalMap 集合。

　　（2）在放入过程中通过 key.threadLocalHashCode & (len-1) 计算当前 ThreadLocal 与 value 的映射放入的下标。同理，使用该算法获取下标处对应的 Entry 映射。

　　源码如下。

```java
public class ThreadLocal<T> {
    // 设置线程本地变量
    public void set(T value) {
        Thread t = Thread.currentThread();           // 获取当前线程
        ThreadLocalMap map = getMap(t);              // 获取线程 map
        if (map != null)                             // map 不为空，直接放入
            map.set(this, value);
        else
```

```
            createMap(t, value);                      // 否则创建线程 map
    }

    // 获取线程 map。这里直接取线程对象的 ThreadLocal.ThreadLocalMap 对象
    ThreadLocalMap getMap(Thread t) {
        return t.threadLocals;
    }

    // 创建线程 map，这里初始化了 ThreadLocalMap 对象
    void createMap(Thread t, T firstValue) {
        t.threadLocals = new ThreadLocalMap(this, firstValue);
    }

    // 用于处理 ThreadLocal 和 value 映射的线程 map
    static class ThreadLocalMap {

        private Entry[] table;                        // 保存映射的数组

        ThreadLocalMap(ThreadLocal<?> firstKey, Object firstValue) {
            table = new Entry[INITIAL_CAPACITY];       // 创建数组
            int i = firstKey.threadLocalHashCode & (INITIAL_CAPACITY - 1); // 计算当前 ThreadLocal 中对应
value 的下标
            table[i] = new Entry(firstKey, firstValue);  // 将映射关系放入下标所在的 entry 中
            size = 1;
            setThreshold(INITIAL_CAPACITY);
        }

        // 将指定的映射关系放入对应的数组下标
        private void set(ThreadLocal<?> key, Object value) {
            Entry[] tab = table;
            int len = tab.length;
            int i = key.threadLocalHashCode & (len-1);  // 根据 hash 值计算该映射关系应该放置的索引下标
                                                        // 省略掉发生冲突二次寻址的过程
            ...
            tab[i] = new Entry(key, value);             // 创建新的 entry
            ...
        }

        // 根据 ThreadLocal 对象获取映射值
        private Entry getEntry(ThreadLocal<?> key) {
            int i = key.threadLocalHashCode & (table.length - 1);
            ...
        }
    }
}
```

首先是 FastThreadLocal 类的原理。关于放入方法，可根据源码得出以下结论。

（1）FastThreadLocal 中的 variablesToRemoveIndex 下标通常为 0，且在该下标处存放着一个 Set 集合，该集合用于保存所有在 InternalThreadLocalMap 中保存值的 FastThreadLocal。

（2）保存 FastThreadLocal 的映射值的核心类是 InternalThreadLocalMap。

（3）生成 FastThreadLocal 对象时会给 FastThreadLocal 生成唯一的下标 index，通过该下标，就不需要像原生 JDK 的 ThreadLocalMap 一样，通过计算 hash 值来取余数找到需要放置的下标。

（4）放置对象时，可以根据是否放置的值为 InternalThreadLocalMap.UNSET 占位符决定是否从

InternalThreadLocalMap 中移除映射。

（5）移除映射时，首先将映射值从 InternalThreadLocalMap 取出，再设置 InternalThreadLocalMap 中对应该 FastThreadLocal index 下标处为 UNSET 占位对象，然后根据是否存在移除的对象值，决定回调 onRemoval 钩子函数

源码如下。

```java
public class FastThreadLocal<V> {
    private static final int variablesToRemoveIndex = InternalThreadLocalMap.nextVariableIndex(); // 该类创建时生成的唯一 id，通常为 0

    private final int index;                      // 当前 FastThreadLocal 在 InternalThreadLocalMap 中的下标

    public FastThreadLocal() {                    // 创建对象时生成唯一的下标
        index = InternalThreadLocalMap.nextVariableIndex();
    }

    public final void set(V value) {
        if (value != InternalThreadLocalMap.UNSET) {   // 如果 value 不是占位对象 UNSET，则直接设置
            set(InternalThreadLocalMap.get(), value);
        } else {                                       // 否则执行清理操作
            remove();
        }
    }

    public final void set(InternalThreadLocalMap threadLocalMap, V value) {
        if (value != InternalThreadLocalMap.UNSET) {   // 该方法是通用方法，这里还得判断一次占位对象
            if (threadLocalMap.setIndexedVariable(index, value)) { // 注意，这里直接根据 index 下标设置值即可
                addToVariablesToRemove(threadLocalMap, this);  // 设置成功，将当前 FastThreadLocal 对象
// 添加到 threadLocalMap 对应的 variablesToRemoveIndex 下标处，表明需要进行清理的 FastThreadLocal（这里
// 通常为 0 下标）
            }
        } else {
            remove(threadLocalMap);
        }
    }

    // 将 FastThreadLocal<?> variable 添加到 InternalThreadLocalMap threadLocalMap 对应的下标处
    private static void addToVariablesToRemove(InternalThreadLocalMap threadLocalMap, <?> variable) {
        Object v = threadLocalMap.indexedVariable(variablesToRemoveIndex); // 获取对应的下标值
        Set<FastThreadLocal<?>> variablesToRemove;
        if (v == InternalThreadLocalMap.UNSET || v == null) {  // 如果不存在集合，则要创建集合
            variablesToRemove = Collections.newSetFromMap(new IdentityHashMap<FastThreadLocal<?>, Boolean>());
            threadLocalMap.setIndexedVariable(variablesToRemoveIndex, variablesToRemove);
        } else { // 否则将已经存在的 v 转为 Set<FastThreadLocal<?>>集合
            variablesToRemove = (Set<FastThreadLocal<?>>)
                                Set<FastThreadLocal<?>>) v;
        }
        // 将 FastThreadLocal 对象放入集合
        variablesToRemove.add(variable);
```

```java
    }
    // 清理当前 FastThreadLocal 在 InternalThreadLocalMap 中的存储
    public final void remove(InternalThreadLocalMap threadLocalMap) {
        if (threadLocalMap == null) {
            return;
        }
        Object v = threadLocalMap.removeIndexedVariable(index);  // 从 threadLocalMap 对应的 index 下标中移除该对象，并设置为 UNSET，返回该对象值
        removeFromVariablesToRemove(threadLocalMap, this);  // 由于该 FastThreadLocal 的映射值已经从 InternalThreadLocalMap 中移除，需要将其从下标为 variablesToRemoveIndex 处的集合中移除

        if (v != InternalThreadLocalMap.UNSET) {  // 如果移除成功，则回调钩子方法（默认为空，可以重写子类完成自己的逻辑）
            try {
                onRemoval((V) v);
            } catch (Exception e) {
                PlatformDependent.throwException(e);
            }
        }
    }

    // 将指定的 FastThreadLocal 从 InternalThreadLocalMap 中对应下标为 variablesToRemoveIndex（通常为 0）中的集合中移除
    private static void removeFromVariablesToRemove(
            InternalThreadLocalMap threadLocalMap, FastThreadLocal<?> variable) {
        // 获取集合并掉用其 remove 方法移除
        Object v = threadLocalMap.indexedVariable(variablesToRemoveIndex);
        if (v == InternalThreadLocalMap.UNSET || v == null) {
            return;
        }
        Set<FastThreadLocal<?>> variablesToRemove = (Set<FastThreadLocal<?>>) v;
        variablesToRemove.remove(variable);
    }

    // 直接移除操作，直接获取当前线程的 InternalThreadLocalMap 对象，然后调用上述的
    // remove(InternalThreadLocalMap threadLocalMap) 方法
    public final void remove() {
        remove(InternalThreadLocalMap.getIfSet());
    }
```

下标为 0 处的集合到底有什么用？为什么在当前线程添加映射时要将 FastThreadLocal 放入该集合？接下来继续看 removeAll 静态方法的移除操作。根据源码得出结论。

（1）在当前线程中可以创建多个 FastThreadLocal，而一个线程只有一个 InternalThreadLocalMap。

（2）当在添加 FastThreadLocal 映射值时，就会将其放入 InternalThreadLocalMap 下标为 0 处的 set 集合。

（3）这时当需要移除该线程所有的 FastThreadLocal 映射时，就可以获取该集合将其中的保存的 InternalThreadLocalMap 的值全部移除。

```java
public class FastThreadLocal<V> {
    public static void removeAll() {
```

```
            InternalThreadLocalMap threadLocalMap = InternalThreadLocalMap.getIfSet(); // 获取当前线程的
InternalThreadLocalMap
        if (threadLocalMap == null) {
            return;
        }
        try {
            Object v = threadLocalMap.indexedVariable(variablesToRemoveIndex); // 获取下标为 0 的 set 集
合,该集合保留了当前线程所有的 FastThreadLocal 变量
            if (v != null && v != InternalThreadLocalMap.UNSET) { // 存在集合时,遍历调用 FastThreadLocal
的 remove 方法移除
                @SuppressWarnings("unchecked")
                Set<FastThreadLocal<?>> variablesToRemove = (Set<FastThreadLocal<?>>) v;
                FastThreadLocal<?>[] variablesToRemoveArray =
                    variablesToRemove.toArray(new FastThreadLocal[variablesToRemove.size()]);
                for (FastThreadLocal<?> tlv: variablesToRemoveArray) {
                    tlv.remove(threadLocalMap);
                }
            }
        } finally { // 最后调用 InternalThreadLocalMap 的 remove 静态方法,完成最终的清理操作(其实就是将
FastThreadLocalThread 的 InternalThreadLocalMap threadLocalMap 置空,因为这个 map 已经没有用了)
            InternalThreadLocalMap.remove();
        }
    }
}
```

FastThreadLocal 的核心是 InternalThreadLocalMap,对于 InternalThreadLocalMap,根据源码得出以下结论。

(1)InternalThreadLocalMap 继承自 UnpaddedInternalThreadLocalMap,对于保存 index 和 value 的核心变量 Object[] indexedVariables 将在 UnpaddedInternalThreadLocalMap 中定义。

(2)保存变量的核心是 InternalThreadLocalMap,可以知道 FastThreadLocalThread 中保存有 InternalThreadLocalMap 的引用,则这时移除和获取都可以直接通过该引用操作。但如果是普通 Thread 线程呢?其实这种情况也可以兼容,只要将 InternalThreadLocalMap 放入普通的 Thread 类的 ThreadLocalMap 即可。通过在 UnpaddedInternalThreadLocalMap 中定义 ThreadLocal slowThreadLocalMap 完成该操作。

```
class UnpaddedInternalThreadLocalMap {
    static final ThreadLocal<InternalThreadLocalMap> slowThreadLocalMap = new
ThreadLocal<InternalThreadLocalMap>();                 // 兼容原生的 JDK ThreadLocal,注意这里的类
型为 InternalThreadLocalMap,这是什么意思?请继续往下看

    static final AtomicInteger nextIndex = new AtomicInteger(); // 原子性生成下一个索引下标

    Object[] indexedVariables;                          // 核心变量,用于保存 ThreadLocal 的变量值(读
者还记得前面的 FastThreadLocal 的 index 下标么)

    UnpaddedInternalThreadLocalMap(Object[] indexedVariables) {
        this.indexedVariables = indexedVariables;
    }
}
```

```java
public final class InternalThreadLocalMap extends UnpaddedInternalThreadLocalMap {
    public static final Object UNSET = new Object(); // 表明未使用的 InternalThreadLocalMap 下标占位符

    // 构造器初始化父类的 Object[] indexedVariables 数组，默认数组长度为 32
    private InternalThreadLocalMap() {
        super(newIndexedVariableTable());
    }

    private static Object[] newIndexedVariableTable() {
        Object[] array = new Object[32];
        Arrays.fill(array, UNSET);
        return array;
    }

    // 获取当前线程所属的 InternalThreadLocalMap 对象。根据当前线程类型是否为 FastThreadLocalThread，
    // 从而调用 fastGet 或 slowGet 获取 InternalThreadLocalMap。因为只有 FastThreadLocalThread 内部拥有
    // InternalThreadLocalMap 引用
    public static InternalThreadLocalMap get() {
        Thread thread = Thread.currentThread();
        if (thread instanceof FastThreadLocalThread) {
            return fastGet((FastThreadLocalThread) thread);
        } else {
            return slowGet();
        }
    }

    // 从当前 FastThreadLocalThread 属性变量中获取 InternalThreadLocalMap 对象。如果不存在，就从这里初始化
    private static InternalThreadLocalMap fastGet(FastThreadLocalThread thread) {
        InternalThreadLocalMap threadLocalMap = thread.threadLocalMap();
        if (threadLocalMap == null) {
            thread.setThreadLocalMap(threadLocalMap = new InternalThreadLocalMap());
        }
        return threadLocalMap;
    }

    // 当前线程不是 FastThreadLocalThread，则要从线程原生的 ThreadLocalMap 中获取 InternalThreadLocalMap。
    // 如果这里不存在，就会初始化 InternalThreadLocalMap 对象，再将其放入原生的 ThreadLocal 中
    private static InternalThreadLocalMap slowGet() {
        ThreadLocal<InternalThreadLocalMap> slowThreadLocalMap =
UnpaddedInternalThreadLocalMap.slowThreadLocalMap;
        InternalThreadLocalMap ret = slowThreadLocalMap.get();
        if (ret == null) {
            ret = new InternalThreadLocalMap();
            slowThreadLocalMap.set(ret);
        }
        return ret;
    }

    // 移除当前线程中的 InternalThreadLocalMap。这里同样分为 FastThreadLocalThread 和普通线程。如果是
    // FastThreadLocalThread，可直接把保存的引用设置为 null；如果是普通 Thread，则直接调用原生 ThreadLocal
    // 的 remove 方法
    public static void remove() {
        Thread thread = Thread.currentThread();
```

```
        if (thread instanceof FastThreadLocalThread) {
            ((FastThreadLocalThread) thread).setThreadLocalMap(null);
        } else {
            slowThreadLocalMap.remove();
        }
    }
}
```

这里主要对比普通的 ThreadLocal 和 FastThreadLocal 的原理。

（1）普通的 ThreadLocalMap 在存储和移除 ThreadLocal 和 Value 的映射时需要进行 hash 取余操作。

（2）FastThreadLocal 在初始化对象时保留了一个 index 下标，存储和移除时不需要取余，只要通过保存的 index 下标获取即可，这就是为什么它叫作 FastThreadLocal。

6.4.7　shutdownGracefully 方法

shutdownGracefully 方法由外部线程调用，用于结束执行器组中的子执行器。该方法的实现较为简单，仅遍历了子执行器数组，并调用它们的 shutdownGracefully 方法，返回 terminationFuture 对象。所有子执行器完成关闭后，该对象会被设置为完成状态。源码如下。

```
public Future<?> shutdownGracefully(long quietPeriod, long timeout, TimeUnit unit) {
    for (EventExecutor l: children) {
        l.shutdownGracefully(quietPeriod, timeout, unit);
        return terminationFuture();
    }
}

public Future<?> terminationFuture() { // 返回 terminationFuture 实例
    return terminationFuture;
}
```

6.4.8　awaitTermination 方法

awaitTermination 方法用于同步等待所有子执行器全部终止。timeout 参数用于指定等待时长。实现过程如下。

（1）根据超时时间和单位计算出基于当前时间退出等待的截止时间 deadline。

（2）遍历所有子执行器，调用它们的 awaitTermination 方法等待。在调用子类方法过程中，修正等待的时间，因为每次等待后会消耗一段等待时间，所以要减掉该时间。

（3）等待过程中判断 deadline，如果发现超时，则退出等待。

（4）调用 isTerminated 方法返回子执行器是否已经终止。

```
public boolean awaitTermination(long timeout, TimeUnit unit)
        throws InterruptedException {
    long deadline = System.nanoTime() + unit.toNanos(timeout);
    loop: for (EventExecutor l: children) {
        for (;;) {
            long timeLeft = deadline - System.nanoTime();
            if (timeLeft <= 0) {
                break loop;
```

```
            }
            if (!.awaitTermination(timeLeft, TimeUnit.NANOSECONDS)) {
                break;
            }
        }
    }
    return isTerminated();
}

public boolean isTerminated() {    // 判断子执行器是否终止，遍历自执行器数组。只要有一个子执行器没有终止，
                                   // 就返回 false
    for (EventExecutor l: children) {
        if (!l.isTerminated()) {
            return false;
        }
    }
    return true;
}
```

6.5　DefaultEventExecutorGroup 类

由于绝大部分工作都在 MultithreadEventExecutorGroup 中实现了，该类只是在构造函数中对父类进行初始化，同时实现了唯一的 newChild 方法。使用该方法创建的子执行器实例是 DefaultEventExecutor 对象。源码如下。

```
public class DefaultEventExecutorGroup extends MultithreadEventExecutorGroup {

    public DefaultEventExecutorGroup(int nThreads) {
        this(nThreads, null);
    }

    // 构造函数，提供了默认的 maxPendingTasks 参数：SingleThreadEventExecutor.DEFAULT_MAX_PENDING_
    EXECUTOR_TASKS 表示等待执行的任务个数
    public DefaultEventExecutorGroup(int nThreads, ThreadFactory threadFactory) {
        this(nThreads, threadFactory,
            SingleThreadEventExecutor.DEFAULT_MAX_PENDING_EXECUTOR_TASKS,
            RejectedExecutionHandlers.reject());
    }

    public DefaultEventExecutorGroup(int nThreads, ThreadFactory threadFactory, int maxPendingTasks,
                                     RejectedExecutionHandler rejectedHandler) {
        super(nThreads, threadFactory, maxPendingTasks, rejectedHandler);
    }

    // 由父类源码可知，传递的参数 args（这里为 maxPendingTasks）会传递到该方法，供子执行器
    DefaultEventExecutor 使用
    // 为什么？父执行器组仅用于管理子执行器（创建、选择、销毁），真实业务逻辑在子执行器中完成
    protected EventExecutor newChild(Executor executor, Object... args) throws Exception {
        return new DefaultEventExecutor(this, executor, (Integer) args[0], (RejectedExecutionHandler) args[1]);
    }
}
```

6.6 AbstractEventExecutor 类

AbstractEventExecutor 类和 AbstractEventExecutorGroup 一样，实现了一些默认的功能方法。读者请注意以下几点。

（1）shutdownGracefully 方法中提供了默认的静默期、关闭超时时间。
（2）提交执行方法由父类 AbstractExecutorService 完成。
（3）默认不支持周期性调度方法。

源码如下：

```java
public abstract class AbstractEventExecutor extends AbstractExecutorService implements EventExecutor {
    static final long DEFAULT_SHUTDOWN_QUIET_PERIOD = 2;    // 静默期
    static final long DEFAULT_SHUTDOWN_TIMEOUT = 15;        // 关闭超时时间

    private final EventExecutorGroup parent;
    private final Collection<EventExecutor> selfCollection = Collections.<EventExecutor>singleton(this);

    protected AbstractEventExecutor() {
        this(null);
    }

    protected AbstractEventExecutor(EventExecutorGroup parent) {
        this.parent = parent;
    }

    @Override
    public EventExecutorGroup parent() {
        return parent;
    }

    // 由于子执行器组只有自己，所以永远返回它自己
    @Override
    public EventExecutor next() {
        return this;
    }

    // 判断当前线程是否为当前 EventExecutor 的执行线程，由子类完成
    @Override
    public boolean inEventLoop() {
        return inEventLoop(Thread.currentThread());
    }

    @Override
    public Iterator<EventExecutor> iterator() {
        return selfCollection.iterator();
    }

    // 关闭方法由子类完成，这里指定了默认的静默周期和关闭超时时间
```

```java
@Override
public Future<?> shutdownGracefully() {
    return shutdownGracefully(DEFAULT_SHUTDOWN_QUIET_PERIOD,
DEFAULT_SHUTDOWN_TIMEOUT, TimeUnit.SECONDS);
}


/* ---- 异步执行创建默认的实例 */

@Override
public <V> Promise<V> newPromise() {
    return new DefaultPromise<V>(this);
}

@Override
public <V> ProgressivePromise<V> newProgressivePromise() {
    return new DefaultProgressivePromise<V>(this);
}

@Override
public <V> Future<V> newSucceededFuture(V result) {
    return new SucceededFuture<V>(this, result);
}

@Override
public <V> Future<V> newFailedFuture(Throwable cause) {
    return new FailedFuture<V>(this, cause);
}

/* ---- 提交方法使用父类 AbstractExecutorService 来实现 */

@Override
public Future<?> submit(Runnable task) {
    return (Future<?>) super.submit(task);
}

@Override
public <T> Future<T> submit(Runnable task, T result) {
    return (Future<T>) super.submit(task, result);
}

@Override
public <T> Future<T> submit(Callable<T> task) {
    return (Future<T>) super.submit(task);
}

/* ---- 默认不支持周期性调度 */
@Override
public ScheduledFuture<?> schedule(Runnable command, long delay,
                                    TimeUnit unit) {
    throw new UnsupportedOperationException();
}

@Override
```

```java
    public <V> ScheduledFuture<V> schedule(Callable<V> callable, long delay, TimeUnit unit) {
        throw new UnsupportedOperationException();
    }

    @Override
    public ScheduledFuture<?> scheduleAtFixedRate(Runnable command, long initialDelay, long period, TimeUnit unit) {
        throw new UnsupportedOperationException();
    }

    @Override
    public ScheduledFuture<?> scheduleWithFixedDelay(Runnable command, long initialDelay, long delay, TimeUnit unit) {
        throw new UnsupportedOperationException();
    }
}
```

6.7　AbstractScheduledEventExecutor 方法

AbstractScheduledEventExecutor 方法中定义了周期性任务队列 scheduledTaskQueue 和周期性任务调度功能方法的实现，用于对队列进行 CRUD。同时提供 ScheduledExecutorService 中的函数实现。该类比较简单且功能单一，方法实现也不复杂，所以笔者直接复制了整个类，并进行了方法注释。读者这里只需要注意以下知识。

（1）队列由 PriorityQueue 类实现。

（2）只有当前执行器所属线程才能对队列进行操作，这时保证线程安全（保证线程安全的两种方法：① 多线程操作上锁；② 单线程操作，其他线程将操作放入单线程的普通任务队列中，由单线程串行化实现，可通过 Redis 实现）。

```java
public abstract class AbstractScheduledEventExecutor extends AbstractEventExecutor {

    // 周期性任务调度队列
    Queue<ScheduledFutureTask<?>> scheduledTaskQueue;

    // 获取基于 ScheduledFutureTask 类加载时间的相对时间
    protected static long nanoTime() {
        return ScheduledFutureTask.nanoTime();
    }

    // 获取或初始化任务调度队列，这里的队列很简单，PriorityQueue 由数组实现的小顶堆优先级队列
    Queue<ScheduledFutureTask<?>> scheduledTaskQueue() {
        if (scheduledTaskQueue == null) {
            scheduledTaskQueue = new PriorityQueue<ScheduledFutureTask<?>>();
        }
        return scheduledTaskQueue;
    }

    // 取消所有周期性调度任务
    protected void cancelScheduledTasks() {
```

```java
        Queue<ScheduledFutureTask<?>> scheduledTaskQueue = this.scheduledTaskQueue;
        if (isNullOrEmpty(scheduledTaskQueue)) {                // 队列为空，则直接退出
            return;
        }
        final ScheduledFutureTask<?>[] scheduledTasks =
            scheduledTaskQueue.toArray(new ScheduledFutureTask<?>[scheduledTaskQueue.size()]); // 导出所有周期性任务

        for (ScheduledFutureTask<?> task: scheduledTasks) {  // 遍历取消这些任务
            task.cancelWithoutRemove(false);
        }
        scheduledTaskQueue.clear();                             // 清空任务队列
    }

    // 从周期性任务队列中获取可以执行的周期性任务
    protected final Runnable pollScheduledTask(long nanoTime) {
        assert inEventLoop();                                   // 执行该方法的线程必须为当前执行器所属线程
        Queue<ScheduledFutureTask<?>> scheduledTaskQueue = this.scheduledTaskQueue;
        // 从队列中看看是否存在任务，如果不存在，则直接返回
        ScheduledFutureTask<?> scheduledTask = scheduledTaskQueue == null ? null : scheduledTaskQueue.peek();
        if (scheduledTask == null) {
            return null;
        }
        // 判断周期性调度任务的等待时间是否小于传入的等待时间。如果小于，则获取该任务返回。注意，这里的 remove 不会导致阻塞
        if (scheduledTask.deadlineNanos() <= nanoTime) {
            scheduledTaskQueue.remove();
            return scheduledTask;
        }
        return null;
    }

    // 从队列中获取到下一次需要执行的周期性任务的等待时间
    protected final long nextScheduledTaskNano() {
        Queue<ScheduledFutureTask<?>> scheduledTaskQueue = this.scheduledTaskQueue;
        // 获取第一个需要执行的任务（由于这里队列是小顶堆优先队列实现，所以第一个任务一定是最近一个需要执行的周期性任务）。如果任务不存在，则返回-1，否则返回需要等待的时间
        ScheduledFutureTask<?> scheduledTask = scheduledTaskQueue == null ? null : scheduledTaskQueue.peek();
        if (scheduledTask == null) {
            return -1;
        }
        return Math.max(0, scheduledTask.deadlineNanos() - nanoTime());
    }

    // 直接获取周期性任务队列中第一个需要执行的任务。不存在则返回 null
    final ScheduledFutureTask<?> peekScheduledTask() {
        Queue<ScheduledFutureTask<?>> scheduledTaskQueue = this.scheduledTaskQueue;
        if (scheduledTaskQueue == null) {
            return null;
        }
        return scheduledTaskQueue.peek();
```

```java
}
// 查看任务队列中是否存在可执行的任务，要求任务存在且已经达到执行的时间
protected final boolean hasScheduledTasks() {
    Queue<ScheduledFutureTask<?>> scheduledTaskQueue = this.scheduledTaskQueue;
    ScheduledFutureTask<?> scheduledTask = scheduledTaskQueue == null ? null : scheduledTaskQueue.peek();
    return scheduledTask != null && scheduledTask.deadlineNanos() <= nanoTime();
}

// 周期性调度实现，该方法对参数校验后直接调用 schedule 方法完成
@Override
public ScheduledFuture<?> schedule(Runnable command, long delay, TimeUnit unit) {
    ObjectUtil.checkNotNull(command, "command");
    ObjectUtil.checkNotNull(unit, "unit");
    if (delay < 0) {
        throw new IllegalArgumentException(
            String.format("delay: %d (expected: >= 0)", delay));
    }
    return schedule(new ScheduledFutureTask<Void>(
        this, command, null, ScheduledFutureTask.deadlineNanos(unit.toNanos(delay))));
}

// 对 callable 接口支持
@Override
public <V> ScheduledFuture<V> schedule(Callable<V> callable, long delay, TimeUnit unit) {
    ObjectUtil.checkNotNull(callable, "callable");
    ObjectUtil.checkNotNull(unit, "unit");
    if (delay < 0) {
        throw new IllegalArgumentException(
            String.format("delay: %d (expected: >= 0)", delay));
    }
    return schedule(new ScheduledFutureTask<V>(
        this, callable, ScheduledFutureTask.deadlineNanos(unit.toNanos(delay))));
}

// 固定周期执行方法
@Override
public ScheduledFuture<?> scheduleAtFixedRate(Runnable command, long initialDelay, long period, TimeUnit unit) {
    ObjectUtil.checkNotNull(command, "command");
    ObjectUtil.checkNotNull(unit, "unit");
    if (initialDelay < 0) {
        throw new IllegalArgumentException(
            String.format("initialDelay: %d (expected: >= 0)", initialDelay));
    }
    if (period <= 0) {
        throw new IllegalArgumentException(
            String.format("period: %d (expected: > 0)", period));
    }

    return schedule(new ScheduledFutureTask<Void>(
        this, Executors.<Void>callable(command, null),
        ScheduledFutureTask.deadlineNanos(unit.toNanos(initialDelay)), unit.toNanos(period)));
```

```java
    }

    // 固定周期执行方法
    @Override
    public ScheduledFuture<?> scheduleWithFixedDelay(Runnable command, long initialDelay, long delay,
TimeUnit unit) {
        ObjectUtil.checkNotNull(command, "command");
        ObjectUtil.checkNotNull(unit, "unit");
        if (initialDelay < 0) {
            throw new IllegalArgumentException(
                String.format("initialDelay: %d (expected: >= 0)", initialDelay));
        }
        if (delay <= 0) {
            throw new IllegalArgumentException(
                String.format("delay: %d (expected: > 0)", delay));
        }

        return schedule(new ScheduledFutureTask<Void>(
            this, Executors.<Void>callable(command, null),
            ScheduledFutureTask.deadlineNanos(unit.toNanos(initialDelay)), -unit.toNanos(delay)));
    }

    // 完整实现周期性调度方法
    <V> ScheduledFuture<V> schedule(final ScheduledFutureTask<V> task) {
        if (inEventLoop()) { // 如果当前线程不位于事件循环组，则加入周期性任务队列（只能由当前执行器的线程完成队列添加操作，这样就可以保证所有操作均由一个线程完成，此时保证了线程安全）
            scheduledTaskQueue().add(task);
        } else { // 否则直接调用 execute 方法执行（因为这里是 Executor 接口的实例，所以由线程执行的方法当然为 Executor 接口中定义的 execute 方法），在该方法中将任务放置到周期性任务队列中
            execute(new Runnable() {
                @Override
                public void run() {
                    scheduledTaskQueue().add(task);
                }
            });
        }
        return task;
    }

    // 将指定的任务 task 从周期性调度任务队列中移除。同样，这里只能有执行器所属的执行线程完成该操作，保证线程安全
    final void removeScheduled(final ScheduledFutureTask<?> task) {
        if (inEventLoop()) {
            scheduledTaskQueue().remove(task);
        } else {
            execute(new Runnable() {
                @Override
                public void run() {
                    removeScheduled(task);
                }
            });
        }
    }
}
```

6.8　SingleThreadEventExecutor 类

AbstractScheduledEventExecutor 实现了周期性任务调度的支持，该类继承自 AbstractScheduledEventExecutor，同时提供了在单线程中执行所有提交任务的功能方法。

6.8.1　核心变量与构造器

通过以下变量构造类。

（1）SingleThreadEventExecutor 中有执行状态 state。

（2）任务执行队列 Queue taskQueue，默认为长度为 maxPendingTasks 的 LinkedBlockingQueue。

（3）Set shutdownHooks 用于在关闭时执行其中的 Runnable 动作。

（4）DefaultThreadProperties 内部类实现了 ThreadProperties，该接口将提供给外部获取线程的属性，默认的线程属性只是包装了 Thread 对象中的方法。

```
// 标识接口表明执行提交的任务具有顺序性
public interface OrderedEventExecutor extends EventExecutor {
}

public abstract class SingleThreadEventExecutor extends AbstractScheduledEventExecutor implements OrderedEventExecutor {
    // 最大挂起执行的任务。这里最小为 16，默认为整型最大值
    static final int DEFAULT_MAX_PENDING_EXECUTOR_TASKS = Math.max(16,
SystemPropertyUtil.getInt("io.Netty.eventexecutor.maxPendingTasks", Integer.MAX_VALUE));

    // 执行器状态
    private static final int ST_NOT_STARTED = 1;
    private static final int ST_STARTED = 2;
    private static final int ST_SHUTTING_DOWN = 3;
    private static final int ST_SHUTDOWN = 4;
    private static final int ST_TERMINATED = 5;

    // 唤醒和执行空操作的任务
    private static final Runnable WAKEUP_TASK = new Runnable() {
        @Override
        public void run() {
            // Do nothing.
        }
    };
    private static final Runnable NOOP_TASK = new Runnable() {
        @Override
        public void run() {
            // Do nothing.
        }
    };

    // JUC 的基础。封装 state 变量和 threadProperties 变量的原子性修改操作
```

```java
    private static final AtomicIntegerFieldUpdater<SingleThreadEventExecutor> STATE_UPDATER;
    private static final AtomicReferenceFieldUpdater<SingleThreadEventExecutor, ThreadProperties> PROPERTIES_UPDATER;

    private final Queue<Runnable> taskQueue;                              // 任务队列
    private volatile Thread thread;                                       // 执行线程对象
    private volatile ThreadProperties threadProperties;                   // 线程属性
    private final Executor executor;                                      // 执行器对象
    private volatile boolean interrupted;                                 // 标识线程是否被中断
    private final Semaphore threadLock = new Semaphore(0);                // 线程锁信号量
    private final Set<Runnable> shutdownHooks = new LinkedHashSet<Runnable>(); // 在关闭时执行的操作链表
    private final boolean addTaskWakesUp;                                 // 是否允许添加任务唤醒线程
    private final int maxPendingTasks;                                    // 最大挂起的任务
    private final RejectedExecutionHandler rejectedExecutionHandler;      // 拒绝函数
    private volatile int state = ST_NOT_STARTED;                          // 当前执行器状态
    private volatile long gracefulShutdownQuietPeriod;                    // 静默期持续时间
    private volatile long gracefulShutdownTimeout;                        // 关闭超时时间
    private long gracefulShutdownStartTime;                               // 关闭开始时间
    private final Promise<?> terminationFuture = new DefaultPromise<Void>(GlobalEventExecutor.INSTANCE);   // 用于外部线程等待关闭的 Promise

    // 构造器，指定了默认的 ThreadPerTaskExecutor 执行器
    protected SingleThreadEventExecutor(
            EventExecutorGroup parent, ThreadFactory threadFactory, boolean addTaskWakesUp) {
        this(parent, new ThreadPerTaskExecutor(threadFactory), addTaskWakesUp);
    }

    // 构造器，指定了默认的最大挂起执行任务和拒绝函数（抛出任务拒绝异常）
    protected SingleThreadEventExecutor(EventExecutorGroup parent, Executor executor, boolean addTaskWakesUp) {
        this(parent, executor, addTaskWakesUp, DEFAULT_MAX_PENDING_EXECUTOR_TASKS,
                RejectedExecutionHandlers.reject());
    }

    // 最终构造器
    protected SingleThreadEventExecutor(EventExecutorGroup parent, Executor executor,
                                        boolean addTaskWakesUp, int maxPendingTasks,
                                        RejectedExecutionHandler rejectedHandler) {
        super(parent);                                                    // 初始化执行器组
        this.addTaskWakesUp = addTaskWakesUp;
        this.maxPendingTasks = Math.max(16, maxPendingTasks);
        this.executor = ObjectUtil.checkNotNull(executor, "executor");
        taskQueue = newTaskQueue(this.maxPendingTasks);                   // 创建任务队列
        rejectedExecutionHandler = ObjectUtil.checkNotNull(rejectedHandler, "rejectedHandler");
    }

    // 默认为指定数量的 LinkedBlockingQueue
    protected Queue<Runnable> newTaskQueue(int maxPendingTasks) {
        return new LinkedBlockingQueue<Runnable>(maxPendingTasks);
    }
```

```java
// 内部类，用于获取当前执行器所属线程对象的属性
private static final class DefaultThreadProperties implements ThreadProperties {
    private final Thread t;

    DefaultThreadProperties(Thread t) {
        this.t = t;
    }

    @Override
    public State state() {
        return t.getState();
    }

    @Override
    public int priority() {
        return t.getPriority();
    }

    @Override
    public boolean isInterrupted() {
        return t.isInterrupted();
    }

    @Override
    public boolean isDaemon() {
        return t.isDaemon();
    }

    @Override
    public String name() {
        return t.getName();
    }

    @Override
    public long id() {
        return t.getId();
    }

    @Override
    public StackTraceElement[] stackTrace() {
        return t.getStackTrace();
    }

    @Override
    public boolean isAlive() {
        return t.isAlive();
    }
}
}
```

6.8.2　execute 核心方法实现

核心的方法便是 execute，因为 Executor 接口定义了如何执行任务，子类会通过该方法向事件执行

器中添加执行的任务。处理流程如下。

（1）判断任务是否为空。

（2）根据当前线程是否为当前执行器的执行线程，选择是直接添加任务，还是调用 startThread 尝试启动该事件执行器的线程执行任务，然后将任务添加到任务队列中。如果线程还未启动的话，则无法执行任务，所以需要尝试 startThread，就需要该方法允许重入。同时，由于任务队列是 LinkedBlockingQueue，具有线程安全，所以可以多任务来操作。

（3）判断当前事件执行器是否已经关闭，从而将任务从队列中移除，同时抛出拒绝异常。

（4）如果设置 addTaskWakesUp 为 false，则根据 wakesUpForTask 方法的返回值选择是否唤醒线程。

```java
public void execute(Runnable task) {
    if (task == null) {
        throw new NullPointerException("task");
    }

    // 根据当前线程是否为当前执行器的执行线程判断是直接添加任务，还是调用 startThread 启动该事件执行器
    // 的线程执行任务，然后将任务添加到任务队列中
    boolean inEventLoop = inEventLoop();
    if (inEventLoop) {
        addTask(task);
    } else {
        startThread();                                    // 启动任务线程
        addTask(task);                                    // 将任务添加到任务队列
        if (isShutdown() && removeTask(task)) {           // 判断当前事件执行器是否已经关闭，从队列中移除任务，同时抛出拒绝异常
            reject();
        }
    }
    if (!addTaskWakesUp && wakesUpForTask(task)) {  // 如果把 addTaskWakesUp 设置为 false，则根据
// wakesUpForTask 方法的返回值选择是否唤醒线程
        wakeup(inEventLoop);
    }
}

// 该方法的当前版本总是返回 true
protected boolean wakesUpForTask(Runnable task) {
    return true;
}
```

6.8.3 addTask 核心方法

addTask 方法相对简单，判断任务为空后，调用 offerTask 方法将任务添加到队列，添加失败时，抛出任务拒绝异常。在 offerTask 中，会判断事件执行器状态，从而决定是否拒绝或添加任务。

```java
protected void addTask(Runnable task) {
    if (task == null) {
        throw new NullPointerException("task");
    }
    if (!offerTask(task)) {
        reject(task);
    }
}
```

```
}
final boolean offerTask(Runnable task) {
    if (isShutdown()) {
        reject();
    }
    return taskQueue.offer(task);
}
```

6.8.4 startThread 核心方法

startThread 方法用于启动事件执行器的线程。流程如下。

（1）尝试原子性的将状态从 ST_NOT_STARTED 修改为 ST_STARTED。如果成功，则调用 doStartThread 启动线程。

（2）创建 Runnable 任务，并执行。

（3）在 Runnable 中获取当前线程并判断中断。如果设置了中断标记位，则中断当前线程。

（4）更新最后执行任务的时间。

（5）执行子类需要重写的 run 方法 SingleThreadEventExecutor.this.run()。通过 try catch 捕捉所有子类执行的异常。保证当前线程不会因为子类的异常而退出。

（6）当执行完毕后将执行器状态修改为 ST_SHUTTING_DOWN。

（7）执行 confirmShutdown 方法，运行所有剩余的任务和关闭钩子函数。

（8）调用 cleanup 钩子函数执行资源清理操作。

（9）把事件执行器的状态修改为终止，同时释放线程锁信号量。

```
private void startThread() {
    if (STATE_UPDATER.get(this) == ST_NOT_STARTED) {
        if (STATE_UPDATER.compareAndSet(this, ST_NOT_STARTED, ST_STARTED)) { // 原子性更新 state
            doStartThread();
        }
    }
}

private void doStartThread() {
    assert thread == null;
    executor.execute(new Runnable() {             // 直接通过执行器执行任务
        @Override
        public void run() {
            // 获取当前线程并判断中断，如果设置了中断标记位，则中断当前线程
            thread = Thread.currentThread();
            if (interrupted) {
                thread.interrupt();
            }

            boolean success = false;
            updateLastExecutionTime(); // 更新最后执行任务的时间，即获取当前时间对 lastExecutionTime 进
行赋值
            try { // 执行子类需要重写的 run 方法。通过 try catch 捕捉所有子类执行 run 的异常。保证当前线程
不会因为子类的异常而退出
                SingleThreadEventExecutor.this.run();
```

```java
                success = true;
            } catch (Throwable t) {
                logger.warn("Unexpected exception from an event executor: ", t);
            } finally {
                // 执行完毕后将执行器状态修改为 ST_SHUTTING_DOWN。由此可见，上述的 run 方法一旦
// 返回，当前执行器就等同关闭，这时很容易推测到子类的 SingleThreadEventExecutor.this.run()是处理业务的核
// 心操作
                for (;;) {
                    int oldState = STATE_UPDATER.get(SingleThreadEventExecutor.this);
                    if (oldState >= ST_SHUTTING_DOWN || STATE_UPDATER.compareAndSet(
                        SingleThreadEventExecutor.this, oldState, ST_SHUTTING_DOWN)) {
                        break;
                    }
                }

                // 如果成功执行 SingleThreadEventExecutor.this.run()，而不是因为抛出异常而退出，则需要
// 判断 gracefulShutdownStartTime 是否为 0。如果 gracefulShutdownStartTime 为 0，则表明子类没有调用
// confirmShutdown 方法，这时会触发一个错误日志
                if (success && gracefulShutdownStartTime == 0) {
                    logger.error("Buggy " + EventExecutor.class.getSimpleName() + " implementation; " +
                        SingleThreadEventExecutor.class.getSimpleName() +
".confirmShutdown() must be called " +
                        "before run() implementation terminates.");
                }

                try {
                    // 执行 confirmShutdown 方法，运行所有剩余的任务和关闭钩子函数
                    for (;;) {
                        if (confirmShutdown()) {
                            break;
                        }
                    }
                } finally {
                    try {
                        cleanup();          // 执行资源清理操作，子类可以实现该钩子函数完成资源释放
                    } finally {
                        // 修改事件执行器状态为终止状态，同时释放线程锁信号量
                        STATE_UPDATER.set(SingleThreadEventExecutor.this, ST_TERMINATED);
                        threadLock.release();
                        if (!taskQueue.isEmpty()) {                 // 任务队列不为空，触发警告日志
                            logger.warn(
                                "An event executor terminated with " +
                                "non-empty task queue (" + taskQueue.size() + ')');
                        }
                        terminationFuture.setSuccess(null);         // 设置终止 Future 成功完成
                    }
                }
            }
        }
    });
}

// 子类重写该方法执行处理操作
```

```java
protected abstract void run();

// 子类实现该钩子方法完成资源的清理
protected void cleanup() {
    // NOOP
}
```

6.8.5 confirmShutdown 核心方法

confirmShutdown 方法用于确保事件处理器处于关闭状态。流程如下。

（1）判断状态是否处于 ST_SHUTTING_DOWN（正在关闭状态）。
（2）判断当前调用该方法的线程是否正在执行该事件执行器的线程。
（3）取消所有周期性调度任务。
（4）更新 gracefulShutdownStartTime 关闭开始时间。
（5）执行所有任务队列中的任务成功或执行钩子函数成功，判断当前状态是否已经处于关闭状态。如果处于关闭状态，直接返回 true。因为此时执行器已经关闭，并且没有任务需要执行。否则调用 wakeup 函数，唤醒线程并返回 false。
（6）获取当前时间并判断状态 state 是否处于 ST_SHUTTING_DOWN（关闭状态）。如果已经关闭，则直接返回 true。否则判断当前 shutdown 的时间是否已经超过关闭超时时间，如果是，则返回 true。
（7）判断最后一次执行任务的时间是否小于设置的关闭静默期时间，调用 wakeup 唤醒线程继续执行任务。睡眠时间为 100ms，表示在静默期内每隔 100ms 判断一次是否有任务放入队列执行。
（8）如果在静默期内，没有任务放入队列执行，则返回 true，安全 shutdown。

```java
protected boolean confirmShutdown() {
    if (!isShuttingDown()) { // 判断状态是否处于正在关闭状态：STATE_UPDATER.get(this)>=ST_SHUTTING_DOWN
        return false;
    }
    if (!inEventLoop()) {           // 当前调用该方法的线程不是正在执行该事件执行器的线程，抛出异常
        throw new IllegalStateException("must be invoked from an event loop");
    }
    cancelScheduledTasks();         // 取消所有周期性调度任务
    if (gracefulShutdownStartTime == 0) { // 更新关闭开始时间
        gracefulShutdownStartTime = ScheduledFutureTask.nanoTime();
    }
    if (runAllTasks() || runShutdownHooks()) { // 执行所有任务队列中的任务成功或执行钩子函数成功，判断当前状态是否已经处于关闭状态。如果处于关闭状态，则直接返回 true，因为此时执行器已经关闭，并且没有任何新任务需要执行
        if (isShutdown()) {
            return true;
        }
        // 如果队列中仍有任务，则等待一段时间，直到静默期时间段内没有任务放入队列
        wakeup(true);
        return false;
    }
    final long nanoTime = ScheduledFutureTask.nanoTime();  // 获取当前时间
    if (isShutdown() || nanoTime - gracefulShutdownStartTime > gracefulShutdownTimeout) { // 如果状态已经关闭，则直接返回 true，否则判断当前关闭的时间是否已经超过关闭超时时间。如果是，则返回 true
        return true;
```

```
        }
        if (nanoTime - lastExecutionTime <= gracefulShutdownQuietPeriod) { // 最后一次执行任务的时间小于设置
的关闭静默期时间，这时调用 wakeup 唤醒线程继续执行任务。这里睡眠 100ms，表示在静默期内每 100ms 判断
是否有任务放入队列执行
            wakeup(true);
            try {
                Thread.sleep(100);
            } catch (InterruptedException e) {
                // Ignore
            }
            return false;
        }
        // 静默期内没有任务放入队列，直接返回 true，安全关闭
        return true;
    }

    // 取消所有周期性调度任务
    protected void cancelScheduledTasks() {
        assert inEventLoop();
        // 获取周期性任务队列。如果队列为空，则直接返回
        Queue<ScheduledFutureTask<?>> scheduledTaskQueue = this.scheduledTaskQueue;
        if (isNullOrEmpty(scheduledTaskQueue)) {
            return;
        }
        // 将周期性任务队列中的任务转为数组，并遍历该数组并取消执行
        final ScheduledFutureTask<?>[] scheduledTasks =
            scheduledTaskQueue.toArray(new ScheduledFutureTask<?>[scheduledTaskQueue.size()]);
        for (ScheduledFutureTask<?> task: scheduledTasks) {
            task.cancelWithoutRemove(false);
        }
        scheduledTaskQueue.clear();                    // 将周期性任务队列清空
    }

    // 唤醒线程函数，判断执行器状态后，向队列放入唤醒线程的空任务。因为线程可能处于阻塞队列获取任务状态，
通过向其中放入空任务，可以将线程唤醒
    protected void wakeup(boolean inEventLoop) {
        if (!inEventLoop || STATE_UPDATER.get(this) == ST_SHUTTING_DOWN) {
            taskQueue.offer(WAKEUP_TASK);
        }
    }
```

6.8.6 runAllTasks 核心方法

runAllTasks 方法用于执行普通任务队列中的所有任务。流程如下。

（1）将当前需要执行的周期性任务队列中的任务放入普通任务队列 taskQueue。

（2）执行 taskQueue 中的所有任务，直到周期性任务队列和 taskQueue 为空。

（3）更新最终执行时间。

（4）调用子类实现的钩子函数 afterRunningAllTasks。

```
protected boolean runAllTasks() {
    assert inEventLoop();
```

```java
        boolean fetchedAll;
        boolean ranAtLeastOne = false;
        do {
            fetchedAll = fetchFromScheduledTaskQueue();
            if (runAllTasksFrom(taskQueue)) {
                ranAtLeastOne = true;
            }
        } while (!fetchedAll);                                  // 继续处理，直到获取所有的计划任务
        if (ranAtLeastOne) {                                    // 更新最终执行时间
            lastExecutionTime = ScheduledFutureTask.nanoTime();
        }
        afterRunningAllTasks();                                 // 调用子类实现的钩子函数
        return ranAtLeastOne;
    }

    // 将当前需要执行的周期性任务从周期性任务队列中移到普通任务队列 taskQueue 中
    private boolean fetchFromScheduledTaskQueue() {
        long nanoTime = AbstractScheduledEventExecutor.nanoTime();
        Runnable scheduledTask = pollScheduledTask(nanoTime);   // 获取当前已经到期的周期性任务
        while (scheduledTask != null) {
            if (!taskQueue.offer(scheduledTask)) {              // 将其放入普通任务队列
                // 如果任务队列中没有剩余空间，则将它放回 scheduledTaskQueue，等待普通任务队列有空间后
    再次提取它
                scheduledTaskQueue().add((ScheduledFutureTask<?>) scheduledTask);
                return false;
            }
            scheduledTask = pollScheduledTask(nanoTime);        // 获取下一个周期性任务
        }
        return true;
    }

    // 执行 taskQueue 中的所有任务
    protected final boolean runAllTasksFrom(Queue<Runnable> taskQueue) {
        Runnable task = pollTaskFrom(taskQueue);                // 从队列中提取任务
        if (task == null) {
            return false;
        }
        for (;;) {
            safeExecute(task);                                  // 执行任务
            task = pollTaskFrom(taskQueue);                     // 提取下一个任务
            if (task == null) {
                return true;
            }
        }
    }

    // 使用 try catch 预防发生异常导致线程退出
    private static void safeExecute(Runnable task) {
        try {
            task.run();
        } catch (Throwable t) {
            logger.warn("A task raised an exception. Task: {}", task, t);
        }
    }
```

6.8.7 runShutdownHooks 核心方法

该方法用于执行所有注册的执行器关闭时执行的钩子函数。源码如下。

```java
private boolean runShutdownHooks() {
    boolean ran = false;
    while (!shutdownHooks.isEmpty()) {    // 执行所有钩子直到队列为空
        List<Runnable> copy = new ArrayList<Runnable>(shutdownHooks);  // 将当前所有钩子复制快照到当前 copy 引用，避免在执行中新增加 hook
        shutdownHooks.clear();              // 清空队列
        for (Runnable task: copy) {         // 遍历钩子队列中的任务
            try {
                task.run();
            } catch (Throwable t) {
                logger.warn("Shutdown hook raised an exception.", t);
            } finally {
                ran = true;
            }
        }
    }
    if (ran) {                              // 执行成功，更新执行器最后一次执行任务的时间
        lastExecutionTime = ScheduledFutureTask.nanoTime();
    }
    return ran;
}
```

6.8.8 awaitTermination 核心方法

awaitTermination 方法用于外部线程等待当前执行器关闭，timeout 指定等待超时时间。源码如下。

```java
public boolean awaitTermination(long timeout, TimeUnit unit) throws InterruptedException {
    if (unit == null) {
        throw new NullPointerException("unit");
    }
    if (inEventLoop()) {    // 不能等待自身完成（即当前线程是当前事件执行器的执行线程）
        throw new IllegalStateException("cannot await termination of the current thread");
    }
    if (threadLock.tryAcquire(timeout, unit)) {  // 通过线程信号量完成等待（Semaphore threadLock = new Semaphore(0)，由于信号量初始化为 0，执行线程调用 release 时会唤醒等待线程）
        threadLock.release();   // 成功后需要释放信号量，如果还有别的线程正在等待信号量，就会被唤醒
    }
    return isTerminated(); // 判断是否处于 ST_TERMINATED 终止状态（doStartThread 方法会在 finally 块中将 state 转变为 ST_TERMINATED）
}
```

6.8.9 takeTask 核心方法

takeTask 方法的权限修饰符是 protected，所以必然是子类进行调用，通过方法名也很容易知道该方法的作用，即从任务队列中获取待执行的任务。流程如下。

（1）判断任务队列必须为阻塞队列，因为保证线程安全，同时可以在没有任务可获取时阻塞等待

任务。

（2）调用 peekScheduledTask 方法，查看周期性任务调度队列的第一个待执行任务。

（3）如果周期性任务不存在，则从普通任务队列中获取任务执行。此时调用 taskQueue.take()，所以当不存在任务时会阻塞。如果存在任务，且任务为唤醒任务，则直接设置为空返回即可。

（4）如果存在周期性任务，则获取等待时间。如果等待时间大于 0，则在等待期间去普通任务队列里获取任务执行。如果等待时间小于或等于 0，表明周期性任务已经可以执行了，则将周期性任务调度队列中已经达到执行时间的任务放入普通任务队列，从队列中获取任务，然后返回。

```java
protected Runnable takeTask() {
    assert inEventLoop();
    if (!(taskQueue instanceof BlockingQueue)) {   // 任务队列必须为阻塞队列，因为保证线程安全，同时可以在没有任务可获取时阻塞等待任务
        throw new UnsupportedOperationException();
    }
    BlockingQueue<Runnable> taskQueue = (BlockingQueue<Runnable>) this.taskQueue;
    for (;;) {
        // 查看周期性任务调度队列的第一个待执行任务
        ScheduledFutureTask<?> scheduledTask = peekScheduledTask();
        if (scheduledTask == null) {              // 如果周期性任务不存在，则从普通任务队列中获取任务执行
            Runnable task = null;
            try {
                task = taskQueue.take(); // 获取普通任务队列中的任务。这里为 take 方法，不存在任务时会阻塞
                if (task == WAKEUP_TASK) { // 如果是唤醒任务，则直接设置为空返回
                    task = null;
                }
            } catch (InterruptedException e) {
                // Ignore
            }
            return task;
        } else { // 周期性任务存在，获取等待时间
            long delayNanos = scheduledTask.delayNanos();
            Runnable task = null;
            if (delayNanos > 0) {    // 等待时间大于 0，在等待期间去普通任务队列里获取任务执行
                try {
                    task = taskQueue.poll(delayNanos, TimeUnit.NANOSECONDS);
                } catch (InterruptedException e) {
                    return null;
                }
            }
            if (task == null) { // 如果任务为空，表明在等待时间段内，没有放入新任务或已经可以执行周期性任务。此时将周期性任务调度队列中已经达到执行时间的任务放入普通任务队列，然后从队列中获取放入的任务执行（注意，这里是 Netty 修复的一个 bug，读者可以考虑下如果不移动周期性任务到普通任务队列，当 taskqueue 中总是存在一个任务执行时，周期性任务将永远得不到执行）
                /** 这是没修复前的逻辑，不管 delayNanos 的值如何，这里都将从 taskQueue 队列中获取任务，如果该队列中始终存在一个任务，就不会执行这个达到执行时间的周期性任务
                Runnable task;
                if (delayNanos > 0) {
                    try {
                        task = taskQueue.poll(delayNanos, TimeUnit.NANOSECONDS);
                    } catch (InterruptedException e) {
```

```
                        // Waken up.
                        return null;
                }
        } else {                // 这里是关键（必须去掉，因为此时 delayNanos<=0, 已经可以执行了）
            task = taskQueue.poll();
        }
        if (task == null) {
            fetchFromScheduledTaskQueue();
            task = taskQueue.poll();
        }
        **/
        fetchFromScheduledTaskQueue();
        task = taskQueue.poll();
    }
    if (task != null) {          // 任务存在，直接返回
        return task;
    }
}
```

6.8.10 shutdownGracefully 核心方法

shutdownGracefully 方法用于关闭事件执行器。当然，事件执行器最终是通过封装 Runnable 对象放入 Executor 中执行，这时 Executor 启动一个线程执行事件执行器中的 run 方法。该方法关闭执行器的流程如下。

（1）对静默期持续时间，等待关闭超时时间，超时时间单位进行参数校验。

（2）如果是关闭中状态，直接返回 terminationFuture。

（3）循环等待状态变为关闭中状态。在循环中执行以下操作。

① 如果状态是关闭中，返回 terminationFuture。

② 如果当前执行器线程为事件执行线程，则直接把 newState 状态修改为 ST_SHUTTING_DOWN，否则根据当前状态来修改。如果处于 ST_NOT_STARTED（未启动）、ST_STARTED（已经启动），则直接修改为 ST_SHUTTING_DOWN，其他状态均不变。

③ 原子性更新状态。

④ 将静默期、超时时间转为实例变量。

⑤ 如果之前的状态为未启动状态，则启动线程处理关闭状态。

⑥ 如果线程是关闭状态，则不需要唤醒，其他状态则需要唤醒线程处理关闭状态。

⑦ 返回 terminationFuture。

```
public Future<?> shutdownGracefully(long quietPeriod, long timeout, TimeUnit unit) {
    // 参数校验
    if (quietPeriod < 0) {
        throw new IllegalArgumentException("quietPeriod: " + quietPeriod + " (expected >= 0)");
    }
    if (timeout < quietPeriod) {
        throw new IllegalArgumentException(
                "timeout: " + timeout + " (expected >= quietPeriod (" + quietPeriod + "))");
```

```java
    }
    if (unit == null) {
        throw new NullPointerException("unit");
    }
    // 已经处于关闭中状态,直接返回 terminationFuture,调用方可以使用 Future 等待状态转为 TS_TERMINATE
    if (isShuttingDown()) {
        return terminationFuture();
    }
    boolean inEventLoop = inEventLoop();         // 判断当前调用线程是否为事件执行器的执行线程
    boolean wakeup;
    int oldState;
    for (;;) {                                    // 循环等待关闭
        if (isShuttingDown()) {                   // 已经处于关闭中，返回 Future
            return terminationFuture();
        }
        int newState;
        wakeup = true;
        oldState = STATE_UPDATER.get(this);       // 获取当前状态
        if (inEventLoop) {                        // 如果当前执行器线程为事件执行线程，则直接修改
// newState 状态为 ST_SHUTTING_DOWN，表明处于关闭状态（注意，这里能直接修改，是因为当前为执行线程，
// 既然执行到该方法，当前状态肯定为 ST_STARTED，直接修改即可）
            newState = ST_SHUTTING_DOWN;
        } else {
            // 否则根据当前状态来修改。如果处于 ST_NOT_STARTED（未启动）、ST_STARTED（已经启
// 动），则直接修改为 ST_SHUTTING_DOWN，其他状态均不变（即如果状态已经处于关闭中、关闭、终止，则
// 不需要修改）
            switch (oldState) {
                case ST_NOT_STARTED:
                case ST_STARTED:
                    newState = ST_SHUTTING_DOWN;
                    break;
                default:
                    newState = oldState;
                    wakeup = false;
            }
        }
        if (STATE_UPDATER.compareAndSet(this, oldState, newState)) { // 原子性更新状态
            break;
        }
    }
    // 将静默期，超时时间转为实例变量
    gracefulShutdownQuietPeriod = unit.toNanos(quietPeriod);
    gracefulShutdownTimeout = unit.toNanos(timeout);
    if (oldState == ST_NOT_STARTED) {             // 如果之前的状态为未启动状态，则启动线程
        doStartThread();
    }
    if (wakeup) {        // 如果线程已经处于关闭状态，则不需要唤醒，其他时候需要唤醒线程来处理关闭状态
        wakeup(inEventLoop);
    }
    return terminationFuture();
}
```

6.9　DefaultEventExecutor

在 SingleThreadEventExecutor 中，执行 execute 方法时会将任务放入普通任务队列中，同时调用 doStartThread 方法启动执行线程，而执行线程会调用子类实现的 run 方法，完成执行器任务的执行。该类作为默认执行器会重写 run 方法完成自身的逻辑，而其他所有的操作，都已经在 SingleThreadEventExecutor 中完成。所以这里直接介绍 run 方法和构造函数实现。

```java
public final class DefaultEventExecutor extends SingleThreadEventExecutor {
    // 构造器：通过指定执行器初始化父类
    public DefaultEventExecutor(Executor executor) {
        this(null, executor);
    }

    // 构造器：通过指定事件执行器组初始化父类
    public DefaultEventExecutor(EventExecutorGroup parent) {
        this(parent, new DefaultThreadFactory(DefaultEventExecutor.class));
    }

    // 构造器：通过指定事件执行器组和线程工厂初始化父类
    public DefaultEventExecutor(EventExecutorGroup parent, ThreadFactory threadFactory) {
        super(parent, threadFactory, true);
    }

    // 重写父类的 run 方法，在该方法中不断循环执行任务队列中的任务
    protected void run() {
        for (;;) {
            // 从队列中获取任务执行，并且更新最后任务执行时间
            Runnable task = takeTask();
            if (task != null) {
                task.run();
                updateLastExecutionTime();
            }
            if (confirmShutdown()) { // 确认是否需要关闭事件执行器。如果关闭，则直接退出循环结束任务执行
                break;
            }
        }
    }
}
```

第 7 章 EventLoop 与 EventLoopGroup 原理

7.1 EventLoopGroup 接口与 EventLoop 接口

7.1.1 EventLoopGroup 接口

EventLoopGroup 是特殊的 EventExecutorGroup 事件执行器组,本书中译为事件循环组。它在 EventExecutorGroup 的基础上增加了对 Channel 通道的注册操作。源码如下。

```java
public interface EventLoopGroup extends EventExecutorGroup {
    /**
     * 返回事件循环组中的下一个事件循环对象
     */
    @Override
    EventLoop next();

    /**
     * 将 channel 注册到事件循环组中,返回的 ChannelFuture 表明执行结果
     */
    ChannelFuture register(Channel channel);

    /**
     * 使用 ChannelPromise 注册通道,只需知道是将 channel 通道包装在 ChannelPromise 中即可
     */
    ChannelFuture register(ChannelPromise promise);
}
```

7.1.2 EventLoop 接口

EventLoop 表示 EventLoopGroup 中的事件循环,该接口是特殊的 EventLoopGroup,很明显,又是一个只有自己的组。所以该组将处理所有注册到其中的 channel 操作,同时由于其包含在 EventLoopGroup 中,所以增加了一个用于获取所属事件循环组的父类方法。源码如下。

```java
public interface EventLoop extends OrderedEventExecutor, EventLoopGroup {
    @Override
    EventLoopGroup parent();
}
```

7.2 MultithreadEventLoopGroup 原理

MultithreadEventLoopGroup 表示一个多线程处理的事件循环组,实现相当简单,因为大部分操作都在 MultithreadEventExecutorGroup 抽象类中实现了,这里只不过对 register 方法进行了默认实现。所有操作均调用 next 方法获取事件循环组中的下一个 EventLoop 事件循环处理。所以更能说明 MultithreadEventLoopGroup 只是管理 EventLoop,实际执行操作的是 EventLoop。源码如下。

```java
public abstract class MultithreadEventLoopGroup extends MultithreadEventExecutorGroup implements EventLoopGroup {
    private static final int DEFAULT_EVENT_LOOP_THREADS; // 默认事件循环线程个数
    static {
        // 默认取 CPU 核心数 * 2
        DEFAULT_EVENT_LOOP_THREADS = Math.max(1, SystemPropertyUtil.getInt(
            "io.Netty.eventLoopThreads", Runtime.getRuntime().availableProcessors() * 2));
    }

    // 构造器:如果线程数为 0,则将使用默认线程数
    protected MultithreadEventLoopGroup(int nThreads, Executor executor, Object... args) {
        super(nThreads == 0 ? DEFAULT_EVENT_LOOP_THREADS : nThreads, executor, args);
    }

    // 默认线程工厂,指定优先级为最高优先级
    @Override
    protected ThreadFactory newDefaultThreadFactory() {
        return new DefaultThreadFactory(getClass(), Thread.MAX_PRIORITY);
    }

    // 直接调用父类的选择器选择下一个事件循环对象
    @Override
    public EventLoop next() {
        return (EventLoop) super.next();
    }

    // 由子类创建事件循环对象
    @Override
    protected abstract EventLoop newChild(Executor executor, Object... args) throws Exception;

    /*****************所有操作均由事件循环对象完成****************/
    @Override
    public ChannelFuture register(Channel channel) {
        return next().register(channel);
    }

    @Override
    public ChannelFuture register(ChannelPromise promise) {
        return next().register(promise);
    }
```

```
    @Deprecated
    @Override
    public ChannelFuture register(Channel channel, ChannelPromise promise) {
        return next().register(channel, promise);
    }
}
```

7.3　DefaultEventLoopGroup 原理

DefaultEventLoopGroup 是默认的事件循环组实现，同样较为简单。既然是默认事件循环组，实际处理的肯定也是默认事件循环对象，即 DefaultEventLoop。源码如下。

```
public class DefaultEventLoopGroup extends MultithreadEventLoopGroup {

    /**
     * 使用默认的线程数
     */
    public DefaultEventLoopGroup() {
        this(0);
    }

    /**
     * 指定线程数创建
     */
    public DefaultEventLoopGroup(int nThreads) {
        this(nThreads, null);
    }

    /**
     * 指定线程数和线程工厂
     */
    public DefaultEventLoopGroup(int nThreads, ThreadFactory threadFactory) {
        super(nThreads, threadFactory);
    }

    // 事件循环对象为默认事件循环
    @Override
    protected EventLoop newChild(Executor executor, Object... args) throws Exception {
        return new DefaultEventLoop(this, executor);
    }
}
```

7.4　NioEventLoopGroup 类

NioEventLoopGroup 用于处理 NIO Channel 操作的事件循环组。这里创建的子事件循环为 NioEventLoop。同理，由于是操作 NIO 的事件，需要初始化 SelectorProvider，创建 Selector 实例。该类也是在生产环境中常用的事件循环组，但是实现较为简单，因为大量的操作也是在 MultithreadEventLoopGroup 中完成。源码如下。

```java
public class NioEventLoopGroup extends MultithreadEventLoopGroup {

    /**
     * 构造器：初始化 SelectorProvider，该对象将初始化选择器
     */
    public NioEventLoopGroup(int nThreads, ThreadFactory threadFactory) {
        this(nThreads, threadFactory, SelectorProvider.provider());
    }


    /**
     * 构造器：使用默认的选择策略 DefaultSelectStrategyFactory.INSTANCE，该选择策略会在事件循环对象
     * 中处理 Selector 通道选择时使用
     */
    public NioEventLoopGroup(
            int nThreads, ThreadFactory threadFactory, final SelectorProvider selectorProvider) {
        this(nThreads, threadFactory, selectorProvider, DefaultSelectStrategyFactory.INSTANCE);
    }

    // 使用默认拒绝策略初始化父类
    public NioEventLoopGroup(int nThreads, ThreadFactory threadFactory,
            final SelectorProvider selectorProvider, final SelectStrategyFactory selectStrategyFactory) {
        super(nThreads, threadFactory, selectorProvider, selectStrategyFactory,
RejectedExecutionHandlers.reject());
    }
    /**
     * 设置事件循环处理线程，处理 Channel IO 的时间占比和执行普通任务的时间占比，默认为 50%
     */
    public void setIoRatio(int ioRatio) {
        for (EventExecutor e: this) {
            ((NioEventLoop) e).setIoRatio(ioRatio);
        }
    }

    /**
     * 用于修复 JDK 的 epoll 100% CPU 的错误（在部分 Linux 的 2.6 的 kernel 中，poll 和 epoll 对于突然中断
     * 的连接 Socket 会对返回的 eventSet 事件集合置为 POLLHUP，也可能是 POLLERR，eventSet 事件集合发生了
     * 变化，可能会导致唤醒 Selector。了解即可），解决办法是让所有事件循环重新创建 Selector 对象
     */
    public void rebuildSelectors() {
        for (EventExecutor e: this) {
            ((NioEventLoop) e).rebuildSelector();
        }
    }

    // 创建的事件循环对象为 NioEventLoop，这里将传递给事件循环组的 SelectorProvider、SelectStrategyFactory、
    // RejectedExecutionHandler 传递给了事件循环对象
    @Override
    protected EventLoop newChild(Executor executor, Object... args) throws Exception {
        return new NioEventLoop(this, executor, (SelectorProvider) args[0],
                ((SelectStrategyFactory) args[1]).newSelectStrategy(),
```

```
(RejectedExecutionHandler) args[2]);
    }
}
```

7.5 ThreadPerChannelEventLoopGroup 原理

ThreadPerChannelEventLoopGroup 为每个通道创建 EventLoop 实例，这种特性就是 BIO 的特性，在后面描述 oio 时会用到该类。注意，在生产线上慎用该类（通常不用），因为无法利用 NIO 的特性，每个通道配一个线程会导致系统性能极度下降。

7.5.1 核心变量与构造器

通过变量的定义，很容易理解以下几点内容。
（1）该类对于每个通道都使用一个 EventLoop 对象处理。
（2）通道数可能非常大，可以通过设置 maxChannels 限制并发的通道数。
（3）通道数达到最大时，则会抛出 ChannelException tooManyChannels 异常。
（4）创建 activeChildren 和 idleChildren 缓存 EventLoop 对象，提升性能。

```
public class ThreadPerChannelEventLoopGroup extends AbstractEventExecutorGroup implements
EventLoopGroup {
    private final Object[] childArgs;              // 传递给管理的 EventLoop 对象使用参数
    private final int maxChannels;                 // 最大通道数
    final Executor executor;                       // 执行器对象
    final Set<EventLoop> activeChildren =
        Collections.newSetFromMap(PlatformDependent.<EventLoop, Boolean>newConcurrentHashMap());
    // 管理活动的 EventLoop 对象
    final Queue<EventLoop> idleChildren = new ConcurrentLinkedQueue<EventLoop>(); // 管理空闲的
EventLoop 对象
    private final ChannelException tooManyChannels;   // 在通道数量超过最大通道数时抛出的异常对象

    private volatile boolean shuttingDown;            // 是否处于关闭状态
    private final Promise<?> terminationFuture = new
DefaultPromise<Void>(GlobalEventExecutor.INSTANCE);  // 关闭结果 Promise，调用方可以根据该对象来判断
当前执行器组是否已经终止
    private final FutureListener<Object> childTerminationListener = new FutureListener<Object>() {
        @Override
        public void operationComplete(Future<Object> future) throws Exception {
            if (isTerminated()) {
                terminationFuture.trySuccess(null);
            }
        }
    }; // EventLoop 对象终止后回调的监听器

    // 构造方法：设置 maxChannels 最大通道数为 0，表示无限制
    protected ThreadPerChannelEventLoopGroup() {
        this(0);
    }
```

```java
// 构造方法：设置 maxChannels 最大通道数为 maxChannels，并提供了默认工厂函数
protected ThreadPerChannelEventLoopGroup(int maxChannels) {
    this(maxChannels, Executors.defaultThreadFactory());
}

// 构造方法：设置 maxChannels 最大通道数为 maxChannels，并提供了默认执行器（每个任务创建一个线程来执行）
protected ThreadPerChannelEventLoopGroup(int maxChannels,ThreadFactory threadFactory, Object... args) {
    this(maxChannels, new ThreadPerTaskExecutor(threadFactory), args);
}

// 完整构造器：对参数进行初始化
protected ThreadPerChannelEventLoopGroup(int maxChannels, Executor executor, Object... args) {
    if (maxChannels < 0) {
        throw new IllegalArgumentException(String.format(
            "maxChannels: %d (expected: >= 0)", maxChannels));
    }
    if (executor == null) {
        throw new NullPointerException("executor");
    }

    if (args == null) {
        childArgs = EmptyArrays.EMPTY_OBJECTS;
    } else {
        childArgs = args.clone();
    }

    this.maxChannels = maxChannels;
    this.executor = executor;

    tooManyChannels = ThrowableUtil.unknownStackTrace(
        new ChannelException("too many channels (max: " + maxChannels + ')'),
        ThreadPerChannelEventLoopGroup.class, "nextChild()");
}
```

7.5.2 newChild 核心方法

newChild 方法创建管理的 EventLoop 对象。创建的类型是 ThreadPerChannelEventLoop 对象。源码如下。

```java
protected EventLoop newChild(@SuppressWarnings("UnusedParameters") Object... args) throws Exception {
    return new ThreadPerChannelEventLoop(this);
}
```

7.5.3 next 核心方法

对于每个通道，该类都有一个 EventLoop 对象用于执行，所以 next 方法不支持操作。源码如下。

```java
public EventLoop next() {
    throw new UnsupportedOperationException();
}
```

7.5.4　shutdownGracefully 核心方法

该方法用于关闭事件循环组。流程如下。

（1）设置 shuttingDown 为 true，标识已经处于关闭中状态。
（2）关闭所有活动的 EventLoop。
（3）关闭所有空闲的 EventLoop。
（4）如果此时状态已经变为终止状态，则设置 terminationFuture 执行成功。

```java
public Future<?> shutdownGracefully(long quietPeriod, long timeout, TimeUnit unit) {
    shuttingDown = true;                           // 标识已经处于关闭中状态
    for (EventLoop l: activeChildren) {            // 关闭所有活动的 EventLoop
        l.shutdownGracefully(quietPeriod, timeout, unit);
    }
    for (EventLoop l: idleChildren) {              // 关闭所有空闲的 EventLoop
        l.shutdownGracefully(quietPeriod, timeout, unit);
    }

    // 如果已变为终止状态，则把 terminationFuture 设置为执行成功，且调用方可以通过 terminationFuture 获取
    // 终止状态
    if (isTerminated()) {
        terminationFuture.trySuccess(null);
    }
    return terminationFuture();
}
```

7.5.5　awaitTermination 核心方法

该方法用于超时等待事件循环组终止。流程如下。

（1）计算超时时间。
（2）遍历所有活动的 EventLoop，调用它们的 awaitTermination。
（3）遍历所有空闲的 EventLoop，调用它们的 awaitTermination。

```java
public boolean awaitTermination(long timeout, TimeUnit unit)
        throws InterruptedException {
    long deadline = System.nanoTime() + unit.toNanos(timeout); // 计算超时时间
    for (EventLoop l: activeChildren) {            // 遍历所有活动的 EventLoop，调用它们的 awaitTermination
        for (;;) {
            long timeLeft = deadline - System.nanoTime();
            if (timeLeft <= 0) {                   // 判断是否超时
                return isTerminated();
            }
            if (l.awaitTermination(timeLeft, TimeUnit.NANOSECONDS)) {
                break;
            }
        }
    }
    for (EventLoop l: idleChildren) {              // 遍历所有空闲的 EventLoop，调用它们的 awaitTermination
        for (;;) {
            long timeLeft = deadline - System.nanoTime();
```

```
            if (timeLeft <= 0) {            // 判断是否超时
                return isTerminated();
            }
            if (!.awaitTermination(timeLeft, TimeUnit.NANOSECONDS)) {
                break;
            }
        }
    }
    return isTerminated();
}
```

7.5.6 register 核心方法

register 系列方法把传递的 Channel 注册到管理的 EventLoop 中。核心方法是 nextChild，该方法返回一个可用的 EventLoop，再调用 register 方法完成注册。区别如下。

（1）register(Channel channel)在注册时创建了 DefaultChannelPromise 对通道进行包装。

（2）register(ChannelPromise promise)直接将 ChannelPromise 传递给 register 方法。

```
public ChannelFuture register(Channel channel) {
    if (channel == null) {
        throw new NullPointerException("channel");
    }
    try {
        EventLoop l = nextChild();
        return l.register(new DefaultChannelPromise(channel, l));
    } catch (Throwable t) {
        return new FailedChannelFuture(channel, GlobalEventExecutor.INSTANCE, t);
    }
}

@Override
public ChannelFuture register(ChannelPromise promise) {
    try {
        return nextChild().register(promise);
    } catch (Throwable t) {
        promise.setFailure(t);
        return promise;
    }
}
```

7.5.7 nextChild 核心方法

nextChild 方法在 register 方法中会返回一个可用的 EventLoop。流程如下。

（1）判断状态是否为关闭中。如果是，则抛出任务拒绝异常。

（2）尝试从空闲的队列中获取 EventLoop。如果获取失败，则判断是否达到最大通道数。如果达到最大通道数，则抛出 tooManyChannels 异常；否则创建一个新的 EventLoop，并将其添加到 activeChildren 活动对象集中，然后返回该 EventLoop。

```
private EventLoop nextChild() throws Exception {
    if (shuttingDown) {
```

```
            throw new RejectedExecutionException("shutting down");
        }

        EventLoop loop = idleChildren.poll();
        if (loop == null) {                                                              // 获取失败
            if (maxChannels > 0 && activeChildren.size() >= maxChannels) {               // 判断是否达到最大通道限制
                throw tooManyChannels;
            }
            loop = newChild(childArgs);                                                  // 创建新的 EventLoop
            loop.terminationFuture().addListener(childTerminationListener);
        }
        activeChildren.add(loop);                                                        // 添加到活动集
        return loop;
    }
```

7.6 OioEventLoopGroup 类

OioEventLoopGroup 类非常简单，实际没有完成任何操作，只是在构造函数中初始化父类。主要用于在命名上符合 Netty 对于事件循环组的定义规范。例如，NIO（NioEventLoopGroup），BIO（OioEventLoopGroup）。源码如下。

```
public class OioEventLoopGroup extends ThreadPerChannelEventLoopGroup {
    public OioEventLoopGroup() {
        this(0);
    }

    public OioEventLoopGroup(int maxChannels) {
        this(maxChannels, Executors.defaultThreadFactory());
    }

    public OioEventLoopGroup(int maxChannels, Executor executor) {
        super(maxChannels, executor);
    }

    public OioEventLoopGroup(int maxChannels, ThreadFactory threadFactory) {
        super(maxChannels, threadFactory);
    }
}
```

7.7 SingleThreadEventLoop 原理

SingleThreadEventLoop 类表示单线程执行的事件循环对象。同样，大部分操作已经在 SingleThreadEventExecutor 中实现。这里主要添加了 Queue tailTasks 队列的处理，同时实现了 EventLoop 中 register 函数。

7.7.1 核心变量与构造器

该类主要新增的操作是 Queue tailTasks 队列操作。源码如下。

```
public abstract class SingleThreadEventLoop extends SingleThreadEventExecutor implements EventLoop {
    protected static final int DEFAULT_MAX_PENDING_TASKS = Math.max(16,
SystemPropertyUtil.getInt("io.Netty.eventLoop.maxPendingTasks", Integer.MAX_VALUE)); // tailTasks 队列中最大放置的任务长度

    private final Queue<Runnable> tailTasks;          // 放置 tail 任务的队列。即在 taskQueue 普通任务队列执行后,需要执行的任务

    // 构造方法:设置了默认为 DEFAULT_MAX_PENDING_TASKS 和拒绝函数
    protected SingleThreadEventLoop(EventLoopGroup parent, ThreadFactory threadFactory, boolean addTaskWakesUp) {
        this(parent, threadFactory, addTaskWakesUp, DEFAULT_MAX_PENDING_TASKS,
RejectedExecutionHandlers.reject());
    }

    // 构造方法:初始化父类同时创建 tailTasks 队列
    protected SingleThreadEventLoop(EventLoopGroup parent, ThreadFactory threadFactory,
                                    boolean addTaskWakesUp, int maxPendingTasks,
                                    RejectedExecutionHandler rejectedExecutionHandler) {
        super(parent, threadFactory, addTaskWakesUp, maxPendingTasks, rejectedExecutionHandler);
        tailTasks = newTaskQueue(maxPendingTasks);
    }
}
```

7.7.2 next 核心方法

next 方法直接调用父类的 next 方法获取下一个 EventLoop 对象。源码如下。

```
public EventLoop next() {
    return (EventLoop) super.next();
}
```

7.7.3 executeAfterEventLoopIteration 核心方法

executeAfterEventLoopIteration 方法用于把任务添加到 tailTasks 队列中。流程如下。
(1)判断 EventLoop 是否已经关闭。如果关闭,则抛出任务拒绝异常。
(2)尝试把任务放入 tailTasks。如果达到最大长度,导致放入失败,则抛出任务拒绝异常。
(3)根据设置唤醒工作任务线程。

```
public final void executeAfterEventLoopIteration(Runnable task) {
    ObjectUtil.checkNotNull(task, "task");
    if (isShutdown()) {                              // 判断 EventLoop 是否已经关闭
        reject();
    }
```

```
        if (!tailTasks.offer(task)) {                    // 尝试将任务放入 tailTasks
            reject(task);
        }

        if (wakesUpForTask(task)) {                      // 根据设置唤醒工作任务线程
            wakeup(inEventLoop());
        }
    }
```

7.7.4　afterRunningAllTasks 核心方法

afterRunningAllTasks 方法是 SingleThreadEventExecutor 类里 runAllTasks 方法的钩子函数。taskQueue 中的所有任务都执行完毕后，会回调该钩子函数。这里重写了该方法，执行 tailTasks 队列中的任务。源码如下。

```
protected void afterRunningAllTasks() {
    runAllTasksFrom(tailTasks);
}
```

7.7.5　register 核心方法

register 方法用于将 Channel 注册到当前 EventLoop 中。在封装 ChannelPromise 对象后，调用 Channel 的 unsafe 对象的 register 方法完成注册。源码如下。

```
public ChannelFuture register(Channel channel) {
    return register(new DefaultChannelPromise(channel, this));
}

@Override
public ChannelFuture register(final ChannelPromise promise) {
    ObjectUtil.checkNotNull(promise, "promise");
    promise.channel().unsafe().register(this, promise);    // 注册通道
    return promise;
}
```

7.7.6　hasTasks 核心方法

hasTasks 方法用于判断是否需要执行任务。注意，要判断父类的 taskQueue 和当前 tailTasks。源码如下。

```
@Override
protected boolean hasTasks() {
    return super.hasTasks() || !tailTasks.isEmpty();
}
```

7.7.7　pendingTasks 核心方法

pendingTasks 方法用于获取当前需要执行的任务个数。任务个数等于父类 taskQueue 的任务数加上 tailTasks 的任务数。源码如下。

```
@Override
```

```
public int pendingTasks() {
    return super.pendingTasks() + tailTasks.size();
}
```

7.8 NioEventLoop

前面描述了 NioEventLoopGroup。实际上它就是一个管理者，用于管理 NioEventLoop 的生命周期，创建并传递设置的参数到 NioEventLoop，关闭 NioEventLoop。实际执行操作的是 NioEventLoop，所以它非常重要，也相对复杂，该类是业务开发中经常使用到的 EventLoop。

7.8.1 核心变量与构造器

Java 中，实现 IO 多路复用的是 Selector，该类必然要与该类耦合。从变量和构造器中可以了解如下几点。

（1）在构造器中创建了 Selector 对象。

（2）对原生的 selectedKeys 和 publicSelectedKeys 进行增强，增强的内容如下。

① JDK 原生自带的选择集的类型为 HashSet。

② 操作 HashSet，就需要计算 hash 值，同时在遍历时需要使用迭代器对象。

③ 增强的集合就是一个数组，不需要使用迭代器，同时也不需要计算 hash 值（内部使用数组下标操作）。

由于内部实现就是两个数组进行切换操作，所以笔者这里将其省略。感兴趣的读者可以打开源码一览究竟。

```
public final class NioEventLoop extends SingleThreadEventLoop {
    private static final int CLEANUP_INTERVAL = 256;                              // 默认清理周期
    private static final boolean DISABLE_KEYSET_OPTIMIZATION =
            SystemPropertyUtil.getBoolean("io.Netty.noKeySetOptimization", false); // 标识是否正确 Selector 进行优化，默认为 false
    private static final int MIN_PREMATURE_SELECTOR_RETURNS = 3;                  // 执行 select 操作时，执行循环的次数（后面在描述 select 方法时会详细介绍）
    private static final int SELECTOR_AUTO_REBUILD_THRESHOLD;                     // 自动重新创建 Selector 的阈值（用于避免 CPU 100%的错误）

    private final IntSupplier selectNowSupplier = new IntSupplier() {
        @Override
        public int get() throws Exception {
            return selectNow();
        }
    }; // 用于封装 selectNow 的 Supplier

    private final Callable<Integer> pendingTasksCallable = new Callable<Integer>() {
        @Override
        public Integer call() throws Exception {
            return NioEventLoop.super.pendingTasks();
        }
    }; // 用于封装获取挂起的任务数的 Callable
```

第 7 章 EventLoop 与 EventLoopGroup 原理

```java
// 静态代码块，用于初始化静态变量：SELECTOR_AUTO_REBUILD_THRESHOLD
static {
    String key = "sun.nio.ch.bugLevel";
    try {
        String buglevel = SystemPropertyUtil.get(key);    // 获取设置的 JDK 错误等级，通常不设置该变量
        if (buglevel == null) {
            System.setProperty(key, "");
        }
    } catch (SecurityException e) {
    }
    int  selectorAutoRebuildThreshold=SystemPropertyUtil.getInt("io.Netty.selectorAutoRebuildThreshold", 512);
                                                        // 默认自动重建 Selector 阈值为 512
    if (selectorAutoRebuildThreshold < MIN_PREMATURE_SELECTOR_RETURNS) {  // 如果小于最小 selector 返回事件个数，则设置为 0
        selectorAutoRebuildThreshold = 0;
    }
    SELECTOR_AUTO_REBUILD_THRESHOLD = selectorAutoRebuildThreshold;
}

Selector selector;                                       // JDK 的 Selector 对象
private SelectedSelectionKeySet selectedKeys;            // 已经选择的 SelectionKey
private final SelectorProvider provider;                 // 用于创建 Selector 对象的提供器
private final AtomicBoolean wakenUp = new AtomicBoolean(); // 用于控制是否调用 Selector.select()对线程阻塞
private final SelectStrategy selectStrategy;             // 选择策略。用于控制执行事件时的方法，例如执行 select 或 selectNow
private volatile int ioRatio = 50;                       // 控制线程在执行 Channel IO 操作的时间和执行任务队列的事件时间占比，默认各占一半
private int cancelledKeys;                               // 保存取消的 SelectionKey 个数
private boolean needsToSelectAgain;                      // 控制是否应该重新进行 select 操作

// 构造方法：初始化父类，设置最大挂起任务为 DEFAULT_MAX_PENDING_TASKS
NioEventLoop(NioEventLoopGroup parent, Executor executor, SelectorProvider selectorProvider,SelectStrategy strategy, RejectedExecutionHandler rejectedExecutionHandler) {
    super(parent, executor, false, DEFAULT_MAX_PENDING_TASKS, rejectedExecutionHandler);
    if (selectorProvider == null) {
        throw new NullPointerException("selectorProvider");
    }
    if (strategy == null) {
        throw new NullPointerException("selectStrategy");
    }
    provider = selectorProvider;
    selector = openSelector();                           // 创建 Selector 实例
    selectStrategy = strategy;
}

// 创建 Selector 实例
private Selector openSelector() {
    final Selector selector;
    try {
        selector = provider.openSelector();              // 直接通过提供器打开 Selector
```

```java
        } catch (IOException e) {
            throw new ChannelException("failed to open a new selector", e);
        }
        if (DISABLE_KEYSET_OPTIMIZATION) {  // 如果优化关闭, 则直接返回
            return selector;
        }
        try {                                           // 否则创建 SelectedSelectionKeySet 对象
            SelectedSelectionKeySet selectedKeySet = new SelectedSelectionKeySet();
            Class<?> selectorImplClass =
                Class.forName("sun.nio.ch.SelectorImpl", false,
                    PlatformDependent.getSystemClassLoader());
            // 确保当前选择器的实现为 SelectorImpl
            if (!selectorImplClass.isAssignableFrom(selector.getClass())) {
                return selector;
            }
            // 获取 SelectorImpl 类的 selectedKeys (已经选择的 SelectionKey) 和 publicSelectedKeys
            //(SelectionKey 的视图, 只允许被移除, 但不允许被添加), 并将它们设置为 SelectedSelectionKeySet selectedKeySet
            Field selectedKeysField = selectorImplClass.getDeclaredField("selectedKeys");
            Field publicSelectedKeysField = selectorImplClass.getDeclaredField("publicSelectedKeys");

            selectedKeysField.setAccessible(true);
            publicSelectedKeysField.setAccessible(true);

            selectedKeysField.set(selector, selectedKeySet);
            publicSelectedKeysField.set(selector, selectedKeySet);
            selectedKeys = selectedKeySet;
        } catch (Throwable t) {
            selectedKeys = null;
            logger.trace("Failed to instrument an optimized java.util.Set into: {}", selector, t);
        }
        return selector;
    }
}
```

7.8.2 run 核心方法

上文已经介绍过,事件循环线程会执行子类复写的 run 方法,所以该方法是分析此类的起点。执行流程如下。

(1) 根据 selectNowSupplier 和任务队列有无任务来计算将要进行的操作。

(2) 根据设置的 ioRatio (默认为 50%) 执行 Selector 中注册 Channel IO 事件和任务队列中的任务 (taskQueue、ScheduledTaskQueue、tailTasks)。

(3) 判断事件循环是否已经关闭。如果关闭,则取消关闭所有 Selectionkey。

```java
// 选择策略接口
public interface SelectStrategy {
    /**
     * 表明需要执行一个阻塞 select 操作
     */
    int SELECT = -1;
    /**
     * 表明执行一个非阻塞的 selectNow 操作
```

```java
    */
    int CONTINUE = -2;

    int calculateStrategy(IntSupplier selectSupplier, boolean hasTasks) throws Exception;
}


// 这里只需要了解默认选择策略即可
final class DefaultSelectStrategy implements SelectStrategy {
    static final SelectStrategy INSTANCE = new DefaultSelectStrategy();       // 单例模式

    private DefaultSelectStrategy() { }

    @Override
    public int calculateStrategy(IntSupplier selectSupplier, boolean hasTasks) throws Exception {
        return hasTasks ? selectSupplier.get() : SelectStrategy.SELECT;       // 如果有任务，则直接返回
selectNow 的准备好的事件个数，否则返回 SELECT，表示执行阻塞的选择操作
    }
}

protected void run() {
    for (;;) {
        try {
            switch (selectStrategy.calculateStrategy(selectNowSupplier, hasTasks())) { // 根据
selectNowSupplier 和任务队列中是否有任务，计算要进行的操作
                case SelectStrategy.CONTINUE:                // 如果是 CONTINUE，则继续循环
                    continue;
                case SelectStrategy.SELECT:
                    select(wakenUp.getAndSet(false));        // 执行 select 选择操作，注意这里把 wakenUp
设置为 false，并使用原来的值进行 select
                    if (wakenUp.get()) {                     // 如果设置了唤醒，则执行唤醒方法（该方法的调用会导致下一
次执行 select 阻塞操作时立即返回）
                        selector.wakeup();
                    }
                default:                                     // 默认不进行选择
            }
            cancelledKeys = 0;
            needsToSelectAgain = false;
            final int ioRatio = this.ioRatio;
            if (ioRatio == 100) {    // 如果设置全部处理 IO 通道事件，则先执行 SelectedKey，再执行任务队列
中的任务
                processSelectedKeys();
                runAllTasks();
            } else {        // 否则计算 processSelectedKeys 执行 IO 事件的时间，再设置执行任务队列的时间
                final long ioStartTime = System.nanoTime();
                processSelectedKeys();
                final long ioTime = System.nanoTime() - ioStartTime;
                runAllTasks(ioTime * (100 - ioRatio) / ioRatio);    // ioRatio 为 50，假如执行 IO 事件的时间为
100ms，则：100ms * (100-50) / 50 = 100ms，正好各占一半
            }

            if (isShuttingDown()) {  // 判断事件循环是否已关闭，如果关闭，则将所有 Selectionkey 取消关闭
                closeAll();
```

```
                if (confirmShutdown()) {        // 确认没任务执行，且状态正确，则结束循环
                    break;
                }
            }
        } catch (Throwable t) {                 // 捕捉所有移除进行日志记录
            logger.warn("Unexpected exception in the selector loop.", t);
            // 避免出现连续抛出异常，导致 CPU 占用过多
            try {
                Thread.sleep(1000);
            } catch (InterruptedException e) {
            }
        }
    }
}
```

7.8.3 select 核心方法

select 方法用于执行 Selector 的 select 函数，获取注册到 Selector 中准备好的事件，参数 oldWakenUp 表示之前 wakeup 的旧值。流程如下。

（1）通过周期性任务调度超时时间，计算 select timeout 超时时间。

（2）判断超时。

（3）判断任务队列中是否存在执行任务。如果存在，则执行 selectNow，避免任务队列中的任务得不到执行。

（4）执行超时 selector.select(timeoutMillis)等待。

（5）判断用户是否唤醒当前线程、队列中是否存在任务、是否有通道事件准备就绪，从而结束当前循环。

（6）根据 selectCnt 判断是否发生 selector 空转的 bug，重新构建 Selector。

```
private void select(boolean oldWakenUp) throws IOException {
    Selector selector = this.selector;
    try {
        int selectCnt = 0;  // 计算 for 循环次数，用于与 MIN_PREMATURE_SELECTOR_RETURNS 变量进行
日志分析（注意，selectCnt 记录的是超时时间内 select 函数被唤醒的次数，详见以下代码）
        long currentTimeNanos = System.nanoTime();
        long selectDeadLineNanos = currentTimeNanos + delayNanos(currentTimeNanos);        // 计算select
timeout 超时时间（这是怎么计算的？考虑下周期性调度任务的超时。）
        for (;;) {
            long timeoutMillis = (selectDeadLineNanos - currentTimeNanos + 500000L) / 1000000L;  // 计算超
时时间毫秒数 1000000L 为纳秒转为毫秒，前面的加 500000L 是为了对计算的毫秒数向上取整
            if (timeoutMillis <= 0) {         // 不存在超时时间，判断是否是第一次循环（通过selectCnt），如果
是，则直接执行 selectNow
                if (selectCnt == 0) {
                    selector.selectNow();
                    selectCnt = 1;
                }
                break;
            }
            if (hasTasks() && wakenUp.compareAndSet(false, true)) {  // 任务队列中存在任务（taskqueue 或
tailTask），且 cas 设置 wakeup 标志位为 true 成功，执行 selectNow 后结束循环
```

```java
                    selector.selectNow();
                    selectCnt = 1;
                    break;
                }
                int selectedKeys = selector.select(timeoutMillis);      // 执行超时 select 等待
                selectCnt ++;                                            // 自增循环等待次数
                if (selectedKeys != 0 || oldWakenUp || wakenUp.get() || hasTasks() || hasScheduledTasks()) {
                    // - 已经由通道事件产生
                    // - 外部函数设置了唤醒操作
                    // - 任务队列中存在任务
                    // - 存在可以周期性调度任务
                    // - 结束循环
                    break;
                }
                if (Thread.interrupted()) {                              // 线程被中断，记录日志结束循环
                    if (logger.isDebugEnabled()) {
                        logger.debug("Selector.select() returned prematurely because " +
                                "Thread.currentThread().interrupt() was called. Use " +
                                "NioEventLoop.shutdownGracefully() to shutdown the NioEventLoop.");
                    }
                    selectCnt = 1;
                    break;
                }

                long time = System.nanoTime();
                if (time - TimeUnit.MILLISECONDS.toNanos(timeoutMillis) >= currentTimeNanos) {
                    // 达到超时时间，则重置 selectCnt（再次强调：selectCnt 记录的是超时时间内 select 函数返
回循环的次数）
                    selectCnt = 1;
                } else if (SELECTOR_AUTO_REBUILD_THRESHOLD > 0 &&
                        selectCnt >= SELECTOR_AUTO_REBUILD_THRESHOLD) {            // 设置重新构建
Selector 的阈值，同时当前循环次数大于 SELECTOR_AUTO_REBUILD_THRESHOLD，则判定出现了 selector
空转的 bug。记录日志重新创建 selector
                    logger.warn(
                            "Selector.select() returned prematurely {} times in a row; rebuilding Selector {}.",
                            selectCnt, selector);

                    rebuildSelector();
                    selector = this.selector;

                    // 上面重新构建了 selector, 立即执行一次 selectNow, 更新 selectedKeys 集合
                    selector.selectNow();
                    selectCnt = 1;
                    break;
                }
                currentTimeNanos = time;
            }
            if (selectCnt > MIN_PREMATURE_SELECTOR_RETURNS) {   // 循环次数大于设置的 PREMATURE 次
数，记录日志
                if (logger.isDebugEnabled()) {
                    logger.debug("Selector.select() returned prematurely {} times in a row for Selector {}.",
                            selectCnt - 1, selector);
                }
```

```
        } catch (CancelledKeyException e) {
            if (logger.isDebugEnabled()) {
                logger.debug(CancelledKeyException.class.getSimpleName()+" raised by a Selector {} - JDK bug?",
                    selector, e);
            }
        }
    }
}

private static final long SCHEDULE_PURGE_INTERVAL = TimeUnit.SECONDS.toNanos(1); // 如果没有周期
性任务,则默认进行 1s 的超时等待

// 计算 select timeout 超时时间
protected long delayNanos(long currentTimeNanos) {
    ScheduledFutureTask<?> scheduledTask = peekScheduledTask();
    if (scheduledTask == null) {
        return SCHEDULE_PURGE_INTERVAL;
    }

    return scheduledTask.delayNanos(currentTimeNanos);
}
```

7.8.4 rebuildSelector 核心方法

rebuildSelector 方法在 select 方法中对 Selector 空转的 bug 进行处理。处理方式是重新构建 Selector 对象。流程如下。

(1) 如果当前执行线程不是事件循环组的线程,则向队列中提交任务来让执行线程执行重新构建操作。

(2) 判断当前不存在选择器,则直接返回。

(3) 使用 SelectorProvider 创建新的选择器实例。

(4) 将旧 Selector 中注册的通道注册到新的 Selector 中。

```
public void rebuildSelector() {
    if (!inEventLoop()) {
        execute(new Runnable() {
            @Override
            public void run() {
                rebuildSelector();
            }
        });
        return;
    }

    final Selector oldSelector = selector;
    final Selector newSelector;

    if (oldSelector == null) {                              // 当前不存在选择器,则直接返回
        return;
    }

    try {
```

```
            newSelector = openSelector();                    // 通过 SelectorProvider 创建新的选择器实例
    } catch (Exception e) {
        logger.warn("Failed to create a new Selector.", e);
        return;
    }
    // 将旧 Selector 中注册的通道注册到新的 Selector 中
    int nChannels = 0;
    for (;;) {
        try {
            for (SelectionKey key: oldSelector.keys()) {      // 遍历之前注册的通道选择集
                Object a = key.attachment();                   // 获取与 SelectionKey 绑定的对象
                try {
                    if (!key.isValid() || key.channel().keyFor(newSelector) != null) {   // 避免重复注册
                        continue;
                    }
                    int interestOps = key.interestOps();       // 获取之前注册的感兴趣事件集
                    key.cancel();                              // 将 key 从旧的选择器中取消
                    SelectionKey newKey = key.channel().register(newSelector, interestOps, a); // 重新将通
道注册到新的 Selector 中，注意这里需要携带：之前的感兴趣事件集、绑定对象
                    if (a instanceof AbstractNioChannel) {
                        // 如果绑定的对象是 AbstractNioChannel 实例，则要将其中保存的选择键更新
                        ((AbstractNioChannel) a).selectionKey = newKey;
                    }
                    nChannels ++;
                } catch (Exception e) {                        // 发生异常，关闭通道
                    logger.warn("Failed to re-register a Channel to the new Selector.", e);
                    if (a instanceof AbstractNioChannel) {
                        AbstractNioChannel ch = (AbstractNioChannel) a;
                        ch.unsafe().close(ch.unsafe().voidPromise());
                    } else {
                        @SuppressWarnings("unchecked")
                        NioTask<SelectableChannel> task = (NioTask<SelectableChannel>) a;
                        invokeChannelUnregistered(task, key, e);
                    }
                }
            }
        } catch (ConcurrentModificationException e) {
            continue;
        }
        break;
    }
    selector = newSelector;                                   // 更新选择器对象
    try {
        // 关闭旧的选择器
        oldSelector.close();
    } catch (Throwable t) {
        if (logger.isWarnEnabled()) {
            logger.warn("Failed to close the old Selector.", t);
        }
    }
}
```

7.8.5 processSelectedKeys 核心方法

processSelectedKeys 方法用于执行 Selector 中注册的通道事件，即可读、可写、新建连接。如果使用了优化的选择集，则调用 processSelectedKeysOptimized 方法执行，否则使用普通的 processSelectedKeysPlain 方法执行准备好的事件。执行流程如下。

（1）获取需要处理的 SelectionKey，并将其从 publicSelectedKeys 中移除。

（2）根据携带对象的类型来选择处理。如果是 AbstractNioChannel 的子类，则直接调用方法 processSelectedKey 执行；否则先将携带对象转为 NioTask 再调用方法执行。

（3）如果设置了 needsToSelectAgain 标志位，则在处理完当前选择键后，需要再次选择。

优化操作主要是对原生的 HashSet，直接操作数组索引下标，不需要计算 hash，也不使用迭代器对象。源码如下。

```java
private void processSelectedKeys() {
    if (selectedKeys != null) {
        processSelectedKeysOptimized(selectedKeys.flip());
    } else {
        processSelectedKeysPlain(selector.selectedKeys());
    }
}

// 使用优化的选择键集执行
private void processSelectedKeysOptimized(SelectionKey[] selectedKeys) {
    for (int i = 0;; i ++) {
        final SelectionKey k = selectedKeys[i];
        if (k == null) {          // 判断数组项是否为空
            break;
        }
        selectedKeys[i] = null;   // 直接将数组项设置为空

        // 根据携带对象的类型来选择处理
        final Object a = k.attachment();
        if (a instanceof AbstractNioChannel) {
            processSelectedKey(k, (AbstractNioChannel) a);
        } else {
            @SuppressWarnings("unchecked")
            NioTask<SelectableChannel> task = (NioTask<SelectableChannel>) a;
            processSelectedKey(k, task);
        }

        // 设置需要再次选择。此时将原有的 selectedKeys 数组中当前 i 后面的 selectedKey 清空，然后执行再
        // 次选择，并将 selectedKeys 中的数组进行切换，此时将获得正确的 selectedKeys 数组并且重置 index 循环下标为
        // -1，当开始循环时设置为 0，表示重新从数组最开始执行
        if (needsToSelectAgain) {
            for (;;) {
                i++;
                if (selectedKeys[i] == null) {
                    break;
                }
                selectedKeys[i] = null;
            }
```

```java
            selectAgain();
            selectedKeys = this.selectedKeys.flip();
            i = -1;
        }
    }
}

// 使用普通的选择键集完成处理
private void processSelectedKeysPlain(Set<SelectionKey> selectedKeys) {
    if (selectedKeys.isEmpty()) {                    // 键集为空，则返回
        return;
    }
    Iterator<SelectionKey> i = selectedKeys.iterator();    // 获取迭代器对象
    for (;;) {
        // 获取需要处理的 SelectionKey，并从 publicSelectedKeys 中移除
        final SelectionKey k = i.next();
        final Object a = k.attachment();
        i.remove();

        if (a instanceof AbstractNioChannel) {           // 处理 AbstractNioChannel 的子类
            processSelectedKey(k, (AbstractNioChannel) a);
        } else {                                          // 处理 NioTask
            NioTask<SelectableChannel> task = (NioTask<SelectableChannel>) a;
            processSelectedKey(k, task);
        }
        if (!i.hasNext()) {                              // 迭代器处理完成，结束循环
            break;
        }
        if (needsToSelectAgain) {  // 如果设置了在处理完选择键后需要再次选择，此时需要重新创建迭代器对
象，避免发生 ConcurrentModificationException 异常
            selectAgain();
            selectedKeys = selector.selectedKeys();
            if (selectedKeys.isEmpty()) {
                break;
            } else {
                i = selectedKeys.iterator();
            }
        }
    }
}

// 重新对选择器进行选择
private void selectAgain() {
    needsToSelectAgain = false;                      // 还原标识变量
    try {
        selector.selectNow();
    } catch (Throwable t) {
        logger.warn("Failed to update SelectionKeys.", t);
    }
}
```

7.8.6 processSelectedKey 核心频道方法

该方法用于处理 SelectionKey 和其携带的 AbstractNioChannel 子类对象。流程如下。

（1）获取通道 NioUnsafe 对象。
（2）判断选择键无效并关闭通道。
（3）获取准备好的事件集。
（4）根据事件集的类型：OP_READ、OP_ACCEPT、OP_WRITE、OP_CONNECT，调用 Channel 的相应方法完成处理。

所有处理方法由 Channel 实现。源码如下。

```java
private void processSelectedKey(SelectionKey k, AbstractNioChannel ch) {
    final AbstractNioChannel.NioUnsafe unsafe = ch.unsafe();        // 获取通道 NioUnsafe 对象
    if (!k.isValid()) {                          // 选择键无效（选择键被 cancel、channel 关闭、Selector 被关闭）
        final EventLoop eventLoop;
        try {
            eventLoop = ch.eventLoop();
        } catch (Throwable ignored) {   // Channel 的实现没有使用事件循环组则返回，如果没有实现该方法，
                                        // 则可能抛出异常，这里不做处理，直接返回
            return;
        }
        if (eventLoop != this || eventLoop == null) { // Channel 已经从事件循环组中解除注册，则直接返回
            return;
        }
        unsafe.close(unsafe.voidPromise());      // 关闭仍在事件循环组中注册的通道
        return;
    }

    try {
        int readyOps = k.readyOps();             // 获取准备好的事件集
        if ((readyOps & (SelectionKey.OP_READ | SelectionKey.OP_ACCEPT)) != 0 || readyOps == 0) { // 如
        // 果为 OP_READ 和 OP_ACCEPT 则处理读操作，同样对 readyOps=0 进行判断，是为了避免发生 JDK 的事件集
        // 空返回 Bug（其实这可能是操作系统的 Bug：在部分 Linux2.6 的 kernel 中，poll 和 epoll 对于突然中断的连接 socket
        // 会将返回的 eventSet 事件集合置为 POLLHUP，也可能是 POLLERR，eventSet 事件集合发生了变化，有可能唤
        // 醒 Selector）
            unsafe.read();                       // 处理读操作
            if (!ch.isOpen()) {                  // 通道已关闭，直接返回
                return;
            }
        }
        if ((readyOps & SelectionKey.OP_WRITE) != 0) {    // 执行通道写操作
            ch.unsafe().forceFlush();
        }
        if ((readyOps & SelectionKey.OP_CONNECT) != 0) {  // 执行连接操作
            int ops = k.interestOps();
            ops &= ~SelectionKey.OP_CONNECT;              // 移除选择集中的 OP_CONNECT,如果不移除，
                                                          // 将总是返回该事件集
            k.interestOps(ops);
            unsafe.finishConnect();
        }
    } catch (CancelledKeyException ignored) {
        unsafe.close(unsafe.voidPromise());
    }
}
```

7.8.7 processSelectedKey 核心任务方法

processSelectedKey(SelectionKey k, NioTask task)方法用于处理 SelectionKey 和其携带的 NioTask 对象。该方法主要回调 task 的 channelReady 方法完成对通道的处理。同时，根据处理结果完成进一步的处理。源码如下。

```java
private static void processSelectedKey(SelectionKey k, NioTask<SelectableChannel> task) {
    int state = 0;              // 标志执行状态：1、task 回调方法正常执行；2、执行发生异常
    try {
        task.channelReady(k.channel(), k);  // 回调 task 方法
        state = 1;
    } catch (Exception e) {     // 发生异常，将选择键删除同时回调任务的 channelUnregistered 方法
        k.cancel();
        invokeChannelUnregistered(task, k, e);
        state = 2;
    } finally {
        switch (state) {
            case 0:             // 发生非 Exception 异常，取消选择键同时回调任务的 channelUnregistered 方法
                k.cancel();
                invokeChannelUnregistered(task, k, null);
                break;
            case 1:             // 正常状态，验证选择键是否有效
                if (!k.isValid()) {
                    invokeChannelUnregistered(task, k, null);
                }
                break;
        }
    }
}
```

7.9 ThreadPerChannelEventLoop 类

ThreadPerChannelEventLoop 类用于处理 OIO 模型的 Channel 通道对象，即由 ThreadPerChannelEventLoop 对象处理 Channel。该类较为简单，大部分方法都在 SingleThreadEventLoop 中实现。这里主要初始化父类，同时，在 run 方法中不断从队列中获取任务处理，直到通道被解除注册。源码如下。

```java
public class ThreadPerChannelEventLoop extends SingleThreadEventLoop {

    private final ThreadPerChannelEventLoopGroup parent;
    private Channel ch;

    public ThreadPerChannelEventLoop(ThreadPerChannelEventLoopGroup parent) {
        super(parent, parent.executor, true);
        this.parent = parent;
    }

    // 实现通道的注册。使用父类 super.register 进行注册，同时添加监听器。注册完成时，回调获取当前 Channel
    // 对象。注册失败时，调用 deregister 方法，将自身从 ThreadPerChannelEventLoopGroup 中移除
```

```java
@Override
public ChannelFuture register(ChannelPromise promise) {
    return super.register(promise).addListener(new ChannelFutureListener() {
        @Override
        public void operationComplete(ChannelFuture future) throws Exception {
            if (future.isSuccess()) {
                ch = future.channel();
            } else {
                deregister();
            }
        }
    });
}

// 主要事件函数
@Override
protected void run() {
    for (;;) {
        // 获取任务执行（后文在介绍 OioChannel 时，可以看到该任务的提交）
        Runnable task = takeTask();
        if (task != null) {
            task.run();
            updateLastExecutionTime();
        }
        // 判断通道是否关闭
        Channel ch = this.ch;
        if (isShuttingDown()) {
            if (ch != null) {
                ch.unsafe().close(ch.unsafe().voidPromise());
            }
            if (confirmShutdown()) {
                break;
            }
        } else {
            if (ch != null) {
                // 如果通道已解除注册，执行队列中所有任务后，结束当前事件循环
                if (!ch.isRegistered()) {
                    runAllTasks();
                    deregister();
                }
            }
        }
    }
}

// 将自身从父 ThreadPerChannelEventLoopGroup 中移除
protected void deregister() {
    ch = null; // 不再使用通道对象，释放引用
    parent.activeChildren.remove(this);
    parent.idleChildren.add(this);
}
}
```

第 8 章

Future 与 Promise 原理

8.1 Future 接口

在 JUC 中，Future 代表异步执行操作。使用 Future 可以获取到异步执行的任务状态，也可以等待任务执行完毕，获取任务执行结果。Netty 中，Future 继承自 JUC 的 Future，但扩展了一些方法，主要添加了监听器操作。读者可以想想，这样做有什么好处？

原来的 Future 没有监听器模式，只能通过 get 阻塞等待任务结束。有了监听器，只要向 Future 异步执行任务中添加一个监听器即可。任务执行完毕时，由执行线程完成该回调任务。还可以从中抽取异同部分：响应中断等待、不可响应中断等待、超时等待、不超时等待。源码如下。

```java
public interface Future<V> extends java.util.concurrent.Future<V> {

    /**
     * 判断当前 IO 操作是否执行完毕
     */
    boolean isSuccess();

    /**
     * 判断是否能够通过 cancel(boolean)方法结束任务执行
     */
    boolean isCancellable();

    /**
     * 获取导致 IO 操作失败的异常
     */
    Throwable cause();

    /**
     * 向当前 Future 添加监听器，Future 代表的异步任务执行完毕时，回调该监听器
     * 如果当前任务已完成，则直接执行该监听器
     */
    Future<V> addListener(GenericFutureListener<? extends Future<? super V>> listener);

    /**
     * 添加一组监听器
     */
    Future<V> addListeners(GenericFutureListener<? extends Future<? super V>>... listeners);

    /**
```

```java
 * 移除监听器
 */
Future<V> removeListener(GenericFutureListener<? extends Future<? super V>> listener);

/**
 * 移除一组监听器
 */
Future<V> removeListeners(GenericFutureListener<? extends Future<? super V>>... listeners);

/**
 * 等待当前 Future 代表的异步任务执行完毕。任务执行失败时，把异常信息抛出给调用方。可响应线程中断
 */
Future<V> sync() throws InterruptedException;

/**
 * 等待当前 Future 代表的异步任务执行完毕。任务执行失败时，把异常信息抛出给调用方。不响应线程中断
 */
Future<V> syncUninterruptibly();

/**
 * 等待当前 Future 代表的异步任务执行完毕。但任务执行失败时，不会把异常信息抛出给调用方。可响应线程中断
 */
Future<V> await() throws InterruptedException;

/**
 * 等待当前 Future 代表的异步任务执行完毕。但任务执行失败时，不会把异常信息抛出给调用方。不响应线程中断
 */
Future<V> awaitUninterruptibly();

/**
 * 指定超时时间，等待当前 Future 代表的异步任务执行完毕。但任务执行失败时，不会把异常信息抛出给调用方。可响应线程中断
 */
boolean await(long timeout, TimeUnit unit) throws InterruptedException;

/**
 * 指定超时时间单位为毫秒，为上一个方法的简化版本。笔者认为这里冗余了。等待当前 Future 代表的异步任务执行完毕。但任务执行失败时，不会把异常信息抛出给调用方。可响应线程中断
 */
boolean await(long timeoutMillis) throws InterruptedException;

/**
 * 指定超时时间，等待当前 Future 代表的异步任务执行完毕。但任务执行失败时，不会把异常信息抛出给调用方。不响应线程中断
 */
boolean awaitUninterruptibly(long timeout, TimeUnit unit);

/**
 * 指定超时时间单位为毫秒，为上一个方法的简化版本。笔者认为这里冗余了。等待当前 Future 代表的异步任务执行完毕。但任务执行失败时，不会把异常信息抛出给调用方。不响应线程中断
 */
```

```
boolean awaitUninterruptibly(long timeoutMillis);

/**
 * 立即返回当前 Future 代表的异步任务的结果，如果还未完成，则返回 null
 */
V getNow();
}
```

8.2　GenericFutureListener 与 FutureListener 接口

8.2.1　GenericFutureListener 接口

GenericFutureListener 接口用于表示监听异步执行任务的 Future 监听器。Future 被标识完成时，会立即回调该接口的 operationComplete 方法，在该方法参数中传入了 Future，就可以使用 Future 获取异步任务的执行结果。源码如下。

```
// JDK 的标记接口。表明一个事件监听器
public interface EventListener {
}

// 通用监听器接口。注意，这里的泛型定义 F extends Future<?>，表明只要是 Future 的子接口都可以传入
public interface GenericFutureListener<F extends Future<?>> extends EventListener {

    /**
     * Future 执行完毕时，回调该方法
     */
    void operationComplete(F future) throws Exception;
}
```

8.2.2　FutureListener 接口

FutureListener 接口较为简单，只是将通用事件监听器 GenericFutureListener 的泛型固定为 Future。源码如下。

```
public interface FutureListener<V> extends GenericFutureListener<Future<V>> { }
```

8.3　AbstractFuture 接口

AbstractFuture 抽象类为 Future 的模板类。该接口主要实现了原生 JDK 中 Future 的 get 方法。源码如下。

```
public abstract class AbstractFuture<V> implements Future<V> {

    // 阻塞等待当前 Future 代表的异步任务完成
    @Override
    public V get() throws InterruptedException, ExecutionException {
```

```
        await();                   // 等待任务执行完毕
        Throwable cause = cause();
        if (cause == null) {        // 没有发生异常，直接返回当前结果
            return getNow();
        }
        // 执行异常，包装异常对象返回
        if (cause instanceof CancellationException) {
            throw (CancellationException) cause;
        }
        throw new ExecutionException(cause);
    }

    @Override
    public V get(long timeout, TimeUnit unit) throws InterruptedException, ExecutionException, TimeoutException {
        // 与 get 方法相同，只不过这里进行了超时等待，且在超时时间到达后，抛出 TimeoutException
        if (await(timeout, unit)) {
            Throwable cause = cause();
            if (cause == null) {
                return getNow();
            }
            if (cause instanceof CancellationException) {
                throw (CancellationException) cause;
            }
            throw new ExecutionException(cause);
        }
        throw new TimeoutException();
    }
}
```

8.4　ChannelGroupFuture 接口

ChannelGroupFuture 是特殊的 Future，代表异步的 ChannelGroup 通道组的操作，由代表 IO 操作的 ChannelFuture 组成。在 Netty 中，因为所有 ChannelGroup 中的 IO 操作都是异步的，这意味着调用任何方法都会立即返回。所以需要一个 ChannelGroupFuture 表明这些异步执行的任务，通过它可以得知 IO 操作是否已完成。同时，通过 Future 的定义得知，还可以在调用方等待该 IO 操作时完成。源码如下。

```
public interface ChannelGroupFuture extends Future<Void>, Iterable<ChannelFuture> {

    /**
     * 返回与之关联的 ChannelGroup 通道组对象
     */
    ChannelGroup group();

    /**
     * 返回代表在 ChannelGroup 中的 Channel 通道的单个 I/O 操作的 ChannelFuture 对象
     */
    ChannelFuture find(Channel channel);

    /**
```

```
 * 与ChannelGroupFuture相关的所有I/O操作都成功,且没有任何失败时,返回true
 */
@Override
boolean isSuccess();

/* 以下两个方法从注释上看起来作用差不多,但是实现却是相反的。isPartialSuccess方法在判断时,首先判
断成功的IO操作个数:successCount != 0 && successCount != futures.size()。isPartialFailure则首先判断失败
的IO操作个数:failureCount != 0 && failureCount != futures.size() */

/**
 * 关联的IO操作中有部分执行成功,部分执行失败时返回true
 */
boolean isPartialSuccess();

/**
 * 关联的IO操作中有部分执行失败,部分执行成功回true
 */
boolean isPartialFailure();
}
```

8.5 GenericProgressiveFutureListener 监听器

GenericProgressiveFutureListener 监听器扩展了 GenericFutureListener,通过扩展的 operationProgressed 方法可以获取任务执行的总进度(total)和当前进度(progress)。源码如下。

```
public interface GenericProgressiveFutureListener<F extends ProgressiveFuture<?>> extends
GenericFutureListener<F> {
    void operationProgressed(F future, long progress, long total) throws Exception;
}
```

8.6 ChannelFuture 接口

ChannelFuture 接口表示异步执行的 Channel 通道的 IO 操作。该接口较为简单,扩展的 Future 接口方法是 channel(),用于获取与之关联的 Channel 通道对象。源码如下。

```
public interface ChannelFuture extends Future<Void> {
    /**
     * 与当前Future关联的通道对象
     */
    Channel channel();

    /**
     * 当前Future代表空的Future,返回true。这时,不允许调用以下方法。
     *      addListener(GenericFutureListener)
     *      addListeners(GenericFutureListener[])
     *      await()}
     *      await(long, TimeUnit)} ()
     *      await(long)} ()
```

```
 *      awaitUninterruptibly()
 *      sync()
 *      syncUninterruptibly()
 */
boolean isVoid();
}
```

8.7 Promise 接口

Promise 接口也是特殊的 Future 接口,用于对 Future 代表的异步任务结果操作。这里面的两大核心方法:设置成功,设置失败同时保存异常结果。注意,设置结果不可重复,如果在多线程间设置执行结果,可以使用 tryXXX 方法,因为该方法不会抛出 IllegalStateException 异常。源码如下。

```java
public interface Promise<V> extends Future<V> {
    /**
     * 设置当前 Future 代表的异步任务执行结果完成,同时使用参数 result 设置结果。如果已设置当前结果,则
     * 抛出无效参数异常
     */
    Promise<V> setSuccess(V result);

    /**
     * 尝试设置当前 Future 代表的异步任务执行结果完成,同时使用参数 result 设置结果。如果已设置当前任务,
     * 则返回 false,否则返回 true
     */
    boolean trySuccess(V result);

    /**
     * 设置当前 Future 代表的异步任务执行失败,同时使用参数 cause 设置异常信息。如果已设置当前结果,
     * 则抛出无效参数异常
     */
    Promise<V> setFailure(Throwable cause);

    /**
     * 设置当前 Future 代表的异步任务执行失败,同时使用参数 cause 设置异常信息。如果已设置当前结果,
     * 则返回 false,否则返回 true
     */
    boolean tryFailure(Throwable cause);

    /**
     * 设置 Future 代表的异步任务在执行后,是否可以取消
     */
    boolean setUncancellable();
}
```

8.8 DefaultPromise 接口

DefaultPromise 是 Promise 接口的默认实现,完成了所有的接口实现。其中涉及的方法较多,所以

分拆讲解。

8.8.1 核心变量与构造器

通过变量定义得知。

（1）result 异步执行结果由 RESULT_UPDATER 原子性更新，从而保证线程安全。

（2）EventExecutor executor 事件执行器用于通知监听器。

```java
public class DefaultPromise<V> extends AbstractFuture<V> implements Promise<V> {
    private static final int MAX_LISTENER_STACK_DEPTH = Math.min(8,
SystemPropertyUtil.getInt("io.Netty.defaultPromise.maxListenerStackDepth", 8)); // 最大监听器栈深度，在调用
监听器时使用，用于限制调用

    private static final AtomicReferenceFieldUpdater<DefaultPromise, Object> RESULT_UPDATER; // 原子性
更新结果引用对象（JUC 的基础）

    private static final Signal SUCCESS = Signal.valueOf(DefaultPromise.class, "SUCCESS"); // 执行结果为
null 时，设置的成功对象

    private static final Signal UNCANCELLABLE = Signal.valueOf(DefaultPromise.class,
"UNCANCELLABLE");                        // 标识当前执行的任务不能取消

    private static final CauseHolder CANCELLATION_CAUSE_HOLDER = new
CauseHolder(ThrowableUtil.unknownStackTrace(
        new CancellationException(), DefaultPromise.class, "cancel(...)")); // 调用 cancel 方法时标识取消的占
位对象

    static {
        // 初始化 result 属性原子操作对象
        AtomicReferenceFieldUpdater<DefaultPromise, Object> updater =
            PlatformDependent.newAtomicReferenceFieldUpdater(DefaultPromise.class, "result");
        RESULT_UPDATER = updater == null ?
AtomicReferenceFieldUpdater.newUpdater(DefaultPromise.class,
                                                                            Object.class,
"result") : updater;
    }

    private volatile Object result;           // 代表执行结果
    private final EventExecutor executor;     // 通知监听器的事件执行器对象
    private Object listeners;                 // 监听器对象
    private short waiters;                    // 等待当前任务执行完毕的线程数
    private boolean notifyingListeners;       // 标识当前正在通知监听器

    public DefaultPromise(EventExecutor executor) {
        this.executor = checkNotNull(executor, "executor");
    }

    // 静态内部类。作为异常信息占位对象
    private static final class CauseHolder {
        final Throwable cause;
        CauseHolder(Throwable cause) {
```

```
            this.cause = cause;
        }
    }
}
```

8.8.2　await 核心方法

await 方法用于在外部线程调用，响应中断的等待当前异步执行任务完成。流程如下。

（1）如果任务已完成，则直接返回。
（2）如果线程被中断，则抛出中断异常。
（3）判断死锁（等待自身）。
（4）在当前 Promise 对象上等待执行完毕。

```
// 用于判断死锁
protected void checkDeadLock() {
    EventExecutor e = executor();            // 获取通过构造函数传入的执行器对象
    if (e != null && e.inEventLoop()) {       // 判断当前线程是否为事件执行器中的线程（等待自身）
        throw new BlockingOperationException(toString());
    }
}

// 判断当前任务是否执行完毕
public boolean isDone() {
    return isDone0(result);
}

private static boolean isDone0(Object result) {  // 结果不为空，且不是不能取消的占位符
    return result != null && result != UNCANCELLABLE;
}

public Promise<V> await() throws InterruptedException {
    if (isDone()) {                           // 已完成，直接返回
        return this;
    }
    if (Thread.interrupted()) {               // 线程被中断，抛出中断异常
        throw new InterruptedException(toString());
    }
    checkDeadLock();                          // 判断死锁
    synchronized (this) {                     // 在当前 Promise 对象上等待
        while (!isDone()) {                   // 仍为完成时等待，同时增加等待线程数
            incWaiters();
            try {
                wait();
            } finally {
                decWaiters();
            }
        }
    }
    return this;
}
```

8.8.3 awaitUninterruptibly 核心方法

该方法用于在外部线程调用，不响应中断，等待当前异步执行任务完成。流程如下。

（1）如果任务已完成，直接返回。

（2）判断死锁。

（3）在当前 Promise 对象上等待任务执行完毕。如果等待过程中线程被中断，则设置中断标识 interrupted。

（4）任务执行后，如果设置了 interrupted 中断标志位，则调用 Thread.currentThread().interrupt 方法重新设置线程标志位。

```java
public Promise<V> awaitUninterruptibly() {
    if (isDone()) {                                  // 已完成，直接返回
        return this;
    }
    checkDeadLock();                                 // 判断死锁
    boolean interrupted = false;                     // 标识线程在等待过程中被中断的标志位
    synchronized (this) {                            // 在当前 Promise 对象上等待
        while (!isDone()) {
            incWaiters();
            try {
                wait();
            } catch (InterruptedException e) {       // 与 await()不同，在捕捉中断异常的同时设置了 interrupted 标志位
                interrupted = true;
            } finally {
                decWaiters();
            }
        }
    }
    if (interrupted) {                               // 由于异常捕捉会清除中断标志位，所以这里重新设置线程标志位
        Thread.currentThread().interrupt();
    }
    return this;
}
```

8.8.4 cancel 核心方法

cancel 方法用于取消异步执行的任务，参数 mayInterruptIfRunning 表示当任务正在执行时，是否通过中断停止执行。这里直接通过 compareAndSet 将结果修改为 CANCELLATION_CAUSE_HOLDER。如果修改成功，则唤醒等待任务执行完毕的线程，同时通知监听器。源码如下。

```java
// 如果有等待线程，则唤醒它们
private synchronized void checkNotifyWaiters() {
    if (waiters > 0) {
        notifyAll();
    }
}

public boolean cancel(boolean mayInterruptIfRunning) {
    if (RESULT_UPDATER.compareAndSet(this, null, CANCELLATION_CAUSE_HOLDER)) { // CAS 原子性
```

更新
```
        checkNotifyWaiters();
        notifyListeners();
        return true;
    }
    return false;
}
```

8.8.5 sync 核心方法

sync 方法用于外部线程调用,响应中断异常的任务完成。如果任务发生了异常,把异常抛出。首先,使用 await 方法等待任务执行完毕,然后,判断是否存在导致当前任务执行失败的异常信息。如果存在,则将其抛出。源码如下。

```
// 如果存在导致当前任务执行失败的异常信息,则将其抛出
private void rethrowIfFailed() {
    Throwable cause = cause();
    if (cause == null) {
        return;
    }
    PlatformDependent.throwException(cause);
}

public Promise<V> sync() throws InterruptedException {
    await(); // 调用 await 等待完成
    rethrowIfFailed();
    return this;
}
```

8.8.6 syncUninterruptibly 核心方法

syncUninterruptibly 方法与 sync 相同,只不过这里调用了 awaitUninterruptibly 而不是 await。源码如下。

```
public Promise<V> syncUninterruptibly() {
    awaitUninterruptibly();
    rethrowIfFailed();
    return this;
}
```

8.8.7 setSuccess 核心方法

setSuccess 方法用于设置成功执行的结果值。result 为 null 时把结果设置为 SUCCESS。由源码可知,如何设置任务不允许被取消:设置 UNCANCELLABLE 为 result,同时在执行完毕后将其替换为真实结果。源码如下。

```
// 根据结果值是否为 null,选择是否使用静态变量 SUCCESS 设置执行结果
private boolean setSuccess0(V result) {
    return setValue0(result == null ? SUCCESS : result);
}
```

```
// 共用方法，用于原子性设置 result 结果值
private boolean setValue0(Object objResult) {
    if (RESULT_UPDATER.compareAndSet(this, null, objResult) ||  // 首先尝试原子性将 null 修改为 objResult
        RESULT_UPDATER.compareAndSet(this, UNCANCELLABLE, objResult)) {  // 上一步失败后，有可能
                                                                // 之前设置了 UNCANCELLABLE 标志位，表示不可取消，这里可以尝试将其替换为真实结果
        checkNotifyWaiters();                    // 成功后唤醒等待线程
        return true;
    }
    return false;
}

public Promise<V> setSuccess(V result) {
    if (setSuccess0(result)) {                   // 直接调用 setSuccess0 完成设置，成功后通知监听器
        notifyListeners();
        return this;
    }
    throw new IllegalStateException("complete already: " + this);  // 如果设置失败，则表明任务已经被其他线
                                                                   // 程设置完成，抛出异常
}
```

8.8.8　setFailure 核心方法

setFailure 方法用于设置执行异常但是已完成的任务。参数 cause 用于指明导致任务结束的异常信息。源码如下。

```
// 将异常信息包装为 CauseHolder 对象以完成设置
private boolean setFailure0(Throwable cause) {
    return setValue0(new CauseHolder(checkNotNull(cause, "cause")));
}

public Promise<V> setFailure(Throwable cause) {
    if (setFailure0(cause)) {
        notifyListeners();                       // 完成后通知监听器
        return this;
    }
    throw new IllegalStateException("complete already: " + this, cause);  // 重复设置抛出异常
}
```

8.8.9　trySuccess 核心方法

trySuccess 方法用于尝试设置任务的结果值，但不抛出 IllegalStateException 异常。流程一致，只不过在末尾不抛出异常，而是返回 false。源码如下。

```
public boolean trySuccess(V result) {
    if (setSuccess0(result)) {
        notifyListeners();
        return true;
    }
    return false;
}
```

8.8.10　tryFailure 核心方法

同 trySuccess 一样，设置失败后，返回 false。源码如下。

```
public boolean tryFailure(Throwabie cause) {
    if (setFailure0(cause)) {
        notifyListeners();
        return true;
    }
    return false;
}
```

8.8.11　addListener 核心方法

addListener 方法用于向 Future 中添加监听器对象。流程如下。

（1）判断监听器不为空。

（2）对当前对象上锁保证线程安全，同时调用 addListener0 添加监听器。

（3）添加完成后，如果任务已经执行完毕，通知监听器。

```
private void addListener0(GenericFutureListener<? extends Future<? super V>> listener) {
    if (listeners == null) {                    // 监听器变量为空，将当前对象设置为 listeners 对象
        listeners = listener;
    } else if (listeners instanceof DefaultFutureListeners) { // 监听器对象为默认监听器，直接将监听器添加到监听器数组中
        ((DefaultFutureListeners) listeners).add(listener);
    } else {  // 否则创建默认监听器对象，同时将当前监听器和目前需要添加的监听器放入其中
        listeners = new DefaultFutureListeners((GenericFutureListener<? extends Future<V>>) listeners, listener);
    }
}

public Promise<V> addListener(GenericFutureListener<? extends Future<? super V>> listener) {
    checkNotNull(listener, "listener");         // 监听器不能为空
    synchronized (this) {                        // 保证线程安全
        addListener0(listener);
    }
    if (isDone()) {                              // 添加完成后，如果任务已执行完毕，则通知监听器
        notifyListeners();
    }
    return this;
}
```

上面介绍了 DefaultFutureListeners 类，接下来介绍该类的实现原理。由源码可知，DefaultFutureListeners 类使用数组保存监听器，数组容量不够时，通过翻倍扩容存放新的监听器。源码如下。

```
final class DefaultFutureListeners {
    private GenericFutureListener<? extends Future<?>>[] listeners;  // 监听器数组
    private int size;                                                 // 总的监听器个数
    private int progressiveSize;                                      // 获取任务执行进度的监听器个数

    DefaultFutureListeners(GenericFutureListener<? extends Future<?>> first, GenericFutureListener<?
```

```
        extends Future<?>> second) {
    // 创建监听器数组，同时将两个监听器放入 0 和 1 下标
    listeners = new GenericFutureListener[2];
    listeners[0] = first;
    listeners[1] = second;
    size = 2;                          // 更新监听器长度
    // 根据监听器类型是否为带进度查询的监听器类型来增加 progressiveSize 计数
    if (first instanceof GenericProgressiveFutureListener) {
        progressiveSize ++;
    }
    if (second instanceof GenericProgressiveFutureListener) {
        progressiveSize ++;
    }
}

// 向监听器数组添加监听器
public void add(GenericFutureListener<? extends Future<?>> l) {
    GenericFutureListener<? extends Future<?>>[] listeners = this.listeners;
    final int size = this.size;
    if (size == listeners.length) {    // 开辟 2 倍的数组，添加监听器
        this.listeners = listeners = Arrays.copyOf(listeners, size << 1);
    }
    listeners[size] = l;
    // 增加监听器计数
    this.size = size + 1;
    if (l instanceof GenericProgressiveFutureListener) {
        progressiveSize ++;
    }
}

// 返回监听器列表
public GenericFutureListener<? extends Future<?>>[] listeners() {
    return listeners;
}
}
```

8.8.12　notifyListeners 核心方法

notifyListeners 方法用于通知所有监听器。流程如下。

（1）获取事件执行器对象。

（2）如果当前线程属于事件执行器的执行线程，则判断调用栈深后调用 notifyListenersNow 通知监听器。

（3）如果是外部线程调用，则向事件执行器提交一个 Runnable 任务完成通知。

```
private void notifyListeners() {
    EventExecutor executor = executor();    // 获取事件执行器
    if (executor.inEventLoop()) {           // 当前线程属于事件执行器的执行线程
        final InternalThreadLocalMap threadLocals = InternalThreadLocalMap.get(); // 获取 TL 中的当前线程执
行监听器的栈深度
        final int stackDepth = threadLocals.futureListenerStackDepth();
        if (stackDepth < MAX_LISTENER_STACK_DEPTH) { // 如果当前线程通知监听器的次数小于设置的最
```

大栈深，则增加计数，并通知监听器
                threadLocals.setFutureListenerStackDepth(stackDepth + 1);
            try {
                notifyListenersNow();
            } finally {   // 调用完成后还原之前的栈深
                threadLocals.setFutureListenerStackDepth(stackDepth);
            }
            return;
        }
        // 如果超过最大调用深度，则不会继续调用监听器
    }

    // 如果是外部线程调用，则通过 Runnalbe 的任务交给时间执行器线程处理
    safeExecute(executor, new Runnable() {
        @Override
        public void run() {
            notifyListenersNow();   // 注意这里不会判断栈深，因为不是同一个线程
        }
    });
}
```

## 8.8.13  notifyListenersNow 核心方法

notifyListenersNow 方法用于通知监听器。流程如下。

（1）如果正在通知监听器，或监听为空，则直接返回。否则设置当前正在通知监听器，同时将全局监听器从成员变量中移入局部变量。

（2）根据监听器类型循环来通知监听器，直到成员变量为空。

```
private void notifyListenersNow() {
    Object listeners;
    synchronized (this) {
        // 如果正在通知监听器，或监听为空，则直接返回
        if (notifyingListeners || this.listeners == null) {
            return;
        }
        // 否则设置正在通知监听器，同时摘下全局监听器
        notifyingListeners = true;
        listeners = this.listeners;
        this.listeners = null;
    }
    for (;;) {   // 循环通知监听器
        if (listeners instanceof DefaultFutureListeners) { // 通知 DefaultFutureListeners 监听器
            notifyListeners0((DefaultFutureListeners) listeners);
        } else {   // 否则通知 GenericFutureListener 监听器
            notifyListener0(this, (GenericFutureListener<? extends Future<V>>) listeners);
        }
        synchronized (this) {   // 通知完毕后，判断是否还存在监听器，如果不存在，则直接返回
            if (this.listeners == null) {
                notifyingListeners = false;
                return;
            }
            listeners = this.listeners;
```
```

```
        this.listeners = null;
    }
  }
}
```

8.8.14 notifyListeners0 核心方法

以下两个方法用于执行实际的通知操作。

(1) 在 DefaultFutureListeners 中，遍历监听器列表，通知监听器（使用 notifyListener0(Future future, GenericFutureListener l)方法）。

(2) 在 GenericFutureListener 中，回调监听器的 operationComplete 方法，同时捕捉执行异常。

```
private void notifyListeners0(DefaultFutureListeners listeners) {
    GenericFutureListener<?>[] a = listeners.listeners();
    int size = listeners.size();
    for (int i = 0; i < size; i ++) {
        notifyListener0(this, a[i]);
    }
}

private static void notifyListener0(Future future, GenericFutureListener l) {
    try {
        l.operationComplete(future);
    } catch (Throwable t) {
        logger.warn("An exception was thrown by " + l.getClass().getName() + ".operationComplete()", t);
    }
}
```

8.8.15 notifyProgressiveListeners 核心方法

notifyProgressiveListeners 方法用于通知可以获取任务进度的监听器。参数 progress 表示当前任务执行进度，total 表示当前任务总进度。流程如下。

(1) 获取进度监听器列表，如果没有监听器，则直接返回。

(2) 判断当前线程是否属于事件执行器的执行线程。如果不是，则将流程包装为 Runnable 任务并提交到事件执行器的任务队列；如果是，则直接通知监听器。

(3) 根据当前监听器列表数组类型调用对应方法完成通知（即泛型是否为 ProgressiveFuture），最终通知方法为 notifyProgressiveListeners0。

```
void notifyProgressiveListeners(final long progress, final long total) {
    final Object listeners = progressiveListeners();    // 获取进度监听器列表
    if (listeners == null) {                             // 没有监听器，直接返回
        return;
    }
    final ProgressiveFuture<V> self = (ProgressiveFuture<V>) this;
    EventExecutor executor = executor();
    if (executor.inEventLoop()) {                        // 当前线程属于事件执行器执行线程
        // 根据当前监听器列表数组类型调用对应方法完成通知（泛型是否为 ProgressiveFuture）
        if (listeners instanceof GenericProgressiveFutureListener[]) {
            notifyProgressiveListeners0(
                self, (GenericProgressiveFutureListener<?>[]) listeners, progress, total);  // 通知
```

```
GenericProgressiveFutureListener 数组类型
        } else {
            notifyProgressiveListener0(
                self, (GenericProgressiveFutureListener<ProgressiveFuture<V>>) listeners, progress, total);
// 通知GenericProgressiveFutureListener<ProgressiveFuture<V>> 数组类型
        }
    } else {   // 若线程不属于事件执行器的执行线程,直接将上述流程包装为 Runnable 任务,由事件执行线程
来执行,保证线程安全
        if (listeners instanceof GenericProgressiveFutureListener[]) {
            final GenericProgressiveFutureListener<?>[] array =
                (GenericProgressiveFutureListener<?>[]) listeners;
            safeExecute(executor, new Runnable() {
                @Override
                public void run() {
                    notifyProgressiveListeners0(self, array, progress, total);
                }
            });
        } else {
            final GenericProgressiveFutureListener<ProgressiveFuture<V>> l =
                (GenericProgressiveFutureListener<ProgressiveFuture<V>>) listeners;
            safeExecute(executor, new Runnable() {
                @Override
                public void run() {
                    notifyProgressiveListener0(self, l, progress, total);
                }
            });
        }
    }
}
```

8.8.16 progressiveListeners 核心方法

progressiveListeners 方法用于获取任务进度监听器。流程如下。

(1) 如果当前 Future 的监听器列表对象为空,则直接返回。

(2) 如果监听器列表对象是 DefaultFutureListeners,则将其中的监听器数组中的进度监听器复制到新的数组中并返回。

(3) 如果监听器列表对象是 GenericProgressiveFutureListener 监听器对象,则直接返回。

```
private synchronized Object progressiveListeners() {
    Object listeners = this.listeners;
    if (listeners == null) {                              // 当前监听器列表为空
        return null;
    }

    if (listeners instanceof DefaultFutureListeners) {    // 监听器列表对象为 DefaultFutureListeners,将其中的监
听器数组中的进度监听器复制到新的数组中并返回
        DefaultFutureListeners dfl = (DefaultFutureListeners) listeners;
        int progressiveSize = dfl.progressiveSize();
        switch (progressiveSize) {                        // 优化了进度监听器为 0 和 1 的情况,不需要创建数组
            case 0:
                return null;
            case 1:
```

```
            for (GenericFutureListener<?> l: dfl.listeners()) {
                if (l instanceof GenericProgressiveFutureListener) {
                    return l;
                }
            }
            return null;
        }
        // 创建进度监听器数组，并遍历 DefaultFutureListeners 监听器列表，找到进度监听器并将其放入数组中
        GenericFutureListener<?>[] array = dfl.listeners();
        GenericProgressiveFutureListener<?>[] copy = new
 GenericProgressiveFutureListener[progressiveSize];
        for (int i = 0, j = 0; j < progressiveSize; i ++) {
            GenericFutureListener<?> l = array[i];
            if (l instanceof GenericProgressiveFutureListener) {
                copy[j ++] = (GenericProgressiveFutureListener<?>) l;
            }
        }
        return copy;
    } else if (listeners instanceof GenericProgressiveFutureListener) {
        // 如果是 GenericProgressiveFutureListener 监听器对象，直接返回
        return listeners;
    } else {
        return null;                                           // 其他监听器类型将返回 null
    }
}
```

8.8.17　notifyProgressiveListener 核心方法

以下两个 notifyProgressiveListeners() 方法，根据监听器类型回调进度监听器。由源码可知。对于 GenericProgressiveFutureListener<?>[] listeners 数组，遍历监听器列表调用 notifyProgressiveListener()。对于 GenericProgressiveFutureListener l，直接在 try catch 块中回调 operationProgressed 方法，使用 try catch 块可以避免调用异常导致当前线程退出。源码如下。

```
private static void notifyProgressiveListeners0(
        ProgressiveFuture<?> future, GenericProgressiveFutureListener<?>[] listeners, long progress, long total) {
    for (GenericProgressiveFutureListener<?> l: listeners) {   // 遍历调用
        if (l == null) {
            break;
        }
        notifyProgressiveListener0(future, l, progress, total);
    }
}

private static void notifyProgressiveListener0(
        ProgressiveFuture future, GenericProgressiveFutureListener l, long progress, long total) {
    try {
        l.operationProgressed(future, progress, total);        // 单个监听器，直接调用
    } catch (Throwable t) {
        logger.warn("An exception was thrown by " + l.getClass().getName() + ".operationProgressed()", t);
    }
}
```

8.9　ChannelPromise 接口

ChannelPromise 接口用于扩展 ChannelFuture 和 Promise 接口，提供了对 ChannelFuture 变量的写操作。该接口主要提供了对 Channel 的处理，设置成功的空参数操作，同时支持返回新的 ChannelPromise 对象。源码如下。

```
public interface ChannelPromise extends ChannelFuture, Promise<Void> {
    // 设置通道执行成功，此时不需要提供参数
    ChannelPromise setSuccess();
    boolean trySuccess();

    // 如果接口的实例 isVoid 方法返回 true，则返回新的 ChannelPromise，否则返回该接口本身
    ChannelPromise unvoid();
}
```

8.9.1　DefaultChannelPromise 类

DefaultChannelPromise 类实现了 ChannelPromise 接口，但是它继承自 DefaultPromise 类，所以大部分工作都在 DefaultPromise 父类中实现，该类主要提供对 ChannelFuture 接口和 ChannelPromise 接口的新增方法进行实现，只需了解这几个方法和变量即可。由源码可知。

（1）DefaultChannelPromise 实现了 FlushCheckpoint 接口，该接口用于刷新还原点，后面会详细介绍该接口和其所属类。

（2）定义了所属通道对象和还原点变量。

（3）setSuccess()设置空结果，通过调用父类 DefaultPromise 的 setSuccess(null)完成设置。

（4）isVoid 总是返回 false，所以 unvoid 方法总是返回当前 DefaultChannelPromise 对象。

```
public class DefaultChannelPromise extends DefaultPromise<Void> implements ChannelPromise, FlushCheckpoint {
    private final Channel channel;     // 所属通道对象
    private long checkpoint;           // 当前还原点

    // 通过通道对象创建构造器
    public DefaultChannelPromise(Channel channel) {
        this.channel = channel;
    }

    // 通过通道对象和执行器创建构造器
    public DefaultChannelPromise(Channel channel, EventExecutor executor) {
        super(executor);
        this.channel = channel;
    }

    // 设置返回空结果
    @Override
    public ChannelPromise setSuccess() {
        return setSuccess(null);
```

```java
    }

    @Override
    public ChannelPromise setSuccess(Void result) {
        super.setSuccess(result);
        return this;
    }

    @Override
    public boolean trySuccess() {
        return trySuccess(null);
    }

    // unvoid 总是返回当前 ChannelPromise
    @Override
    public ChannelPromise unvoid() {
        return this;
    }

    // isVoid 总是返回 false
    @Override
    public boolean isVoid() {
        return false;
    }

    // 刷新和设置当前还原点
    @Override
    public long flushCheckpoint() {
        return checkpoint;
    }

    @Override
    public void flushCheckpoint(long checkpoint) {
        this.checkpoint = checkpoint;
    }
}
```

以下源码是 ChannelFlushPromiseNotifier 类。FlushCheckpoint 属于该类的内部接口。ChannelFlushPromiseNotifier 类允许向其中注册 ChannelFuture 实例，一旦写入一定数量的数据并到达判断点，就会通知 ChannelFuture。该类在 Netty 中不用，了解下即可。源码如下。

```java
public final class ChannelFlushPromiseNotifier {
    private long writeCounter;          // 数据写入量
    private final Queue<FlushCheckpoint> flushCheckpoints = new ArrayDeque<FlushCheckpoint>(); // 保存注册的 ChannelFuture 实例
    private final boolean tryNotify;    // 用于控制通知方式。true：调用 ChannelFuture 的 tryFailure 或 trySuccess。false：调用 ChannelFuture 的 setFailure 或 setSuccess

    // 构造器。用于设置 tryNotify 变量
    public ChannelFlushPromiseNotifier(boolean tryNotify) {
        this.tryNotify = tryNotify;
    }
```

```java
// 向当前 ChannelFlushPromiseNotifier 中添加 ChannelPromise 实例,它将在指定的 pendingDataSize 到达后通知
public ChannelFlushPromiseNotifier add(ChannelPromise promise, long pendingDataSize) {
    if (promise == null) {
        throw new NullPointerException("promise");
    }
    if (pendingDataSize < 0) {
        throw new IllegalArgumentException("pendingDataSize must be >= 0 but was " + pendingDataSize);
    }
    long checkpoint = writeCounter + pendingDataSize;
    if (promise instanceof FlushCheckpoint) {     // 如果已经实现了 FlushCheckpoint 接口,设置新的 checkpoint 同时将其添加到队列中
        FlushCheckpoint cp = (FlushCheckpoint) promise;
        cp.flushCheckpoint(checkpoint);
        flushCheckpoints.add(cp);
    } else { // 否则包装到默认的 FlushCheckpoint 接口实现类,然后将其添加到队列中
        flushCheckpoints.add(new DefaultFlushCheckpoint(checkpoint, promise));
    }
    return this;
}

// 通知注册的 ChannelPromise 实例
public ChannelFlushPromiseNotifier notifyPromises() {
    notifyPromises0(null);
    return this;
}

// 实际通知操作
private void notifyPromises0(Throwable cause) {
    if (flushCheckpoints.isEmpty()) {           // 没有需要通知的实例,直接返回
        writeCounter = 0;
        return;
    }
    final long writeCounter = this.writeCounter;
    for (;;) {                                   // 遍历处理注册的 ChannelPromise 实例
        FlushCheckpoint cp = flushCheckpoints.peek();   // 获取队列头部的 ChannelPromise 实例
        if (cp == null) {                        // 如果通知列表中没有任何内容,则重置计数器
            this.writeCounter = 0;
            break;
        }
        if (cp.flushCheckpoint() > writeCounter) {   // ChannelPromise 实例的还原点大于 writeCounter
            if (writeCounter > 0 && flushCheckpoints.size() == 1) { // 队列中只有一个 ChannelPromise 实例
                this.writeCounter = 0;           // 还原写入数量
                cp.flushCheckpoint(cp.flushCheckpoint() - writeCounter); // 重新设置 Checkpoint
            }
            break; // 由于是 FIFO 的队列,所以如果头部的还原点未达到 pendingDataSize,则结束循环
        }
        // 从队列中移除一个 ChannelPromise 实例,根据 tryNotify 变量调用不同方法完成通知
        flushCheckpoints.remove();
        ChannelPromise promise = cp.promise();
        if (cause == null) {                     // 异常为空,调用空成功方法
            if (tryNotify) {
```

```java
                    promise.trySuccess();
                } else {
                    promise.setSuccess();
                }
            } else {
                if (tryNotify) {
                    promise.tryFailure(cause);
                } else {
                    promise.setFailure(cause);
                }
            }
        }
        // 避免写入数量发生溢出
        final long newWriteCounter = this.writeCounter;
        if (newWriteCounter >= 0x8000000000L) {
            // 只有当计数器变得非常大时才重置计数器
            this.writeCounter = 0;
            for (FlushCheckpoint cp: flushCheckpoints) {    // 遍历列表修正 Checkpoint
                cp.flushCheckpoint(cp.flushCheckpoint() - newWriteCounter);
            }
        }
    }
}

// 内部接口，用于操作 Checkpoint
interface FlushCheckpoint {
    long flushCheckpoint();
    void flushCheckpoint(long checkpoint);
    ChannelPromise promise();
}

// 默认 FlushCheckpoint 包装类，传入的 ChannelPromise 没有实现 FlushCheckpoint 接口时，包装到其中
private static class DefaultFlushCheckpoint implements FlushCheckpoint {
    private long checkpoint;
    private final ChannelPromise future;

    DefaultFlushCheckpoint(long checkpoint, ChannelPromise future) {
        this.checkpoint = checkpoint;
        this.future = future;
    }

    @Override
    public long flushCheckpoint() {
        return checkpoint;
    }

    @Override
    public void flushCheckpoint(long checkpoint) {
        this.checkpoint = checkpoint;
    }

    @Override
    public ChannelPromise promise() {
        return future;
    }
}
```

        }
    }

8.9.2 DefaultChannelGroupFuture 类

DefaultChannelGroupFuture 类用于实现 ChannelGroupFuture 接口。ChannelGroupFuture 接口代表了一组 Channel 的异步的 IO 操作。以下介绍默认实现如何管理这些操作。由源码可知。

（1）该类继承自 DefaultPromise，大部分操作均在 DefaultPromise 类中完成。

（2）successCount 变量记录成功数量。

（3）failureCount 变量记录失败数量。

（4）Map<Channel, ChannelFuture> futures 保存管理的通道异步 IO 操作集合。

（5）构建 DefaultChannelGroupFuture 实例时，向管理的所有 ChannelFuture 中添加 childListener 监听器。异步操作完成后，回调其中的 operationComplete 方法，在该方法中完成如下操作。

① 记录成功和失败的 IO 异步任务个数。

② 所有异步 IO 操作均完成时（callSetDone 为 true），根据管理的 IO 异步任务是否存在失败，选择调用 setFailure0 或 setSuccess0。

③ 调用 setFailure0 时，遍历所有 ChannelFuture，找到失败的 ChannelFuture，然后将其和 Channel 封装为 DefaultEntry 元组数组，作为参数调用 setFailure0 方法。

```
final class DefaultChannelGroupFuture extends DefaultPromise<Void> implements ChannelGroupFuture {
    private final ChannelGroup group;                                    // 所属通道组
    private final Map<Channel, ChannelFuture> futures;                   // 通道的异步 IO 操作集合
    private int successCount;                                            // 成功数量
    private int failureCount;                                            // 失败数量
    private final ChannelFutureListener childListener = new ChannelFutureListener() {  // 用于向管理的 ChannelFuture 添加的监听器
        @Override
        public void operationComplete(ChannelFuture future) throws Exception {
            boolean success = future.isSuccess();
            boolean callSetDone;                                         // 表示所有异步 IO 操作均完成
            synchronized (DefaultChannelGroupFuture.this) {              // 记录成功和失败个数
                if (success) {
                    successCount ++;
                } else {
                    failureCount ++;
                }
                callSetDone = successCount + failureCount == futures.size();
                assert successCount + failureCount <= futures.size();
            }
            if (callSetDone) {                        // 所有异步操作完成后，根据结果来调用 setFailure0 或 setSuccess0
                if (failureCount > 0) {               // 有执行失败的通道 IO 任务
                    List<Map.Entry<Channel, Throwable>> failed =
                            new ArrayList<Map.Entry<Channel, Throwable>>(failureCount);
                    for (ChannelFuture f: futures.values()) {// 遍历添加失败的 ChannelFuture 到 failed 列表中
                        if (!f.isSuccess()) {
                            failed.add(new DefaultEntry<Channel, Throwable>(f.channel(), f.cause())); // 包装通道和所属异常信息到 DefaultEntry
                        }
```

```
                setFailure0(new ChannelGroupException(failed));
            } else {
                setSuccess0();
            }
        }
    }
};

// 构造器，用于初始化成员变量
DefaultChannelGroupFuture(ChannelGroup group, Map<Channel, ChannelFuture> futures, EventExecutor executor) {
    super(executor);
    this.group = group;
    this.futures = Collections.unmodifiableMap(futures);
    for (ChannelFuture f: this.futures.values()) {       // 遍历 ChannelFuture，向其中添加监听器
        f.addListener(childListener);
    }
    if (this.futures.isEmpty()) {                         // Future 为空，直接设置成功
        setSuccess0();
    }
}

// 以下两个方法的判断顺序：先判断成功个数，还是先判断失败个数
@Override
public synchronized boolean isPartialSuccess() {
    return successCount != 0 && successCount != futures.size();
}

@Override
public synchronized boolean isPartialFailure() {
    return failureCount != 0 && failureCount != futures.size();
}

// 内部类，用于封装通道和异常对象的元组
private static final class DefaultEntry<K, V> implements Map.Entry<K, V> {
    private final K key;
    private final V value;

    DefaultEntry(K key, V value) {
        this.key = key;
        this.value = value;
    }
}
}
```

小结

事件驱动层篇

本篇内容可以用 Netty NIO 事件驱动框架的示意图进行总结描述。

（1）EventLoopGroup 中管理一组 EventLoop，数量为 CPU 核心数乘以 2。

（2）异步操作由 EventExecutorChooser 对象选择的 EventLoop 执行。

（3）在 EventLoop 中存在三个队列：scheduledTaskQueue、taskQueue、tailTasks，分别用于保存周期性调度任务、普通任务、尾部任务，调用顺序在 runAllTask 方法中定义。

（4）Channel 注册到每个 EventLoop 的 Selector 中，由 NioEventLoop 线程调用 select 方法选择准备好事件的通道对象，对其进行 OPWRITE（写）、OPREAD（读）、OP_ACCEPT（接收客户端连接）操作。

示意图如图 8-1 所示。

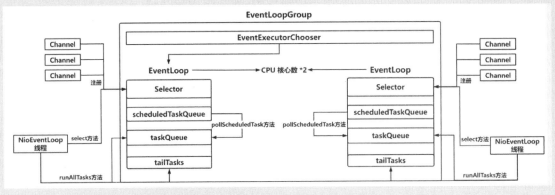

图 8-1　Netty NIO 事件驱动框架示意图

第 3 篇 内存管理层

首先，介绍逻辑推理。

（1）计算机磁盘上存放的是二进制程序：数据和代码。

（2）CPU 只能从内存中提取数据和执行指令，所以要从磁盘中加载程序，放入内存，才能执行。

（3）在用户空间（C 语言编程、Java 编程）申请内存时，可以用 malloc/free（C 语言）、new/delete（C++）、new/gc（Java）等功能函数分配内存。

（4）内存由操作系统管理，分配时要通过操作系统提供的接口（把操作系统当作 B/S 架构中的 S（Server），它提供了 Controller 接口，只要按照协议格式传递数据，获取结果即可）：mmap（内存映射）、brk（推动程序数据段上方的 edata 指针分配内存）。

（5）在程序中直接使用系统调用分配内存时，要进行上下文切换：用户态转为内核态，这将消耗程序性能。所以，在操作系统的分配内存函数中进行包装，可以让进程不直接与 OS 的 Contoller 对接，因为程序需要平台无关化（只需要调用库函数，不需要管不同 OS 的 Contoller 协议格式，因为每个操作系统提供的分配函数都不一样）。

（6）为了避免频繁发生进程上下文切换，库函数在用户空间层面实现了内存池，可以直接从操作系统中分配一大片空间，然后在用户库函数层面提供内存管理操作。

（7）进程分配内存时，就可以直接从内存池中分配，不需要再经历上下文切换到内核态由操作系统来处理，极大地增加了进程的性能。目前常用的 C 内存管理函数库为 ptmalloc（glibc 自带）、tcmalloc（谷歌）、jemalloc（facebook）。

（8）Java 语言只能通过 Java 的编程函数进行操作，不能直接使用 C 的内存管理函数，这就意味着如果需要保证程序的分配性能，只能在 Java 语言层面自己实现一个内存池。

综上所述，Netty 为了保证高性能，在提供事件循环组的异步处理基础上，自身实现了一套内存管理系统。通过这套内存系统可以极大地提升 Netty 的性能，因为除了首次从 JVM 中分配，其他时间均在 Java 层面进行内存分配和回收。本篇将详细介绍 Netty 对于原生的 ByteBuffer 类的改造，如何设计 Netty 自己的 ByteBuf 类，以及非池化内存管理和池化内存管理的原理。

第 9 章

ByteBuf 与衍生类原理

9.1 ByteBuf 原理

首先，回顾原生 JDK 的 ByteBuffer 的设计。

（1）在 Java 中内存类型分为两类：堆内内存、堆外内存。

（2）使用堆内内存时，可以新建一个 byte 数组，在 ByteBuffer 中就需要 byte[] hb。

（3）使用堆外内存时，可以通过 Unsafe 内分配一个 byte 数组，在 ByteBuffer 中要使用 long address 保存分配的内存首地址。

（4）对于 byte 数组的操作，读或写需要使用下标表明当前读写操作处理的位置。

（5）进行读或写时，需要一个限制位 limit。该限制位用于表明读或写的结束下标，这时可以约束下标不会超过 limit。

（6）进行读或写时，需要一个限制位 capacity，该限制位用于表明数组的容量，可以约束 limit 不会超过 capacity。

（7）需要重复进行处理时，可以通过标志位，对当前下标进行标记，然后在需要重复读取时将下标还原到标记位的位置。

```
private int mark = -1;      // 用于标记当前处理位置，需要重复处理时，可以通过 reset 将 position 还原到 mark
                            标记的下标处
private int position = 0;   // 用于表示当前处理的 index 下标
private int limit;          // 用于表示当前读或写的限制下标
private int capacity;       // 用于表示当前可以容纳数据的容量

long address;               // 使用堆外内存时使用，保存堆外内存地址
final byte[] hb;            // 使用堆内内存时使用，保存存放数据的数组
```

在简单回顾原生的 ByteBuffer 设计后，不难看出，这种设计异常复杂，用户需要感知这四个标记位，学习成本较高。所以 Netty 对 ByteBuffer 进行了重新设计，提供了 ByteBuf 类。

（1）该类代表了一个随机和顺序的字节数组。

（2）封装了对于堆内内存 byte[]数组和原生 JDK ByteBuffer（堆外内存）的操作。

（3）提供了两个索引下标来支持顺序读和写操作：readerIndex（读操作下标）、writerIndex（写操作下标）。

（4）使用 capacity 变量和 maxCapacity 作为容量限制。

（5）这两个下标满足 0 <= readerIndex <= writerIndex <= capacity。

（6）ByteBuf 实际存放信息的对象是 byte[]字节数组时，可直接通过 array 方法获取该数组。

（7）ByteBuf 实际存放信息的对象是 ByteBuffer 对象时，可直接通过 nioBuffer 方法获取该对象。

9.1.1 构造器与核心变量

```java
public abstract class ByteBuf implements ReferenceCounted, Comparable<ByteBuf> {
    public abstract byte[] array();                              // 包装对象是 byte 数组时，获取该数组本身
    public abstract ByteBuffer nioBuffer();                      // 包装对象是 ByteBuffer 对象时，获取该对象本身
    public abstract int writableBytes();                         // 获取当前可写的字节数：this.capacity - this.writerIndex
    public abstract int readableBytes();                         // 获取当前可读的字节数：this.writerIndex - this.readerIndex
    public abstract int writerIndex();                           // 获取当前写下标
    public abstract ByteBuf writerIndex(int writerIndex);        // 设置当前写下标
    public abstract int readerIndex();                           // 获取当前读下标
    public abstract ByteBuf readerIndex(int readerIndex);        // 获取当前写下标
    public abstract int capacity();                              // 获取当前容量
    public abstract ByteBuf capacity(int newCapacity);           // 设置当前容量
    public abstract int maxCapacity();                           // 获取最大容量
    public abstract ByteBuf discardReadBytes();                  // 丢弃已经读取的字节
}
```

9.1.2 ReferenceCounted 接口

ByteBuf 分配时会从内存池中获取，不使用时需要把内存放回内存池，这时可以使用引用计数法标识当前 ByteBuf 是否已经被释放，如果引用计数为 0，就可以将其放回内存池。新创建 ReferenceCounted 实例时，它的 reference count 引用计数值被初始化为 1；使用 retain 方法时，会增加一个 reference count 引用计数；调用 release 方法时，会减少一个 reference count 引用计数。如果引用计数变为 0，会将其表示的空间进行释放。源码如下。

```java
public interface ReferenceCounted {
    /**
     * 返回当前引用计数
     */
    int refCnt();

    /**
     * 当前引用计数加 1
     */
    ReferenceCounted retain();

    /**
     * 当前引用计数加 increment
     */
    ReferenceCounted retain(int increment);

    /**
     * 记录该结点当前的访问位置，用于调试。如果确定该对象为内存泄露对象，则将该操作记录的信息提供给
     * 用户
     */
    ReferenceCounted touch();

    /**
```

```
     * 类似touch()，只不过这里记录了hint对象，该对象同样用于记录信息，用于在发生内存泄漏时给用户提示
     */
    ReferenceCounted touch(Object hint);

    /**
     * 当前引用计数减1
     */
    boolean release();

    /**
     * 当前引用计数减decrement
     */
    boolean release(int decrement);
}
```

9.2 AbstractByteBuf 原理

AbstractByteBuf 类是 ByteBuf 抽象类的骨架实现，将其中的大量模板方法完成实现。由于方法较多，本书不再赘述，只讲解了一些难点的方法。例如 readerIndex()、writerIndex()等直接返回变量的方法均省略不提。

9.2.1 核心变量与构造器

由源码可知，AbstractByteBuf定义了读写下标变量，同时定义了用于回退操作的mark变量，最大长度变量。在当前骨架类中保存了判断内存泄漏的ResourceLeakDetector对象。源码如下。

```
public abstract class AbstractByteBuf extends ByteBuf {
    static final ResourceLeakDetector<ByteBuf> leakDetector =
        ResourceLeakDetectorFactory.instance().newResourceLeakDetector(ByteBuf.class); // 全局静态常量
对象，用于监测内存泄漏，稍后讲解
    int readerIndex;                          // 读下标
    int writerIndex;                          // 写下标
    private int markedReaderIndex;            // 标记读下标，方便回退
    private int markedWriterIndex;            // 标记写下标，方便回退
    private int maxCapacity;                  // 数组容量

    // 构造器。用于初始化 maxCapacity
    protected AbstractByteBuf(int maxCapacity) {
        if (maxCapacity < 0) {
            throw new IllegalArgumentException("maxCapacity: " + maxCapacity + " (expected: >= 0)");
        }
        this.maxCapacity = maxCapacity;
    }
}
```

9.2.2 writeByte 核心方法

writeByte 方法用于向 ByteBuf 中放入 value 值，由源码可知，实际放入操作由子类来实现。这是为什么呢？因为当前骨架模板类并不知道子类保存数据的地方，不能确定是在 byte 数组，还是在

ByteBuffer 的堆外内存。源码如下。

```java
public ByteBuf writeByte(int value) {
    ensureAccessible();
    ensureWritable0(1);                             // 确保当前有足够的空间可以写入 value 值
    _setByte(writerIndex++, value);                 // 完成实际写入并增加写下标
    return this;
}

private void ensureWritable0(int minWritableBytes) {
    if (minWritableBytes <= writableBytes()) {      // 需要写入的数量小于或等于可以写入的下标值, 直接
可写, 为覆盖写入
        return;
    }

    if (minWritableBytes > maxCapacity - writerIndex) {  // 写入数量大于最大剩余的写入空间, 则抛出异常
        throw new IndexOutOfBoundsException(String.format(
            "writerIndex(%d) + minWritableBytes(%d) exceeds maxCapacity(%d): %s",
            writerIndex, minWritableBytes, maxCapacity, this));
    }

    // 否则需要调整 capacity

    // 将当前写入后的新容量对齐到 2 的倍数
    int newCapacity = alloc().calculateNewCapacity(writerIndex + minWritableBytes, maxCapacity);

    // 设置容量为放入后的新容量
    capacity(newCapacity);
}
protected abstract void _setByte(int index, int value);    // 实际写入操作由子类完成
```

9.2.3 writeBytes 核心方法

上述方法用于写入单个值, writeBytes 方法用于写入多个值。入参 src 表明需要写入的字节数组, 入参 srcIndex 表明开始写入的下标, 入参 length 表示写入的个数。同样, 写入操作由子类完成。源码如下。

```java
public ByteBuf writeBytes(byte[] src, int srcIndex, int length) {
    ensureAccessible();
    ensureWritable(length);
    setBytes(writerIndex, src, srcIndex, length);   // 完成实际写入, 传入 writerIndex, 表明从 writerIndex
下标处开始写入
    writerIndex += length;
    return this;
}
public abstract ByteBuf setBytes(int index, byte[] src, int srcIndex, int length);    // 由子类完成
```

9.2.4 readByte 核心方法

readByte 方法用于从 ByteBuf 中读取一个字节。实际读取操作由子类完成。源码如下。

```java
public byte readByte() {
    checkReadableBytes0(1);
```

```java
        int i = readerIndex;
        byte b = _getByte(i);
        readerIndex = i + 1;                                    // 增加读下标
        return b;
}

private void checkReadableBytes0(int minimumReadableBytes) {
    ensureAccessible();
    if (readerIndex > writerIndex - minimumReadableBytes) {    // 读下标大于写下标,若此时没有可读数据,抛出异常
        throw new IndexOutOfBoundsException(String.format(
            "readerIndex(%d) + length(%d) exceeds writerIndex(%d): %s",
            readerIndex, minimumReadableBytes, writerIndex, this));
    }
}

protected abstract byte _getByte(int index);                    // 由于不知道数据在哪,实际读操作由子类完成
```

9.2.5　readBytes 核心方法

readBytes 方法用于批量读取数据。dst 用于存放读取数据的数组,dstIndex 表明存放下标,length 表明存放长度。源码如下。

```java
public ByteBuf readBytes(byte[] dst, int dstIndex, int length) {
    checkReadableBytes(length);
    getBytes(readerIndex, dst, dstIndex, length);               // 传入 readerIndex 表明从该下标开始读取
    readerIndex += length;                                      // 增加读下标的值
    return this;
}
public abstract ByteBuf getBytes(int index, byte[] dst, int dstIndex, int length);  // 实际读取操作由子类完成
```

9.2.6　writeZero 核心方法

writeZero 方法用于向数组中写入 0。length 表明写入长度。由源码可知。

（1）为了避免循环的次数,将写入的长度分为 long 8B 长整型写以及 int 4B 整型写。

（2）根据 length 计算写入的 long 个数、byte 个数。

（3）写入 long 0。

（4）根据剩余的 byte 数写入 int 和剩余字节数。

```java
public ByteBuf writeZero(int length) {
    if (length == 0) {                                          // 写入长度为 0,直接返回
        return this;
    }
    ensureWritable(length);
    int wIndex = writerIndex;
    checkIndex(wIndex, length);                                 // 判断写入后是否发生越界
    int nLong = length >>> 3;                                   // 计算写入长度为 long 个数。除以 8B
    int nBytes = length & 7;                                    // 计算写入长度为 byte 的个数。截断三位,即 111,考虑下 2^3 等于 8
    for (int i = nLong; i > 0; i --) {                          // 写入 long
        _setLong(wIndex, 0);
        wIndex += 8;
```

```
    }
    if (nBytes == 4) {                                          // 写入整型
        _setInt(wIndex, 0);
        wIndex += 4;
    } else if (nBytes < 4) {                                    // 写入小于 4B 的长度
        for (int i = nBytes; i > 0; i --) {
            _setByte(wIndex, (byte) 0);
            wIndex++;
        }
    } else {                        // nBytes 大于 4B 小于 8B（因为 8B 为 long 长整型将计算在 nLong 中写入）
        _setInt(wIndex, 0);
        wIndex += 4;                                            // 先写入一个整型 4B 的 0
        for (int i = nBytes - 4; i > 0; i --) {
            _setByte(wIndex, (byte) 0);                         // 写入剩余 byte 数
            wIndex++;
        }
    }
    writerIndex = wIndex;                                       // 更新写下标
    return this;
}
```

9.2.7 discardReadBytes 核心方法

discardReadBytes 方法用于丢弃已经读取的字节。由源码可知，所谓丢弃，就是将未读的数据挪到下标为 0 处，然后更新 writerIndex、readerIndex 和标记下标。源码如下。

```
public ByteBuf discardReadBytes() {
    ensureAccessible();
    if (readerIndex == 0) {                                     // 没有读取过数据，直接返回
        return this;
    }
    if (readerIndex != writerIndex) {
        setBytes(0, this, readerIndex, writerIndex - readerIndex);  // 否则将 writerIndex - readerIndex 后面未
读的值挪到数组的首部
        writerIndex -= readerIndex;                             // 更新写下标
        adjustMarkers(readerIndex);                             // 更新标记位
        readerIndex = 0;                                        // 重置写下标
    } else {                        // 读下标等于写下标，表明全部读完，重置 writerIndex 和 readerIndex 即可
        adjustMarkers(readerIndex);
        writerIndex = readerIndex = 0;
    }
    return this;
}
```

9.3 AbstractReferenceCountedByteBuf 类

AbstractReferenceCountedByteBuf 类用于定义引用计数 ByteBuf 的模板类。实现了 ReferenceCounted 接口的方法。由于实现较为简单，笔者这里将其全部放到一个类中讲解。由源码可知，其实就是对 refCnt 引用计数变量进行原子性操作，该值初始化为 1，变为 0 后会调用由子类实现的 deallocate 方法完成对

占用内存的释放。源码如下。

```java
public abstract class AbstractReferenceCountedByteBuf extends AbstractByteBuf {

    private static final AtomicIntegerFieldUpdater<AbstractReferenceCountedByteBuf> refCntUpdater;  // 原子性更新 refCnt 变量

    static {                                                    // 创建原子性更新 refCnt 变量的属性更新对象
        AtomicIntegerFieldUpdater<AbstractReferenceCountedByteBuf> updater =
                PlatformDependent.newAtomicIntegerFieldUpdater(AbstractReferenceCountedByteBuf.class, "refCnt");
        if (updater == null) {
            updater = AtomicIntegerFieldUpdater.newUpdater(AbstractReferenceCountedByteBuf.class, "refCnt");
        }
        refCntUpdater = updater;
    }

    private volatile int refCnt = 1;                            // 引用计数，初始化为 1

    // 构造器。更新父类 maxCapacity
    protected AbstractReferenceCountedByteBuf(int maxCapacity) {
        super(maxCapacity);
    }

    // 获取引用计数
    @Override
    public int refCnt() {
        return refCnt;
    }

    /**
     * 更新引用计数，非线程安全，由子类调用
     */
    protected final void setRefCnt(int refCnt) {
        this.refCnt = refCnt;
    }

    @Override
    public ByteBuf retain() {                                   // 对引用计数加 1
        for (;;) {                                              // 循环更新
            int refCnt = this.refCnt;
            if (refCnt == 0) {                                  // 计数值为 0 表明已经回收
                throw new IllegalReferenceCountException(0, 1);
            }
            if (refCnt == Integer.MAX_VALUE) {                  // 达到最大值
                throw new IllegalReferenceCountException(Integer.MAX_VALUE, 1);
            }
            if (refCntUpdater.compareAndSet(this, refCnt, refCnt + 1)) {   // 原子性更新引用计数值
                break;
            }
        }
        return this;
```

```java
}
// 更新引用计数值为 increment
@Override
public ByteBuf retain(int increment) {
    if (increment <= 0) {                                                    // 已经被回收
        throw new IllegalArgumentException("increment: " + increment + " (expected: > 0)");
    }
    for (;;) {                                                               // 同 retain()
        int refCnt = this.refCnt;
        if (refCnt == 0) {
            throw new IllegalReferenceCountException(0, increment);
        }
        if (refCnt > Integer.MAX_VALUE - increment) {
            throw new IllegalReferenceCountException(refCnt, increment);
        }
        if (refCntUpdater.compareAndSet(this, refCnt, refCnt + increment)) { // 这里增量为 increment
            break;
        }
    }
    return this;
}

// 默认实现返回当前对象
@Override
public ByteBuf touch() {
    return this;
}

// 默认实现返回当前对象
@Override
public ByteBuf touch(Object hint) {
    return this;
}

// 对引用计数-1
@Override
public boolean release() {
    for (;;) {                                                               // 循环处理
        int refCnt = this.refCnt;
        if (refCnt == 0) {                                                   // 已经被回收
            throw new IllegalReferenceCountException(0, -1);
        }

        if (refCntUpdater.compareAndSet(this, refCnt, refCnt - 1)) {         // 原子性更新
            if (refCnt == 1) {   // 如果更新前的值为 1，此时更新成功，值为 0，则要释放当前 ByteBuf 占用的内存
                deallocate();
                return true;
            }
            return false;
        }
    }
}
```

```java
}
// 对引用计数减 decrement
@Override
public boolean release(int decrement) {
    if (decrement <= 0) {
        throw new IllegalArgumentException("decrement: " + decrement + " (expected: > 0)");
    }

    for (;;) { // 同 release()
        int refCnt = this.refCnt;
        if (refCnt < decrement) {
            throw new IllegalReferenceCountException(refCnt, -decrement);
        }

        if (refCntUpdater.compareAndSet(this, refCnt, refCnt - decrement)) {   // 这里原子性减 decrement
            if (refCnt == decrement) {
                deallocate();
                return true;
            }
            return false;
        }
    }
}

/**
 * 引用计数为 0 时调用，由子类实现内存释放。该方法只会调用一次
 */
protected abstract void deallocate();
}
```

9.4 UnpooledHeapByteBuf 类

UnpooledHeapByteBuf 类用于实现大端序存储在 Java 堆内存的 byte[] 数组实现，从名字可以看出，ByteBuf 不提供内存池的功能。

9.4.1 核心变量与构造器

由源码可知，该类保存了 byte[] array 数组，该数组用于存储数据，ByteBuf 则提供读写下标的支持。源码如下。

```java
public class UnpooledHeapByteBuf extends AbstractReferenceCountedByteBuf {
    private final ByteBufAllocator alloc;   // 内存分配器。表明当前 ByteBuf 所属分配器
    byte[] array;                           // 保存数据的内存数组
    private ByteBuffer tmpNioBuf;           // 支持 internalNioBuffer 方法，将 byte[] array 包装为 ByteBuffer 返回

    // 构造器用于初始化 UnpooledHeapByteBuf 成员变量
    private UnpooledHeapByteBuf(
        ByteBufAllocator alloc, byte[] initialArray, int readerIndex, int writerIndex, int maxCapacity) {
```

```
        super(maxCapacity);                    // 初始化父类
    // 校验参数
    if (alloc == null) {
        throw new NullPointerException("alloc");
    }
    if (initialArray == null) {
        throw new NullPointerException("initialArray");
    }
    if (initialArray.length > maxCapacity) {
        throw new IllegalArgumentException(String.format(
            "initialCapacity(%d) > maxCapacity(%d)", initialArray.length, maxCapacity));
    }
    this.alloc = alloc;
    setArray(initialArray);                    // 初始化 byte[] array 数组
    setIndex(readerIndex, writerIndex);        // 初始化读写下标
    }
}
```

9.4.2 getByte 核心方法

getByte 方法用于从数组中读取下标为 index 的值，直接通过 memory[index]读取对应下标值返回。源码如下。

```
protected byte _getByte(int index) {
    return HeapByteBufUtil.getByte(array, index);
}

static byte getByte(byte[] memory, int index) {
    return memory[index];
}
```

9.4.3 getBytes 核心方法

getBytes 方法用于从数组中读取数据放入 dst 数组。由于数据就存放在 byte[]数组中，所以这里直接通过 System.arraycopy 方法将 array 数组中的数据复制到 byte[] dst 即可。源码如下。

```
public ByteBuf getBytes(int index, byte[] dst, int dstIndex, int length) {
    checkDstIndex(index, length, dstIndex, dst.length);
    System.arraycopy(array, index, dst, dstIndex, length);
    return this;
}
```

9.4.4 setByte 核心方法

setByte 方法用于在数组下标为 index 处放入 value 值。直接通过 memory[index] = (byte) value 设置对应下标值返回。源码如下。

```
protected void _setByte(int index, int value) {
    HeapByteBufUtil.setByte(array, index, value);
}

static void setByte(byte[] memory, int index, int value) {
```

```
    memory[index] = (byte) value;
}
```

9.4.5　setBytes 核心方法

setBytes 方法用于从 src 数组中读取数据放入 array 数组。由于数据就存放在 byte[]数组中,所以直接通过 System.arraycopy 方法将 src 数组中的数据复制到 byte[] array 即可。源码如下。

```
public ByteBuf setBytes(int index, byte[] src, int srcIndex, int length) {
    checkSrcIndex(index, length, srcIndex, src.length);
    System.arraycopy(src, srcIndex, array, index, length);
    return this;
}
```

9.4.6　setShort 核心方法

本来该方法的底层就是调用 setByte 完成操作,但是由于该类使用大端序进行存储,这里通过 short 2B 的存储给读者展示大端序的存储过程,读者在使用 long 4byte 时,可以举一反三。

```
protected void _setShort(int index, int value) {
    HeapByteBufUtil.setShort(array, index, value);
}

// 大端存储
static void setShort(byte[] memory, int index, int value) {
    memory[index] = (byte) (value >>> 8);               // 低八位在后
    memory[index + 1] = (byte) value;                    // 高八位在前
}
```

9.4.7　capacity 核心方法

capacity 方法用于调整当前 capacity 的容量,ByteBuf 用 maxCapacity 和 capacity 表明当前容量,capacity 可以当作 ByteBuffer 的 limit。由源码可知,capacity 只是一个逻辑概念,相当于当前 array 数组的 length,设置新的 newCapacity 大于 oldCapacity 时,就进行扩容,否则进行缩容。源码如下。

```
public ByteBuf capacity(int newCapacity) {
    ensureAccessible();
    if (newCapacity < 0 || newCapacity > maxCapacity()) {   // 新容量小于 0,或大于最大容量
        throw new IllegalArgumentException("newCapacity: " + newCapacity);
    }
    int oldCapacity = array.length;
    if (newCapacity > oldCapacity) {                         // 新容量大于旧容量,进行数据复制
        byte[] newArray = new byte[newCapacity];
        System.arraycopy(array, 0, newArray, 0, array.length);
        setArray(newArray);
    } else if (newCapacity < oldCapacity) {                  // 新容量小于旧容量,进行数据裁切
        byte[] newArray = new byte[newCapacity];
        int readerIndex = readerIndex();
        if (readerIndex < newCapacity) {                     // 读下标小于新容量,根据写下标裁切
            int writerIndex = writerIndex();
            if (writerIndex > newCapacity) {
```

```
                    writerIndex(writerIndex = newCapacity);
                }
                System.arraycopy(array, readerIndex, newArray, readerIndex, writerIndex - readerIndex);
            } else {                                                    // 否则直接将读写下标设置为新容量位置
                setIndex(newCapacity, newCapacity);
            }
            setArray(newArray);
        }
        return this;
    }
```

9.4.8 nioBuffer 核心方法

nioBuffer 方法用于获取 JDK 的原生 ByteBuffer 对象。由源码可知，该方法将内部的数组包装放入 ByteBuffer 并返回。源码如下。

```
public ByteBuffer nioBuffer(int index, int length) {
    ensureAccessible();
    return ByteBuffer.wrap(array, index, length).slice();   // 这里的 slice 方法，读者可以参考 HeapByteBuffer
源码的构造器，其实这里是对 pos 和 limit 重置
}
```

9.5 UnpooledUnsafeHeapByteBuf 类

UnpooledUnsafeHeapByteBuf 类用于在拥有 Unsafe 类的 JDK 版本中，通过 Unsafe 的方式获取 byte 数组中的值。由于该类不提供新的变量，仅是获取数组的方法不一样而已，笔者这里只做简要分析。

由源码可知，最终通过 PlatformDependent0 类的 Unsafe 类完成。读者这里不用深究 PlatformDependent 类和 PlatformDependent0 类，因为这两个类仅是封装对 Unsafe 的操作（因为不同的 JDK 版本对于 JDK 原生类库进行了改造，所以需要根据版本运行时动态调整一些方法的调用和属性获取）。注意，一般来说，HeapByteBuf 直接通过数组 byte[N]的方式访问，PlatformDependent 类和 PlatformDependent0 类主要用于服务堆外内存。源码如下。

```
final class UnpooledUnsafeHeapByteBuf extends UnpooledHeapByteBuf {
    // 构造器初始父类
    UnpooledUnsafeHeapByteBuf(ByteBufAllocator alloc, int initialCapacity, int maxCapacity) {
        super(alloc, initialCapacity, maxCapacity);
    }

    // 获取 byte 中值通过 UnsafeByteBufUtil
    protected byte _getByte(int index) {
        return UnsafeByteBufUtil.getByte(array, index);
    }

    // 同样通过 UnsafeByteBufUtil
    protected void _setByte(int index, int value) {
        UnsafeByteBufUtil.setByte(array, index, value);
    }
```

```java
    public ByteBuf writeZero(int length) {
        if (PlatformDependent.javaVersion() >= 7) {  // 只支持 JDK 7 以上版本
            ensureWritable(length);
            int wIndex = writerIndex;
            _setZero(wIndex, length);                // 从写下标处写入
            writerIndex = wIndex + length;           // 修正写入下标索引
            return this;
        }
        return super.writeZero(length);              // JDK 7 以下通过父类来写入
    }

    // 最终通过 UnsafeByteBufUtil 来写入
    private void _setZero(int index, int length) {
        checkIndex(index, length);
        UnsafeByteBufUtil.setZero(array, index, length);
    }
}

// 该工具类用于包装 Unsafe 的操作。可使用该类的方法直接调用 PlatformDependent 类完成操作
final class UnsafeByteBufUtil {

    static byte getByte(byte[] array, int index) {
        return PlatformDependent.getByte(array, index);
    }

    static void setByte(byte[] array, int index, int value) {
        PlatformDependent.putByte(array, index, (byte) value);
    }

    static void setZero(byte[] array, int index, int length) {
        if (length == 0) {
            return;
        }
        PlatformDependent.setMemory(array, index, length, ZERO);
    }
}

// 该类会探测当前运行时的环境，同时设置相关变量，提供平台相关的功能支持。这些方法调用
// PlatformDependent0 类完成调用
public final class PlatformDependent {
    public static byte getByte(byte[] data, int index) {
        return PlatformDependent0.getByte(data, index);
    }
    public static void putByte(byte[] data, int index, byte value) {
        PlatformDependent0.putByte(data, index, value);
    }
    public static void setMemory(byte[] dst, int dstIndex, long bytes, byte value) {
        PlatformDependent0.setMemory(dst, BYTE_ARRAY_BASE_OFFSET + dstIndex, bytes, value);
    }
}

// 该类用于封装 sun.misc.* 包中的操作。最终通过调用 Unsafe 类完成操作
final class PlatformDependent0 {
    static final Unsafe UNSAFE;
```

```java
    static byte getByte(byte[] data, int index) {
        return UNSAFE.getByte(data, BYTE_ARRAY_BASE_OFFSET + index);
    }
    static void putByte(byte[] data, int index, byte value) {
        UNSAFE.putByte(data, BYTE_ARRAY_BASE_OFFSET + index, value);
    }
    static void setMemory(Object o, long offset, long bytes, byte value) {
        UNSAFE.setMemory(o, offset, bytes, value);
    }
}
```

9.6 UnpooledDirectByteBuf 原理

该类用于封装一个非池化的基于原生 JDK 的 ByteBuffer(准确来说是 DirectByteBuffer)的 ByteBuf 对象。通常使用 Unpooled.directBuffer(int)分配该缓冲区对象。

9.6.1 核心变量与构造器

从源码中该类将在构造器分配 JDK 原生的 DirectByteBuffer 对象。源码如下。

```java
public class UnpooledDirectByteBuf extends AbstractReferenceCountedByteBuf {
    private final ByteBufAllocator alloc;    // 所属内存分配器
    private ByteBuffer buffer;               // 存放数据的 DirectByteBuffer 对象
    private ByteBuffer tmpNioBuf;            // 用于支持 internalNioBuffer 的临时缓冲区
    private int capacity;    // 当前容量(当前存储数据的不再是堆内存的 byte 数组, 所以需要保存容量)
    private boolean doNotFree;               // 设置当前在设置新的 Buffer 后是否释放旧的 Buffer

    // 构造器用于初始化父类和当前成员
    protected UnpooledDirectByteBuf(ByteBufAllocator alloc, int initialCapacity, int maxCapacity) {
        super(maxCapacity);
        // 校验参数
        if (alloc == null) {
            throw new NullPointerException("alloc");
        }
        if (initialCapacity < 0) {
            throw new IllegalArgumentException("initialCapacity: " + initialCapacity);
        }
        if (maxCapacity < 0) {
            throw new IllegalArgumentException("maxCapacity: " + maxCapacity);
        }
        if (initialCapacity > maxCapacity) {
            throw new IllegalArgumentException(String.format(
                "initialCapacity(%d) > maxCapacity(%d)", initialCapacity, maxCapacity));
        }
        this.alloc = alloc;
        setByteBuffer(ByteBuffer.allocateDirect(initialCapacity));    // 分配 JDK 原生的直接缓冲区
    }

    // 初始化成员变量
    private void setByteBuffer(ByteBuffer buffer) {
```

```
        ByteBuffer oldBuffer = this.buffer;
        if (oldBuffer != null) {           // 之前设置过 ByteBuffer
            if (doNotFree) {    // 当前不允许释放旧的 ByteBuffer，这时可能在其他地方使用该旧的缓冲区
                doNotFree = false;
            } else {                       // 否则释放旧的缓冲区
                freeDirect(oldBuffer);
            }
        }
        // 初始化成员变量
        this.buffer = buffer;
        tmpNioBuf = null;
        capacity = buffer.remaining();     // 容量等于当前可写数量（因为这里是新创建的 ByteBuffer，所以 pos
0 limit initialCapacity capacity initialCapacity）
    }
}
```

9.6.2　setByte 核心方法

setByte 方法直接通过 DirectByteBuffer 对象存放数据。源码如下。

```
protected void _setByte(int index, int value) {
    buffer.put(index, (byte) value);
}
```

9.6.3　setBytes 核心方法

setBytes 方法首先获取临时缓冲区对象，然后将该对象的 position 设置为当前写下标，limit 设置为写入的内容长度，最后调用 put 方法，将 src 数组中 srcIndex 起始下标处长度为 length 的内容写入 ByteBuffer。注意，由于 tmpBuf 临时缓冲区的内容与原始缓冲区的内容共享，只是不共享下标，所以这里写入临时缓冲区等于写入原始缓冲区。源码如下。

```
public ByteBuf setBytes(int index, byte[] src, int srcIndex, int length) {
    checkSrcIndex(index, length, srcIndex, src.length);
    ByteBuffer tmpBuf = internalNioBuffer();
    tmpBuf.clear().position(index).limit(index + length);
    tmpBuf.put(src, srcIndex, length);
    return this;
}
```

9.6.4　getByte 核心方法

getByte 方法直接通过 DirectByteBuffer 对象获取数据。

```
protected byte _getByte(int index) {
    return buffer.get(index);
}
```

9.6.5　getBytes 核心方法

getBytes 方法通过原始缓冲区 Buffer 进行复制，此时两者的 position、limit、mark 将互相分离，但是数据共享，所以可以安全地设置临时缓冲区的 position 和 limit 进行读取操作。源码如下。

```java
public ByteBuf getBytes(int index, byte[] dst, int dstIndex, int length) {
    getBytes(index, dst, dstIndex, length, false);
    return this;
}

private void getBytes(int index, byte[] dst, int dstIndex, int length, boolean internal) {
    checkDstIndex(index, length, dstIndex, dst.length);
    ByteBuffer tmpBuf;
    if (internal) {                                          // 指定 internal 则使用内部缓冲区
        tmpBuf = internalNioBuffer();
    } else {
        tmpBuf = buffer.duplicate();
    }
    tmpBuf.clear().position(index).limit(index + length);    // 设置 position 和 limit 为下面的读操作做准备
    tmpBuf.get(dst, dstIndex, length);
}
```

9.6.6 capacity 核心方法

capacity 方法用于调整新的容量。从源码中可知，其中的处理过程同 UnpooledHeapByteBuf 的 capacity(int newCapacity)，只不过这里把 byte 数组换成了 DirectByteBuffer 缓冲区的操作，读者在想不明白时，可以回顾下 position、limit 的作用。源码如下。

```java
public ByteBuf capacity(int newCapacity) {
    ensureAccessible();
    if (newCapacity < 0 || newCapacity > maxCapacity()) {     // 新容量小于 0 或容量超出最大容量抛出异常
        throw new IllegalArgumentException("newCapacity: " + newCapacity);
    }

    int readerIndex = readerIndex();
    int writerIndex = writerIndex();
    int oldCapacity = capacity;
    if (newCapacity > oldCapacity) {                          // 新容量大于旧容量，进行扩容并复制数据
        ByteBuffer oldBuffer = buffer;
        ByteBuffer newBuffer = allocateDirect(newCapacity);   // 分配新的缓冲区
        oldBuffer.position(0).limit(oldBuffer.capacity());    // 复制所有数据（pos - limit 可读）
        newBuffer.position(0).limit(oldBuffer.capacity());    // 读取所有数据（pos - limit 可写）
        newBuffer.put(oldBuffer);                             // 开始复制
        newBuffer.clear();                                    // 还原新缓冲区的标志位
        setByteBuffer(newBuffer);
    } else if (newCapacity < oldCapacity) {                   // 新容量小于旧容量，进行缩容并复制数据
        ByteBuffer oldBuffer = buffer;
        ByteBuffer newBuffer = allocateDirect(newCapacity);   // 分配新的缓冲区
        if (readerIndex < newCapacity) {                      // 读下标小于新容量，说明有数据还未读取
            if (writerIndex > newCapacity) {                  // 写下标大于新容量，直接将 writerIndex 限制为 newCapacity
                writerIndex(writerIndex = newCapacity);
            }
            // 设置好 postion 和 limit，准备读写
            oldBuffer.position(readerIndex).limit(writerIndex);
            newBuffer.position(readerIndex).limit(writerIndex);
            newBuffer.put(oldBuffer);                         // 开始复制
            newBuffer.clear();
```

```
        } else {    // readerIndex>=newCapacity, 说明没有数据需要读取，直接设置 readerIndex 和 writerIndex
为新容量下标
            setIndex(newCapacity, newCapacity);
        }
        setByteBuffer(newBuffer);
    }
    return this;
}
```

9.6.7　freeDirect 核心方法

freeDirect 方法用于释放分配的 DirectByteBuffer 对象。这里直接通过调用 JDK 原生的 Cleaner 对象的 clean 方法完成释放。源码如下。

```
public static void freeDirectBuffer(ByteBuffer buffer) {
    if (hasUnsafe() && !isAndroid()) {
        PlatformDependent0.freeDirectBuffer(buffer);
    }
}

static void freeDirectBuffer(ByteBuffer buffer) {
    Cleaner0.freeDirectBuffer(buffer);
}

private static final long CLEANER_FIELD_OFFSET;         // cleaner 对象偏移量
private static final Method CLEAN_METHOD;               // cleanner 清理方法对象

static {                                                // 初始化 DirectByteBuffer 的 cleaner 对象偏移量和方法偏移量
    Field cleanerField = direct.getClass().getDeclaredField("cleaner");
    cleanerField.setAccessible(true);
    CLEANER_FIELD_OFFSET = PlatformDependent0.objectFieldOffset(cleanerField); // cleaner 对象偏移量
    Object cleaner = cleanerField.get(direct);
    try {                                               // Cleaner implements Runnable 支持 JDK 9 以上
        Runnable runnable = (Runnable) cleaner;
        CLEAN_METHOD = Runnable.class.getDeclaredMethod("run");
    } catch (ClassCastException ignored) {              // 否则调用原来的 clean 方法
        CLEAN_METHOD = cleaner.getClass().getDeclaredMethod("clean");
    }
}

static void freeDirectBuffer(ByteBuffer buffer) {
    if (CLEANER_FIELD_OFFSET == -1 || !buffer.isDirect()) {
        return;
    }
    try {                                               // 直接调用 Cleaner 的方法完成释放
        Object cleaner = PlatformDependent0.getObject(buffer, CLEANER_FIELD_OFFSET);
        if (cleaner != null) {
            CLEAN_METHOD.invoke(cleaner);
        }
    } catch (Throwable t) {
        // Nothing we can do here.
    }
}
```

9.7　UnpooledUnsafeDirectByteBuf 方法

UnpooledUnsafeDirectByteBuf 方法同 UnpooledDirectByteBuf 一样，只不过操作数据的方法由原生的 ByteBuffer 变为使用 Unsafe 操作。源码如下。

```java
public class UnpooledUnsafeDirectByteBuf extends AbstractReferenceCountedByteBuf {
    private final ByteBufAllocator alloc;     // 所属内存分配器
    private long memoryAddress;               // 内存地址
    private ByteBuffer tmpNioBuf;             // 临时缓冲区
    private int capacity;                     // 容量
    private boolean doNotFree;
    ByteBuffer buffer;                        // 封装的 DirectByteBuffer

    protected UnpooledUnsafeDirectByteBuf(ByteBufAllocator alloc, int initialCapacity, int maxCapacity) {
        super(maxCapacity);
        if (alloc == null) {
            throw new NullPointerException("alloc");
        }
        if (initialCapacity < 0) {
            throw new IllegalArgumentException("initialCapacity: " + initialCapacity);
        }
        if (maxCapacity < 0) {
            throw new IllegalArgumentException("maxCapacity: " + maxCapacity);
        }
        if (initialCapacity > maxCapacity) {
            throw new IllegalArgumentException(String.format(
                "initialCapacity(%d) > maxCapacity(%d)", initialCapacity, maxCapacity));
        }

        this.alloc = alloc;
        setByteBuffer(allocateDirect(initialCapacity), false);
    }

    // 初始化成员变量，同 UnpooledDirectByteBuf
    final void setByteBuffer(ByteBuffer buffer, boolean tryFree) {
        if (tryFree) {
            ByteBuffer oldBuffer = this.buffer;
            if (oldBuffer != null) {
                if (doNotFree) {
                    doNotFree = false;
                } else {
                    freeDirect(oldBuffer);
                }
            }
        }
        this.buffer = buffer;
        memoryAddress = PlatformDependent.directBufferAddress(buffer);
        tmpNioBuf = null;
        capacity = buffer.remaining();
```

```java
}

/* 分配与释放内存同 UnpooledDirectByteBuf */
protected ByteBuffer allocateDirect(int initialCapacity) {
    return ByteBuffer.allocateDirect(initialCapacity);
}

protected void freeDirect(ByteBuffer buffer) {
    PlatformDependent.freeDirectBuffer(buffer);
}

/* 操作内存不同于 UnpooledDirectByteBuf,这里直接使用 Unsafe 来操作 */
protected byte _getByte(int index) {
    return UnsafeByteBufUtil.getByte(addr(index));
}

protected byte _getByte(int index) {
    return UnsafeByteBufUtil.getByte(addr(index));
}

}
```

9.8　UnpooledUnsafeNoCleanerDirectByteBuf 类

UnpooledUnsafeNoCleanerDirectByteBuf 类操作数据的方法与 UnpooledUnsafeDirectByteBuf 相同，通过 Unsafe 来操作。只不过这里不通过 Cleaner 释放内存，而是手动释放内存。源码如下。

```java
final class UnpooledUnsafeNoCleanerDirectByteBuf extends UnpooledUnsafeDirectByteBuf {

    UnpooledUnsafeNoCleanerDirectByteBuf(ByteBufAllocator alloc, int initialCapacity, int maxCapacity) {
        super(alloc, initialCapacity, maxCapacity);
    }

    // 分配内存
    @Override
    protected ByteBuffer allocateDirect(int initialCapacity) {
        return PlatformDependent.allocateDirectNoCleaner(initialCapacity);
    }

    // 释放内存
    @Override
    protected void freeDirect(ByteBuffer buffer) {
        PlatformDependent.freeDirectNoCleaner(buffer);
    }
}
```

先来看分配原理。这里使用反射获取到了 DirectByteBuffer 的构造器。DirectByteBuffer(long addr(堆外内存地址), int cap(容量))，本来在 DirectByteBuffer 中会自己生成 long address 来调用该构造器，但由于没有 Cleaner，则要通过 Unsafe.allocateMemory 分配了一段堆外内存初始化 DirectByteBuffer 对象。源码如下。

```java
public static ByteBuffer allocateDirectNoCleaner(int capacity) {
    incrementMemoryCounter(capacity);
    try {
        return PlatformDependent0.allocateDirectNoCleaner(capacity);
    } catch (Throwable e) {
        decrementMemoryCounter(capacity);
        throwException(e);
        return null;
    }
}

static ByteBuffer allocateDirectNoCleaner(int capacity) {
    return newDirectBuffer(UNSAFE.allocateMemory(capacity), capacity);   // 注意，这里直接使用 Unsafe 分配了一段堆外内存
}

private static final Constructor<?> DIRECT_BUFFER_CONSTRUCTOR;   // DirectByteBuffer 的构造器

static { // 初始化 DirectByteBuffer 的构造器：DirectByteBuffer(long addr(堆外内存地址), int cap(容量))
    DIRECT_BUFFER_CONSTRUCTOR = direct.getClass().getDeclaredConstructor(long.class, int.class);
    DIRECT_BUFFER_CONSTRUCTOR.setAccessible(true);
}

// 这里通过自己创建的堆外内存初始化 DirectByteBuffer 对象
private static ByteBuffer newDirectBuffer(long address, int capacity) {
    try {
        return (ByteBuffer) DIRECT_BUFFER_CONSTRUCTOR.newInstance(address, capacity);
    } catch (Throwable cause) {
        // Not expected to ever throw!
        if (cause instanceof Error) {
            throw (Error) cause;
        }
        throw new Error(cause);
    }
}
```

然后是释放原理。这里不使用 Cleaner 释放内存，而是手动调用 unsafe 类的 freeMemory 方法完成堆外内存的释放。源码如下。

```java
public static void freeDirectNoCleaner(ByteBuffer buffer) {
    int capacity = buffer.capacity();
    PlatformDependent0.freeMemory(PlatformDependent0.directBufferAddress(buffer));
    decrementMemoryCounter(capacity);
}

static void freeMemory(long address) {
    UNSAFE.freeMemory(address);
}
```

9.9 PooledByteBuf 原理

前面讲的都是非池化的 ByteBuf，现在来学习使用了池化的 PooledByteBuf。注意，这里的池化说

的是内存池，先回顾一下前面的知识。

（1）ByteBuf 维护了 readIndex 和 writeIndex，分别用于表示读下标和写下标。

（2）UnPooledByteBuf 可以使用 byte[]数组和 DirectByteBuffer 对象保存数据。

（3）所以得出结论，ByteBuf 维护了操作数据的基本方法，而实际的存储将由存放数据的载体实现。

由此可知，PooledByteBuf 也是使用 byte[]数组或 DirectByteBuffer 对象，只不过这些保存数据的载体被内存池管理，即分配时从内存池中获取，释放时返还给内存池。注意，由于内存池是另外一块内容，这里只需要关注 PooledByteBuf 自身对于操作数据的原理即可，内存池的相关介绍，笔者会在后面紧跟着内存分配器一起讲解。由源码可知。

（1）Recycler 类用于回收 PooledByteBuf 的对象池，即对于 PooledByteBuf 对象也进行了缓存。

（2）memory 为实际保存数据的内存。

（3）由于 memory 属于内存池管理的所有内存，所以需要使用 offset 和 length 说明当前从内存池中分配可以使用的区间。

（4）PoolThreadCache 用于缓存从共用的 PoolChunk 中分配的内存，这时可以让当前线程无锁化分配。

```java
abstract class PooledByteBuf<T> extends AbstractReferenceCountedByteBuf {
    private final Recycler.Handle<PooledByteBuf<T>> recyclerHandle;  // 用于回收 PooledByteBuf 的对象池
    protected PoolChunk<T> chunk;         // 当前所属内存池，这里了解即可，后面会详细讲解
    protected long handle;                // 用于指向 PoolChunk 内存池中的分配空间数据，同样了解即可
    protected T memory;                   // 保存数据的内存
    protected int offset;                 // 保存数据内存的偏移量
    protected int length;                 // 保存数据内存的长度
    int maxLength;                        // 支持的最大长度
    PoolThreadCache cache;   // 所属线程本地内存缓存（用于加速分配，即 ThreadLocal 放置内存，用于无锁化分配）
    private ByteBuffer tmpNioBuf;         // 临时缓冲区

    // 构造器。用于初始化 maxCapacity 和对象池 recyclerHandle
    protected PooledByteBuf(Recycler.Handle<? extends PooledByteBuf<T>> recyclerHandle, int maxCapacity) {
        super(maxCapacity);
        this.recyclerHandle = (Handle<PooledByteBuf<T>>) recyclerHandle;
    }
}
```

9.9.1 init 核心方法

init 方法由 PoolChunk 内存管理器调用，使用分配的内存信息初始化成员变量。源码如下。

```java
void init(PoolChunk<T> chunk, long handle, int offset, int length, int maxLength, PoolThreadCache cache) {
    this.chunk = chunk;
    this.handle = handle;
    memory = chunk.memory;
    this.offset = offset;
    this.length = length;
    this.maxLength = maxLength;
    tmpNioBuf = null;
    this.cache = cache;
}
```

9.9.2　initUnpooled 核心方法

当分配的内存太大了，这时会导致这片空间无法管理。因为内存池的设计是针对小内存进行管理，如果内存太大，会导致内存资源紧缺，大内存无法被利用。所以，initUnpooled 方法就是用于在 PoolChunk 内存管理器认为分配的内存太大时使用，将这片大空间采用不池化的技术来分配，所以这里只需要 PoolChunk 和 length，表明所属内存池和长度即可。源码如下。

```
void initUnpooled(PoolChunk<T> chunk, int length) {
    this.chunk = chunk;
    handle = 0;
    memory = chunk.memory;
    offset = 0;
    this.length = maxLength = length;
    tmpNioBuf = null;
    cache = null;
}
```

9.9.3　reuse 核心方法

reuse 方法用于复用当前 PooledByteBuf 对象。内存分配器在从对象池中取出该对象进行复用时调用，用 maxCapacity 初始化父类同时还原标志位。源码如下。

```
final void reuse(int maxCapacity) {
    maxCapacity(maxCapacity);
    setRefCnt(1);                              // 重置引用计数
    setIndex0(0, 0);                           // 还原 writeIndex 和 readIndex 为 0
    discardMarks();                            // 还原读写标记位为 0
}
```

9.9.4　capacity 核心方法

capacity 方法用于调整容量。流程如下。

（1）如果当前为非池化内存，则重新分配。

（2）如果为池化内存，则对比新容量和当前长度；如果新容量大于当前长度且小于最大长度，则更新 length 为新容量即可。如果新容量小于当前长度，则进一步进行空间判断，以避免持有较大的不使用空间，导致内存浪费；如果新容量等于当前长度，则直接返回。

（3）如果以上条件均不满足，则调用分配器重新分配内存。

```
public final ByteBuf capacity(int newCapacity) {
    ensureAccessible();
    if (chunk.unpooled) {                      // 非池化内存
        if (newCapacity == length) {           // 新容量等于当前长度，则直接返回
            return this;
        }
    } else {                                   // 池化内存
        if (newCapacity > length) {            // 新容量大于当前长度
            if (newCapacity <= maxLength) {    // 新容量小于支持的最大长度，更新 length 为新容量
                length = newCapacity;
```

```
                return this;
            }
        } else if (newCapacity < length) {                  // 新容量小于当前长度
            if (newCapacity > maxLength >>> 1) {            // 新容量大于当前长度的一半，需要进行空间判
断，避免持有较大的不使用空间，导致内存浪费
                if (maxLength <= 512) {                     // 最大长度小于 512B 时，在更新时判断新容量
与最大长度的差值是否在 16B 以内
                    if (newCapacity > maxLength - 16) {    // 新容量比最大长度大 16，即后面有小于 16B 的
空间，则直接将 length 更新为新容量即可
                        length = newCapacity;
                        setIndex(Math.min(readerIndex(), newCapacity), Math.min(writerIndex(),
newCapacity));
                                                            // 更新读和写下标
                        return this;
                    }
                } else {        // > 512（例如>= 1024），直接更新长度为新容量，这时不会造成过多内存浪费
                    length = newCapacity;
                    setIndex(Math.min(readerIndex(), newCapacity), Math.min(writerIndex(), newCapacity));
                    return this;
                }
            }
        } else {                                            // 新容量等于当前长度，则直接返回
            return this;
        }
    }
    // 重新分配内存（后面学习内存池时会详细讲解）
    chunk.arena.reallocate(this, newCapacity, true);
    return this;
}
```

9.9.5　deallocate 核心方法

deallocate 方法用于将从内存池中分配的内存归还给内存池。这里通过保留有内存池信息的 handle 变量，将其传递给所属内存池，并调用 free 方法完成释放。释放完成时，将重置变量，并调用 recycle 方法将当前对象放回对象池中。源码如下。

```
protected final void deallocate() {
    if (handle >= 0) {
        final long handle = this.handle;
        this.handle = -1;
        memory = null;
        chunk.arena.free(chunk, handle, maxLength, cache);
        recycle();
    }
}
```

9.9.6　recycle 核心方法

recycle 方法用于将当前 PooledByteBuf 放回内存池，直接调用 recyclerHandle 完成回收。源码如下。

```
private void recycle() {
    recyclerHandle.recycle(this);
}
```

第 10 章

Netty 对象池原理

10.1 Recycler 原理

Recycler 类在 Netty 中起对象池的作用。推理如下。
（1）Java 中创建对象分两步，分配内存和初始化对象。
（2）在 GC 中，如果分配的对象生命周期较短，则会在新生代中清理。
（3）为了减少分配对象、销毁对象和 GC 标记对程序性能的影响，创建对象池来缓存分配的对象。

所以对象池的存在会减少对 YGC（Young Garbage Collection，年轻代垃圾回收）的影响，因为缓存的对象本身会被移动到老年代中。而该类就作为 Netty 架构中的内存池存在。

10.1.1 核心构造器与变量

由源码可知。
（1）NOOP_HANDLE 表示对 Handle 接口的空实现，表示不做回收操作。
（2）可以通过对应的系统变量名设置各个缓存容量参数。
（3）FastThreadLocal<Stack> threadLocal 用于保存线程保存对象的栈。
（4）实际对象池实现为栈对象，Recycler 作为栈对象的门面。

```
public abstract class Recycler<T> {
    private static final Handle NOOP_HANDLE = new Handle() {
        @Override
        public void recycle(Object object) {
            // NOOP
        }
    }; // 默认不回收对象的处理对象

    private static final AtomicInteger ID_GENERATOR = new AtomicInteger(Integer.MIN_VALUE); // ID 生成器
    private static final int OWN_THREAD_ID = ID_GENERATOR.getAndIncrement(); // 生成 Recycler 的默认 ID
    private static final int DEFAULT_INITIAL_MAX_CAPACITY = 32768; // 默认保存 32K 大小的对象
    private static final int DEFAULT_MAX_CAPACITY;                 // 保存当前允许的最大容量
    private static final int INITIAL_CAPACITY;                     // 初始容量
    private static final int MAX_SHARED_CAPACITY_FACTOR;           // 最大共享容量因子（用于支持 Stack 类与 WeakOrderQueue 类，后面会描述）
    private static final int LINK_CAPACITY; // 连接容量（用于支持 WeakOrderQueue 类中的 Link 大小，后面会描述）
```

```java
static {
    // 设置最大容量。默认为 DEFAULT_INITIAL_MAX_CAPACITY 32K
    int maxCapacity = SystemPropertyUtil.getInt("io.Netty.recycler.maxCapacity",
DEFAULT_INITIAL_MAX_CAPACITY);
    if (maxCapacity < 0) {
        maxCapacity = DEFAULT_INITIAL_MAX_CAPACITY;
    }
    DEFAULT_MAX_CAPACITY = maxCapacity;
    MAX_SHARED_CAPACITY_FACTOR = max(2,
SystemPropertyUtil.getInt("io.Netty.recycler.maxSharedCapacityFactor",2));   // 共享容量因子默认为 2
    LINK_CAPACITY = MathUtil.findNextPositivePowerOfTwo(
        max(SystemPropertyUtil.getInt("io.Netty.recycler.linkCapacity", 16), 16));  // 连接容量默认为 16
    INITIAL_CAPACITY = min(DEFAULT_MAX_CAPACITY, 256);   // 初始容量为 256
}
private final int maxCapacity;
private final int maxSharedCapacityFactor;

// ThreadLocal 用于保存当前线程的 Stack 对象
private final FastThreadLocal<Stack<T>> threadLocal = new FastThreadLocal<Stack<T>>() {
    @Override
    protected Stack<T> initialValue() {
        return new Stack<T>(Recycler.this, Thread.currentThread(), maxCapacity,
maxSharedCapacityFactor);
    }
};

// ThreadLocal 用于保存当前线程的 Map<Stack<?>, WeakOrderQueue>对象（了解即可，后面会详细解释）
private static final FastThreadLocal<Map<Stack<?>, WeakOrderQueue>> DELAYED_RECYCLED =
    new FastThreadLocal<Map<Stack<?>, WeakOrderQueue>>() {
    @Override
    protected Map<Stack<?>, WeakOrderQueue> initialValue() {
        return new WeakHashMap<Stack<?>, WeakOrderQueue>();
    }
};

// 使用默认最大容量的构造器
protected Recycler() {
    this(DEFAULT_MAX_CAPACITY);
}

// 使用指定 maxCapacity 最大容量的构造器
protected Recycler(int maxCapacity) {
    this(maxCapacity, MAX_SHARED_CAPACITY_FACTOR);
}

// 使用指定 maxCapacity 最大容量和 maxSharedCapacityFactor 共享容量因子的构造器
protected Recycler(int maxCapacity, int maxSharedCapacityFactor) {
    if (maxCapacity <= 0) {
        this.maxCapacity = 0;
        this.maxSharedCapacityFactor = 1;
    } else {
        this.maxCapacity = maxCapacity;
        this.maxSharedCapacityFactor = max(1, maxSharedCapacityFactor);
```

```
            }
        }
}
```

10.1.2 Handle 核心接口

Handle 接口用于定义如何回收对象，同时也包装了缓存对象，在 Stack 中缓存的便是 Handle 接口的实例。上面已经看到了 NOOP_HANDLE，现在来看 DefaultHandle 类对于 recycle 方法的实现。流程如下。

（1）判断当前回收的对象与当前对象不等，表明两者不属于同一个 Handle，抛出异常。

（2）如果当前线程为 DefaultHandle 所属的对象栈，则直接压入对象栈中。

（3）否则获取当前线程所属的 Map<Stack<?>, WeakOrderQueue>集合，将对象放入所属 Stack 的 WeakOrderQueue 中。

```java
public interface Handle<T> {
    void recycle(T object);
}

static final class DefaultHandle<T> implements Handle<T> {
    private int lastRecycledId;         // 最后回收的 id
    private int recycleId;              // 回收 id
    private Stack<?> stack;             // 所属栈（对象池）
    private Object value;               // 缓存的对象

    DefaultHandle(Stack<?> stack) {
        this.stack = stack;
    }

    @Override
    public void recycle(Object object) {
        if (object != value) {          // 回收的对象与当前对象不等，表明两者不属于同一个 Handle（其实不难看
出，归还对象时，归还的为 DefaultHandle）
            throw new IllegalArgumentException("object does not belong to handle");
        }
        Thread thread = Thread.currentThread();
        if (thread == stack.thread) {   // 当前线程为 DefaultHandle 所属的对象栈，则直接压入栈中即可
            stack.push(this);
            return;
        }
        // 获取当前线程所属的 Map<Stack<?>, WeakOrderQueue> 集合，将对象放入所属 Stack 的
WeakOrderQueue 中
        Map<Stack<?>, WeakOrderQueue> delayedRecycled = DELAYED_RECYCLED.get();
        WeakOrderQueue queue = delayedRecycled.get(stack);
        if (queue == null) {            // 初始化 WeakOrderQueue
            queue = WeakOrderQueue.allocate(stack, thread);
            if (queue == null) {        // 分配失败不缓存该对象
                return;
            }
            delayedRecycled.put(stack, queue);
        }
```

```
            queue.add(this);        // 将当前 DefaultHandler 保存到 WeakOrderQueue 中
    }
}
```

10.1.3 Stack 核心类

Stack 类作为 Recycler 的内部类，提供了对于对象池的支持。这里使用了栈的特性来构建对象池。栈为先进后出，反过来说后进先出，很好地利用了缓存命中的特性。如果一个线程刚好释放了一个对象，这时该对象还在缓存中，直接 pop 该对象，会直接利用缓存的特性加快对象的访问。

1．核心变量与构造器

由源码可知。

（1）DefaultHandle<?>[] elements 用于作为对象池保存对象。

（2）maxCapacity 和 availableSharedCapacity 将从所属 Recycler 中继承。

（3）对象数组将在构造器中创建。

```
static final class Stack<T> {
    final Recycler<T> parent;                       // 所属回收器
    final Thread thread;                            // 所属线程对象
    private DefaultHandle<?>[] elements;            // 用于保存对象（这里就是对象池）
    private final int maxCapacity;                  // 最大容量
    private int size;                               // 当前存放的对象个数
    final AtomicInteger availableSharedCapacity;    // 可用共享容量
    private volatile WeakOrderQueue head;           // WeakOrderQueue 头结点
    private WeakOrderQueue cursor, prev; // 当前处理的 WeakOrderQueue 和前一个 WeakOrderQueue 引用

    // 构造器用于初始化成员
    Stack(Recycler<T> parent, Thread thread, int maxCapacity, int maxSharedCapacityFactor) {
        this.parent = parent;
        this.thread = thread;
        this.maxCapacity = maxCapacity;
        availableSharedCapacity = new AtomicInteger(max(maxCapacity / maxSharedCapacityFactor,
LINK_CAPACITY));          // 通过 maxCapacity、maxSharedCapacityFactor 和 LINK_CAPACITY 计算
availableSharedCapacity
        elements = new DefaultHandle[min(INITIAL_CAPACITY, maxCapacity)];   // 创建对象数组
    }
}
```

2．newHandle 核心方法

该方法用于创建 DefaultHandle。仅在构造其中传入当前引用，表明 DefaultHandle 所属栈。源码如下。

```
DefaultHandle<T> newHandle() {
    return new DefaultHandle<T>(this);
}
```

3．push 核心方法

push 方法用于将 DefaultHandle 对象压入对象栈中。流程如下。

（1）判断 recycleId 和 lastRecycledId，回收 id 不为 0 表明已经被回收，抛出异常。

（2）初始化 recycleId 和 lastRecycledId 为 Recycle 类的默认线程 ID。

(3）如果对象池中的对象达到最大容量，则不缓存当前对象。
(4）如果对象池中的对象达到 elements 数组的长度，则对数组进行 2 倍扩容。
(5）将对象放入数组中，并增加 size 大小。

```java
void push(DefaultHandle<?> item) {
    if ((item.recycleId | item.lastRecycledId) != 0) {      // 回收 id 不为 0 表明已经被回收，不允许重复回收
        throw new IllegalStateException("recycled already");
    }
    item.recycleId = item.lastRecycledId = OWN_THREAD_ID; // 初始化 recycleId 和 lastRecycledId 为 Recycle 类的默认线程 ID
    int size = this.size;
    if (size >= maxCapacity) {                              // 达到最大容量，不缓存当前对象
        return;
    }
    if (size == elements.length) {                          // 对象数组 2 倍扩容
        elements = Arrays.copyOf(elements, min(size << 1, maxCapacity));
    }
    elements[size] = item;
    this.size = size + 1;
}
```

4. pop 核心方法

pop 方法用于从对象池中获取缓存的对象。流程如下。

(1）可用对象为 0，尝试从 WeakOrderQueue 中获取，获取失败，直接返回 null。
(2）从 elements 数组中对应 size 下标处取出对象，并赋值为 null，重置 recycleId 和 lastRecycledId。

```java
DefaultHandle<T> pop() {
    int size = this.size;
    if (size == 0) {            // 可用对象为 0，尝试从 WeakOrderQueue 中获取，获取失败，直接返回 null
        if (!scavenge()) {
            return null;
        }
        size = this.size;
    }
    size --;                    // 减少 size
    // 取出对象并置空
    DefaultHandle ret = elements[size];
    elements[size] = null;
    if (ret.lastRecycledId != ret.recycleId) {
        throw new IllegalStateException("recycled multiple times");
    }
    ret.recycleId = 0;
    ret.lastRecycledId = 0;
    this.size = size;
    return ret;
}
```

5. scavenge 核心方法

scavenge 方法用于从 WeakOrderQueue 中获取缓存的对象。流程如下。

(1）如果 scavengeSome 方法返回为 true，则表明获取成功。

（2）否则重置 prev 和 cursor 并返回失败。

在 scavengeSome 中完成真正获取对象的方法。流程如下。

（1）设置当前处理的 WeakOrderQueue cursor 对象。

（2）调用 WeakOrderQueue cursor 的 transfer 方法完成对象的迁移，如果迁移成功，则结束循环并返回。

（3）否则判断获取下一个 WeakOrderQueue 对象。

（4）如果当前与 WeakOrderQueue cursor 绑定的线程对象已经被回收，则判断是否缓存对象，如果没有，则转移其中的对象。

（5）如果线程仍存在，则直接保存当前 cursor 为 prev。

（6）cursor 更新为 next，继续循环处理。

读者可能对于 WeakOrderQueue 不太熟悉。后面内容就是对该类的讲解，读者可先把 WeakOrderQueue 当作一个单向链表，里面保存有 DefaultHandler 对象，这里需要将其转移到 Stack 的 elements 数组中处理。源码如下。

```
boolean scavenge() {
    if (scavengeSome()) {
        return true;
    }
    prev = null;
    cursor = head;
    return false;
}

boolean scavengeSome() {
    // 设置当前处理的 WeakOrderQueue 对象
    WeakOrderQueue cursor = this.cursor;
    if (cursor == null) {                                // cursor 为空，取 head 作为当前操作对象
        cursor = head;
        if (cursor == null) {
            return false;
        }
    }
    boolean success = false;
    WeakOrderQueue prev = this.prev;                     // 保存前一个处理的 WeakOrderQueue
    do {
        if (cursor.transfer(this)) {                     // 调用 WeakOrderQueue 的 transfer 方法完成对象的迁移
            // （后面会讲到，转移到当前 Stack 的 elements 中）
            success = true;
            break;
        }
        WeakOrderQueue next = cursor.next;               // 获取下一个 WeakOrderQueue 对象
        if (cursor.owner.get() == null) {                // 当前与 WeakOrderQueue 绑定的线程对象已经被回收，
            // 则判断缓存中是否还有对象，如果没有，则将其中的对象进行转移
            if (cursor.hasFinalData()) {
                for (;;) {                               // 死循环保证 WeakOrderQueue 中对象全部转移
                    if (cursor.transfer(this)) {
                        success = true;
                    } else {
```

```
                    break;
                }
            }
        }
        if (prev != null) {              // 前一个 WeakOrderQueue 不为空,则将前一个 WeakOrderQueue
的 next 指向当前 next,因为 cursor 已经被回收,所以断开链表中的连接
            prev.next = next;
        } else {                         // 线程仍存在,直接保存当前 cursor 为 prev
            prev = cursor;
        }
        cursor = next;                   // cursor 更新为 next
    } while (cursor != null && !success);  // 循环处理直到 cursor 为空(WeakOrderQueue 链表遍历完毕)或处
理失败
    // 更新变量为最新值
    this.prev = prev;
    this.cursor = cursor;
    return success;
}
```

6. increaseCapacity 核心方法

increaseCapacity 方法在 WeakOrderQueue 中调用,用于对 elements 进行扩容,因为在 WeakOrderQueue 的 transfer 方法中,将保存的对象传递到 elements 数组时,如果容量不够,需要扩容。源码如下。

```
int increaseCapacity(int expectedCapacity) {
    int newCapacity = elements.length;
    int maxCapacity = this.maxCapacity;
    do { // 对 newCapacity 进行 2 倍递增,直到 newCapacity 大于 expectedCapacity 且小于等于 maxCapacity
        newCapacity <<= 1;
    } while (newCapacity < expectedCapacity && newCapacity < maxCapacity);
    newCapacity = min(newCapacity, maxCapacity);        // 取新容量与最大容量的最小值(因为上面以 2 倍扩
容,有可能在最后一次循环对容量进行了 2 倍扩容后大于最大容量)
    if (newCapacity != elements.length) {               // 开辟新数组同时将旧容量中数据复制到新数组中
        elements = Arrays.copyOf(elements, newCapacity);
    }
    return newCapacity;
}
```

10.1.4 WeakOrderQueue 核心类

WeakOrderQueue 类设计相当巧妙,由 Stack 的 push 方法可知:需要回收对象的所属线程不是 Stack 的所属线程时,需要依赖该类完成对象的暂时存储。为什么如此设计,推理如下。

(1)对象池中的对象只能被所属的线程获取(在后面的 get 方法中会看到,每次拿到的 Stack 都是通过 ThreadLocal 获取的),但是获取到的对象可以传递到其他任何一个线程。

(2)当在别的线程中调用回收方法放回对象池中,会和 Stack 所属线程发生竞争。

(3)竞争的方法有两种,上锁互斥,由一个线程处理该对象,或者是无锁化处理。

(4)Netty 采用的是第二种无锁化处理。

(5)这时就很容易知道 WeakOrderQueue 的作用了,它保存了某个 Stack 需要回收的对象,由 Stack 所属线程在执行 pop 操作时进行转移即可。

（6）这时不难得出 FastThreadLocal<Map<Stack<?>, WeakOrderQueue>> DELAYED_RECYCLED 变量的作用：每个线程有一个 Map<Stack<?>, WeakOrderQueue>结构，用于存放栈和 WeakOrderQueue 的映射。

1. 核心变量与构造器

由源码可知。

（1）Link 内部类用于实际保存回收的对象数组 elements，其中 value 为写下标，readIndex 为读下标。

（2）Link 对象将通过变量 next 形成一个单链表。

（3）在构造器中初始化了 Link 首尾引用，并将 WeakOrderQueue 线程安全地链入 Stack 的 WeakOrderQueue 链表中。

```
private static final class WeakOrderQueue {

    // 内部类用于实际保存回收的对象数组 elements，这里直接继承 AtomicInteger，使用其内部的 value 作为写
    下标，这时由于 value 为 volatile 修饰，本身具备线程可见性
    private static final class Link extends AtomicInteger {
        private final DefaultHandle<?>[] elements = new DefaultHandle[LINK_CAPACITY]; // 对象数组
        private int readIndex;                // 读下标
        private Link next;                    // 连接到下一个 link 对象
    }
    private Link head, tail;                  // Link 链表的首尾引用
    private WeakOrderQueue next;              // WeakOrderQueue 链表
    private final WeakReference<Thread> owner; // 所属线程对象（这里使用弱引用，不会导致 Thread 对象
    不能被回收）
    private final int id = ID_GENERATOR.getAndIncrement();   // 获取 id
    private final AtomicInteger availableSharedCapacity;     // 可用共享容量

    // 构造器用于初始化变量
    private WeakOrderQueue(Stack<?> stack, Thread thread) {
        head = tail = new Link();
        owner = new WeakReference<Thread>(thread);
        synchronized (stack) {       // 对所属 stack 上锁，用于将自身链入 Stack 的 WeakOrderQueue 链
        表（由于多线程可以并发创建 WeakOrderQueue，所以需要保证线程安全）
            next = stack.head;
            stack.head = this;
        }
        availableSharedCapacity = stack.availableSharedCapacity; // 直接继承 stack 的设置
    }
}
```

2. allocate 核心方法

该函数用于创建 WeakOrderQueue 对象。直接通过 reserveSpace 方法判断是否需要创建 WeakOrderQueue 对象，在 reserveSpace 方法中对 availableSharedCapacity 与 LINK_CAPACITY 进行判断，如果 availableSharedCapacity 小于 LINK_CAPACITY，则不创建 WeakOrderQueue。源码如下。

```
static WeakOrderQueue allocate(Stack<?> stack, Thread thread) {
    return reserveSpace(stack.availableSharedCapacity, LINK_CAPACITY)
        ? new WeakOrderQueue(stack, thread) : null;
}
```

```java
private static boolean reserveSpace(AtomicInteger availableSharedCapacity, int space) {
    assert space >= 0;
    for (;;) {
        int available = availableSharedCapacity.get();
        if (available < space) {
            return false;
        }
        if (availableSharedCapacity.compareAndSet(available, available - space)) { // 对总 availableSharedCapacity 减 LINK_CAPACITY 大小
            return true;
        }
    }
}
```

3．add 核心方法

add 方法用于向 WeakOrderQueue 中添加需要回收的对象 DefaultHandle。流程如下。

（1）把 lastRecycledId 设置为当前 WeakOrderQueue 的 id。

（2）获取当前 link 的写下标，并判断是否达到最大容量，如果达到最大容量，则调用 reserveSpace 方法判断当前存储的所有 Link 是否到达了设置的 availableSharedCapacity 大小。如果达到该大小，则不缓存该对象；如果还有空间存放，但是由于当前 Link 已经满了，则创建一个新的 Link 存放对象，同时将该 Link 放入全局的 Link 链表。

（3）将对象放入 Link 对象中。

（4）更新 writeIndex 下标。

```java
void add(DefaultHandle<?> handle) {
    handle.lastRecycledId = id;                  // 设置 lastRecycledId 为当前 WeakOrderQueue 的 id
    Link tail = this.tail;
    int writeIndex;
    if ((writeIndex = tail.get()) == LINK_CAPACITY) {  // Link 存放的对象达到最大容量（单个 Link 的大小达到单个 Link 的限制）
        if (!reserveSpace(availableSharedCapacity, LINK_CAPACITY)) {
            // 所有 link 占用的空间，达到最大容量，丢弃该对象
            return;
        }
        // 分配一个新的 Link 用于存放数据，并放入全局 Link 链表
        this.tail = tail = tail.next = new Link();
        writeIndex = tail.get();
    }
    tail.elements[writeIndex] = handle;          // 将对象放入 Link 对象中
    handle.stack = null;                          // 此时不需要再缓存 stack 的引用，因为当前 DefaultHandle 已经放入了 WeakOrderQueue，同时 WeakOrderQueue 与 Stack 的关系会绑定在 Stack 的 WeakOrderQueue 链表中
    tail.lazySet(writeIndex + 1);                 // 采用 unsafe.putOrderedInt 方法来优化对于 Link 的 volatile 类型的 value 进行更新操作的全屏障性能损耗，因为当前 Link 的写下标只在所属线程内使用，所以不需要保证可见性，因为在写入该变量的线程中，该变量本身可见（可参考下笔者的《深入理解 Java 高并发编程》中的 Volatile 篇）
}
```

4．hasFinalData 核心方法

hasFinalData 方法用于判断 WeakOrderQueue 中 Link 链表是否还有对象。只需要判断最后一个 Link

中 readIndex 读下标是否等于写下标。源码如下。

```
boolean hasFinalData() {
    return tail.readIndex != tail.get();
}
```

5. transfer 核心方法

transfer 方法用于将 WeakOrderQueue 中保存的对象转移到 Stack 中。流程如下。

（1）判定当前WeakOrderQueue没有缓存对象，直接返回false。

（2）从 Link 链表中的头结点开始转移对象，如果当前 head 已经读取完毕，则获取 Link 链表的下一个结点；如果没有，则返回 false。

（3）计算当前 Link head 的开始读取的下标和结束下标。

（4）计算 stack 中存放对象的容量，如果容量不够，则进行扩容。

（5）如果 Stack 中有空间存放对象，则循环复制 head 中的对象放入 Stack。

（6）如果 Stack 中没有空间存放对象，则返回 false。

```
boolean transfer(Stack<?> dst) {
    Link head = this.head;
    if (head == null) {                                 // 没有缓存对象，直接返回
        return false;
    }
    // 从 Link 链表中的头结点开始转移对象
    if (head.readIndex == LINK_CAPACITY) { // 当前 head 已经读取完毕，获取 Link 链表的下一个结点；如果没有，则返回 false
        if (head.next == null) {
            return false;
        }
        this.head = head = head.next;
    }
    // 计算开始读取的下标和结束下标
    final int srcStart = head.readIndex;
    int srcEnd = head.get();
    final int srcSize = srcEnd - srcStart;
    if (srcSize == 0) {
        return false;
    }

    // 计算 stack 中存放对象的容量，如果容量不够，则进行扩容
    final int dstSize = dst.size;
    final int expectedCapacity = dstSize + srcSize;

    if (expectedCapacity > dst.elements.length) {
        final int actualCapacity = dst.increaseCapacity(expectedCapacity);
        srcEnd = min(srcStart + actualCapacity - dstSize, srcEnd); // 扩容后重新计算 srcEnd，因为此时可能扩容后，空间仍不能放入 Link 中所有的对象
    }
    if (srcStart != srcEnd) {                           // Stack 中有空间存放对象
        final DefaultHandle[] srcElems = head.elements;
        final DefaultHandle[] dstElems = dst.elements;
        int newDstSize = dstSize;
```

```java
        for (int i = srcStart; i < srcEnd; i++) {            // 循环复制数据
            DefaultHandle element = srcElems[i];
            if (element.recycleId == 0) {                    // 把 recycleId 更新为 lastRecycledId
                element.recycleId = element.lastRecycledId;
            } else if (element.recycleId != element.lastRecycledId) {
                throw new IllegalStateException("recycled already");
            }
            element.stack = dst;                             // 重新关联 DefaultHandle 与 Stack
            dstElems[newDstSize ++] = element;               // 放入 Stack 中
            srcElems[i] = null;                              // 清空 Link 的引用
        }
        dst.size = newDstSize;
        if (srcEnd == LINK_CAPACITY && head.next != null) {  // 转移当前 Link 的全部对象,且 Link 链表中还
存在 Link(head.next),则更新空间后将 head 设置为链表中的下一个 Link
            reclaimSpace(LINK_CAPACITY);                     // 更新 availableSharedCapacity 空间
            this.head = head.next;
        }
        head.readIndex = srcEnd;                             // 更新当前 Link 的读下标为 srcEnd,注意,这里的 head
仍然是发生转移的 Link,而不是 this.head
        return true;
    } else {                                                 // 栈已经满了,不进行回收
        return false;
    }
}
```

6. finalize 核心方法

finalize 方法重写了 Object 类中的 finalize 方法。用于在 WeakOrderQueue 对象被回收时,释放内存,这里的 super.finalize 不会产生实际操作,本身父类 Object 的该方法实现为空方法,为了规范编写,关键点在于 finally 代码块。由源码可知,在该代码块中遍历 Link 链表,增加 stack.availableSharedCapacity 变量的大小。源码如下。

```java
protected void finalize() throws Throwable {
    try {
        super.finalize();
    } finally {
        Link link = head;
        while (link != null) {  // 遍历链表回收每个 Link 的大小,即 LINK_CAPACITY
            reclaimSpace(LINK_CAPACITY);
            link = link.next;
        }
    }
}

// 回收空间,但是仅对 availableSharedCapacity 增加大小(在构造器中 availableSharedCapacity 将从
stack.availableSharedCapacity 中获取,所以这里增加大小,等于增加 Stack 的 availableSharedCapacity 大小)
private void reclaimSpace(int space) {
    assert space >= 0;
    availableSharedCapacity.addAndGet(space);
}
```

7. get 核心方法

get 方法用于从对象池中获取对象。流程如下。

（1）判断最大容量是否为 0。如果为 0，则表明没有实现对象池，直接调用子类完成对象的创建。

（2）获取当前线程的 stack 对象，并尝试从 stack 对象中获取到缓存的对象。

（3）获取失败，从栈中创建一个新的 DefaultHandle，然后调用子类创建对象，并把对象包装到 DefaultHandle 中。

（4）返回 DefaultHandle 中包装的对象。

```
public final T get() {
    if (maxCapacity == 0) {              // 最大容量为 0，表明没有实现对象池，直接调用子类完成对象
的创建（这里使用 NOOP_HANDLE，当子类回收对象时，这里什么都不做，仅作为一个空策略（策略模式））
        return newObject((Handle<T>) NOOP_HANDLE);
    }
    Stack<T> stack = threadLocal.get();  // 获取当前线程的 stack 对象
    DefaultHandle<T> handle = stack.pop(); // 尝试从 stack 对象中获取到缓存的对象
    if (handle == null) {                // 获取失败，从栈中创建一个新的 DefaultHandle
        handle = stack.newHandle();
        handle.value = newObject(handle);
    }
    return (T) handle.value;
}

// 唯一抽象方法，子类完成对象的创建
protected abstract T newObject(Handle<T> handle);
```

10.2 PooledHeapByteBuf 原理

PooledByteBuf 类定义了初始化、回收 ByteBuf 的操作，定义了池化的 ByteBuf 的所有功能。它唯一不能做的就是创建对象和回收器，因为它并不知道具体实现是什么，是 byte 数组，还是 ByteBuffer？所以关于具象化的实现由子类自身完成。PooledHeapByteBuf 类定义了一个特殊的 PooledByteBuf，即使用堆内内存数组 byte[]作为管理对象。由源码可知。

（1）在泛型中指定了 byte[]数组作为数据载体。

（2）Recycler RECYCLER 对象定义了对象池的对象创建。

（3）操作数据的方法通过 HeapByteBufUtil 类完成，在该类中将直接操作数组获取对应下标值。

```
class PooledHeapByteBuf extends PooledByteBuf<byte[]> {
    private static final Recycler<PooledHeapByteBuf> RECYCLER = new Recycler<PooledHeapByteBuf>() {
        @Override
        protected PooledHeapByteBuf newObject(Handle<PooledHeapByteBuf> handle) {
            return new PooledHeapByteBuf(handle, 0); // 这里将 PooledHeapByteBuf 与 Handle 对象关联
        }
    };                                   // PooledHeapByteBuf 对象池

    // 获取对象池中的 PooledHeapByteBuf 对象实例
    static PooledHeapByteBuf newInstance(int maxCapacity) {
        PooledHeapByteBuf buf = RECYCLER.get();
        buf.reuse(maxCapacity);          // 使用 maxCapacity 重置内部变量
        return buf;
```

```java
}
// 构造器初始化父类
PooledHeapByteBuf(Recycler.Handle<? extends PooledHeapByteBuf> recyclerHandle, int maxCapacity) {
    super(recyclerHandle, maxCapacity);
}

// 获取下标为 index 的值
protected byte _getByte(int index) {
    return HeapByteBufUtil.getByte(memory, idx(index));
}

// 设置下标为 index 的值
protected void _setByte(int index, int value) {
    HeapByteBufUtil.setByte(memory, idx(index), value);
}
/*省略其他操作数组的方法，同 setByte 和 getByte*/
}
// 封装数组的操作
final class HeapByteBufUtil {

    static byte getByte(byte[] memory, int index) {
        return memory[index];
    }

    static void setByte(byte[] memory, int index, int value) {
        memory[index] = (byte) value;
    }
}
```

10.3　PooledUnsafeHeapByteBuf 类

PooledUnsafeHeapByteBuf 类同 PooledHeapByteBuf，只不过是把内部获取数据的方法转换为使用 Unsafe 来操作 byte 数组的方法。UnsafeByteBufUtil 类前面过，其中封装了 PlatformDependent 操作 Unsafe，由于封装的较为简单，这里就不占用过多篇幅阐述。源码如下。

```java
final class PooledUnsafeHeapByteBuf extends PooledHeapByteBuf {

    private static final Recycler<PooledUnsafeHeapByteBuf> RECYCLER = new Recycler<PooledUnsafeHeapByteBuf>() {
        @Override
        protected PooledUnsafeHeapByteBuf newObject(Handle<PooledUnsafeHeapByteBuf> handle) {
            return new PooledUnsafeHeapByteBuf(handle, 0);
        }
    };                    // PooledUnsafeHeapByteBuf 对象池

    // 提供获取 PooledUnsafeHeapByteBuf 对象方法
    static PooledUnsafeHeapByteBuf newUnsafeInstance(int maxCapacity) {
        PooledUnsafeHeapByteBuf buf = RECYCLER.get();
```

```
        buf.reuse(maxCapacity);
        return buf;
    }

    // 构造器初始化父类
    private PooledUnsafeHeapByteBuf(Handle<PooledUnsafeHeapByteBuf> recyclerHandle, int maxCapacity){
        super(recyclerHandle, maxCapacity);
    }

    // 获取下标为 index 的值
    @Override
    protected byte _getByte(int index) {
        return UnsafeByteBufUtil.getByte(memory, idx(index));
    }

    // 设置下标为 index 的值
    protected void _setByte(int index, int value) {
        UnsafeByteBufUtil.setByte(memory, idx(index), value);
    }
    /*省略其他操作数组的方法，同 setByte 和 getByte*/
}
```

10.4　PooledDirectByteBuf 类

PooledDirectByteBuf 类用于封装原生 JDK 的 DirectByteBuffer 对象。由源码可知。
（1）泛型中指定父类泛型 T 的类型为 ByteBuffer（实际类型为 DirectByteBuffer）。
（2）实际操作数据的方法调用 ByteBuffer 类的 get 和 put 方法完成。

```
final class PooledDirectByteBuf extends PooledByteBuf<ByteBuffer> {

    private static final Recycler<PooledDirectByteBuf> RECYCLER = new Recycler<PooledDirectByteBuf>({
        @Override
        protected PooledDirectByteBuf newObject(Handle<PooledDirectByteBuf> handle) {
            return new PooledDirectByteBuf(handle, 0);
        }
    }; // PooledDirectByteBuf 对象池

    // 外部调用，获取对象池中的对象
    static PooledDirectByteBuf newInstance(int maxCapacity) {
        PooledDirectByteBuf buf = RECYCLER.get();
        buf.reuse(maxCapacity);
        return buf;
    }

    // 构造器初始化父类
    private PooledDirectByteBuf(Recycler.Handle<PooledDirectByteBuf> recyclerHandle, int maxCapacity{
        super(recyclerHandle, maxCapacity);
    }
```

```java
    protected byte _getByte(int index) {
        return memory.get(idx(index));
    }

    protected void _setByte(int index, int value) {
        memory.put(idx(index), (byte) value);
    }
}
```

10.5　CompositeByteBuf 原理

前面描述的都是单个 ByteBuf，它的类型可能属于如下类别。

（1）基于堆内存的 byte[]数组的 HeapByteBuf，同时该类型又分为：使用 byte[index]下标的普通方式获取，使用 Unsafe 操作数组偏移量来操作。

（2）基于堆外内存的 DirectByteBuffer 的 HeapByteBuf，该类型也分为：使用 Unsafe，直接操作 DirectByteBuffer。

本节介绍的 CompositeByteBuf 是一个虚拟的 ByteBuf 缓冲区，它可以由上述的缓冲区组成，即混合缓冲区。

10.5.1　核心变量与构造器

由源码可知。

（1）Component 对象表示缓冲区。

（2）List components 用于保存所有缓冲区。

```java
public class CompositeByteBuf extends AbstractReferenceCountedByteBuf implements Iterable<ByteBuf> {
    private static final ByteBuffer EMPTY_NIO_BUFFER = Unpooled.EMPTY_BUFFER.nioBuffer(); // 用于没
有缓冲区时的空缓冲区对象
    private static final Iterator<ByteBuf> EMPTY_ITERATOR = Collections.<ByteBuf>emptyList().iterator();
    // 没有缓冲区时，需要迭代的空迭代器

    private final ByteBufAllocator alloc;              // 所属分配器
    private final List<Component> components;          // 包含的缓冲区组件
    private final int maxNumComponents;                // 支持组合的最大缓冲区个数
    private boolean freed;                             // 是否需要释放

    // 构造器，通常由 ByteBufAllocator 分配调用创建对象
    public CompositeByteBuf(ByteBufAllocator alloc, boolean direct, int maxNumComponents, ByteBuf...
buffers) {
        super(Integer.MAX_VALUE);                      // 初始化最大容量为整型最大值
        if (alloc == null) {                           // 必须包含分配器
            throw new NullPointerException("alloc");
        }
        if (maxNumComponents < 2) { // 支持的最大缓冲区组件必须大于或等于 2（这就是叫组合缓冲区的原因）
            throw new IllegalArgumentException(
                    "maxNumComponents: " + maxNumComponents + " (expected: >= 2)");
```

```java
        }
        // 初始化成员变量
        this.alloc = alloc;
        this.direct = direct;
        this.maxNumComponents = maxNumComponents;
        components = newList(maxNumComponents);
        addComponents0(false, 0, buffers);   // 将缓冲区添加到组件数组中
        consolidateIfNeeded(); // 判断缓冲区容量，如果组件的数量超过当前操作允许的最大数量，则将它们合并
        setIndex(0, capacity()); // 设置读下标为 0，写下标为缓冲区列表中的最后一个缓冲区的结束偏移量
    }

    // 内部类，用于保证 ByteBuf 的属性信息
    private static final class Component {
        final ByteBuf buf;
        final int length;                      // 长度
        int offset;                            // 当前操作数据偏移量
        int endOffset;                         // 缓冲区结束偏移量

        Component(ByteBuf buf) {
            this.buf = buf;
            length = buf.readableBytes();
        }

        void freeIfNecessary() {
            buf.release();                     // 释放缓冲区
        }
    }
}
```

10.5.2 addComponents 核心方法

addComponents 方法用于向 CompositeByteBuf 中添加缓冲区，由于缓冲区的包装类为 Component，所以这里称为添加组件。increaseWriterIndex 参数表示在添加后是否增加写索引下标，cIndex 为 buffers 中开始添加的下标，通常该下标为 0。源码如下。

```java
private int addComponents0(boolean increaseWriterIndex, int cIndex, ByteBuf... buffers) {
    checkNotNull(buffers, "buffers");
    int i = 0;
    try {
        checkComponentIndex(cIndex);
        while (i < buffers.length) {           // 遍历 buffers 数组，调用 addComponent0 完成添加
            ByteBuf b = buffers[i++];
            if (b == null) {
                break;
            }
            cIndex = addComponent0(increaseWriterIndex, cIndex, b) + 1; // 每次递增 1
            int size = components.size();
            if (cIndex > size) {               // cIndex 达到最大的 components 的长度，重置 cIndex 为 list 长
度（实际插入操作会在 addComponent0 中处理）
                cIndex = size;
            }
        }
```

```
        return cIndex;
    } finally {
        for (; i < buffers.length; ++i) {        // 添加完成后释放缓冲区
            ByteBuf b = buffers[i];
            if (b != null) {
                try {
                    b.release();
                } catch (Throwable ignored) {
                    // ignore
                }
            }
        }
    }
}
```

10.5.3　addComponent 核心方法

addComponents0 方法批量添加缓冲区时，调用 addComponent 方法完成实际添加操作。流程如下。

（1）判断插入下标 cIndex 是否在合法范围内，否则会引起数组越界。

（2）设置 ByteBuf 字节序为大端序，同时调用 slice 方法创建一个新的 ByteBuf，然后包装为 Component 对象。

（3）如果插入下标达到 list 的长度，即插入位置为末尾，则将组件添加到列表中，判断当前是否为第一个组件，如果不是，则通过下标获取到列表的前一个组件，通过其偏移量来设置当前组件的 offset 和 endOffset。

（4）如果插入位置为中间，首先将组件放入缓冲区列表对应下标处，然后更新组件偏移量。

（5）如果需要更新写下标索引，则将当前写下标索引加上缓冲区的可读字节数。

```
private int addComponent0(boolean increaseWriterIndex, int cIndex, ByteBuf buffer) {
    assert buffer != null;
    boolean wasAdded = false;
    try {
        checkComponentIndex(cIndex);        // 判断插入下标：cIndex > 0 && cIndex < components.size()
        int readableBytes = buffer.readableBytes();
        @SuppressWarnings("deprecation")
        Component c = new Component(buffer.order(ByteOrder.BIG_ENDIAN).slice()); // 设置 ByteBuf 字节序
为大端序（因为在网络传输中，将采用大端序来传输字节），同时调用 slice 方法创建一个新的 ByteBuf，然后包
装为 Component 对象
        if (cIndex == components.size()) {   // 插入下标达到 list 的长度，即插入位置为末尾
            wasAdded = components.add(c);    // 将组件添加到列表中
            if (cIndex == 0) {               // 如果插入下标为 0，表明此时列表中没有组件，设置 endOffset
偏移量为当前组件的可读字节数
                c.endOffset = readableBytes;
            } else {                         // 否则获取前一个 Component 的 endOffset，设置其作为当前组
件的 offset，然后将 endOffset 设置为 offset+readableBytes（累计总和值）
                Component prev = components.get(cIndex - 1);
                c.offset = prev.endOffset;
                c.endOffset = c.offset + readableBytes;
            }
        } else {        // 插入下标不是在列表末尾，为中间插入，首先将组件放入缓冲区列表对应下标处，然后更
新组件偏移量（很耗时）
```

```
            components.add(cIndex, c);
            wasAdded = true;
            if (readableBytes != 0) {
                updateComponentOffsets(cIndex);        // 更新偏移量
            }
        }
        if (increaseWriterIndex) {        // 如果更新写下标索引，则将当前写下标索引加上缓冲区的可读字节数
            writerIndex(writerIndex() + buffer.readableBytes());
        }
        return cIndex;
    } finally {
        if (!wasAdded) {                              // 添加失败，释放缓冲区
            buffer.release();
        }
    }
}

// 为中间插入时，需要更新当前组件和后面的组件的缓冲区偏移量
private void updateComponentOffsets(int cIndex) {
    int size = components.size();
    if (size <= cIndex) {                             // 末尾插入，忽略
        return;
    }
    Component c = components.get(cIndex);
    if (cIndex == 0) {        // 如果为头部插入，则设置当前 offset 为 0，同时 endOffset 为当前可用的字节数即可
        c.offset = 0;
        c.endOffset = c.length;
        cIndex ++;
    }
    for (int i = cIndex; i < size; i ++) {            // 遍历当前索引下标后面的所有缓冲区组件，依靠前一个
components.get(i - 1)组件的 offset 和 endOffset 修正当前缓冲区的对应值
        Component prev = components.get(i - 1);
        Component cur = components.get(i);
        cur.offset = prev.endOffset;
        cur.endOffset = cur.offset + cur.length;
    }
}
```

10.5.4 removeComponent 核心方法

removeComponent 方法用于从缓冲区列表中移除对应 cIndex 下标处的组件。方法实现很简单，先移除下标为 cIndex 的组件，然后修正偏移量。源码如下。

```
public CompositeByteBuf removeComponent(int cIndex) {
    checkComponentIndex(cIndex);
    Component comp = components.remove(cIndex);
    comp.freeIfNecessary();
    if (comp.length > 0) {
        updateComponentOffsets(cIndex);
    }
    return this;
}
```

10.5.5　removeComponents 核心方法

removeComponents 方法用于从缓冲区列表中批量移除组件，从对应 cIndex 下标处开始，数量为 numComponents。流程如下。

（1）如果移除数量为 0，则直接返回。
（2）将需要移除的组件通过 subList 方法获取到子列表。
（3）遍历列表，判断缓冲区中是否有可用字节数设置 needsUpdate。
（4）如果移除的这一批缓冲区中，包含了可用字节数，即 needsUpdate 为 true，则需要从删除的下标处开始，修正后面的缓冲区的 offset 和 endOffset。

```
public CompositeByteBuf removeComponents(int cIndex, int numComponents) {
    checkComponentIndex(cIndex, numComponents);
    if (numComponents == 0) {        // 移除数量为 0，直接返回
        return this;
    }
    List<Component> toRemove = components.subList(cIndex, cIndex + numComponents); // 将需要移除的组件通过 subList 方法获取到子列表
    boolean needsUpdate = false;
    for (Component c: toRemove) { // 遍历列表，判断缓冲区中是否有可用字节数来设置 needsUpdate
        if (c.length > 0) {
            needsUpdate = true;
        }
        c.freeIfNecessary();        // 释放缓冲区
    }
    toRemove.clear();

    if (needsUpdate) {              // 如果移除的这一批缓冲区中，包含了可用字节数，则需要从删除的下标处开始，修正后面的缓冲区的 offset 和 endOffset
        updateComponentOffsets(cIndex);
    }
    return this;
}
```

10.5.6　capacity 核心方法

capacity 方法用于获取组件列表中的可用容量。由于已经在列表的末尾的 endOffset 中记录了所有待使用的数据，所以直接获取到列表末尾组件的 endOffset 即可。源码如下。

```
public int capacity() {
    final int numComponents = components.size();
    if (numComponents == 0) {
        return 0;
    }
    return components.get(numComponents - 1).endOffset;
}
```

10.5.7　decompose 核心方法

decompose 方法用于将偏移量为 offset，长度为 length 的缓冲区进行分解，并返回 ByteBuf 列表。

流程如下。

（1）长度为 0，直接返回。
（2）将 offset 转为对应的 component 下标（使用二分查找法查找）。
（3）处理第一个需要处理的组件，将该组件的缓冲区进行复制并更新读下标（缓冲区可用字节数）。
（4）循环遍历 offset-length 区间所涉及的组件，并将它们对应的 ByteBuf 添加到 slice 数组中。
（5）遍历列表，将所有的 ByteBuf 进行 slice 方法调用。此时会设置读下标为 0，写下标为当前可读字节数。

```java
public List<ByteBuf> decompose(int offset, int length) {
    checkIndex(offset, length);
    if (length == 0) {                                     // 长度为 0，直接返回
        return Collections.emptyList();
    }
    int componentId = toComponentIndex(offset);            // 将 offset 转为对应的 component 下标
    List<ByteBuf> slice = new ArrayList<ByteBuf>(components.size());
    Component firstC = components.get(componentId);        // 第一个需要处理的组件
    ByteBuf first = firstC.buf.duplicate();                // 将当前组件的缓冲区进行复制
    first.readerIndex(offset - firstC.offset);             // 更新读下标（缓冲区可用字节数）
    ByteBuf buf = first;
    int bytesToSlice = length;
    do {                  // 循环遍历 offset-length 区间所涉及的组件，并将它们对应的 ByteBuf 添加到 slice 数组中
        int readableBytes = buf.readableBytes();
        if (bytesToSlice <= readableBytes) {   // 缓冲区的可读字节数大于需要切割的 length 长度，表明为最后一个组件。修正 writerIndex 后，添加到 slice 数组中并结束循环
            buf.writerIndex(buf.readerIndex() + bytesToSlice); // 修正实际写下标为当前 length 的值
            slice.add(buf);
            break;
        } else {                                           // 不是最后一个组件，添加并获取到下一个组件循环
            slice.add(buf);
            bytesToSlice -= readableBytes;                 // 更新 bytesToSlice，此时为当前已经处理的可读字节数
            componentId ++;
            buf = components.get(componentId).buf.duplicate();
        }
    } while (bytesToSlice > 0);

    // 遍历列表，将所有的 ByteBuf 进行 slice 方法调用。此时会设置读下标为 0，写下标为当前可读字节数
    for (int i = 0; i < slice.size(); i ++) {
        slice.set(i, slice.get(i).slice());
    }
    return slice;
}

// 根据提供的 offset，计算出该偏移量字节所在的 Component 下标
public int toComponentIndex(int offset) {
    checkIndex(offset);
    for (int low = 0, high = components.size(); low <= high;) {   // 遍历二分查找，因为此时的 offset 到 endOffset 是按顺序递增的
        int mid = low + high >>> 1;
        Component c = components.get(mid);
        if (offset >= c.endOffset) {
            low = mid + 1;
```

```
        } else if (offset < c.offset) {
            high = mid - 1;
        } else {
            return mid;
        }
    }
    throw new Error("should not reach here");
}
```

10.5.8　findComponent 核心方法

findComponent 方法用于根据提供的 offset 找到包含该 offset 的组件。实现原理同 toComponentIndex 方法，只不过这里返回的不是下标，而是 Component 对象。源码如下。

```
private Component findComponent(int offset) {
    checkIndex(offset);
    for (int low = 0, high = components.size(); low <= high;) { // 二分查找
        int mid = low + high >>> 1;
        Component c = components.get(mid);
        if (offset >= c.endOffset) {
            low = mid + 1;
        } else if (offset < c.offset) {
            high = mid - 1;
        } else {
            assert c.length != 0;
            return c;
        }
    }
    throw new Error("should not reach here");
}
```

10.5.9　getByte 核心方法

getByte 方法用于获取 index 下标指定的字节，通过 findComponent 方法获取该下标所在的组件，同时调用内部的 ByteBuf 获取字节。源码如下。

```
protected byte _getByte(int index) {
    Component c = findComponent(index);
    return c.buf.getByte(index - c.offset);          // offset 为起始下标，index 与之相减可得到对应字节数
}
```

10.5.10　setByte 核心方法

setByte 方法用于设置 index 下标指定的字节，通过 findComponent 方法获取该下标所在的组件，同时调用内部的 ByteBuf 设置字节。源码如下。

```
public CompositeByteBuf setByte(int index, int value) {
    Component c = findComponent(index);
    c.buf.setByte(index - c.offset, value);
    return this;
}
```

10.6 WrappedByteBuf 类

WrappedByteBuf 类作为 SimpleLeakAwareByteBuf 和 AdvancedLeakAwareByteBuf 的父类，实现了基本的装饰者模式。该类内部包含一个 ByteBuf，buf 变量用于保存包装的 ByteBuf 对象，其中定义了一些 final 修饰的方法，硬性规定了这些方法不能被子类重写。源码如下。

```java
class WrappedByteBuf extends ByteBuf {

    protected final ByteBuf buf;    // 包装的 ByteBuf 对象

    protected WrappedByteBuf(ByteBuf buf) {
        if (buf == null) {
            throw new NullPointerException("buf");
        }
        this.buf = buf;
    }

    /*所有方法均是调用内部封装的 ByteBuf 完成。笔者这里就不列出所有方法了。注意，有些方法增加了 final 修饰符，而有些没有增加，这意味着子类可以复写该方法完成自己的额外逻辑*/

    public final int capacity() {
        return buf.capacity();
    }

    public final int readerIndex() {
        return buf.readerIndex();
    }

    public final int writerIndex() {
        return buf.writerIndex();
    }

    public byte getByte(int index) {
        return buf.getByte(index);
    }

    public ByteBuf setByte(int index, int value) {
        buf.setByte(index, value);
        return this;
    }
}
```

10.7 SimpleLeakAwareByteBuf 类

SimpleLeakAwareByteBuf 类继承自 WrappedByteBuf 类，该类重写了部分方法，实现了内存泄漏判断机制。推理如下。

（1）在原生 DirectByteBuffer 中，通过生成一个虚引用 Cleaner，在 DirectByteBuffer 对象被垃圾回收（GC）后，把 Cleaner 对象也回收。因为 Cleaner 的强引用 DirectByteBuffer 已经被回收。

（2）Cleaner 对象被回收时，会调用 remove 方法将自身从 Cleaner 链表中移除，同时调用 run 方法将原生 DirectByteBuffer 分配的堆外内存释放。

（3）ByteBuf 如果使用池化的实例 PooledByteBuf。则这片内存将由 Netty 自身进行管理：byteC:\Users\Daniel\AppData\Roaming\Microsoft\Word\堆内内存、DirectByteBuffer（堆外内存）。

（4）如果分配 PooledByteBuf 后，没有将这部分内存归还给 Netty 的内存池，当 PooledByteBuf 对象被回收后，这片内存将不可用，这时发生了内存泄漏。

（5）Netty 有提示用户发生内存泄漏的机制，用户可以根据该提示对内存泄漏的代码进行修改。

SimpleLeakAwareByteBuf 和 AdvancedLeakAwareByteBuf 类便是这样一种机制：判断 Netty 应用的内存泄漏，提供报告信息。而 SimpleLeakAwareByteBuf 是简单地进行抽样的判断，AdvancedLeakAwareByteBuf 对每一个 PooledByteBuf 都进行判断，所以性能上 SimpleLeakAwareByteBuf 较好，精确度上 AdvancedLeakAwareByteBuf 较好。该方法的核心原理在于内部变量 ResourceLeak。释放时会操作 ResourceLeak 进行关闭。源码如下。

```java
final class SimpleLeakAwareByteBuf extends WrappedByteBuf {
    private final ResourceLeak leak; // 内存泄漏包装对象

    SimpleLeakAwareByteBuf(ByteBuf buf, ResourceLeak leak) {
        super(buf);
        this.leak = leak;
    }
    /* 由于是简单的泄漏判断，所以 touch 方法不会保存任何信息 */
    @Override
    public ByteBuf touch() {
        return this;
    }

    @Override
    public ByteBuf touch(Object hint) {
        return this;
    }

    /* release 方法将在内存释放后对 ResourceLeak 管理对象关闭。读者可以思考：如果没有关闭，移除该对象后会怎样？请结合原生的 Cleaner 进行思考 */
    @Override
    public boolean release() {
        boolean deallocated = super.release();
        if (deallocated) {
            leak.close();
        }
        return deallocated;
    }

    @Override
    public boolean release(int decrement) {
        boolean deallocated = super.release(decrement);
        if (deallocated) {
```

```
            leak.close();
        }
        return deallocated;
}

/* 重写了需要生成新 ByteBuf 的方法，方法较多，笔者这里不再一一解释。读者只需要明白：只要是生成新
的 ByteBuf，都需要是该类的实例 */

    @Override
    public ByteBuf order(ByteOrder endianness) {
        leak.record();
        if (order() == endianness) {
            return this;
        } else {
            return new SimpleLeakAwareByteBuf(super.order(endianness), leak);
        }
    }


    @Override
    public ByteBuf slice() {
        return new SimpleLeakAwareByteBuf(super.slice(), leak);
    }

    @Override
    public ByteBuf duplicate() {
        return new SimpleLeakAwareByteBuf(super.duplicate(), leak);
    }
}
```

10.8　AdvancedLeakAwareByteBuf 类

AdvancedLeakAwareByteBuf 类会进一步加大对 PooledByteBuf 操作的内存泄漏信息的判断。由源码可知，该类记录了所有的方法，发生内存泄漏时可以提供完整详细的信息。源码如下。

```
final class AdvancedLeakAwareByteBuf extends WrappedByteBuf {
    private static final String PROP_ACQUIRE_AND_RELEASE_ONLY =
"io.Netty.leakDetection.acquireAndReleaseOnly";
    private static final boolean ACQUIRE_AND_RELEASE_ONLY;              // 标识是否需要在
recordLeakNonRefCountingOperation 方法中记录当前调用信息

    static {                              // 从系统变量初始化 ACQUIRE_AND_RELEASE_ONLY，默认为 false
        ACQUIRE_AND_RELEASE_ONLY =
SystemPropertyUtil.getBoolean(PROP_ACQUIRE_AND_RELEASE_ONLY, false);
    }

    private final ResourceLeak leak;          // 内存泄漏包装对象

    AdvancedLeakAwareByteBuf(ByteBuf buf, ResourceLeak leak) {
        super(buf);
        this.leak = leak;
```

```java
}

// 核心方法，用于记录当前调用栈信息
static void recordLeakNonRefCountingOperation(ResourceLeak leak) {
    if (!ACQUIRE_AND_RELEASE_ONLY) {
        leak.record();
    }
}

// 向内存泄漏的管理对象 ResourceLeak 提供 hint 对象，用于判断内存是否泄漏
@Override
public ByteBuf touch(Object hint) {
    leak.record(hint);
    return this;
}

/* 当内存释放到内存池时（引用计数为 1），调用 leak.close 方法将内存泄漏关闭 */

@Override
public boolean release() {
    boolean deallocated = super.release();
    if (deallocated) {
        leak.close();
    } else {
        leak.record();
    }
    return deallocated;
}

@Override
public boolean release(int decrement) {
    boolean deallocated = super.release(decrement);
    if (deallocated) {
        leak.close();
    } else {
        leak.record();
    }
    return deallocated;
}

/* 将需要生成新 ByteBuf 的方法进行重写，方法较多，笔者这里不再一一解释。读者只需要明白：只要是生成新的 ByteBuf，都需要是该类的实例。同时，会在调用 recordLeakNonRefCountingOperation 方法时对当前调用栈进行记录 */

public ByteBuf slice() {
    recordLeakNonRefCountingOperation(leak);
    return new AdvancedLeakAwareByteBuf(super.slice(), leak);
}

public ByteBuf duplicate() {
    recordLeakNonRefCountingOperation(leak);
    return new AdvancedLeakAwareByteBuf(super.duplicate(), leak);
}
}
```

10.9 WrappedCompositeByteBuf 类

WrappedCompositeByteBuf 类是用于实现组合缓冲区 CompositeByteBuf 的包装类，其中所有方法均是调用包装对象 CompositeByteBuf wrapped 来完成操作。源码如下。

```
class WrappedCompositeByteBuf extends CompositeByteBuf {
    private final CompositeByteBuf wrapped;

    // 构造器保存包装对象
    WrappedCompositeByteBuf(CompositeByteBuf wrapped) {
        super(wrapped.alloc());    // 使用包装的组合缓冲区分配器来初始化父类
        this.wrapped = wrapped;
    }

    /* 所有方法均是调用包装的 CompositeByteBuf 对象操作，这里只列出部分 */
    @Override
    public boolean release() {
        return wrapped.release();
    }

    @Override
    public boolean release(int decrement) {
        return wrapped.release(decrement);
    }
}
```

10.10 SimpleLeakAwareCompositeByteBuf 类

SimpleLeakAwareCompositeByteBuf 类继承自 WrappedCompositeByteBuf，对其中的部分方法进行了实现，实现了对内存泄漏的简单抽样判断机制。该类的核心为 ResourceLeak 变量，在构造器中对其进行初始化，在调用 release 方法将包装的内存释放后，会调用 close 方法完成关闭，此时不会导致内存泄漏。源码如下。

```
final class SimpleLeakAwareCompositeByteBuf extends WrappedCompositeByteBuf {

    private final ResourceLeak leak;

    SimpleLeakAwareCompositeByteBuf(CompositeByteBuf wrapped, ResourceLeak leak) {
        super(wrapped);
        this.leak = leak;
    }

    @Override
    public boolean release() {
        boolean deallocated = super.release();
        if (deallocated) {           // 占用的内存释放后，关闭泄漏判断
```

```java
            leak.close();
        }
        return deallocated;
    }

    @Override
    public boolean release(int decrement) {
        boolean deallocated = super.release(decrement);
        if (deallocated) {             // 释放占用的内存后，关闭泄漏判断
            leak.close();
        }
        return deallocated;
    }

    /*所有产生新 ByteBuf 的方法都被包装为 SimpleLeakAwareByteBuf 对象，这里不再一一列出*/
    @Override
    public ByteBuf slice() {
        return new SimpleLeakAwareByteBuf(super.slice(), leak);
    }

    @Override
    public ByteBuf retainedSlice() {
        return new SimpleLeakAwareByteBuf(super.retainedSlice(), leak);
    }
}
```

10.11 AdvancedLeakAwareCompositeByteBuf 类

AdvancedLeakAwareCompositeByteBuf 类继承自 WrappedCompositeByteBuf，并对其中的部分方法进行了实现。该类实现了对内存泄漏的增强抽样判断机制，核心为 ResourceLeak 变量，在构造器中对其进行初始化，然后调用 release 方法释放内存，再调用 close 方法完成关闭，此时不会导致内存泄漏。注意，该类与 SimpleLeakAwareCompositeByteBuf 不同的地方在于：所有调用方法都对调用栈进行了记录，而后者不会记录栈信息。源码如下。

```java
final class AdvancedLeakAwareCompositeByteBuf extends WrappedCompositeByteBuf {

    private final ResourceLeak leak;

    AdvancedLeakAwareCompositeByteBuf(CompositeByteBuf wrapped, ResourceLeak leak) {
        super(wrapped);
        this.leak = leak;
    }

    @Override
    public boolean release() {
        boolean deallocated = super.release();
        if (deallocated) {             // 释放占用的内存后，关闭泄漏判断
            leak.close();
        } else {
            leak.record();
```

```java
        }
        return deallocated;
    }

    @Override
    public boolean release(int decrement) {
        boolean deallocated = super.release(decrement);
        if (deallocated) {          // 释放占用的内存后，关闭泄漏判断
            leak.close();
        } else {
            leak.record();
        }
        return deallocated;
    }

    /* 所有产生新 ByteBuf 的方法都被包装为 SimpleLeakAwareByteBuf 对象，同时所有操作都调用
recordLeakNonRefCountingOperation 记录当前调用栈信息 */

    @Override
    public ByteBuf slice() {
        recordLeakNonRefCountingOperation(leak);
        return new AdvancedLeakAwareByteBuf(super.slice(), leak);
    }
    @Override
    public ByteBuf duplicate() {
        recordLeakNonRefCountingOperation(leak);
        return new AdvancedLeakAwareByteBuf(super.duplicate(), leak);
    }
}
```

10.12　ResourceLeak 接口

ResourceLeak 接口定义了内存泄漏包装对象的核心功能。源码如下。

```java
public interface ResourceLeak {
    // 记录调用栈信息
    void record();

    // 记录调用栈信息，同时指定 hint 对象，增加输出内存泄漏时的额外信息
    void record(Object hint);

    // 关闭内存泄漏判断。资源释放时，正常情况下应该调用该方法。此时不会导致内存泄漏
    boolean close();
}
```

10.13　ResourceLeakDetector 类

ResourceLeakDetector 类是 Netty 的内存泄漏管理器。该类提供了对 ResourceLeak 接口的实现类定

义，也提供了创建 ResourceLeak 实现类对象的方法，同时也提供了判断与报告内存泄漏的机制。

10.13.1 核心变量与构造器

由源码可知。

（1）Level 枚举类用于定义判断等级。

（2）默认判断等级为 Level.SIMPLE。

（3）DefaultResourceLeak 作为 ResourceLeak 接口的实现类，继承自 PhantomReference。该类同 JDK 原生的 Cleaner 机制一样，均采用虚引用判断内存泄漏。

```java
public class ResourceLeakDetector<T> {
    /* 定义系统变量配置项的名称 */
    private static final String PROP_LEVEL_OLD = "io.Netty.leakDetectionLevel";
    private static final String PROP_LEVEL = "io.Netty.leakDetection.level";
    private static final Level DEFAULT_LEVEL = Level.SIMPLE;     // 默认判断等级

    private static final String PROP_MAX_RECORDS = "io.Netty.leakDetection.maxRecords";
    private static final int DEFAULT_MAX_RECORDS = 4;            // 默认最大记录个数
    private static final int MAX_RECORDS;

    /**
     * 内存泄漏判断等级枚举
     */
    public enum Level {
        /**
         * 关闭内存泄漏
         */
        DISABLED,
        /**
         * 对内存资源进行简单抽样判断
         */
        SIMPLE,
        /**
         * 更进一步采用高级采样资源进行泄漏判断，以高开销为代价，报告泄漏对象最近被访问的位置
         */
        ADVANCED,
        /**
         * 对所有内存资源进行判断，报告更详细的信息。该功能消耗较大，建议在测试环境中使用
         */
        PARANOID
    }

    private static Level level;                                  // 判断等级

    static {                                                     // 从环境变量中读取信息初始化的内存泄漏管理器变量
        // 是否开启资源判断
        final boolean disabled;
        if (SystemPropertyUtil.get("io.Netty.noResourceLeakDetection") != null) {
            disabled = SystemPropertyUtil.getBoolean("io.Netty.noResourceLeakDetection", false);
        } else {
```

```java
            disabled = false;
        }
        Level defaultLevel = disabled ? Level.DISABLED : DEFAULT_LEVEL; // 没有被关闭，初始化为默认等级
        // 首先尝试从旧属性名中读取
        String levelStr = SystemPropertyUtil.get(PROP_LEVEL_OLD,defaultLevel.name()).trim().toUpperCase();
        // 使用新的属性名
        levelStr = SystemPropertyUtil.get(PROP_LEVEL, levelStr).trim().toUpperCase();
        Level level = DEFAULT_LEVEL;
        for (Level l : EnumSet.allOf(Level.class)) {
            if (levelStr.equals(l.name()) || levelStr.equals(String.valueOf(l.ordinal()))) {
                level = l;
            }
        }
        MAX_RECORDS = SystemPropertyUtil.getInt(PROP_MAX_RECORDS,DEFAULT_MAX_RECORDS);
        ResourceLeakDetector.level = PARANOID;
    }

    private static final String[] STACK_TRACE_ELEMENT_EXCLUSIONS = {
            "io.Netty.util.ReferenceCountUtil.touch(",
            "io.Netty.buffer.AdvancedLeakAwareByteBuf.touch(",
            "io.Netty.buffer.AbstractByteBufAllocator.toLeakAwareBuffer(",
            "io.Netty.buffer.AdvancedLeakAwareByteBuf.recordLeakNonRefCountingOperation("
    }; // 用于记录进行过滤的堆栈信息，这里过滤掉了资源判断本身的调用栈信息

    static final int DEFAULT_SAMPLING_INTERVAL = 128;    // 抽样判断时使用（对其取余运算）

    private final DefaultResourceLeak head = new DefaultResourceLeak(null);    // 内存泄漏资源的头结点
    private final DefaultResourceLeak tail = new DefaultResourceLeak(null);    // 内存泄漏资源的尾结点

    private final ReferenceQueue<Object> refQueue = new ReferenceQueue<Object>(); // JDK 引用队列
    private final ConcurrentMap<String,Boolean> reportedLeaks =PlatformDependent.newConcurrentHashMap();
                                                            // 用于标识已经报告的内存泄漏

    private final String resourceType;                      // 资源类型字符串
    private final int samplingInterval;                     // 抽样判断值
    private final int mask;                                 // 抽样判断值掩码，用于优化取模运算
    private final long maxActive;                           // 最大活动实例个数
    private long active;                                    // 当前活动实例个数
    private final AtomicBoolean loggedTooManyActive = new AtomicBoolean(); // 标识是否记录过多的活动实例
    private long leakCheckCnt;                              // 泄漏判断数量

    // 内部类，实现了 ResourceLeak 接口，注意，这里继承自虚引用完成对内存资源的包装
    private final class DefaultResourceLeak extends PhantomReference<Object> implements ResourceLeak {
        private final String creationRecord;                // 实例创建时的记录字符串
        private final Deque<String> lastRecords = new ArrayDeque<String>(); // 实例操作时记录字符串队列
        private final AtomicBoolean freed;                  // 标识是否已经释放资源
        private DefaultResourceLeak prev;                   // 泄漏资源双向链表的前驱结点
        private DefaultResourceLeak next;                   // 泄漏资源双向链表的后继结点
        private int removedRecords;                         // 达到最大记录个数时，移除的记录个数

        // 构造函数，用于初始化父类及成员变量
        DefaultResourceLeak(Object referent) {
```

```java
            super(referent, referent != null ? refQueue : null);// 初始化 PhantomReference
            if (referent != null) {                            // 引用队列存在
                Level level = getLevel();
                if (level.ordinal() >= Level.ADVANCED.ordinal()) { // 等级为 ADVANCED 或 PARANOID，在
创建时记录创建信息字符串
                    creationRecord = newRecord(null, 3);       // 调用 ResourceLeakDetector 的静态方法创建
新的记录字符串
                } else {
                    creationRecord = null;
                }

                synchronized (head) {                          // 将其链入全局内存泄漏链表
                    prev = head;
                    next = head.next;
                    head.next.prev = this;
                    head.next = this;
                    active++;                                   // 增加活动计数
                }
                freed = new AtomicBoolean();
            } else {                                           // 引用队列不存在，freed 标识当前已经被释放
                creationRecord = null;
                freed = new AtomicBoolean(true);
            }
        }

        // 记录调用栈
        @Override
        public void record() {
            record0(null, 3);
        }

        // 记录调用栈并携带 hint 对象
        @Override
        public void record(Object hint) {
            record0(hint, 3);
        }

        private void record0(Object hint, int recordsToSkip) {
            if (creationRecord != null) {                      // 存在创建记录，进行记录
                String value = newRecord(hint, recordsToSkip); // 调用 ResourceLeakDetector 的静态方法创
建新的记录字符串
                synchronized (lastRecords) {                   // 将记录放入 lastRecords 双端队列中
                    int size = lastRecords.size();
                    if (size == 0 || !lastRecords.getLast().equals(value)) { // 判断是否放入重复的记录
                        lastRecords.add(value);
                    }
                    if (size > MAX_RECORDS) {                  // 达到最大记录个数，丢弃最先放入的记录，同时增加移除
记录计数
                        lastRecords.removeFirst();
                        ++removedRecords;
                    }
                }
            }
        }
```

```java
        }
        // 释放内存资源时调用，用于解除对该资源的内存泄漏判断
        @Override
        public boolean close() {
            if (freed.compareAndSet(false, true)) { // 多线程安全，CAS 将资源标识为已经释放。同时将内存
                                                   // 泄漏包装对象从全局链表中移除
                synchronized (head) {
                    active--;
                    prev.next = next;
                    next.prev = prev;
                    prev = null;
                    next = null;
                }
                return true;
            }
            return false;
        }

        // 重写父类方法，完整打印记录的信息
        @Override
        public String toString() {
            if (creationRecord == null) {           // 创建记录不存在，直接返回空字符串
                return EMPTY_STRING;
            }

            final Object[] array;
            final int removedRecords;
            synchronized (lastRecords) {            // 获取记录的字符串数组
                array = lastRecords.toArray();
                removedRecords = this.removedRecords;
            }

            StringBuilder buf = new StringBuilder(16384).append(NEWLINE);
            if (removedRecords > 0) {               // 保存移除的记录数
                buf.append("WARNING: ")
                    .append(removedRecords)
                    .append(" leak records were discarded because the leak record count is limited to ")
                    .append(MAX_RECORDS)
                    .append(". Use system property ")
                    .append(PROP_MAX_RECORDS)
                    .append(" to increase the limit.")
                    .append(NEWLINE);
            }
            buf.append("Recent access records: ")
                .append(array.length)
                .append(NEWLINE);

            if (array.length > 0) {                 // 保存记录的字符串信息
                for (int i = array.length - 1; i >= 0; i--) {
                    buf.append('#')
                        .append(i + 1)
                        .append(':')
                        .append(NEWLINE)
```

```
                    .append(array[i]);
            }
        }
        buf.append("Created at:")
            .append(NEWLINE)
            .append(creationRecord);

        buf.setLength(buf.length() - NEWLINE.length());
        return buf.toString();
    }
}
```

10.13.2 newRecord 核心方法

newRecord 方法用于创建新的记录字符串信息。流程如下。

（1）创建 StringBuilder 字符缓冲区。
（2）写入 hint 对象信息。
（3）写入堆栈信息。

```
// 内存泄漏 hint 对象实现接口，用于获取 hint 对象更加详细的信息
public interface ResourceLeakHint {
    /**
     * 返回 hint 对象用于记录的字符串
     */
    String toHintString();
}

static String newRecord(Object hint, int recordsToSkip) {
    StringBuilder buf = new StringBuilder(4096);                    // 字符缓冲区
    // 写入 hint 对象信息
    if (hint != null) {
        buf.append("\tHint: ");
        if (hint instanceof ResourceLeakHint) {
            buf.append(((ResourceLeakHint) hint).toHintString());
        } else {
            buf.append(hint);
        }
        buf.append(NEWLINE);
    }

    // 写入堆栈信息
    StackTraceElement[] array = new Throwable().getStackTrace();
    for (StackTraceElement e : array) {
        if (recordsToSkip > 0) {                                    // 略过指定数量的栈信息
            recordsToSkip--;
        } else {
            String estr = e.toString();                             // 堆栈信息转为字符串
            boolean excluded = false;
            for (String exclusion : STACK_TRACE_ELEMENT_EXCLUSIONS) {  // 过滤冗余信息
                if (estr.startsWith(exclusion)) {
```

```
                    excluded = true;
                    break;
                }
            }
            if (!excluded) {                    // 将堆栈字符串信息写入缓冲区
                buf.append('\t');
                buf.append(estr);
                buf.append(NEWLINE);
            }
        }
    }
    return buf.toString();
}
```

10.13.3 open 核心方法

open 方法用于根据传入的 obj 资源对象,创建 ResourceLeak 包装类的实例。流程如下。

(1)如果内存泄漏处于关闭,直接返回。

(2)如果泄漏等级为 SIMPLE 或 ADVANCED,则进行抽样判断,在包装对象时报告内存泄漏。

(3)如果泄漏等级为 PARANOID,则对每个资源对象都判断内存泄漏,在包装对象时报告内存泄漏。

```
public final ResourceLeak open(T obj) {
    Level level = ResourceLeakDetector.level;
    if (level == Level.DISABLED) {              // 内存泄漏处于关闭,直接返回
        return null;
    }
    if (level.ordinal() < PARANOID.ordinal()) { // 泄漏等级为 SIMPLE 或 ADVANCED,进行抽样判断
        if ((leakCheckCnt++ & mask) == 0) {     // 使用掩码来加快取模运算,默认为 128,表示每 128 个对象抽样判断一次
            reportLeak(level);                  // 包装的同时,进行内存泄漏报告
            return new DefaultResourceLeak(obj);
        } else {
            return null;
        }
    } else {                                    // 泄漏等级为 PARANOID,对每个资源对象都判断内存泄漏
        reportLeak(level);                      // 包装的同时,进行内存泄漏报告
        return new DefaultResourceLeak(obj);
    }
}
```

10.13.4 reportLeak 核心方法

reportLeak 方法用于报告发生内存泄漏的资源对象。流程如下。

(1)如果关闭错误日志等级,则仅判断关闭资源。

(2)如果最大实例个数达到最大限制,则警告当前需要报告的实例个数过多,会导致严重性能损耗。

(3)循环遍历虚引用队列的对象,对其判断内存泄漏:是否调用 close 方法。如果发生内存泄漏,则报告内存泄漏信息。

```java
private void reportLeak(Level level) {
    if (!logger.isErrorEnabled()) {          // 错误日志等级关闭，仅判断资源关闭，因为此时无法输出错误日志
        for (;;) {
            @SuppressWarnings("unchecked")
            DefaultResourceLeak ref = (DefaultResourceLeak) refQueue.poll();
            if (ref == null) {
                break;
            }
            ref.close();
        }
        return;
    }
    int samplingInterval = level == PARANOID ? 1 : this.samplingInterval;
    if (active * samplingInterval > maxActive && loggedTooManyActive.compareAndSet(false, true)) { // 最大
    // 实例个数达到最大，警告当前需要报告的实例个数过多，会导致严重性能损耗
        reportInstancesLeak(resourceType);
    }

    // 循环遍历虚引用队列的对象，对其判断内存泄漏
    for (;;) {
        DefaultResourceLeak ref = (DefaultResourceLeak) refQueue.poll();
        if (ref == null) {                   // 没有回收的对象，直接返回
            break;
        }
        ref.clear();                         // 清空保存的 reference 对象
        if (!ref.close()) {  // 调用 close 方法尝试将 DefaultResourceLeak 从全局链表中删除，如果删除失败，说
    // 明之前已经调用过 close 方法，则当前对象没有发生内存泄漏，继续循环
            continue;
        }
        String records = ref.toString();     // 生成字符串对象
        if (reportedLeaks.putIfAbsent(records, Boolean.TRUE) == null) {    // 标识当前记录已经被报告
            if (records.isEmpty()) {
            } else {                         // 报告泄漏信息
                reportUntracedLeak(resourceType);
                reportTracedLeak(resourceType, records);
            }
        }
    }
}
```

10.13.5　reportXLeak 核心方法

reportXLeak 系列方法用于打印实际的内存泄漏错误提示信息。

（1）reportTracedLeak 方法用于打印内存泄漏的追踪的详细信息。

（2）reportUntracedLeak 方法用于打印内存泄漏的类型信息。

（3）reportInstancesLeak 方法用于打印当前判断的实例数过多的提示信息。

```java
protected void reportTracedLeak(String resourceType, String records) {
    logger.error(
```

```java
            "LEAK: {}.release() was not called before it's garbage-collected. " +
            "See http://Netty.io/wiki/reference-counted-objects.html for more information.{}",
            resourceType, records);
}

protected void reportUntracedLeak(String resourceType) {
    logger.error("LEAK: {}.release() was not called before it's garbage-collected. " +
            "Enable advanced leak reporting to find out where the leak occurred. " +
            "To enable advanced leak reporting, " +
            "specify the JVM option '-D{}={}' or call {}.setLevel() " +
            "See http://Netty.io/wiki/reference-counted-objects.html for more information.",
            resourceType, PROP_LEVEL, Level.ADVANCED.name().toLowerCase(),
simpleClassName(this));
}

protected void reportInstancesLeak(String resourceType) {
    logger.error("LEAK: You are creating too many " + resourceType + " instances.   " +
            resourceType + " is a shared resource that must be reused across the JVM," +
            "so that only a few instances are created.");
}
```

第 11 章

Netty 内存池原理

11.1 ByteBufAllocator

前面章节涉及的都是内存操作对象 ByteBuf 的实现，其中定义了读写下标、容量等信息。从本章节开始，将了解到用于分配的数据载体：堆内内存 byte[]数组、堆外内存 DirectByteBuffer 等数据载体，即内存分配器。其实不难得出结论，ByteBuf 用于定义操作数据载体的功能函数、ByteBufAllocator 接口用于定义实际分配的内存。注意，由于工具类包含方法太多，且均为工具方法，所以就同 PlatformDependent 类一样，笔者会在涉及时一并讲解，而不再另起章节讲解。从源码中可知，分配的内存分为堆外缓冲区、堆内缓冲区、堆内组合缓冲区、堆外组合缓冲区。源码如下。

```java
public final class ByteBufUtil {
    static final ByteBufAllocator DEFAULT_ALLOCATOR;                  // 默认实例
    static {
        String allocType = SystemPropertyUtil.get(
            "io.Netty.allocator.type", PlatformDependent.isAndroid() ? "unpooled" : "pooled"); // 如果不是安卓平台，则默认为池化的分配器（安卓平台中的内存很珍贵，需要及时给 OS 归还内存）
        allocType = allocType.toLowerCase(Locale.US).trim();
        ByteBufAllocator alloc;
        // 根据类型创建实例
        if ("unpooled".equals(allocType)) {
            alloc = UnpooledByteBufAllocator.DEFAULT;
        } else if ("pooled".equals(allocType)) {
            alloc = PooledByteBufAllocator.DEFAULT;
        } else {
            alloc = PooledByteBufAllocator.DEFAULT;
        }
        DEFAULT_ALLOCATOR = alloc;
    }
}

public interface ByteBufAllocator {
    ByteBufAllocator DEFAULT = ByteBufUtil.DEFAULT_ALLOCATOR;          // 保存默认内存分配器实例

    // 分配一个具备默认容量的 ByteBuf，是否为堆外内存，取决于具体实现
    ByteBuf buffer();

    // 分配一个具备指定 initialCapacity 容量的 ByteBuf，是否为堆外内存，取决于具体实现
    ByteBuf buffer(int initialCapacity);
```

```java
// 分配一个具备指定 initialCapacity 容量且最大容量为 maxCapacity 的 ByteBuf, 是否为堆外内存, 取决于具体实现
ByteBuf buffer(int initialCapacity, int maxCapacity);

// 分配一个具备默认容量的 ByteBuf, 优先分配堆外内存, 因为对于 IO 缓冲区来说, 堆外内存可以减少复制次数
ByteBuf ioBuffer();

// 分配一个具备初始 initialCapacity 容量的 ByteBuf, 优先分配堆外内存, 对于 IO 缓冲区来说, 堆外内存可以减少复制次数
ByteBuf ioBuffer(int initialCapacity);

// 分配具备初始 initialCapacity 容量且最大容量为 maxCapacity 的 ByteBuf, 优先分配堆外内存, 对于 IO 缓冲区来说, 堆外内存可以减少复制次数
ByteBuf ioBuffer(int initialCapacity, int maxCapacity);

// 分配具备默认容量的堆内内存 ByteBuf
ByteBuf heapBuffer();

// 分配具备初始容量 initialCapacity 的堆内内存 ByteBuf
ByteBuf heapBuffer(int initialCapacity);

// 分配具备初始容量 initialCapacity 且最大容量为 maxCapacity 的堆内内存 ByteBuf
ByteBuf heapBuffer(int initialCapacity, int maxCapacity);

// 分配具备默认容量的堆外内存 ByteBuf
ByteBuf directBuffer();

// 分配具备初始容量 initialCapacity 的堆外内存 ByteBuf
ByteBuf directBuffer(int initialCapacity);

// 分配具备初始容量 initialCapacity 且最大容量为 maxCapacity 的堆外内存 ByteBuf
ByteBuf directBuffer(int initialCapacity, int maxCapacity);

// 分配具备默认最大组件的组合缓冲区 CompositeByteBuf, 是否为堆外内存, 取决于具体实现
CompositeByteBuf compositeBuffer();

// 分配具备指定最大组件 maxNumComponents 的组合缓冲区 CompositeByteBuf, 是否为堆外内存, 取决于具体实现
CompositeByteBuf compositeBuffer(int maxNumComponents);

// 分配具备默认最大组件的组合堆内缓冲区 CompositeByteBuf
CompositeByteBuf compositeHeapBuffer();

// 分配具备指定最大组件 maxNumComponents 的组合堆内缓冲区 CompositeByteBuf
CompositeByteBuf compositeHeapBuffer(int maxNumComponents);

// 分配具备默认最大组件的组合堆外缓冲区 CompositeByteBuf
CompositeByteBuf compositeDirectBuffer();

// 分配具备指定最大组件 maxNumComponents 的组合堆外缓冲区 CompositeByteBuf
CompositeByteBuf compositeDirectBuffer(int maxNumComponents);

// 获取堆外内存缓冲区是否进行了池化
```

```
        boolean isDirectBufferPooled();

        // 在对 ByteBuf 进行扩容时，计算新的容量值
        int calculateNewCapacity(int minNewCapacity, int maxCapacity);
}
```

11.2 AbstractByteBufAllocator 类

众所周知，一旦定义了接口，就可以定义一个模板类，将子类共用方法进行实现。同理，这里的 AbstractByteBufAllocator 抽象类就对 ByteBufAllocator 接口的部分方法提供了实现，子类只需要重写方法或实现模板方法即可。

11.2.1 核心变量与构造器

由源码可知。
（1）定义了默认初始容量和最大组件数。
（2）定义了 directByDefault 变量，用于表示是否优先分配堆外内存，在构造器中需要存在 Unsafe 类才能优先分配堆外内存。

```
public abstract class AbstractByteBufAllocator implements ByteBufAllocator {
    private static final int DEFAULT_INITIAL_CAPACITY = 256;    // 默认初始容量
    static final int DEFAULT_MAX_COMPONENTS = 16;               // 默认最大组件数，用于支持组合缓冲区
    private final boolean directByDefault;                      // 是否默认创建堆外内存缓冲区
    private final ByteBuf emptyBuf;                             // 表示容量为 0 的空缓冲区

    // 指定是否优先分配堆外内存 preferDirect 来创建实例
    protected AbstractByteBufAllocator(boolean preferDirect) {
        directByDefault = preferDirect && PlatformDependent.hasUnsafe(); // 必须有 Unsafe 实例才允许默认分配堆外内存
        emptyBuf = new EmptyByteBuf(this);
    }
}
```

11.2.2 buffer 核心方法

buffer 系列方法用于分配内存，只不过根据参数 directByDefault 决定是否优先分配堆外内存，均是调用 heapBuffer(int initialCapacity, int maxCapacity) 和 directBuffer(int initialCapacity, int maxCapacity) 方法完成创建，而该方法对初始容量和最大容量校验后，调用子类实现的 newHeapBuffer 方法和 newDirectBuffer 方法完成创建。源码如下。

```
@Override
public ByteBuf buffer() {
    if (directByDefault) {
        return directBuffer();
    }
    return heapBuffer();
```

```java
}

@Override
public ByteBuf buffer(int initialCapacity) {
    if (directByDefault) {
        return directBuffer(initialCapacity);
    }
    return heapBuffer(initialCapacity);
}

@Override
public ByteBuf buffer(int initialCapacity, int maxCapacity) {
    if (directByDefault) {
        return directBuffer(initialCapacity, maxCapacity);
    }
    return heapBuffer(initialCapacity, maxCapacity);
}

/* 分配堆外内存 */
@Override
public ByteBuf directBuffer() {
    return directBuffer(DEFAULT_INITIAL_CAPACITY, Integer.MAX_VALUE);
}

@Override
public ByteBuf directBuffer(int initialCapacity) {
    return directBuffer(initialCapacity, Integer.MAX_VALUE);
}

@Override
public ByteBuf directBuffer(int initialCapacity, int maxCapacity) {
    if (initialCapacity == 0 && maxCapacity == 0) {
        return emptyBuf;
    }
    validate(initialCapacity, maxCapacity);
    return newDirectBuffer(initialCapacity, maxCapacity);
}

// 子类完成实现
protected abstract ByteBuf newDirectBuffer(int initialCapacity, int maxCapacity);

/* 分配堆内内存 */

// 如果没有指定初始容量和最大容量,则使用默认值: 256 和整型最大值
public ByteBuf heapBuffer() {
    return heapBuffer(DEFAULT_INITIAL_CAPACITY, Integer.MAX_VALUE);
}

public ByteBuf heapBuffer(int initialCapacity, int maxCapacity) {
    if (initialCapacity == 0 && maxCapacity == 0) {
        return emptyBuf;
    }
    validate(initialCapacity, maxCapacity);
    return newHeapBuffer(initialCapacity, maxCapacity);
```

```
}
// 子类完成实现
protected abstract ByteBuf newHeapBuffer(int initialCapacity, int maxCapacity);
```

11.2.3 ioBuffer 核心方法

ioBuffer 方法用于分配 IO 使用的内存，优先分配堆外内存，减少复制次数，但是分配的前提是具备 Unsafe 类。最终通过调用 directBuffer 方法完成创建。源码如下。

```
@Override
public ByteBuf ioBuffer() {
    if (PlatformDependent.hasUnsafe()) {
        return directBuffer(DEFAULT_INITIAL_CAPACITY);
    }
    return heapBuffer(DEFAULT_INITIAL_CAPACITY);
}

@Override
public ByteBuf ioBuffer(int initialCapacity) {
    if (PlatformDependent.hasUnsafe()) {
        return directBuffer(initialCapacity);
    }
    return heapBuffer(initialCapacity);
}

@Override
public ByteBuf ioBuffer(int initialCapacity, int maxCapacity) {
    if (PlatformDependent.hasUnsafe()) {
        return directBuffer(initialCapacity, maxCapacity);
    }
    return heapBuffer(initialCapacity, maxCapacity);
}
```

11.2.4 compositeBuffer 核心方法

该系列方法用于分配组合缓冲区，最终的方法会调用 compositeDirectBuffer(int maxNumComponents) 和 compositeHeapBuffer(int maxNumComponents)，分别用于分配堆外组合缓冲区和堆内组合缓冲区，在这两个方法中分别创建 new CompositeByteBuf(this, true, maxNumComponents) 和 new CompositeByteBuf(this, false, maxNumComponents) 对象。源码如下。

```
public CompositeByteBuf compositeBuffer() {
    if (directByDefault) {
        return compositeDirectBuffer();
    }
    return compositeHeapBuffer();
}

@Override
public CompositeByteBuf compositeBuffer(int maxNumComponents) {
    if (directByDefault) {
        return compositeDirectBuffer(maxNumComponents);
```

```java
    }
    return compositeHeapBuffer(maxNumComponents);
}

@Override
public CompositeByteBuf compositeHeapBuffer() {
    return compositeHeapBuffer(DEFAULT_MAX_COMPONENTS);
}

@Override
public CompositeByteBuf compositeHeapBuffer(int maxNumComponents) {
    return toLeakAwareBuffer(new CompositeByteBuf(this, false, maxNumComponents));
}

@Override
public CompositeByteBuf compositeDirectBuffer() {
    return compositeDirectBuffer(DEFAULT_MAX_COMPONENTS);
}

@Override
public CompositeByteBuf compositeDirectBuffer(int maxNumComponents) {
    return toLeakAwareBuffer(new CompositeByteBuf(this, true, maxNumComponents));
}
```

11.2.5 toLeakAwareBuffer 核心方法

该系列方法用于将创建的 ByteBuf 和 CompositeByteBuf 包装为 XLeakAwareByteBuf，用于判断泄漏的包装对象。通过源码实现较为简单，根据 ResourceLeakDetector 的等级来创建对应的包装对象。源码如下。

```java
protected static ByteBuf toLeakAwareBuffer(ByteBuf buf) {
    ResourceLeak leak;
    switch (ResourceLeakDetector.getLevel()) {
        case SIMPLE:          // 简单采样
            leak = AbstractByteBuf.leakDetector.open(buf);
            if (leak != null) {
                buf = new SimpleLeakAwareByteBuf(buf, leak);
            }
            break;
        case ADVANCED:        // 加强判断
        case PARANOID:
            leak = AbstractByteBuf.leakDetector.open(buf);
            if (leak != null) {
                buf = new AdvancedLeakAwareByteBuf(buf, leak);
            }
            break;
        default:
            break;
    }
    return buf;
}

protected static CompositeByteBuf toLeakAwareBuffer(CompositeByteBuf buf) {
```

```
ResourceLeak leak;
switch (ResourceLeakDetector.getLevel()) {
    case SIMPLE:
        leak = AbstractByteBuf.leakDetector.open(buf);
        if (leak != null) {
            buf = new SimpleLeakAwareCompositeByteBuf(buf, leak);    // 简单组合缓冲区内存判断对象
        }
        break;
    case ADVANCED:
    case PARANOID:
        leak = AbstractByteBuf.leakDetector.open(buf);
        if (leak != null) {
            buf = new AdvancedLeakAwareCompositeByteBuf(buf, leak); // 加强组合缓冲区内存判断对象
        }
        break;
    default:
        break;
}
return buf;
}
```

11.2.6　calculateNewCapacity 核心方法

calculateNewCapacity 方法用于根据传入的最小新容量和最大容量进行计算，返回计算好的新容量值。这里最大为 4MB，如果大于该容量将取该容量的倍数值，小于该容量将从 64B 开始以 2 倍递增计算新容量值。源码如下。

```
public int calculateNewCapacity(int minNewCapacity, int maxCapacity) {
    // 参数校验
    if (minNewCapacity < 0) {
        throw new IllegalArgumentException("minNewCapacity: " + minNewCapacity + " (expectd: 0+)");
    }
    if (minNewCapacity > maxCapacity) {
        throw new IllegalArgumentException(String.format(
            "minNewCapacity: %d (expected: not greater than maxCapacity(%d)",
            minNewCapacity, maxCapacity));
    }
    final int threshold = 1048576 * 4;
    if (minNewCapacity == threshold) {           // 新容量等于容量限制（4MB），则返回限制值
        return threshold;
    }
    // 如果超过限制值，则按照 4MB 的倍数进行增加
    if (minNewCapacity > threshold) {
        int newCapacity = minNewCapacity / threshold * threshold;
        if (newCapacity > maxCapacity - threshold) {    // 新容量大于 maxCapacity，则设置为 maxCapacity
            newCapacity = maxCapacity;
        } else {
            newCapacity += threshold;    // 否则增加一个容量限制值（除法运算可能会导致分配的数量较少，所以加一个限制值）
        }
        return newCapacity;
    }
```

```
// 没有超过最大容量，从 64B 开始以 2 倍增加（64、128、256…），直到找到正好大于 minNewCapacity 的值
int newCapacity = 64;
while (newCapacity < minNewCapacity) {
    newCapacity <<= 1;
}
return Math.min(newCapacity, maxCapacity);    // 计算的新容量不得大于最大容量
}
```

11.3 UnpooledByteBufAllocator 类

UnpooledByteBufAllocator 类继承自 AbstractByteBufAllocator，实现了其中的分配核心函数。该类不会对内存进行池化，所以实现很简单，仅根据内存类型创建 UnpooledHeapByteBuf 或 UnpooledDirectByteBuf。同时 disableLeakDetector 变量用于控制是否对该内存进行泄漏探测，默认为开启状态。可能读者会有疑惑：没有池化内存，为什么还要打开该选项呢？当不用该内存时，垃圾回收（GC）会帮助解决。不管是堆内内存 byte 数组或堆外内存 DirectByteBuffer。但内存泄漏本身就是一个程序错误。这里需要提醒开发者注意，对于 newUnsafeDirectByteBuf 方法创建的 UnpooledUnsafeNoCleanerDirectByteBuf 来说，没有使用 Cleaner，就意味着开辟的内存不会引用 Cleaner 释放。源码如下。

```
public final class PlatformDependent {
    private static final boolean DIRECT_BUFFER_PREFERRED =
        HAS_UNSAFE && !SystemPropertyUtil.getBoolean("io.Netty.noPreferDirect", false);  // 通常为 true

    public static boolean directBufferPreferred() {
        return DIRECT_BUFFER_PREFERRED;
    }
}

public final class UnpooledByteBufAllocator extends AbstractByteBufAllocator {
    private final boolean disableLeakDetector;   // 是否使用内存泄漏探测，true 表示不使用内存泄漏探测，false 表示使用内存泄漏探测
    public static final UnpooledByteBufAllocator DEFAULT =
        new UnpooledByteBufAllocator(PlatformDependent.directBufferPreferred());    // 默认实例，使用 PlatformDependent 决定是否优先分配堆外内存，通常为 true

    // 构造器，设置是否优先分配堆外内存，false 表示使用内存泄漏探测机制
    public UnpooledByteBufAllocator(boolean preferDirect) {
        this(preferDirect, false);
    }

    // 构造器，preferDirect 设置是否优先分配堆外内存，disableLeakDetector 设置是否使用内存泄漏探测机制
    public UnpooledByteBufAllocator(boolean preferDirect, boolean disableLeakDetector) {
        super(preferDirect);
        this.disableLeakDetector = disableLeakDetector;
    }

    // 分配堆内内存，根据是否存在 Unsafe 来选择创建 UnpooledUnsafeHeapByteBuf 或 UnpooledHeapByteBuf
    （由此可见 Netty 非常倾向于自己管理内存）
```

```java
    @Override
    protected ByteBuf newHeapBuffer(int initialCapacity, int maxCapacity) {
        return PlatformDependent.hasUnsafe() ? new UnpooledUnsafeHeapByteBuf(this, initialCapacity,
maxCapacity): new UnpooledHeapByteBuf(this, initialCapacity, maxCapacity);
    }

    // 分配堆外内存，根据是否存在 Unsafe 选择创建 newUnsafeDirectByteBuf 或 UnpooledDirectByteBuf
    @Override
    protected ByteBuf newDirectBuffer(int initialCapacity, int maxCapacity) {
        ByteBuf buf = PlatformDependent.hasUnsafe() ?
            UnsafeByteBufUtil.newUnsafeDirectByteBuf(this, initialCapacity, maxCapacity) :
            new UnpooledDirectByteBuf(this, initialCapacity, maxCapacity);
        return disableLeakDetector ? buf : toLeakAwareBuffer(buf); // 如果使用内存泄漏探测，则对其进行包装
    }

    // 分配堆内组合内存 CompositeByteBuf
    @Override
    public CompositeByteBuf compositeHeapBuffer(int maxNumComponents) {
        CompositeByteBuf buf = new CompositeByteBuf(this, false, maxNumComponents);
        return disableLeakDetector ? buf : toLeakAwareBuffer(buf);
    }

    // 分配堆外组合内存 CompositeByteBuf
    @Override
    public CompositeByteBuf compositeDirectBuffer(int maxNumComponents) {
        CompositeByteBuf buf = new CompositeByteBuf(this, true, maxNumComponents);
        return disableLeakDetector ? buf : toLeakAwareBuffer(buf);
    }

    // 非池化，返回 false
    @Override
    public boolean isDirectBufferPooled() {
        return false;
    }
}
```

11.4　PooledByteBufAllocator 类

UnpooledByteBufAllocator 类很简单，仅实现了简单的创建操作，直接创建 UnpooledByteBuf 实例即可。在这些实例的内部使用自己开辟 byte[]数组或 DirectByteBuffer 对象完成数据的承载。本节介绍的内容是 Netty 的一大难点，即内存池化，也就是内部实现了完整的内存池管理。该类相对复杂，笔者尽量详细拆分，帮助读者更好地进行理解。

11.4.1　核心变量与构造器

由源码可知。

（1）Netty 默认使用 8KB 作为基本管理页。什么是基础管理页，即分配内存的最小单位。

（2）Netty 默认使用 11 作为基本管理块的深度。什么是基本管理块，即管理上述的基本管理页的

内存块。

（3）Netty 的默认基本内存管理块大小为：8192 << 11 = 16MB。

（4）内存区可以给所有线程分配内存。而又由于分配线程有多个，为了更好地进行锁粒度细化，就引入了 ARENA 内存管理区域，减少锁竞争。因为内存分为堆内内存和堆外内存，所以 ARENA 也分为 HEAP_ARENA（堆内内存区域）和 DIRECT_ARENA（堆外内存区域）。

（5）默认的 DEFAULT_NUM_HEAP_ARENA 和 DEFAULT_NUM_DIRECT_ARENA 数量均为：CPU 核心数×2。内存分配器主要是谁在使用，答案一定是 Netty 的事件循环组。通过第 1 篇的介绍可知，事件循环组的个数也就是 CPU 核心数×2，当两者匹配时，是不是可以最大限度地减少锁竞争，即每个线程一个 area。

（6）虽然上述已经极大地减少了锁竞争，但是如果业务线程也参与到分配过程中呢？由于内存区域有限，必然会导致内存竞争，那么如何摆脱竞争呢？这必然是通过 ThreadLocal 实现，它将分配的内存缓存在线程本地的缓存中，再次需要分配时不必再前往内存区域分配，PoolThreadLocalCache 类便是起到了这样的作用。

（7）在 static 静态代码块中，可以通过设定对应的系统变量值修改这些值。

```java
public class PooledByteBufAllocator extends AbstractByteBufAllocator {
    private static final int DEFAULT_NUM_HEAP_ARENA;          // 默认堆内内存区域数量
    private static final int DEFAULT_NUM_DIRECT_ARENA;        // 默认堆外内存区域数量

    private static final int DEFAULT_PAGE_SIZE;               // 默认基本管理页大小
    private static final int DEFAULT_MAX_ORDER;  // 默认基本管理内存块深度，用于 PAGE_SIZE 计算基本内存块大小
    private static final int DEFAULT_TINY_CACHE_SIZE;         // 默认微型内存块管理数量
    private static final int DEFAULT_SMALL_CACHE_SIZE;        // 默认小型内存块管理数量
    private static final int DEFAULT_NORMAL_CACHE_SIZE;       // 默认普通内存块管理数量
    private static final int DEFAULT_MAX_CACHED_BUFFER_CAPACITY;  // 默认最大缓存 Buffer 容量
    private static final int DEFAULT_CACHE_TRIM_INTERVAL;     // 默认缓存清理大小限制

    private static final int MIN_PAGE_SIZE = 4096;            // 最小基本管理页大小是 4k
    private static final int MAX_CHUNK_SIZE = (int) (((long) Integer.MAX_VALUE + 1) / 2); // 最大基本管理内存块大小，取整型值最大值的一半

    static { // 静态代码块，从系统变量中获取设定值，初始化 static final 默认变量
        // 初始化默认基本管理页大小，默认一页大小为 8KB
        int defaultPageSize = SystemPropertyUtil.getInt("io.Netty.allocator.pageSize", 8192);
        Throwable pageSizeFallbackCause = null;
        try {
            validateAndCalculatePageShifts(defaultPageSize);
        } catch (Throwable t) {
            pageSizeFallbackCause = t;
            defaultPageSize = 8192;
        }
        DEFAULT_PAGE_SIZE = defaultPageSize;

        // 初始化默认基本管理块深度，默认是 11
        int defaultMaxOrder = SystemPropertyUtil.getInt("io.Netty.allocator.maxOrder", 11);
        Throwable maxOrderFallbackCause = null;
```

```java
    try {
        validateAndCalculateChunkSize(DEFAULT_PAGE_SIZE, defaultMaxOrder);
    } catch (Throwable t) {
        maxOrderFallbackCause = t;
        defaultMaxOrder = 11;
    }
    DEFAULT_MAX_ORDER = defaultMaxOrder;

    // 初始化默认堆内内存管理区域和堆外内存管理区域个数，默认为 CPU 核心数×2
    final Runtime runtime = Runtime.getRuntime();
    final int defaultMinNumArena = runtime.availableProcessors() * 2;
    final int defaultChunkSize = DEFAULT_PAGE_SIZE << DEFAULT_MAX_ORDER;
    DEFAULT_NUM_HEAP_ARENA = Math.max(0,
                            SystemPropertyUtil.getInt(
                                "io.Netty.allocator.numHeapArenas",
                                (int) Math.min(
                                    defaultMinNumArena,
                                    runtime.maxMemory()/defaultChunkSize/2/3)));
    DEFAULT_NUM_DIRECT_ARENA = Math.max(0,
                            SystemPropertyUtil.getInt(
                                "io.Netty.allocator.numDirectArenas",
                                (int) Math.min(
                                    defaultMinNumArena,
                                    PlatformDependent.maxDirectMemory()/defaultChunkSize/2/3)));

    // 初始化默认微型内存、小内存、普通内存的管理数量，分别默认为：512B（微型内存）、256B（小内存）、64B（普通内存）
    DEFAULT_TINY_CACHE_SIZE = SystemPropertyUtil.getInt("io.Netty.allocator.tinyCacheSize", 512);
    DEFAULT_SMALL_CACHE_SIZE = SystemPropertyUtil.getInt("io.Netty.allocator.smallCacheSize", 256);
    DEFAULT_NORMAL_CACHE_SIZE=SystemPropertyUtil.getInt("io.Netty.allocator.normalCacheSize", 64);

    // 默认最大缓冲区容量为 32KB
    DEFAULT_MAX_CACHED_BUFFER_CAPACITY = SystemPropertyUtil.getInt(
        "io.Netty.allocator.maxCachedBufferCapacity", 32 * 1024);

    // 默认缓存清理周期为 8KB。若长时间不用，超过该阈值会释放
    DEFAULT_CACHE_TRIM_INTERVAL = SystemPropertyUtil.getInt(
        "io.Netty.allocator.cacheTrimInterval", 8192);
}

public static final PooledByteBufAllocator DEFAULT =
    new PooledByteBufAllocator(PlatformDependent.directBufferPreferred()); // 全局默认池化分配器实例

private final PoolArena<byte[]>[] heapArenas;           // 堆内内存管理区域数组，直接指定泛型为 byte[]
private final PoolArena<ByteBuffer>[] directArenas;     // 堆内内存管理区域数组，直接指定泛型为
                                                        // ByteBuffer，实际为 DirectByteBuffer
private final int tinyCacheSize;                        // 微型内存管理数量
private final int smallCacheSize;                       // 小型内存管理数量
private final int normalCacheSize;                      // 普通内存管理数量
private final PoolThreadLocalCache threadCache;         // 线程本地内存缓存，用于无锁化分配

// 构造器，设置不优先分配堆外内存
public PooledByteBufAllocator() {
```

```java
            this(false);
        }

        // 构造器，设置如下：默认堆内内存区个数为 CPU 核心数×2，默认堆外内存区个数为 CPU 核心数×2，默认
        // 页大小为 8KB，默认内存管理深度是 11
        public PooledByteBufAllocator(boolean preferDirect) {
            this(preferDirect, DEFAULT_NUM_HEAP_ARENA, DEFAULT_NUM_DIRECT_ARENA,
DEFAULT_PAGE_SIZE, DEFAULT_MAX_ORDER);
        }

        public PooledByteBufAllocator(int nHeapArena, int nDirectArena, int pageSize, int maxOrder) {
            this(false, nHeapArena, nDirectArena, pageSize, maxOrder);
        }

        // 构造器，设置默认微型内存管理数量为 512B，默认小型内存管理数量为 256B，默认普通内存管理数量为 64B
        public PooledByteBufAllocator(boolean preferDirect, int nHeapArena, int nDirectArena, int pageSize, int
maxOrder) {
            this(preferDirect, nHeapArena, nDirectArena, pageSize, maxOrder,
                    DEFAULT_TINY_CACHE_SIZE, DEFAULT_SMALL_CACHE_SIZE,
DEFAULT_NORMAL_CACHE_SIZE);
        }

        // 构造器，通过入参初始化成员变量
        public PooledByteBufAllocator(boolean preferDirect, int nHeapArena, int nDirectArena, int pageSize, int
maxOrder, int tinyCacheSize, int smallCacheSize, int normalCacheSize) {
            super(preferDirect);
            threadCache = new PoolThreadLocalCache();              // 创建线程本地内存缓存
            this.tinyCacheSize = tinyCacheSize;
            this.smallCacheSize = smallCacheSize;
            this.normalCacheSize = normalCacheSize;
            final int chunkSize = validateAndCalculateChunkSize(pageSize, maxOrder); // 验证并计算基本内存块
大小：8192 << 11 = 16 MB
            if (nHeapArena < 0) {
                throw new IllegalArgumentException("nHeapArena: " + nHeapArena + " (expected: >= 0)");
            }
            if (nDirectArena < 0) {
                throw new IllegalArgumentException("nDirectArea: " + nDirectArena + " (expected: >= 0)");
            }
            int pageShifts = validateAndCalculatePageShifts(pageSize); // 验证并计算 page 页大小基于 2 的偏移
量：2^N = 8192
            if (nHeapArena > 0) {                                          // 创建堆内内存管理区域
                heapArenas = newArenaArray(nHeapArena);                    // 创建堆内内存管理区数组
                List<PoolArenaMetric> metrics = new ArrayList<PoolArenaMetric>(heapArenas.length); // 内存度
量数组，用于做性能监测，由于不重要，忽略即可
                for (int i = 0; i < heapArenas.length; i ++) {             // 分配堆内管理区
                    PoolArena.HeapArena arena = new PoolArena.HeapArena(this, pageSize, maxOrder,
pageShifts, chunkSize);
                    heapArenas[i] = arena;
                    metrics.add(arena);
                }
                heapArenaMetrics = Collections.unmodifiableList(metrics);
            } else {
                heapArenas = null;
```

```
            heapArenaMetrics = Collections.emptyList();
        }

        if (nDirectArena > 0) {                          // 创建堆外内存管理区域
            directArenas = newArenaArray(nDirectArena);  // 创建堆外内存管理区数组
            List<PoolArenaMetric> metrics = new ArrayList<PoolArenaMetric>(directArenas.length);
            for (int i = 0; i < directArenas.length; i ++) {   // 分配堆外管理区
                PoolArena.DirectArena arena = new PoolArena.DirectArena(
                    this, pageSize, maxOrder, pageShifts, chunkSize);
                directArenas[i] = arena;
                metrics.add(arena);
            }
            directArenaMetrics = Collections.unmodifiableList(metrics);
        } else {
            directArenas = null;
            directArenaMetrics = Collections.emptyList();
        }
    }
}
```

11.4.2　newHeapBuffer 核心方法

newHeapBuffer 方法用于分配需要容量的堆内缓冲区。流程如下。

（1）获取内存线程缓存对象 PoolThreadCache。

（2）获取到所属线程的堆内内存 PoolArena<byte[]>内存区对象。

（3）内存区不为空，则直接调用 heapArena.allocate 从内存池分配。

（4）内存区为空，则直接创建非池化的 HeapByteBuf。

（5）调用 AbstractByteBufAllocator 的 toLeakAwareBuffer 方法，根据内存泄漏判断设置选择是否包装 ByteBuf。

```
protected ByteBuf newHeapBuffer(int initialCapacity, int maxCapacity) {
    PoolThreadCache cache = threadCache.get();
    PoolArena<byte[]> heapArena = cache.heapArena;

    ByteBuf buf;
    if (heapArena != null) {              // 从内存池分配
        buf = heapArena.allocate(cache, initialCapacity, maxCapacity);
    } else {
        buf = new UnpooledHeapByteBuf(this, initialCapacity, maxCapacity);
    }

    return toLeakAwareBuffer(buf);        // 前面介绍过有关 AbstractByteBufAllocator 的方法
}
```

11.4.3　newDirectBuffer 核心方法

newDirectBuffer 方法用于分配需要容量的堆外缓冲区。流程如下。

（1）获取内存线程缓存对象 PoolThreadCache。

（2）获取到所属线程的堆内内存 PoolArena 内存区对象。

（3）内存区不为空，则直接调用 directArena.allocate 从内存池分配。

（4）内存区为空，则直接创建非池化的 DirectByteBuf。

（5）调用 AbstractByteBufAllocator 的 toLeakAwareBuffer 方法，根据内存泄漏判断设置选择是否包装 ByteBuf。

```java
protected ByteBuf newDirectBuffer(int initialCapacity, int maxCapacity) {
    PoolThreadCache cache = threadCache.get();
    PoolArena<ByteBuffer> directArena = cache.directArena;
    ByteBuf buf;
    if (directArena != null) {
        buf = directArena.allocate(cache, initialCapacity, maxCapacity);
    } else {            // 分配非池化的堆外内存，根据是否存在 Unsafe 来分配前面介绍的内存对象
        if (PlatformDependent.hasUnsafe()) {
            buf = UnsafeByteBufUtil.newUnsafeDirectByteBuf(this, initialCapacity, maxCapacity);
        } else {
            buf = new UnpooledDirectByteBuf(this, initialCapacity, maxCapacity);
        }
    }
    return toLeakAwareBuffer(buf); // 前面介绍过有关 AbstractByteBufAllocator 的方法
}
```

11.4.4 PoolThreadLocalCache 核心内部类

PoolThreadLocalCache 内部类实现了第 1 篇介绍的 FastThreadLocal，内部缓存了每个线程自己的 PoolThreadCache 对象。该类中的方法均重写了 FastThreadLocal 保留的钩子和模板方法。PooledByteBufAllocator 类的微型内存、小型内存、普通内存中的管理数量、最大缓存数量、缓存清理大小限制值均是服务于 PoolThreadCache 类。源码如下。

```java
final class PoolThreadLocalCache extends FastThreadLocal<PoolThreadCache> {

    // 当线程不存在 PoolThreadCache 时，进行初始化
    @Override
    protected synchronized PoolThreadCache initialValue() {
        // 获取最少线程使用的内存区进行初始化
        final PoolArena<byte[]> heapArena = leastUsedArena(heapArenas);
        final PoolArena<ByteBuffer> directArena = leastUsedArena(directArenas);
        return new PoolThreadCache(
            heapArena, directArena, tinyCacheSize, smallCacheSize, normalCacheSize,
            DEFAULT_MAX_CACHED_BUFFER_CAPACITY, DEFAULT_CACHE_TRIM_INTERVAL);
    }

    // 当 PoolThreadLocalCache 被移除时回调，用于释放 PoolThreadCache 占用的内存
    @Override
    protected void onRemoval(PoolThreadCache threadCache) {
        threadCache.free();
    }

    // 用于计算当前最少线程使用的 PoolArena，用于做负载均衡
    private <T> PoolArena<T> leastUsedArena(PoolArena<T>[] arenas) {
        if (arenas == null || arenas.length == 0) {
```

```
            return null;
        }
        PoolArena<T> minArena = arenas[0];
        for (int i = 1; i < arenas.length; i++) { // 从 arenas 的第一个数组开始遍历，找到最少线程占用的 arena
            PoolArena<T> arena = arenas[i];
            if (arena.numThreadCaches.get() < minArena.numThreadCaches.get()) {
                minArena = arena;
            }
        }
        return minArena;
    }
}
```

11.5　PoolThreadCache 类

PoolThreadCache 类用于每一个线程的内存缓存管理类，其中管理了当前线程分配的微型内存、小型内存、普通内存中的缓存，并持有了线程所分配到的 heapArena 和 directArena，用于加速内存分配。同时，该类的分配函数会在后面的 PoolArena 中进行调用，了解其中的原理将有助于更好地理解 PoolArena 的实现过程。

11.5.1　核心变量和构造器

由源码可知。MemoryRegionCache 数组用于保存内存子页。那么什么是 subpage？一个基础页大小为 8KB，但有时候用不了这么大的空间，这时应怎么办呢？这时可将一页切割为不同大小的内存子页进行管理，例如，8KB 切割为两个 4KB。而这些数组就是用于将微型、小型、正常页（N×8KB）进行缓存，所以当线程再次需要分配时，不用再到 PoolArena 中上锁分配（无锁化分配）。源码如下。

```
final class PoolThreadCache {
    // 在 PoolThreadLocalCache 类的 initialValue 方法中分配的内存区
    final PoolArena<byte[]> heapArena;
    final PoolArena<ByteBuffer> directArena;
    // 保存不同大小的内存缓存数组：tiny、small、normal
    private final MemoryRegionCache<byte[]>[] tinySubPageHeapCaches;      // 微型堆内内存子页缓存
    private final MemoryRegionCache<byte[]>[] smallSubPageHeapCaches;     // 小型堆内内存子页缓存
    private final MemoryRegionCache<ByteBuffer>[] tinySubPageDirectCaches; // 微型堆外内存子页缓存
    private final MemoryRegionCache<ByteBuffer>[] smallSubPageDirectCaches;// 小型堆外内存子页缓存
    private final MemoryRegionCache<byte[]>[] normalHeapCaches;           // 正常堆内内存页缓存
    private final MemoryRegionCache<ByteBuffer>[] normalDirectCaches;     // 正常堆外内存页缓存
    private final int numShiftsNormalDirect;                              // 正常堆内内存页偏移量
    private final int numShiftsNormalHeap;                                // 正常堆外内存页偏移量
    private final int freeSweepAllocationThreshold;                       // 缓存页释放阈值
    private int allocations;                                              // 分配次数
    private final Thread thread = Thread.currentThread();                 // 所属线程对象
    private final Runnable freeTask = new Runnable() {                    // 用于封装释放操作的 Runnable
        @Override
        public void run() {
            free0();
```

```java
        }
    };

    // 构造器，初始化成员变量
    PoolThreadCache(PoolArena<byte[]> heapArena, PoolArena<ByteBuffer> directArena,
            int tinyCacheSize, int smallCacheSize, int normalCacheSize,
            int maxCachedBufferCapacity, int freeSweepAllocationThreshold) {
        if (maxCachedBufferCapacity < 0) {
            throw new IllegalArgumentException("maxCachedBufferCapacity: "
                    + maxCachedBufferCapacity + " (expected: >= 0)");
        }
        if (freeSweepAllocationThreshold < 1) {
            throw new IllegalArgumentException("freeSweepAllocationThreshold: "
                    + freeSweepAllocationThreshold + " (expected: > 0)");
        }
        this.freeSweepAllocationThreshold = freeSweepAllocationThreshold;
        this.heapArena = heapArena;
        this.directArena = directArena;
        if (directArena != null) {                                    // 堆外内存域存在，创建子页管理数组
            tinySubPageDirectCaches = createSubPageCaches(
                    tinyCacheSize, PoolArena.numTinySubpagePools, SizeClass.Tiny);     // 指定管理个数：
static final int numTinySubpagePools = 512 >>> 4; => 微型子页内存单个 SubPageMemoryRegionCache 管理的
数量是 32 个，即 MemoryRegionCache<byte[]/ByteBuffer>[] tinySubPage Heap/Direct Caches 数组长度为 32
            smallSubPageDirectCaches = createSubPageCaches(
                    smallCacheSize, directArena.numSmallSubpagePools, SizeClass.Small); // 指定管理个数：
final int numSmallSubpagePools = pageShifts - 9; =>  13 - 9 = 4 个，即
MemoryRegionCache<byte[]/ByteBuffer>[] smallSubPage Heap/Direct Caches 数组长度为 4
            numShiftsNormalDirect = log2(directArena.pageSize);    // 计算偏移量默认为 2^N=8192 => 13
            normalDirectCaches = createNormalCaches(
                    normalCacheSize, maxCachedBufferCapacity, directArena);
            directArena.numThreadCaches.getAndIncrement();         // 增加占用的内存区域线程计数
        } else {                                                    // 没有直接内存区域，不使用缓存
            tinySubPageDirectCaches = null;
            smallSubPageDirectCaches = null;
            normalDirectCaches = null;
            numShiftsNormalDirect = -1;
        }
        if (heapArena != null) {                                    // 存在堆内内存域，创建子页管理数组
            tinySubPageHeapCaches = createSubPageCaches(
                    tinyCacheSize, PoolArena.numTinySubpagePools, SizeClass.Tiny);
            smallSubPageHeapCaches = createSubPageCaches(
                    smallCacheSize, heapArena.numSmallSubpagePools, SizeClass.Small);

            numShiftsNormalHeap = log2(heapArena.pageSize);        // 计算偏移量默认为 2^N=8192 => 13
            normalHeapCaches = createNormalCaches(
                    normalCacheSize, maxCachedBufferCapacity, heapArena);

            heapArena.numThreadCaches.getAndIncrement();           // 增加占用的内存区域线程计数
        } else { // 没有堆内内存区域，则不使用缓存
            tinySubPageHeapCaches = null;
            smallSubPageHeapCaches = null;
            normalHeapCaches = null;
```

```
            numShiftsNormalHeap = -1;
        }
        // 线程关闭时，调用 freeTask 释放内存
        ThreadDeathWatcher.watch(thread, freeTask);
    }
}
```

11.5.2　createSubPageCaches 核心方法

createSubPageCaches 方法用于创建 Tiny 微型和 Small 小型子页的内存缓存。cacheSize 参数指明了单个 MemoryRegionCache 的管理数量，numCaches 表示缓存指定类型的数组长度，SizeClass 表示子页内存类型。源码如下。

```
private static <T> MemoryRegionCache<T>[] createSubPageCaches(
    int cacheSize, int numCaches, SizeClass sizeClass) {
    if (cacheSize > 0) {
        MemoryRegionCache<T>[] cache = new MemoryRegionCache[numCaches];    // 根据 numCaches 创建 cache 数组
        for (int i = 0; i < cache.length; i++) {                // 初始化数组项为 SubPageMemoryRegionCache 类
            cache[i] = new SubPageMemoryRegionCache<T>(cacheSize, sizeClass); // 注意，这里传入了 cacheSize（微型类型数量为 512，小型类型数量为 256）
        }
        return cache;
    } else {
        return null;
    }
}
```

11.5.3　createNormalCaches 核心方法

createNormalCaches 方法用于创建普通页（8KB）的页内存缓存。cacheSize 参数指明单个 MemoryRegionCache 管理的数量，maxCachedBufferCapacity 表示最大缓存容量、area 表示页内存所属内存区域。源码如下。

```
private static <T> MemoryRegionCache<T>[] createNormalCaches(
    int cacheSize, int maxCachedBufferCapacity, PoolArena<T> area) {
    if (cacheSize > 0) {
        int max = Math.min(area.chunkSize, maxCachedBufferCapacity); // 取内存块与最大缓冲容量的最小值
        int arraySize = Math.max(1, log2(max / area.pageSize) + 1); // 计算缓存普通页的 MemoryRegionCache 数组长度，最小为 1，最大为 log2(maxCachedBufferCapacity / 8192)【2 的 N 次方】+ 1 个。注意，取值 maxCachedBufferCapacity 是由于通常不会设置缓存普通页的大小为 chunkSize 16M，否则就太大了
        MemoryRegionCache<T>[] cache = new MemoryRegionCache[arraySize]; // 创建数组
        for (int i = 0; i < cache.length; i++) {                // 初始化数组项为 NormalMemoryRegionCache 类
            cache[i] = new NormalMemoryRegionCache<T>(cacheSize);
        }
        return cache;
    } else {
        return null;
    }
}
```

11.5.4 MemoryRegionCache 核心内部类

MemoryRegionCache 类表示单个线程的页缓存类。由源码可知。

（1）Queue<Entry> queue 根据传入的 cacheSize 创建容纳管理内存的子页容器。

（2）管理子页的个数为 2 的倍数。

（3）allocate 方法是核心分配方法，外部将通过该方法分配当前缓存的页，但初始化操作由子类完成。

```java
private abstract static class MemoryRegionCache<T> {
    private final int size;                         // 管理页缓存大小
    private final Queue<Entry<T>> queue;            // 放置页的队列
    private final SizeClass sizeClass;              // 所属尺寸大小（Tiny、Small、Normal）
    private int allocations;                        // 已经分配的页数量

    // 构造器，初始化变量
    MemoryRegionCache(int size, SizeClass sizeClass) {
        this.size = MathUtil.findNextPositivePowerOfTwo(size);
        queue = PlatformDependent.newFixedMpscQueue(this.size);
        this.sizeClass = sizeClass;
    }

    // 子页从缓存中取出后，由子类实现初始化指定 PooledByteBuf
    protected abstract void initBuf(PoolChunk<T> chunk, long handle,
                                    PooledByteBuf<T> buf, int reqCapacity);

    // 向队列中添加子页
    public final boolean add(PoolChunk<T> chunk, long handle) {
        Entry<T> entry = newEntry(chunk, handle); // 创建内存包装对象
        boolean queued = queue.offer(entry);
        if (!queued) {                            // 队列满了，添加失败，则释放该 Entry，因为没有对其进行缓存
            entry.recycle();
        }
        return queued;
    }

    // 尝试从缓存中分配指定 reqCapacity 容量大小的 PooledByteBuf
    public final boolean allocate(PooledByteBuf<T> buf, int reqCapacity) {
        Entry<T> entry = queue.poll();
        if (entry == null) {                      // 队列中没有缓存的页，则返回 false
            return false;
        }
        initBuf(entry.chunk, entry.handle, buf, reqCapacity);
        entry.recycle();                          // 将子页包装对象回收
        // 增加分配计数值
        ++ allocations;
        return true;
    }

    // 用于清除缓存的页
    public final int free() {
        return free(Integer.MAX_VALUE);           // 指定释放所有的页
```

```java
    }

    private int free(int max) {
        int numFreed = 0;
        for (; numFreed < max; numFreed++) {          // 遍历队列，取出 entry 并将其释放
            Entry<T> entry = queue.poll();
            if (entry != null) {
                freeEntry(entry);
            } else {
                return numFreed;
            }
        }
        return numFreed;
    }

    // 该方法用于当内存分配不紧张时，将释放缓存的页放回到所属的 PoolChunk 内存块中
    public final void trim() {
        int free = size - allocations;                // 计算未分配页的释放数量
        allocations = 0;
        if (free > 0) {                               // 存在未分配的页，将其释放
            free(free);
        }
    }

    // 将缓存的页实体 Entry 释放，调用 entry.recycle 回收 Entry 对象，同时调用 chunk.arena.freeChunk 对 handle
    // 指定的内存完成释放（前面说过 handle，保留了内存池中的内存信息，后面会详细描述）
    private void freeEntry(Entry entry) {
        PoolChunk chunk = entry.chunk;
        long handle = entry.handle;
        entry.recycle();
        chunk.arena.freeChunk(chunk, handle, sizeClass);
    }

    // 内部类，用于封装 handle 内存信息
    static final class Entry<T> {
        final Handle<Entry<?>> recyclerHandle;
        PoolChunk<T> chunk;
        long handle = -1;
        Entry(Handle<Entry<?>> recyclerHandle) {
            this.recyclerHandle = recyclerHandle;
        }

        void recycle() {                    // 释放对 PoolChunk 和 handle 的引用，同时将该对象放回对象池
            chunk = null;
            handle = -1;
            recyclerHandle.recycle(this);
        }
    }

    // 从 Entry 对象池中获取一个 Entry 对象，作为 handle 的包装对象
    private static Entry newEntry(PoolChunk<?> chunk, long handle) {
        Entry entry = RECYCLER.get();
        entry.chunk = chunk;
        entry.handle = handle;
```

```
        return entry;
    }

    // 定义对象池，用于缓存 Entry 对象
    private static final Recycler<Entry> RECYCLER = new Recycler<Entry>() {
        @Override
        protected Entry newObject(Handle<Entry> handle) {
            return new Entry(handle);
        }
    };
}
```

11.5.5 SubPageMemoryRegionCache 核心内部类

SubPageMemoryRegionCache 类继承自 MemoryRegionCache 类，作为 Tiny、Small 类型的内存子页缓存，实现了 initBuf 方法。该方法直接调用内存子页所属 PoolChunk 内存页的 initBufWithSubpage 方法完成初始化。源码如下。

```
private static final class SubPageMemoryRegionCache<T> extends MemoryRegionCache<T> {
    SubPageMemoryRegionCache(int size, SizeClass sizeClass) {
        super(size, sizeClass);
    }

    @Override
    protected void initBuf(
        PoolChunk<T> chunk, long handle, PooledByteBuf<T> buf, int reqCapacity) {
        chunk.initBufWithSubpage(buf, handle, reqCapacity);
    }
}
```

11.5.6 NormalMemoryRegionCache 核心内部类

NormalMemoryRegionCache 类继承自 MemoryRegionCache 类，作为 Normal 类型的内存子页缓存，实现了 initBuf 方法。该方法直接调用内存子页所属 PoolChunk 内存页的 initBuf 方法完成初始化。该类与 SubPageMemoryRegionCache 类的区别仅在于初始化方法不同。源码如下。

```
private static final class NormalMemoryRegionCache<T> extends MemoryRegionCache<T> {
    NormalMemoryRegionCache(int size) {
        super(size, SizeClass.Normal);
    }

    @Override
    protected void initBuf(
        PoolChunk<T> chunk, long handle, PooledByteBuf<T> buf, int reqCapacity) {
        chunk.initBuf(buf, handle, reqCapacity);
    }
}
```

11.5.7 allocateTiny 核心方法

allocateTiny 方法用于分配微型内存块所属的 MemoryRegionCache。流程如下。

（1）通过传入的 area 和 normCapacity（规格化后的内存，会在 PoolArena 中看到）获得所属的 MemoryRegionCache 对象。

（2）根据获取到的 MemoryRegionCache 对象，调用 allocate 方法完成实际分配。

```
boolean allocateTiny(PoolArena<?> area, PooledByteBuf<?> buf, int reqCapacity, int normCapacity) {
    return allocate(cacheForTiny(area, normCapacity), buf, reqCapacity);
}

private MemoryRegionCache<?> cacheForTiny(PoolArena<?> area, int normCapacity) {
    int idx = PoolArena.tinyIdx(normCapacity);      // normCapacity >>> 4 tiny 以 16B 为界限分割为 32 个
数组，所以用需要分配的值除以 4 得到对应分配的数组下标
    if (area.isDirect()) {                          // 堆外内存分配
        return cache(tinySubPageDirectCaches, idx);
    }
    return cache(tinySubPageHeapCaches, idx);       // 堆内内存分配
}

// 从 MemoryRegionCache 数组中取出下标为 idx 的 MemoryRegionCache 对象并返回
private static <T> MemoryRegionCache<T> cache(MemoryRegionCache<T>[] cache, int idx) {
    if (cache == null || idx > cache.length - 1) {
        return null;
    }
    return cache[idx];
}

// 计算容量所属数组下标
static int tinyIdx(int normCapacity) {
    return normCapacity >>> 4;
}
```

11.5.8 cacheForSmall 核心方法

cacheForSmall 方法用于分配小型内存块所属的 MemoryRegionCache。流程如下。

（1）通过传入的 area 和 normCapacity（规格化后的内存，会在 PoolArena 中看到）获得所属的 MemoryRegionCache 对象。

（2）根据获取到的 MemoryRegionCache 对象，调用 allocate 方法完成实际分配。

```
boolean allocateSmall(PoolArena<?> area, PooledByteBuf<?> buf, int reqCapacity, int normCapacity) {
    return allocate(cacheForSmall(area, normCapacity), buf, reqCapacity);
}

private MemoryRegionCache<?> cacheForSmall(PoolArena<?> area, int normCapacity) {
    int idx = PoolArena.smallIdx(normCapacity);
    if (area.isDirect()) {                          // 堆外内存分配
        return cache(smallSubPageDirectCaches, idx);
    }
    return cache(smallSubPageHeapCaches, idx);      // 堆内内存分配
}

// small 从 tiny 界限 512B 开始分割为 4 个数组：512B（数组下标 0）、1024B（数组下标 1）、2048B（数组下标 2）、4096B（数组下标 3）
```

```
static int smallIdx(int normCapacity) {
    int tableIdx = 0;                    // 512 为 0 下标处
    int i = normCapacity >>> 10;         // 除以 1024，计算 1、2、3 下标
    while (i != 0) {
        i >>>= 1;
        tableIdx ++;
    }
    return tableIdx;
}
```

11.5.9　cacheForNormal 核心方法

cacheForNormal 方法用于获取正常页（8KB）的所属的 MemoryRegionCache。通过传入的 area 和 normCapacity（规格化后的内存，会在 PoolArena 中看到）获得所属的 MemoryRegionCache 对象。源码如下。

```
boolean allocateNormal(PoolArena<?> area, PooledByteBuf<?> buf, int reqCapacity, int normCapacity) {
    return allocate(cacheForNormal(area, normCapacity), buf, reqCapacity);
}

private MemoryRegionCache<?> cacheForNormal(PoolArena<?> area, int normCapacity) {
    if (area.isDirect()) {                                   // 堆外内存分配
        int idx = log2(normCapacity >> numShiftsNormalDirect);   // Netty 将正常页内存（8KB-16MB）分割
// 为正常页缓存，以 2 的倍数递增，这里计算 2 的 N 次方，用于计算 index 下标。numShiftsNormalDirect 通常为
// 2^N=8192 0=> 13。一言以蔽之：用给定的容量除以 13，然后取 2 的倍数，获取下标
        return cache(normalDirectCaches, idx);
    }
    // 堆内内存分配
    int idx = log2(normCapacity >> numShiftsNormalHeap);
    return cache(normalHeapCaches, idx);
}

// 计算 val 2 的倍数
private static int log2(int val) {
    int res = 0;
    while (val > 1) {                     // 将 val 右移增加 res 的大小，直到 val<=1
        val >>= 1;
        res++;
    }
    return res;
}
```

11.5.10　allocate 核心方法

allocate 方法用在上述方法中找到对应的 MemoryRegionCache，用于从缓存中分配子页。主要通过调用 MemoryRegionCache 的 allocate 方法完成分配，同时判断是否达到清理阈值。如果达到，则调用 trim 进行空间释放。源码如下。

```
private boolean allocate(MemoryRegionCache<?> cache, PooledByteBuf buf, int reqCapacity) {
    if (cache == null) {                  // 没有缓存，则直接返回分配失败
        return false;
    }
```

```
        boolean allocated = cache.allocate(buf, reqCapacity);   // 调用缓存分配方法分配
        if (++ allocations >= freeSweepAllocationThreshold) {    // 如果分配次数大于清理阈值,则将分配次数重
置为 0,同时进行清理操作
            allocations = 0;
            trim();
        }
        return allocated;
}
```

11.5.11　trim 核心方法

trim 方法用于释放缓存的页。由源码可知,存在清理堆内、堆外两种类型的页,都是调用 MemoryRegionCache 的 trim 方法。源码如下。

```
void trim() {
    // 释放堆外内存页
    trim(tinySubPageDirectCaches);
    trim(smallSubPageDirectCaches);
    trim(normalDirectCaches);
    // 释放堆内内存页
    trim(tinySubPageHeapCaches);
    trim(smallSubPageHeapCaches);
    trim(normalHeapCaches);
}

private static void trim(MemoryRegionCache<?>[] caches) {
    if (caches == null) {
        return;
    }
    for (MemoryRegionCache<?> c: caches) {
        trim(c);
    }
}

private static void trim(MemoryRegionCache<?> cache) {
    if (cache == null) {
        return;
    }
    cache.trim();
}
```

11.5.12　free 核心方法

在所属线程关闭时,调用 free 方法,用于释放所有管理的内存页。由源码可知,主要调用 cache.free 方法,将管理的所有堆内、堆外两种类型的页归还给所属的 PoolChunk 内存块。源码如下。

```
void free() {
    ThreadDeathWatcher.unwatch(thread, freeTask);   // 解除线程死亡判断,避免重复调用该方法
    free0();                                         // 实际释放
}

private void free0() {
```

```
        int numFreed = free(tinySubPageDirectCaches) +
            free(smallSubPageDirectCaches) +
            free(normalDirectCaches) +
            free(tinySubPageHeapCaches) +
            free(smallSubPageHeapCaches) +
            free(normalHeapCaches);              // 在释放所有页同时，记录释放的数量

        if (numFreed > 0 && logger.isDebugEnabled()) {    // 打印日志
            logger.debug("Freed {} thread-local buffer(s) from thread: {}", numFreed, thread.getName());
        }
        // 减少对堆外、堆内内存区域的引用计数
        if (directArena != null) {
            directArena.numThreadCaches.getAndDecrement();
        }
        if (heapArena != null) {
            heapArena.numThreadCaches.getAndDecrement();
        }
    }

    private static int free(MemoryRegionCache<?>[] caches) {
        if (caches == null) {
            return 0;
        }

        int numFreed = 0;
        for (MemoryRegionCache<?> c: caches) {         // 循环释放并记录释放数量
            numFreed += free(c);
        }
        return numFreed;
    }

    private static int free(MemoryRegionCache<?> cache) {
        if (cache == null) {
            return 0;
        }
        return cache.free();
    }
```

11.6　PoolArena 类

PoolArena 类用于表示内存池的区域，即内部管理内存块 PoolChunk 和小于 8KB 的 PoolSubpage 内存子页。前面在 PooledByteBufAllocator 类中看过该类的参数定义。Netty 为了减少线程分配内存竞争，将 PoolArena 的数量与事件循环组的数量设置为 CPU 核心数的 2 倍，这样有助于提高 CPU 的利用率并且减少竞争导致的性能损耗。本节将详细介绍该类对于 PoolChunk 和 PoolSubpage 的管理方式和参数定义。

11.6.1　核心变量与构造器

由源码可知。
（1）该类定义了类型 SizeClass 枚举，定义了微型类型、小型类型、正常类型。

（2）PoolSubpage[] tinySubpagePools 微型子页数组用于保存从 PoolChunk 中分配的由基础内存页 8KB 切割成的 16B 至 496B 的微型子页。

（3）PoolSubpage[] smallSubpagePools 微型子页数组用于保存从 PoolChunk 中分配的由基础内存页 8KB 切割成的 512B 至 4096B 的小型子页。

（4）如图 11-1 所示，PoolChunkList 用于保存不同占用率的基本内存块（由基础内存页 8KB 组成）链表，这样有助于对不同占用率的内存优先进行分配，增加内存利用率。同时不同百分比占用率的 PoolChunkList 有重叠部分，并不是严格区分。这样设计有利于在占用率提升或降低后，减少 PoolChunk 在不同的 PoolChunkList 中移动。

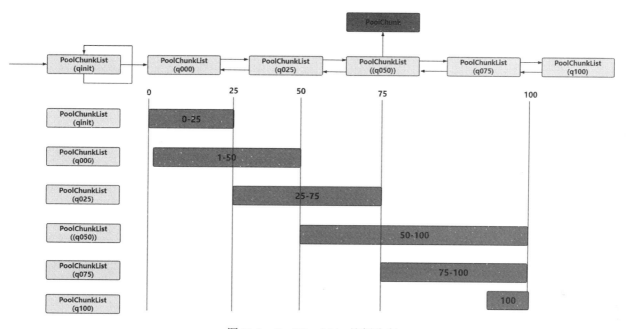

图 11-1　PoolChunkList 比例分布

```
public final class PlatformDependent {
    private static final boolean HAS_UNSAFE = hasUnsafe0();
    private static boolean hasUnsafe0() {
        boolean noUnsafe = SystemPropertyUtil.getBoolean("io.Netty.noUnsafe", false);  // 系统变量获取设置值
        if (isAndroid()) {           // 安卓系统不存在 Unsafe
            return false;
        }
        if (noUnsafe) {              // 系统变量设置没有 Unsafe
            return false;
        }
        boolean tryUnsafe;
        if (SystemPropertyUtil.contains("io.Netty.tryUnsafe")) { // 系统变量设置是否尝试使用 Unsafe
            tryUnsafe = SystemPropertyUtil.getBoolean("io.Netty.tryUnsafe", true);
        } else {
            tryUnsafe = SystemPropertyUtil.getBoolean("org.jboss.Netty.tryUnsafe", true);
        }
        if (!tryUnsafe) {            // 不尝试使用 Unsafe 操作，直接返回
            return false;
```

```java
        }
        try {
            boolean hasUnsafe = PlatformDependent0.hasUnsafe();
            return hasUnsafe;
        } catch (Throwable ignored) {
            // 存在启动 PlatformDependent0 失败的可能
            return false;
        }
    }

    public static boolean hasUnsafe() {
        return HAS_UNSAFE;
    }
}

final class PlatformDependent0 {
    static final Unsafe UNSAFE;
    static {                                    // 初始化 Unsafe
        Unsafe unsafe;
        if (addressField != null) {
            try {
                Field unsafeField = Unsafe.class.getDeclaredField("theUnsafe");
                unsafeField.setAccessible(true);
                unsafe = (Unsafe) unsafeField.get(null);
            } catch (Throwable cause) {
                unsafe = null;
            }
        } else {
            unsafe = null;
        }
        UNSAFE = unsafe;
    }

    static boolean hasUnsafe() {
        return UNSAFE != null;
    }
}

abstract class PoolArena<T> implements PoolArenaMetric {
    static final boolean HAS_UNSAFE = PlatformDependent.hasUnsafe(); // 保存当前运行时是否包含 Unsafe 类
    enum SizeClass {                            // 内存类型
        Tiny,                                   // 微型内存
        Small,                                  // 小型内存
        Normal                                  // 正常内存
    }
    static final int numTinySubpagePools = 512 >>> 4;  // 微型子页内存数量为 32 个
    final PooledByteBufAllocator parent;        // 所属内存分配器
    private final int maxOrder;                 // PoolChunk 内存块的管理深度默认是 11
    final int pageSize;                         // 基本内存管理页大小为 8KB
    final int pageShifts;                       // 基本内存管理页偏移为 13（2^N=8KB）
    final int chunkSize;                        // 基本内存块大小（PoolChunk）为 8KB * maxOrder(11) = 16M
    final int subpageOverflowMask;              // 不属于 subpage 子页的位掩码
    final int numSmallSubpagePools;             // 小型子页内存数量是 4
```

```java
    private final PoolSubpage<T>[] tinySubpagePools;         // 微型子页数组
    private final PoolSubpage<T>[] smallSubpagePools;        // 小型子页数组

    private final PoolChunkList<T> q050;    // 基本内存块（PoolChunk 16M）占用率为 50%~100%的列表
    private final PoolChunkList<T> q025;    // 基本内存块（PoolChunk 16M）占用率为 25%~75%的列表
    private final PoolChunkList<T> q000;    // 基本内存块（PoolChunk 16M）占用率为 1%~50%的列表
    private final PoolChunkList<T> qInit;   // 基本内存块（PoolChunk 16M）占用率为 0%~25%的列表
    private final PoolChunkList<T> q075;    // 基本内存块（PoolChunk 16M）占用率为 75%~100%的列表
    private final PoolChunkList<T> q100;    // 基本内存块（PoolChunk 16M）占用率为 100%的列表

    private final List<PoolChunkListMetric> chunkListMetrics;    // 用于做性能分析，忽略即可
    /* 用于记录分配的 normal 页、tiny 页、small 页、huge 页的次数，用于性能分析，忽略即可 */
    private long allocationsNormal;
    private final LongCounter allocationsTiny = PlatformDependent.newLongCounter();
    private final LongCounter allocationsSmall = PlatformDependent.newLongCounter();
    private final LongCounter allocationsHuge = PlatformDependent.newLongCounter();
    private final LongCounter activeBytesHuge = PlatformDependent.newLongCounter();
    /* 记录释放的 normal 页、tiny 页、small 页、huge 页的次数，用于性能分析，忽略即可 */
    private long deallocationsTiny;
    private long deallocationsSmall;
    private long deallocationsNormal;
    private final LongCounter deallocationsHuge = PlatformDependent.newLongCounter();
    final AtomicInteger numThreadCaches = new AtomicInteger(); // 记录当前内存区域被多少个线程引用计数
（之前涉及过，用该参数来负载均衡 PoolArena）

    // 构造器，从 PooledByteBufAllocator 中继承过来的参数初始化变量
    protected PoolArena(PooledByteBufAllocator parent, int pageSize, int maxOrder, int pageShifts, int chunkSize) {
        this.parent = parent;
        this.pageSize = pageSize;
        this.maxOrder = maxOrder;
        this.pageShifts = pageShifts;
        this.chunkSize = chunkSize;
        subpageOverflowMask = ~(pageSize - 1);
        tinySubpagePools = newSubpagePoolArray(numTinySubpagePools);    // 创建微型子页数组：new PoolSubpage[size]
        for (int i = 0; i < tinySubpagePools.length; i ++) {             // 初始化数组项
            tinySubpagePools[i] = newSubpagePoolHead(pageSize);
        }

        numSmallSubpagePools = pageShifts - 9;
        smallSubpagePools = newSubpagePoolArray(numSmallSubpagePools);  // 创建小型子页数组：new PoolSubpage[size]
        for (int i = 0; i < smallSubpagePools.length; i ++) {            // 初始化数组项
            smallSubpagePools[i] = newSubpagePoolHead(pageSize);
        }

        // 初始化不同占用率的 PoolChunk 链表
        q100 = new PoolChunkList<T>(null, 100, Integer.MAX_VALUE, chunkSize);
        q075 = new PoolChunkList<T>(q100, 75, 100, chunkSize);
        q050 = new PoolChunkList<T>(q075, 50, 100, chunkSize);
        q025 = new PoolChunkList<T>(q050, 25, 75, chunkSize);
        q000 = new PoolChunkList<T>(q025, 1, 50, chunkSize);
```

```java
        qInit = new PoolChunkList<T>(q000, Integer.MIN_VALUE, 25, chunkSize);

        // 将链表互相关联（双向链表）
        q075.prevList(q050);
        q050.prevList(q025);
        q025.prevList(q000);
        q000.prevList(null);
        qInit.prevList(qInit);
        q100.prevList(q075);

        // 性能分析记录，了解即可
        List<PoolChunkListMetric> metrics = new ArrayList<PoolChunkListMetric>(6);
        metrics.add(qInit);
        metrics.add(q000);
        metrics.add(q025);
        metrics.add(q050);
        metrics.add(q075);
        metrics.add(q100);
        chunkListMetrics = Collections.unmodifiableList(metrics);
    }
}
```

11.6.2 HeapArena 核心内部类

HeapArena 内部类继承自 PoolArena，实现了对于堆内存 byte[]数组的功能函数。源码较为简单，一切都围绕字节数组来操作。源码如下。

```java
static final class HeapArena extends PoolArena<byte[]> {

    // 由 PooledByteBufAllocator 调用初始化堆内存管理区域
    HeapArena(PooledByteBufAllocator parent, int pageSize, int maxOrder, int pageShifts, int chunkSize) {
        super(parent, pageSize, maxOrder, pageShifts, chunkSize);
    }

    // 堆内存管理区域不是堆外内存
    @Override
    boolean isDirect() {
        return false;
    }

    // 创建堆内存 byte[]数组的 PoolChunk
    @Override
    protected PoolChunk<byte[]> newChunk(int pageSize, int maxOrder, int pageShifts, int chunkSize) {
        return new PoolChunk<byte[]>(this, new byte[chunkSize], pageSize, maxOrder, pageShifts, chunkSize);
    }

    @Override
    protected PoolChunk<byte[]> newUnpooledChunk(int capacity) {
        return new PoolChunk<byte[]>(this, new byte[capacity], capacity);
    }

    // 堆内内存的释放，由垃圾回收（GC）自动完成，不需要手动参与
    @Override
```

```
    protected void destroyChunk(PoolChunk<byte[]> chunk) {
    }

    // 根据容量创建 PooledHeapByteBuf 对象
    @Override
    protected PooledByteBuf<byte[]> newByteBuf(int maxCapacity) {
        return HAS_UNSAFE ? PooledUnsafeHeapByteBuf.newUnsafeInstance(maxCapacity)
                : PooledHeapByteBuf.newInstance(maxCapacity);
    }

    // 内存复制，只需要使用 System.arraycopy 方法完成复制即可
    @Override
    protected void memoryCopy(byte[] src, int srcOffset, byte[] dst, int dstOffset, int length) {
        if (length == 0) {
            return;
        }
        System.arraycopy(src, srcOffset, dst, dstOffset, length);
    }
}
```

11.6.3 DirectArena 核心内部类

DirectArena 内部类继承自 PoolArena，实现了对于堆外内存 DirectByteBuffer 的功能函数。源码较为简单，一切都围绕堆外内存来操作，源码如下。

```
static final class DirectArena extends PoolArena<ByteBuffer> {

    // 由 PooledByteBufAllocator 调用初始化堆内存管理区域
    DirectArena(PooledByteBufAllocator parent, int pageSize, int maxOrder, int pageShifts, int chunkSize) {
        super(parent, pageSize, maxOrder, pageShifts, chunkSize);
    }

    // 堆内存管理区域为堆外内存
    @Override
    boolean isDirect() {
        return true;
    }

    // 创建堆内存 ByteBuffer 对象的 PoolChunk
    @Override
    protected PoolChunk<ByteBuffer> newChunk(int pageSize, int maxOrder, int pageShifts, int chunkSize) {
        return new PoolChunk<ByteBuffer>(
                this, allocateDirect(chunkSize),
                pageSize, maxOrder, pageShifts, chunkSize);
    }

    @Override
    protected PoolChunk<ByteBuffer> newUnpooledChunk(int capacity) {
        return new PoolChunk<ByteBuffer>(this, allocateDirect(capacity), capacity);
    }

    // 根据运行时是否存在 Unsafe 分配
    private static ByteBuffer allocateDirect(int capacity) {
```

```java
        return PlatformDependent.useDirectBufferNoCleaner() ?
            PlatformDependent.allocateDirectNoCleaner(capacity) : ByteBuffer.allocateDirect(capacity);
    }

    // 由于堆外内存不是堆内内存，不属于垃圾回收（GC）活动范畴，所以这里需要手动释放。前面说过，没有
Cleaner，则手动操作 Unsafe 的 freeMemory 方法释放，否则直接调用 Cleaner 的 run 方法完成清理。
    @Override
    protected void destroyChunk(PoolChunk<ByteBuffer> chunk) {
        if (PlatformDependent.useDirectBufferNoCleaner()) {
            PlatformDependent.freeDirectNoCleaner(chunk.memory);
        } else {
            PlatformDependent.freeDirectBuffer(chunk.memory);
        }
    }

    // 根据是否存在 Unsafe 创建 PooledDirectByteBuf
    @Override
    protected PooledByteBuf<ByteBuffer> newByteBuf(int maxCapacity) {
        if (HAS_UNSAFE) {
            return PooledUnsafeDirectByteBuf.newInstance(maxCapacity);
        } else {
            return PooledDirectByteBuf.newInstance(maxCapacity);
        }
    }

    // 内存复制，需要自己实现
    @Override
    protected void memoryCopy(ByteBuffer src, int srcOffset, ByteBuffer dst, int dstOffset, int length) {
        if (length == 0) {
            return;
        }
        if (HAS_UNSAFE) {              // 运行时存在 Unsafe，通过 UNSAFE.copyMemory 方法操作
            PlatformDependent.copyMemory(
                PlatformDependent.directBufferAddress(src) + srcOffset,
                PlatformDependent.directBufferAddress(dst) + dstOffset, length);
        } else {                       // 否则直接操作 ByteBuffer 类完成复制
            // 需要复制对象操作，因为可能有别的地方在使用 ByteBuffer，不能因为复制导致 position 变量和
limit 发生修改
            src = src.duplicate();
            dst = dst.duplicate();
            // 开始复制
            src.position(srcOffset).limit(srcOffset + length);
            dst.position(dstOffset);
            dst.put(src);           // 通过该方法直接写入 src 即可
        }
    }
}
```

11.6.4 allocate 核心方法

allocate 方法由分配器调用，用于分配需要的内存。流程如下。

（1）分配用于操作实际内存的 PooledByteBuf 对象。

第 11 章 Netty 内存池原理

（2）规格化需要分配内存的大小（这里面其实就可以定义 tiny、small、normal 的表示范围）。

（3）根据分配内存的大小来选择如下范围：微型（16B<= normCapacity < 512B）、小型（512B<= normCapacity <= 4096B）、正常页（8KB 倍数<=16MB（PoolChunk 大小））、大页分配。在微型、小型、正常页分配前均通过缓存分配，如果缓存不存在可用页，则进行上锁分配。大页分配由于超过基础管理内存块，所以不进行池化管理。

```java
// 由子类 DirectArena 或 HeapArena 实现分配
protected abstract PooledByteBuf<T> newByteBuf(int maxCapacity);

PooledByteBuf<T> allocate(PoolThreadCache cache, int reqCapacity, int maxCapacity) {
    PooledByteBuf<T> buf = newByteBuf(maxCapacity);
    allocate(cache, buf, reqCapacity);
    return buf;
}

// 判断规格化后的内存是否为微型或小型内存
boolean isTinyOrSmall(int normCapacity) {
    return (normCapacity & subpageOverflowMask) == 0;        // 一页为 8KB，判断是否小于 1 页，直接取
~(pageSize - 1)与容量相与即可。因为与操作等于二进制截断，相当于截断 8KB 的高位（pageSize - 1 等于取低
12 位为 1，取反后相当于高位 20 位为 1，低 12 位为 0），如果为 0，则表明该内存小于 8KB
}

// 根据规格化后的内存所处的 tinySubpagePools 的下标
static int tinyIdx(int normCapacity) {
    return normCapacity >>> 4;                                // 除以 16 即可，微型类型以 16B 递增
}

// 根据规格化后的内存所处的 smallSubpagePools 的下标
static int smallIdx(int normCapacity) {
    int tableIdx = 0;
    int i = normCapacity >>> 10;    // 除以 1024，因为 small 数组大小有 4 个：0、1、2、3，前面已经介绍过
    while (i != 0) {
        i >>>= 1;
        tableIdx ++;
    }
    return tableIdx;
}

private void allocate(PoolThreadCache cache, PooledByteBuf<T> buf, final int reqCapacity) {
    final int normCapacity = normalizeCapacity(reqCapacity);   // 规格化分配内存大小
    if (isTinyOrSmall(normCapacity)) {                          // 内存分配为微型或小型内存
        int tableIdx;
        PoolSubpage<T>[] table;
        boolean tiny = isTiny(normCapacity);
        if (tiny) {                                             // 16B<= normCapacity < 512B，分配微型内存
            if (cache.allocateTiny(this, buf, reqCapacity, normCapacity)) {   // 优先从线程缓存中分配
                return;
            }
            tableIdx = tinyIdx(normCapacity);              // 缓存分配失败，从当前 tinySubpagePools 中分配
            table = tinySubpagePools;
        } else {                                                // 512B<= normCapacity <= 4096B，分配小型内存
```

```
            if (cache.allocateSmall(this, buf, reqCapacity, normCapacity)) {  // 优先从线程缓存中分配
                return;
            }
            tableIdx = smallIdx(normCapacity);                  // 缓存分配失败，从当前 smallSubpagePools 中分配
            table = smallSubpagePools;
        }
        final PoolSubpage<T> head = table[tableIdx];            // 获取需要分配的子页头结点
        synchronized (head) {                                    // 对头结点上锁分配
            final PoolSubpage<T> s = head.next;
            if (s != head) {                                     // 存在下一个结点
                long handle = s.allocate();
                s.chunk.initBufWithSubpage(buf, handle, reqCapacity); // 初始化分配到的 PooledByteBuf（将
在 PoolChunk 中详细描述）
                // 增加对应计数
                if (tiny) {
                    allocationsTiny.increment();
                } else {
                    allocationsSmall.increment();
                }
                return;
            }
        }
        // 子页内存分配失败，只能先分配一个正常页（8KB），再切割为子页
        allocateNormal(buf, reqCapacity, normCapacity);
        return;
    }
    if (normCapacity <= chunkSize) {                             // 正常容量分配（8KB 倍数）
        if (cache.allocateNormal(this, buf, reqCapacity, normCapacity)) {  // 优先从缓存分配
            return;
        }
        allocateNormal(buf, reqCapacity, normCapacity);
    } else { // 超过 16MB 的内存为大页内存（pageSize * marOrder = 8KB * 11 = 16M），不进行缓存，直接分配
        allocateHuge(buf, reqCapacity);
    }
}
```

11.6.5　normalizeCapacity 核心方法

normalizeCapacity 方法用于对需要分配的容量进行规格化。流程如下。

（1）如果分配的内存为大页内存，则需要多少分配多少，不进行规格化，因为不需要进行池化管理。

（2）如果分配的内存大于等于 512B 且不是微型内存，则需要找到容量最近的 2 的倍数即可。

（3）如果分配的内存正好是 16 的倍数，则取当前需求容量即可。

（4）否则取当前值最近的 16B 的倍数即可（tiny 子页的每一项以 16B 递增直到 496B）。

```
int normalizeCapacity(int reqCapacity) {
    if (reqCapacity < 0) {
        throw new IllegalArgumentException("capacity: " + reqCapacity + " (expected: 0+)");
    }
    if (reqCapacity >= chunkSize) { // 大页内存，需要多少分配多少，不进行规格化，因为不需要进行池化管理
        return reqCapacity;
    }
```

```
    if (!isTiny(reqCapacity)) {              // >= 512 不是微型内存,需要找到容量最近的 2 的倍数即可
        int normalizedCapacity = reqCapacity;
        normalizedCapacity --;                // -1 可以使计算完毕后等于 2^n,否则为 2^n + 1
        normalizedCapacity |= normalizedCapacity >>>  1;
        normalizedCapacity |= normalizedCapacity >>>  2;
        normalizedCapacity |= normalizedCapacity >>>  4;
        normalizedCapacity |= normalizedCapacity >>>  8;
        normalizedCapacity |= normalizedCapacity >>> 16;
        normalizedCapacity ++;
        if (normalizedCapacity < 0) {         // 符号溢出,无符号右移将符号位变为正数
            normalizedCapacity >>>= 1;
        }
        return normalizedCapacity;
    }

    // 正好为 16 的倍数,取当前需求容量即可;小于 16 也没关系,因为在 tinyIdx 方法中会取到 0 下标处的 16B
    if ((reqCapacity & 15) == 0) {
        return reqCapacity;
    }
    return (reqCapacity & ~15) + 16; // 否则取当前值最近的 16B 的倍数即可(tiny 子页的每一项以 16B 递增直到 496B)
}
```

11.6.6 allocateNormal 核心方法

allocateNormal 方法用于分配正常内存页的大小(8KB * N)。流程如下。

(1)分配顺序为:q050、q025、q000、qInit、q075,这样进行分配可以有效增加 PoolChunk 的内存利用率,同时可以减少分配时间。总是优先在占用率较高的 PoolChunk 进行分配,这样可以优先利用空间。至于为什么会这样分配,后续在描述 PoolChunk 时会详细介绍。

(2)如果在 PoolChunkList 中分配成功,则直接返回即可。

(3)否则尝试分配一个新的 PoolChunk 内存管理块,再从该管理块中分配内存,然后尝试将其放入 qInit 链表中。

```
private synchronized void allocateNormal(PooledByteBuf<T> buf, int reqCapacity, int normCapacity) {
    // 在 PoolChunkList 中分配
    if (q050.allocate(buf, reqCapacity, normCapacity) || q025.allocate(buf, reqCapacity, normCapacity) ||
        q000.allocate(buf, reqCapacity, normCapacity) || qInit.allocate(buf, reqCapacity, normCapacity) ||
        q075.allocate(buf, reqCapacity, normCapacity)) {
        ++allocationsNormal;
        return;
    }
    // 分配新的 PoolChunk,然后从中分配
    PoolChunk<T> c = newChunk(pageSize, maxOrder, pageShifts, chunkSize);
    long handle = c.allocate(normCapacity);
    ++allocationsNormal;
    c.initBuf(buf, handle, reqCapacity);    // 分配成功并初始化该 PooledByteBuf 对象。初始化时,并没有计算子页的相关信息,只需要使用低 32 位保留对 PoolChunk 的使用即可(后续将在 PoolChunk 中介绍该方法)
    qInit.add(c);                            // 添加到 qInit 链表中即可
}
```

```
protected abstract PoolChunk<T> newChunk(int pageSize, int maxOrder, int pageShifts, int chunkSize);  // 由
子类实现
```

11.6.7　allocateHuge 核心方法

allocateHuge 方法用于分配大于 16MB 的内存。由于这种内存占用空间较大，所以规格化不做对齐。因为不需要进行池化，所以需要多少就分配多少，在其中通过 newUnpooledChunk 创建不需要池化的 PoolChunk。源码如下。

```
private void allocateHuge(PooledByteBuf<T> buf, int reqCapacity) {
    PoolChunk<T> chunk = newUnpooledChunk(reqCapacity);
    activeBytesHuge.add(chunk.chunkSize());                       // 增加活动大页内存计数
    buf.initUnpooled(chunk, reqCapacity);
    allocationsHuge.increment();
}
protected abstract PoolChunk<T> newUnpooledChunk(int capacity);  // 子类实现
```

11.6.8　findSubpagePoolHead 核心方法

findSubpagePoolHead 方法用于根据传入的 elemSize 分配尺寸，找到对应的 PoolSubpage[]数组下标的头结点并返回。通过源码可知，该方法只用于寻找 tiny、small 的内存页。流程如下。

（1）如果内存小于 512B，子页内存为微型类型。因为微型类型内存以 16B 递增，这里直接除 16 即可。

（2）如果内存大于 512B 且小于 8KB，子页内存为小型类型。小型类型只有 4 个，只需要除 2 递增下标即可。

```
PoolSubpage<T> findSubpagePoolHead(int elemSize) {
    int tableIdx;
    PoolSubpage<T>[] table;
    if (isTiny(elemSize)) {                      // < 512B，微型类型
        tableIdx = elemSize >>> 4;               // 直接除 16 即可
        table = tinySubpagePools;
    } else {                                     // 小型类型
        tableIdx = 0;
        elemSize >>>= 10;
        while (elemSize != 0) {
            elemSize >>>= 1;
            tableIdx ++;
        }
        table = smallSubpagePools;
    }
    return table[tableIdx];
}
```

11.6.9　free 核心方法

free 方法用于释放分配的内存，内存信息由 handle 表示。流程如下。

（1）如果 PoolChunk 为非池化类型，则调用子类实现的 destroyChunk 方法完成内存释放，同时减

少内存使用计数。前面讲过在 HeapArena 中由于内存为 byte[]数组，所以依靠垃圾回收（GC）完成释放，而 DirectArena 为堆外内存，需要根据有无 Cleaner 对象选择不同方式进行清理。

（2）如果 PoolChunk 为池化类型，则优先将该 handle 放到线程缓存中；如果放入失败，则调用 freeChunk 将内存归还给内存池。

```
void free(PoolChunk<T> chunk, long handle, int normCapacity, PoolThreadCache cache) {
    if (chunk.unpooled) {                                       // 非池化类型
        int size = chunk.chunkSize();
        destroyChunk(chunk);
        activeBytesHuge.add(-size);
        deallocationsHuge.increment();
    } else {                                                    // 池化类型
        SizeClass sizeClass = sizeClass(normCapacity);
        if (cache != null && cache.add(this, chunk, handle, normCapacity, sizeClass)) { // 存在缓存，且添加缓存成功
            return;
        }
        freeChunk(chunk, handle, sizeClass);
    }
}
protected abstract void destroyChunk(PoolChunk<T> chunk);       // 子类实现（DirectArena 或 HeapArena）
```

11.6.10　freeChunk 核心方法

freeChunk 方法用于释放 handle 所属的内存到内存池中。流程如下。

（1）根据内存大小信息，更新计数值。

（2）调用 chunk.parent，即用 PoolChunkList 的 free 方法完成释放。

（3）如果 PoolChunkList 决定需要释放该内存，则调用 destroyChunk 将内存归还给 JVM，而不是归还给内存池。

```
void freeChunk(PoolChunk<T> chunk, long handle, SizeClass sizeClass) {
    final boolean destroyChunk;
    synchronized (this) {                                       // 更新计数值
        switch (sizeClass) {
            case Normal:
                ++deallocationsNormal;
                break;
            case Small:
                ++deallocationsSmall;
                break;
            case Tiny:
                ++deallocationsTiny;
                break;
            default:
                throw new Error();
        }
        destroyChunk = !chunk.parent.free(chunk, handle);       // PoolChunkList 完成释放。若释放失败，则返回 false，这里取反将其释放
    }
    if (destroyChunk) {                                         // 归还内存给 JVM
```

```
            destroyChunk(chunk);
        }
    }
```

11.7 PoolChunkList 类

PoolChunkList 类用于管理不同占用率的 PoolChunk 内存块。由于源码实现较为简单，笔者将源码去掉了普通的 get/set/toString 等常用方法，并在这里一起讲解。注意，涉及 Metric 相关的性能计数将不会提及，因为除了计数之外获取计数毫无意义。由源码可知。

（1）PoolChunkList nextList 与 PoolChunkList prevList 变量用于将 PoolChunkList 形成双向链表。

（2）PoolChunk head 用于保存 PoolChunk 链表。

（3）在 PoolChunk 被分配和归还时，都会判断归还的 chunk 的最高和最低利用率，当超过当前 PoolChunkList 的容量限制后，则将其移动到正确的利用率区间链表。

（4）其他所有操作均为标准的双向链表的移除和添加操作（添加时使用头插法，移除时从头结点移除。标准的链表栈结构能够更好地利用缓存空间局部性（后进先出））。

```java
final class PoolChunkList<T> implements PoolChunkListMetric {
    private static final Iterator<PoolChunkMetric> EMPTY_METRICS =
Collections.<PoolChunkMetric>emptyList().iterator();    // 迭代器，为空时放回
    private final PoolChunkList<T> nextList;            // 关联到下一个 PoolChunkList
    private final int minUsage;                         // 管理的 PoolChunk 最小用例
    private final int maxUsage;                         // 管理的 PoolChunk 最大用例
    private final int maxCapacity;                      // 能分配的最大容量
    private PoolChunk<T> head;                          // 管理的 PoolChunk 链表头结点
    private PoolChunkList<T> prevList;                  // 关联到上一个 PoolChunkList

    // 构造器用于初始化成员变量
    PoolChunkList(PoolChunkList<T> nextList, int minUsage, int maxUsage, int chunkSize) {
        this.nextList = nextList;
        this.minUsage = minUsage;
        this.maxUsage = maxUsage;
        maxCapacity = calculateMaxCapacity(minUsage, chunkSize); // 计算当前 PoolChunkList 能分配的最大容量
    }

    // 计算当前 PoolChunkList 能分配的最大容量
    private static int calculateMaxCapacity(int minUsage, int chunkSize) {
        minUsage = minUsage0(minUsage);                 // max(1, value) 获取最小用例值
        if (minUsage == 100) {                          // 如果 minUsage 是 100，则不能分配任何内存
            return 0;
        }

        // 计算可以从 PoolChunkList 中管理的 PoolChunk 分配的最大字节数
        // 例如：如果 PoolChunkList 的 minUsage 等于 25，则可以最多分配 75%的 chunkSize
        return (int) (chunkSize * (100L - minUsage) / 100L); // 16M（默认 maxorder *pageSize）*可以分配的占用率
    }
```

```java
// 用于从当前列表中分配内存
boolean allocate(PooledByteBuf<T> buf, int reqCapacity, int normCapacity) {
    if (head == null || normCapacity > maxCapacity) {  // 不存在 PoolChunk，同时需要分配的容量大于最大可以分配的容量
        return false;
    }
    for (PoolChunk<T> cur = head;;) {                  // 尝试在所有管理的 PoolChunk 链表结点中分配内存
        long handle = cur.allocate(normCapacity);      // 调用 PoolChunk 分配
        if (handle < 0) {                              // 分配失败，继续寻找下一个结点
            cur = cur.next;
            if (cur == null) {                         // 遍历了所有结点仍未分配到内存，则直接返回
                return false;
            }
        } else {                                       // 分配成功
            cur.initBuf(buf, handle, reqCapacity);     // 初始化 ByteBuf 变量，handle 中保存有实际操作的内存信息（高低 32 位）
            if (cur.usage() >= maxUsage) {             // 当前 Chunk 使用率超过当前 PoolChunkList 对象的最大限制值，则将 PoolChunk 从当前列表中移除，同时将其添加到下一个最小用例的 PoolChunkList 中
                remove(cur);                           // 从当前链表中移除
                nextList.add(cur);                     // 添加到指定链表中
            }
            return true;
        }
    }
}

// 该方法用于释放 handle 所表示的内存
boolean free(PoolChunk<T> chunk, long handle) {
    chunk.free(handle);
    if (chunk.usage() < minUsage) {                    // 当前 Chunk 使用率小于当前 PoolChunkList 对象的最小限制值，则将 PoolChunk 从当前列表中移除，同时将其添加到上一个拥有最小用例的 PoolChunkList
        remove(chunk);
        return move0(chunk);
    }
    return true;
}

// 该方法用于将指定内存块从当前链表中移除并将其添加到指定链表中
private boolean move(PoolChunk<T> chunk) {
    if (chunk.usage() < minUsage) {                    // 小于最小用例，向上一个链表移动
        return move0(chunk);
    }
    // 添加到指定链表
    add0(chunk);
    return true;
}

// 将指定 PoolChunk 对象向 PoolChunkList 的前驱结点移动，即利用率较低的链表
private boolean move0(PoolChunk<T> chunk) {
    if (prevList == null) {
        // 前驱链表不存在，即 q000，只有 q000 没有前驱结点
        return false;
```

```java
        return prevList.move(chunk);          // 移动到前驱链表中
    }

    // 指定 PoolChunk 对象在判断利用率没问题后,将其添加到当前 PoolChunkList 链表中
    void add(PoolChunk<T> chunk) {
        if (chunk.usage() >= maxUsage) {// 利用率超过当前链表最大利用率,将其移动到后继链表中
            nextList.add(chunk);
            return;
        }
        add0(chunk);                          // 利用率在区间内,将其添加到当前链表中
    }

    // 将指定的 PoolChunk 添加到当前 PoolChunkList 链表中
    void add0(PoolChunk<T> chunk) {
        chunk.parent = this;                  // 在 PoolChunk 中保存当前 PoolChunkList 引用
        if (head == null) {                   // 不存在头结点时,当前 chunk 为头结点即可
            head = chunk;
            chunk.prev = null;
            chunk.next = null;
        } else {                              // 标准双向链表头插法
            chunk.prev = null;
            chunk.next = head;
            head.prev = chunk;
            head = chunk;
        }
    }

    // 将指定 PoolChunk 对象从当前链表中移除
    private void remove(PoolChunk<T> cur) {
        if (cur == head) {                    // 结点为头结点时,从头结点移除
            head = cur.next;
            if (head != null) {
                head.prev = null;
            }
        } else {                              // 否则从链表中间移除,修正前驱结点和后继结点的 next、prev 引用
            PoolChunk<T> next = cur.next;
            cur.prev.next = next;
            if (next != null) {
                next.prev = cur.prev;
            }
        }
    }
}
```

11.8 PoolSubpage 类

PoolSubpage 类用于保存从 pageSize(默认 8KB)中切割出的对应大小的子页信息,用于服务 Tiny 和 Small 类型的内存信息。在 PoolArena 类中 PoolSubpage[] tinySubpagePools、PoolSubpage[]

smallSubpagePools 数组保存从 8KB 中按照对应数组下标规定的尺寸切割后，生成 PoolSubpage 链表来服务小页分配。例如，tinySubpagePools[0]为 16B，这时分配 Normal Page 8KB，切割为 8KB/16B 个子页，这些子页都是 PoolSubpage 的实例放入 0 下标处。同样，由于该类实现较为简单，笔者也直接将其源码统一讲解。由源码可知。

（1）PoolSubpage 使用 next 和 prev 变量形成双向链表。
（2）初始化时，在 PoolSubpage[]数组中每一项通过 PoolSubpage(int pageSize)构造器生成头结点。
（3）分配时采用头插法放入子页。
（4）所有子页都放入链表时，表明没有线程正在使用该子页链表，需要将其从链表中移除，组成一页放回 PoolChunk。由于使用头插法，直接移除最后一个放入的 chunk 结点即可。

```java
final class PoolSubpage<T> implements PoolSubpageMetric {    // 实现了 PoolSubpageMetric 的性能统计，了解即可
    final PoolChunk<T> chunk;              // 所属内存块（毕竟是从正常页进行切割的,正常页又属于某个内存块）
    private final int memoryMapIdx;        // 指向 PoolChunk 中的 memoryMap 下标
    private final int runOffset;           // 基于内存地址的 0 下标处的偏移量。例如，用 byte[]表示内存，该偏移量就是基于 byte[0]下标处到需要使用的偏移量 byte[runOffset]
    private final int pageSize;            // 被分割的页大小，通常为 8KB
    private final long[] bitmap;           // 页被分割成 N 个子页后，用于标识当前已经被分配的子页
    PoolSubpage<T> prev;                   // 前驱结点
    PoolSubpage<T> next;                   // 后继结点

    boolean doNotDestroy;                  // 标识当前子页是否被归还
    int elemSize;                          // 子页大小
    private int maxNumElems;               // 被切割的数量
    private int bitmapLength;              // 需要的位图长度
    private int nextAvail;                 // 指向下一个可用的位图索引
    private int numAvail;                  // 当前可用的子页数

    // 构造器，用于初始化 PoolSubpage<T>[] tinySubpagePools、PoolSubpage<T>[] smallSubpagePools 数组的头结点
    PoolSubpage(int pageSize) {
        chunk = null;
        memoryMapIdx = -1;
        runOffset = -1;
        elemSize = -1;
        this.pageSize = pageSize;
        bitmap = null;
    }

    // 构造器，用于生成新的子页，携带内存信息
    PoolSubpage(PoolSubpage<T> head, PoolChunk<T> chunk, int memoryMapIdx, int runOffset, int pageSize, int elemSize) {
        this.chunk = chunk;
        this.memoryMapIdx = memoryMapIdx;
        this.runOffset = runOffset;
        this.pageSize = pageSize;
        bitmap = new long[pageSize >>> 10];         // 创建位图数组：pageSize / 16 / 64
        init(head, elemSize);
    }
```

```java
// 初始化成员变量
void init(PoolSubpage<T> head, int elemSize) {
    doNotDestroy = true;
    this.elemSize = elemSize;
    if (elemSize != 0) { // 需要的子页大小存在，直接用 pageSize/elemSize 得到切割子页数。如 8KB/16B
        maxNumElems = numAvail = pageSize / elemSize;
        nextAvail = 0;
        bitmapLength = maxNumElems >>> 6; // 位图长度：子页个数/$2^6$（64 位，一个 long 类型大小，不要忘记用 long 数组表示位图）
        if ((maxNumElems & 63) != 0) {          // 子页数量超过 64 位，增加一个数组长度
            bitmapLength ++;
        }
        for (int i = 0; i < bitmapLength; i ++) {  // 初始化位图
            bitmap[i] = 0;
        }
    }
    addToPool(head);                            // 将头结点添加到子页链表中
}

// 对当前子页分配内存，返回携带子页信息的 handleId（高 32 位表示位图信息，低 32 位表示 PoolChunk 的内存信息）
long allocate() {
    if (elemSize == 0) {                        // 子页的大小为 0
        return toHandle(0);
    }
    if (numAvail == 0 || !doNotDestroy) {       // 没有可用的子页，返回-1，表示分配失败
        return -1;
    }
    final int bitmapIdx = getNextAvail();       // 获取下一个可以分配的位图索引
    // 对位图置 1，表示当前的 bitmapIdx 已经被使用
    int q = bitmapIdx >>> 6;
    bitmap[q] |= 1L << r;
    if (-- numAvail == 0) { // 减少一个可用的子页数，如果没有可用子页，则将其从链表中移除（因为使用位图来标识子页是否已经被使用，所有的子页都已经被分配时，没必要再保存这一串子页链表）
        removeFromPool();
    }
    return toHandle(bitmapIdx);                 // 组合 PoolChunk 的内存信息与子页信息
}

// 将当前子页放入子页链表中
boolean free(PoolSubpage<T> head, int bitmapIdx) {
    if (elemSize == 0) {                        // 分页数量为 0，直接返回
        return true;
    }
    // 将对应下标 bitmapIdx 的位图置 0，异或相同为 0
    int q = bitmapIdx >>> 6;
    bitmap[q] ^= 1L << r;
    setNextAvail(bitmapIdx);                    // 设置下一个可用下标
    if (numAvail ++ == 0) {                     // 在添加当前结点后链表中有可用子页时，将当前子页放入链表，即与 head 进行关联（头插法）
        addToPool(head);
        return true;
```

```java
        }
        if (numAvail != maxNumElems) {  // 可用数量不等于最大数量，说明当前放入的子页不是最后一个，证明
还在使用该子页内存，直接返回成功即可
            return true;
        } else {                          // 否则当前 numAvail == maxNumElems，表明所有子页都已经归还，
说明切割的子页都没有被使用。对其进行释放
            if (prev == next) {           // 如果此子页是池中唯一剩下的一个子页，则不要删除
                return true;
            }
            // 当前页为最后一个子页，将其从链表中移除（由于采用头插法，所以这里只要移除当前结点，就
等于移除了所有子页，这些内存将合并为一整页归还给 PoolChunk 内存块，后续在描述 PoolChunk 时会详细介绍，
这里了解即可）
            doNotDestroy = false;
            removeFromPool();
            return false;
        }
    }

    // 将当前 chunk 结点用头插法插入子页链表
    private void addToPool(PoolSubpage<T> head) {
        prev = head;
        next = head.next;
        next.prev = this;
        head.next = this;
    }

    // 将当前 chunk 从 head 结点上移除链表（该方法由最后一个归还到子页链表的子页调用，因为此时当前子
页链表中没有线程正在使用这些子页，所以需要将其组成一页并释放到 PoolChunk 中）
    private void removeFromPool() {
        prev.next = next;
        next.prev = prev;
        next = null;
        prev = null;
    }

    // 获取下一个可用的链表位图下标
    private int getNextAvail() {
        int nextAvail = this.nextAvail;
        if (nextAvail >= 0) {
            this.nextAvail = -1;
            return nextAvail;
        }
        return findNextAvail();
    }

    // 遍历位图找到可用下标
    private int findNextAvail() {
        final long[] bitmap = this.bitmap;
        final int bitmapLength = this.bitmapLength;
        for (int i = 0; i < bitmapLength; i ++) {  // 循环遍历每个 long 中的位图找到可用下标
            long bits = bitmap[i];
            if (~bits != 0) {                      // 位图中存在可用位
                return findNextAvail0(i, bits);
            }
```

```
            }
            return -1;
        }

        // 遍历每一位找到可用位，即为 0 的那一位
        private int findNextAvail0(int i, long bits) {
            final int maxNumElems = this.maxNumElems;
            final int baseVal = i << 6;         // 左移 6 位，等于乘以 64，即找到基础下标（使用 long 数组保
存位图，这里 i 为 long 的索引下标，需要返回的是基于 long 数组从 0-long 的长度的位索引，所以这里需要基础下
标。例如，i 为 1，基础下标为 1 * 64 = 64，再加上当前 long 64 位中找到的为 0 的下标）
            for (int j = 0; j < 64; j ++) {
                if ((bits & 1) == 0) {
                    int val = baseVal | j;
                    if (val < maxNumElems) {
                        return val;
                    } else {
                        break;
                    }
                }
                bits >>>= 1;
            }
            return -1;
        }

        // 使用 PoolSubpage 的下标和指向 PoolChunk memoryMap 的下标进行组合，形成 handleId。bitmapIdx 放
在高 32 位，memoryMapIdx 放在低 32 位，同时这里是一个特殊值，用于与 0 进行标识
        private long toHandle(int bitmapIdx) {
            return 0x4000000000000000L | (long) bitmapIdx << 32 | memoryMapIdx;
        }
    }
}
```

11.9　PoolChunk 原理

PoolArea 为内存域，用于减少线程并发分配内存时上锁的性能损耗，通常将 PoolArea 和 EventLoopGroup 中的 EventLoop 数量设置为 CPU 核心数×2，这样可以在 IO 和普通事件中充分利用 CPU 资源。PoolArea 中管理的基础内存块为 PoolChunk，而 PoolChunk 中管理的基础页大小为 8KB，所以称一个 PoolChunk 内存块默认大小为：pageSize×maxOrder = 8KB×11 = 16MB。本节详细介绍内存域，说明它是如何管理这些 8KB 内存页的。

在开始之前，有必要进行知识推理。

内存碎片类型分为如下两类。

内碎片。具有 16MB 内存，分配了 4MB，但却只用了 1MB，称剩下的 3MB 为内碎片。

外碎片。具有 16MB 内存，A 线程分配了 4MB，B 线程分配了 3MB，C 线程分配了 2MB，这时内存剩余 7MB，此时内存布局为 A(4MB)→B(3MB)→C(2MB)→空闲(7MB)。如果此时 B 线程释放了内存，内存布局为 A(4MB)→空闲(3MB)→C(2MB)→空闲(7MB)，这时 D 线程需要分配 10MB，这是无法分配的，因为两块空闲线程之间存在着 C 线程的 2MB 内存占用，这时称空闲(3MB)与空闲(7MB)为内存外碎片。

假如一块 16MB 的内存需要进行内存管理，内存分配时的策略有以下几种。

第一种，首次适应策略。空闲分区按内存地址从低到高的顺序以双向链表形式连接在一起。内存分配时，从低地址开始查找并分配，因此容易造成低地址使用率较高，而高地址使用率却较低的情况，同时会造成较小的内存外碎片。

第二种，循环首次适应策略。该策略基于首次适应策略进行改进。在第二次分配时，从下一个空闲分区开始查找。该策略可以将内存分配得更加均匀，查找效率有所提升，但是由于进行了分区跳跃，会在不同分区导致严重内存外碎片。

第三种，最佳适应策略。空间分区链按从小到大的递增顺序关联。内存分配时，从开头开始查找适合的空间内存并分配，完成分配请求后，空闲分区链重新按分区大小排序。该策略的空间利用率高，但同样会有难以利用的小空间分区。这是由于空闲内存块大小在不断发生变化，并没有针对内存大小做优化分类，只有当内存大小刚好等于空闲内存块的大小，此时才会有空间利用率 100%的情况。每次分配完后需要重新排序，存在时间复杂度消耗。

第四种，伙伴分配策略。将内存划分为分区，以最合适的大小满足内存请求。假如把 16MB 的内存切割为 8KB 一页，同时，11 个链表中每一个块链表分别包含大小为 1、2、4、8、16、32、64、128、256、512 和 1024 个连续的页框，最大内存分配为 16MB，此时只需要按照需要的内存大小在不同链表中分配即可。例如，16MB 可以分割为两个 8MB。同理，一个 8MB 可以分割为两个 4MB。如此，将内存分为最细粒度是 8KB。此时，这样的管理就是一颗满二叉树，由于兄弟结点之间的内存大小相同，同时组成了 2 倍兄弟结点大小的父结点，所以称为伙伴算法。该算法有效减少了外碎片，因为可以将兄弟结点整合为父结点的整个空间，但是最小分配粒度还是 pageSize（这里为 8KB），因此有可能造成非常严重的内碎片，因为根本不需要使用这么大的内存页。

第五种，slab 策略。该策略将内存按照需要的小内存尺寸进行分割。例如，需要 16B 的内存，这时可以将从上述伙伴策略中分配到的一页 8KB 分割为 8KB/16B 个小内存，这时可以满足小内存的分配。其实核心思想就是将大页内存进行均分。该算法有助于减少内存的内碎片。

第六种，slab 策略与伙伴分配搭配，效果就很完美。在 Linux 内核、Netty 中均是搭配这两个组合进行内存管理。

Netty 中使用 slab 算法与伙伴算法来进行内存管理，这里读者先理解以下几个术语。

☑ page：伙伴算法内存管理的最小分配单位，通常为 8KB。
☑ chunk：一些列页的组合，即管理 page，也称为内存块。
☑ chunkSize：块大小，通常为 $2^{maxOrder} \times pageSize = 16MB$。
☑ memoryMap：内存集，伙伴算法本身就是一颗满二叉树，所以采用该数组来构建一颗满二叉树。

```tex
树深度              内存大小
depth=0         1 结点 (chunkSize 16MB)
depth=1         2 结点 (chunkSize 16MB / 2 = 8MB 两个 8MB 的内存块)
..
..
depth=d         2^d 结点 (chunkSize/2^d)
..
depth=maxOrder 2^maxOrder 结点(chunkSize/2^maxOrder = pageSize)
```

此时，不难看出，在满二叉树的叶子结点处，depth=maxOrder 由一些列的 page 组成（这里为 8KB）。这时在分配内存时，只需要按照规格化的内存大小（8KB * N），搜索这棵树便可得到想要的内存。

☑ depthMap：深度集，一旦初始化，不会发生改变，与初始值的 memoryMap 相等

可能读者会问：为什么这里要弄出 memoryMap 和 depthMap 两个集，这是与分配算法相关的，下面来看推理。

假定 pageSize 为 8KB，maxOrder 为 11。初始时，memoryMap = depthMap = [0,0,1,1,2,2,2,2,,,,,,11] 这里的数组项为了方便计算，不使用 0 下标，所以下标为 1 处的内存大小为 16MB，即深度为 0 时的大小，后面紧接着就是深度为 1 和 2，一直到深度为 11 的大小。

假如此时在深度为 1 处分配了一个 8MB 的内存，此时由于 16MB 可以分割为两个 8MB，则需要标识当前分配的情况，此时就可以按照以下约定来修改 memoryMap 的下标值，同时与 depthMap 进行对比来查看当前结点分配情况。

若 memoryMap[id] = depth_of_id，则当前 id 下标处的结点及其子结点都未被分配。

若 memoryMap[id] > depth_of_id，则当前 id 下标处的结点的子结点至少有一个被分配，所以如果需要分配当前结点的内存大小时，由于子结点已经部分被分配，所以不能再分配。

若 memoryMap[id] = maxOrder + 1，则当前结点的所有子结点都已经被分配。

通过上述描述，很容易可以得出分配算法。根据规格化的内存 N*pageSize 计算出需要分配的深度 d。从根结点（深度为 0，即 memoryMap 下标为 1）开始遍历。如果 memoryMap[1] >d，则当前 chunk 不能分配该内存。如果当前深度的结点左边的值 memoryMap[left] <= d，则左子树可以进行分配，跳转到左子树分配。否则，尝试右子树，直到找到可以分配的结点。

11.9.1 核心变量与构造器

该类的实现在上文已经详细说明，构造器与变量的定义，均按照上述的算法和逻辑来构建。源码如下。

```
final class PoolChunk<T> implements PoolChunkMetric {
    final PoolArena<T> arena;                  // 所属内存域
    final T memory;                            // 实际管理内存块地址，byte[]或 DirectByteBuffer
    final boolean unpooled;                    // 是否为 Huge 大页内存不进行池化的内存块
    private final byte[] memoryMap;            // 内存集（实现满二叉树）
    private final byte[] depthMap;             // 内存深度（进行原始满二叉树深度保存）
    private final PoolSubpage<T>[] subpages;   // 所有子页数组，用于支持 allocateSubpage 方法分配小页内存
    (在满二叉树的叶子结点上，保存着所有可以分配的 8KB 基础管理页，该数组便保存这些页用于切割为小页内存)
    private final int subpageOverflowMask;     // 子页大小掩码，用于确定请求的容量是否等于或大于 pageSize
    private final int pageSize;                // 基础管理页大小
    private final int pageShifts;              // 基础管理页偏移（2^N = pageSize）
    private final int maxOrder;                // 满二叉树最大深度
    private final int chunkSize;               // 块大小
    private final int log2ChunkSize;           // 2^N =chunkSize
    private final int maxSubpageAllocs;        // 最大能分配的基础管理页数量 1 << maxOrder => 2^maxOrder
    private final byte unusable;               // 用于标记 memoryMap 不可用的 index 下标值，即 memoryMap[id] = maxOrder + 1
    private int freeBytes;                     // 当前 chunk 空闲内存字节数
    PoolChunkList<T> parent;                   // 所属内存块链表
    PoolChunk<T> prev;                         // 在 PoolChunkList 中的前驱结点
```

```java
    PoolChunk<T> next;                                  // 在 PoolChunkList 中的后继结点

    // 构造器用于使用提供的参数信息，初始化成员变量
    PoolChunk(PoolArena<T> arena, T memory, int pageSize, int maxOrder, int pageShifts, int chunkSize) {
        unpooled = false;                               // 通过构造器创建为池化管理内存
        this.arena = arena;
        this.memory = memory;
        this.pageSize = pageSize;
        this.pageShifts = pageShifts;
        this.maxOrder = maxOrder;
        this.chunkSize = chunkSize;
        unusable = (byte) (maxOrder + 1);               // 当前结点已经被分配完毕
        log2ChunkSize = log2(chunkSize);                // 取 16MB 的对数值
        subpageOverflowMask = ~(pageSize - 1);          // pageSize - 1 得到低位掩码，按位取反得到高位掩码
        freeBytes = chunkSize;
        maxSubpageAllocs = 1 << maxOrder;
        // 创建 memorymap 集
        memoryMap = new byte[maxSubpageAllocs << 1];
        depthMap = new byte[memoryMap.length];          // 初始深度集与 memorymap 集相同
        int memoryMapIndex = 1;                         // 不使用数组 0 下标，从 1 下标开始
        for (int d = 0; d <= maxOrder; ++d) {           // 一层一层地向下移动满二叉树，初始化内部值
            int depth = 1 << d;                         // 当前深度。每循环一次，深度增加一倍
            for (int p = 0; p < depth; ++p) {           // 初始化当前深度中的结点数据
                memoryMap[memoryMapIndex] = (byte) d;
                depthMap[memoryMapIndex] = (byte) d;
                memoryMapIndex++;
            }
        }
        subpages = newSubpageArray(maxSubpageAllocs);   // 初始化子页数组，数量为 2^maxOrder
    }

    // 创建子页数组
    private PoolSubpage<T>[] newSubpageArray(int size) {
        return new PoolSubpage[size];
    }

    // 构造器，用于创建 Huge 大页不需要池化的内存块
    PoolChunk(PoolArena<T> arena, T memory, int size) {
        unpooled = true;
        this.arena = arena;
        this.memory = memory;
        memoryMap = null;
        depthMap = null;
        subpages = null;
        subpageOverflowMask = 0;
        pageSize = 0;
        pageShifts = 0;
        maxOrder = 0;
        unusable = (byte) (maxOrder + 1);
        chunkSize = size;
        log2ChunkSize = log2(chunkSize);
        maxSubpageAllocs = 0;
    }
}
```

11.9.2　allocate 核心方法

allocate 方法用于分配规格化后的内存 normCapacity。可以很容易看到，通过 subpageOverflowMask 来判断分配内存的尺寸是否大于或等于 pageSize，来选择合适的分配方法。当为 pageSize 的倍数时，调用 allocateRun 方法，否则调用 allocateSubpage 分配子页。源码如下。

```
long allocate(int normCapacity) {
    if ((normCapacity & subpageOverflowMask) != 0) {   // 判断是否大于或等于 pageSize。其实就是查看规格
                                                        // 化的内存的高位是否为 0，如果不为 0，则高位的值正好为 normCapacity/pageSize（pageSize 的倍数）
        return allocateRun(normCapacity);
    } else {                                            // 分配子页
        return allocateSubpage(normCapacity);
    }
}
```

11.9.3　allocateRun 核心方法

allocateRun 方法用于分配由一个或一组页组成的内存，即 page * N 的内存。流程如下。
（1）计算待分配内存的深度。
（2）调用 allocateNode 分配对应深度的内存。
（3）如果分配失败，即 id 小于 0，则直接返回 id。
（4）否则减少空闲内存计数，并返回 id。

```
private long allocateRun(int normCapacity) {
    int d = maxOrder - (log2(normCapacity) - pageShifts);   // 首先计算分配的深度为 pageSize 的几次方，然后用
                                                              // 最大深度相减即可
    int id = allocateNode(d);                                 // 分配深度为 d 的页
    if (id < 0) {
        return id;
    }
    // 分配成功
    freeBytes -= runLength(id);                               // 减少可用字节数
    return id;
}
// 用于计算指定 id 下标所代表的内存大小：即总大小减掉当前深度
private int runLength(int id) {
    return 1 << log2ChunkSize - depth(id);
}

private byte depth(int id) {
    return depthMap[id];
}
```

11.9.4　allocateNode 核心方法

allocateNode 方法用于分配深度 d 处的内存值，这里需要从 1 下标开始，即从根结点开始遍历查找空闲内存块，因为任何一个块都可能已经被分配。流程如下。

(1) 获取 id 为 1 处的值。
(2) 如果最大内存 16MB 都不能满足需要分配的尺寸，则直接返回-1。
(3) 深度优先遍历，直到找到 id 的深度索引为 d 时才会结束。
(4) 当前下标为 id 处的空间已经被分配，所以将其标明为已使用状态。
(5) 当前结点可能有父结点，也需要更新父结点所在的 memeoryMap 中的值。

```
private byte value(int id) {
    return memoryMap[id];
}

private int allocateNode(int d) {
    int id = 1;
    int initial = -(1 << d);              // 用于判断 id 所在索引区间是否在正确范围
    byte val = value(id);                 // 获取 id 为 1 处的值
    if (val > d) {                        // 最大内存 16MB，都不能满足需要分配的尺寸，则直接返回-1
        return -1;
    }
    while (val < d || (id & initial) == 0) { // 深度优先遍历，直到找到 memoryMap 中深度等于 d 的结点（val>=d），
                                             // 且找到的索引下标在正确区间内（(id & initial) == 0。例如，d 为 1，表示需要的结点在深度为 1 处，由于
                                             // memoryMap=[0,0,1,1,2,2,2,2,,,,,,,,11]，这时 initial = 11111111111111111111111111111110 = -(1 << d)，找到
                                             // 的 val 为 1，id 为 1，此时(id & initial) == 0 满足分配要求）
        id <<= 1;                         // 先获取左结点
        val = value(id);                  // 获取当前左结点的值
        if (val > d) {                    // 左结点没有可分配的空间，则跳转到右结点
            id ^= 1;                      // 用异或做无进位加法，找到当前 id 处的兄弟结点
            val = value(id);              // 更新当前值为兄弟结点的值后继续循环
        }
    }
    setValue(id, unusable);               // 当前下标为 id 处的空间已被分配，将其标明为已使用状态
    updateParentsAlloc(id);               // 当前结点可能有父结点，也需要更新父结点在 memeoryMap 中的值
    return id;
}
```

11.9.5 updateParentsAlloc 核心方法

updateParentsAlloc 方法用于从指定下标处开始更新父结点在 memeoryMap 中的值，因为此时当前结点已经被分配，所以需要更新父类对于结点的分配信息。流程如下。

(1) 循环更新一直到根结点。
(2) 获取当前结点的父结点。
(3) 获取当前结点的值与兄弟结点的值。
(4) 更新父结点的值为兄弟结点之间 val 的最小值。
(5) 更新当前 id 为父结点 id，继续循环。

例如，分配 id 为 1 的结点（8MB），memoryMap=[0,0,1,1,2,2,2,2,,,,,,,11]，此时由于已经分配了 8MB，所以 memoryMap=[0,0,12(maxOrder + 1),1,2,2,2,2,,,,,,,11]。调用该方法时，需要将父结点更新为 1，因为 12>1，所以取最小值，memoryMap=[0,1,12(maxOrder + 1),1,2,2,2,2,,,,,,,11]。再次分配 8MB 时，d 也为 1，这时先找到 id 为 2，由于 memoryMap[2]的值 12 大于 d1，所以 id 为 2 的结点分配失败。同时找到 id

为 3，由于 1=1，所以分配成功。此时 memoryMap=[0,0,12(maxOrder + 1),12(maxOrder + 1,2,2,2,2,,,,,,11)]。调用该方法时更新为 memoryMap=[0,12,12(maxOrder + 1),12(maxOrder + 1,2,2,2,2,,,,,,11)]，再次分配会失败，因为所有结点都已经分配完毕。源码如下。

```
private void updateParentsAlloc(int id) {
    while (id > 1) {                      // 一直更新到根结点
        int parentId = id >>> 1;          // 当前结点整除 2，得到父结点 id
        byte val1 = value(id);            // 获取当前结点的值
        byte val2 = value(id ^ 1);        // 获取到兄弟结点的值
        // 更新父结点的值为兄弟结点之间的 val 的最小值
        byte val = val1 < val2 ? val1 : val2;
        setValue(parentId, val);
        id = parentId;                    // 更新当前 id 为父结点 id，继续循环，此时跳转到父结点 id 来执行
    }
}
```

11.9.6　allocateSubpage 核心方法

allocateSubpage 方法用于分配指定大小 normCapacity 的子页。流程如下。

（1）获取到当前内存容量所处内存区域的 PoolSubpage[]数组的对应下标的头结点 PoolSubpage head。

（2）从满二叉树的叶子结点处开始分配一页，如果分配失败，直接返回-1。

（3）如果分配成功，则将可用内存数减掉一页的大小。

（4）根据当前 id 计算出所在子页数组 this.subpages 的下标。

（5）获取到对应下标处的子页对象，如果不存在，则创建子页。

（6）调用 subpage.init(head, normCapacity)初始化子页。

（7）调用 subpage.allocate()从子页链表中分配一个空闲的子页内存。

```
private long allocateSubpage(int normCapacity) {
    PoolSubpage<T> head = arena.findSubpagePoolHead(normCapacity);   // 获取到当前内存容量所处内存
    区域的 PoolSubpage[]数组的对应下标的头结点（small 或 tiny 数组的头结点）
    synchronized (head) {                 // 头结点上锁保证线程安全
        // 直接从叶子结点处开始分配一页（只有叶子结点的内存页为基础管理页 pageSize 的大小，从这里获取
        一页然后进行切分）
        int d = maxOrder;
        int id = allocateNode(d);
        if (id < 0) {                     // 分配失败，直接返回-1
            return id;
        }
        final PoolSubpage<T>[] subpages = this.subpages;
        final int pageSize = this.pageSize;
        freeBytes -= pageSize;            // 可用内存数减掉一页的大小
        int subpageIdx = subpageIdx(id);  // 根据当前 id 计算出所在子页数组 this.subpages 中的下标：
        memoryMapIdx ^ maxSubpageAllocs, maxSubpageAllocs 为 100000000000，此时相当于将 memoryMapIdx 在
        2048 处的子页开始反转到 PoolSubpage[]数组，从零下标处的 PoolSubpage（memoryMap 中，满二叉树的最左
        结点，读者需要理解：异或等于无进位的加法）
        // 获取到对应下标处的子页对象，如果不存在，则创建子页
        PoolSubpage<T> subpage = subpages[subpageIdx];
        if (subpage == null) {
            subpage = new PoolSubpage<T>(head, this, id, runOffset(id), pageSize, normCapacity);
```

```
            subpages[subpageIdx] = subpage;
        } else {                          // 初始化该子页，进行指定内存 normCapacity 切分
            subpage.init(head, normCapacity);
        }
        return subpage.allocate();        // 从子页链表中分配一个空闲的子页内存
    }
}
```

11.9.7　free 核心方法

free 方法用于释放指定 handle 的内存。流程如下。

（1）获取 memoryMap 下标 memoryMapIdx。
（2）获取内存子页位图下标 bitmapIdx。
（3）如果存在位图信息，则先释放子页内存。
（4）释放 pageSize 一整页内存。
（5）更新指定下标 memoryMapIdx 的深度值。
（6）更新父结点的深度值。

```
void free(long handle) {
    int memoryMapIdx = memoryMapIdx(handle);   // (int) handle 获取 memoryMap 下标（低 32 位）
    int bitmapIdx = bitmapIdx(handle);         // (int) (handle >>> Integer.SIZE) 获取内存子页位图下标（高 32 位）
    if (bitmapIdx != 0) {                      // 位图信息存在，释放子页内存
        PoolSubpage<T> subpage = subpages[subpageIdx(memoryMapIdx)];    // 从子页数组中获取到该子页对象
        // 获取到该子页尺寸大小所对应的头部信息
        PoolSubpage<T> head = arena.findSubpagePoolHead(subpage.elemSize);
        synchronized (head) {                  // 对头部上锁保证释放时线程安全
            if (subpage.free(head, bitmapIdx & 0x3FFFFFFF)) {  // 开始释放子页，如果返回 true，则说明还
有线程占有该内存代表的子页，否则将其从子页链表中摘除，并返回 false，继续执行将一整页数据归还给内存池
                return;
            }
        }
    }
    // 释放 pageSize 一整页内存
    freeBytes += runLength(memoryMapIdx);
    setValue(memoryMapIdx, depth(memoryMapIdx));   // 从 depthMap[id]中取出原有的深度信息设置到指定
memoryMapIdx 下标处
    updateParentsFree(memoryMapIdx);           // 当前结点可用，更新父结点对于子结点分配信息，即深度信息
}
```

11.9.8　updateParentsFree 核心方法

updateParentsFree 方法用于更新指定 id 结点的父结点的内存分配信息，即深度值。流程如下。

（1）获取 id 下标代表的深度信息加 1，该操作是为了在循环中方便计算。
（2）获取父结点下标。
（3）获取当前 id 结点的深度值。
（4）获取当前 id 结点兄弟结点的深度值。
（5）每次迭代减 1，等于当前结点的原始深度。

（6）如果当前结点的兄弟结点的深度都相同，都等于当前 id 的原有深度，则把父结点深度设置为原始深度，因为兄弟结点均完成了释放。

（7）否则取兄弟结点最小值进行更新父结点（深度最小，表明还有空间可分配）。

（8）转移到父结点继续更新。

```
private void updateParentsFree(int id) {
    int logChild = depth(id) + 1;                  // 获取 id 下标代表的深度信息加 1
    while (id > 1) {                               // 一直更新到父结点
        int parentId = id >>> 1;                   // 获取父结点下标
        byte val1 = value(id);                     // 获取当前 id 结点的深度值
        byte val2 = value(id ^ 1);                 // 获取当前 id 结点兄弟结点的深度值
        logChild -= 1; // 每次迭代减 1，等于当前结点的深度（每次迭代都将从当前结点转移到父结点，即深度减 1）
        if (val1 == logChild && val2 == logChild) { // 当前结点的兄弟结点的深度都相同，等于当前 id 的原有深度，把父结点深度设置为原始深度
            setValue(parentId, (byte) (logChild - 1));
        } else {                                   // 否则取兄弟结点最小值进行更新父结点
            byte val = val1 < val2 ? val1 : val2;
            setValue(parentId, val);
        }
        id = parentId;                             // 转移到父结点继续更新
    }
}
```

11.9.9 initBuf 核心方法

initBuf 方法用于初始化分配内存页 PooledByteBuf，此时 handle 中包含了已经分配的内存信息。流程如下。

（1）获取 memoryMap 下标 memoryMapIdx。

（2）获取内存子页下标 bitmapIdx。

（3）如果位图为 0，则初始化正常内存页。

（4）否则初始化内存子页。

```
void initBuf(PooledByteBuf<T> buf, long handle, int reqCapacity) {
    int memoryMapIdx = memoryMapIdx(handle);      // 获取 memoryMap 下标
    int bitmapIdx = bitmapIdx(handle);            // 获取内存子页下标
    if (bitmapIdx == 0) {                         // 位图为 0，初始化正常内存页
        buf.init(this, handle, runOffset(memoryMapIdx), reqCapacity, runLength(memoryMapIdx),
            arena.parent.threadCache());
    } else {                                      // 否则初始化内存子页
        initBufWithSubpage(buf, handle, bitmapIdx, reqCapacity);
    }
}
```

11.9.10 initBufWithSubpage 核心方法

initBufWithSubpage 方法用于初始化从 PoolSubpage 中分配的子页内存。通过源码可知 long handle 64 位变量分为高低 32 位。高 32 位表示 bitmapIdx 信息，即内存子页信息；而低 32 位保留了对 PoolChunk 的内存页的信息，这时需要根据这两个信息计算出 PooledByteBuf 所操作的内存数组的起始偏移量

offset。源码如下。

```java
void initBufWithSubpage(PooledByteBuf<T> buf, long handle, int reqCapacity) {
    initBufWithSubpage(buf, handle, bitmapIdx(handle), reqCapacity);
}

// 从 64 位的 handle 变量中获取到高 32 位的 bitmapIdx 索引下标
private static int bitmapIdx(long handle) {
    return (int) (handle >>> Integer.SIZE);
}

// 从 64 位的 handle 变量中获取到低 32 位的指向 PoolChunk 的 memoryMap 的下标
private static int memoryMapIdx(long handle) {
    return (int) handle;
}

private final int maxSubpageAllocs = 1 << maxOrder; // 在 PoolChunk 中定义,这里粘贴过来方便阅读($2^{11}$ 个子页,一页 8KB)

// 用于获取当前子页在分割后的多个子页的下标,直接异或子页数,子页数通常为 $2^{11}$ = 2048
private int subpageIdx(int memoryMapIdx) {
    return memoryMapIdx ^ maxSubpageAllocs;
}

// 参数 id 表示该内存页所在 PoolChunk 中的 memoryMap 的下标(基于数组 0 的下标),这里通过 id 获取到深
// 度,同时根据深度来计算出所在偏移量,然后乘以该内存块大小即可
private int runOffset(int id) {
    int shift = id ^ 1 << depth(id);
    return shift * runLength(id);
}

// 获取 id 代表的内存块的大小
private int runLength(int id) {
    return 1 << log2ChunkSize - depth(id);
}

private void initBufWithSubpage(PooledByteBuf<T> buf, long handle, int bitmapIdx, int reqCapacity) {
    int memoryMapIdx = memoryMapIdx(handle); // 获取到低 32 位指向 PoolChunk 的 memoryMap 的下标
    PoolSubpage<T> subpage = subpages[subpageIdx(memoryMapIdx)]; // 根据该下标计算出内存页所处的
    subpages 数组的下标
    buf.init(
        this, handle,
        runOffset(memoryMapIdx) + (bitmapIdx & 0x3FFFFFFF) * subpage.elemSize, reqCapacity,
        subpage.elemSize,
        arena.parent.threadCache());// 直接调用 PooledByteBuf 的 init 方法完成初始化。runOffset(memoryMapIdx)+
(bitmapIdx & 0x3FFFFFFF) * subpage.elemSize 用于计算出该子页内存所处的 PoolChunk 管理的内存数组中的
起始下标 memoryMapIdx 指向的内存块的所在内存数组的起始偏移量加上当前子页所处内存块的偏移量。读者可
以这么想:byte[]数组被切割为 N 个 8KB 的页,而 memoryMapIdx 可以获取到这个页所在数组的起始下标,而 8KB
页被切割为 N 个子页大小,如果想要得到真正的偏移量,则需要在 memoryMapIdx 获取到这个页所在数组的起始
下标加上该子页在被切割的该内存页的偏移量,而(bitmapIdx & 0x3FFFFFFF)便是获取到了该内存子页所在内存
页的下标,subpage.elemSize 便是获取到该内存子页切割的字节数(如 16B)
}
```

小结

内存管理篇

本篇内容可以用 Netty 内存管理示意图进行总结描述。

（1）Netty 的分配器分为 PooledByteBufAllocator（池化管理）和 UnpooledByteBufAllocator（非池化管理）。

（2）PooledByteBufAllocator 中默认分配 Cpu 核心数二倍的 PoolArea 内存域，数量与 EventLoopGroup 管理的事件循环线程数保持一致，这是为了减少多个线程进入 PoolArea 时发生锁竞争。

（3）PoolArea 中默认管理多个大小为 16 MB 的 PoolChunk 内存块，它的大小由 chunkSize 决定，内部使用伙伴算法完成 8KB 一页的内存页管理。

（4）SubpagePool 主要用于存放 tiny 和 small 类型的子页，这些子页将从 8KB 基础内存页中切割放入。

（5）在 PoolArena 中的分配大小种类为 Tiny、Small、Normal，大小如图 11-2 所示。

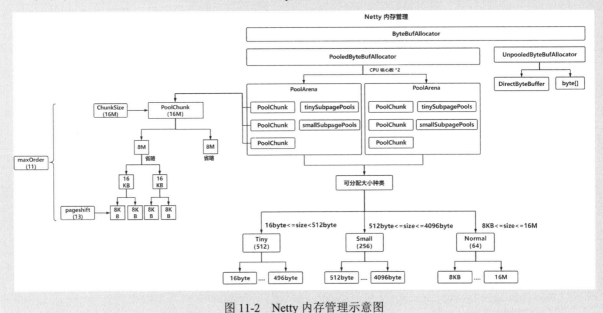

图 11-2　Netty 内存管理示意图

第 4 篇

通道管理层

本篇详细介绍 Netty 中 Channel 通道相关概念，第 1 篇介绍了 Netty 的事件循环组，用于支持 Netty 的线程模型。第 2 篇介绍了内存池原理，用于自定义 Netty 的内存池功能，用于增加内存分配速度。但似乎遗忘了最重要的 IO 细节：Channel 通道对于底层 IO 的支持。本篇将详细讲解编码器、解码器、通道等原理，帮助读者把 Netty 最后一个部分的知识补全。同时，由于底层 IO 细节较多，定义的接口方法也很多，这里也利用了第 1 篇和第 2 篇的知识进行编码，特别是在涉及 Java IO 细节库时，会用到 Java 的 ServerSocket、Socket、ServerSocketChannel、SocketChannel 的知识。所以在学习本篇内容时，一定要先根据接口定义描述纵观全局进行学习。

第 12 章 Netty Channel 层原理

12.1 ChannelOutboundInvoker 接口

ChannelOutboundInvoker 接口用于定义通道输出调用方法。通过该方法定义可知。
（1）该接口主要定义通道对外操作的方法 bind、connect 等。
（2）所有方法均为异步执行，同时返回代表该异步执行结果的 ChannelFuture 实例。
（3）提供创建 ChannelPromise 实例的方法，可以通过这些方法创建成功或失败的 ChannelPromise 实例。

```
public interface ChannelOutboundInvoker {
    // 异步对当前通道执行绑定地址操作，返回代表该异步执行结果的 ChannelFuture 实例
    ChannelFuture bind(SocketAddress localAddress);

    // 异步对当前通道执行连接地址操作，返回代表该异步执行结果的 ChannelFuture 实例
    ChannelFuture connect(SocketAddress remoteAddress);

    // 异步对当前通道执行绑定地址操作，同时在绑定后连接到对应地址并返回代表该异步执行结果的 ChannelFuture 实例
    ChannelFuture connect(SocketAddress remoteAddress, SocketAddress localAddress);

    // 异步对当前通道执行断开地址操作，返回代表该异步执行结果的 ChannelFuture 实例
    ChannelFuture disconnect();

    // 异步对当前通道执行关闭操作，返回代表该异步执行结果的 ChannelFuture 实例
    ChannelFuture close();

    // 异步对当前通道执行解除对 EventExecutor 的注册操作，返回代表该异步执行结果的 ChannelFuture 实例
    ChannelFuture deregister();

    // 以下方法定义，同没有 ChannelPromise promise 一样，只不过在操作完成时，会回调 ChannelPromise 实例
    ChannelFuture bind(SocketAddress localAddress, ChannelPromise promise);
    ChannelFuture connect(SocketAddress remoteAddress, ChannelPromise promise);
    ChannelFuture connect(SocketAddress remoteAddress, SocketAddress localAddress, ChannelPromise promise);
    ChannelFuture disconnect(ChannelPromise promise);
    ChannelFuture close(ChannelPromise promise);
    ChannelFuture deregister(ChannelPromise promise);

    // 对当前通道执行读操作
    ChannelOutboundInvoker read();
```

```java
// 对当前通道执行写操作，数据放入写缓冲区
ChannelFuture write(Object msg);

// 对当前通道执行写操作，数据放入写缓冲区，并在完成后回调 ChannelPromise promise
ChannelFuture write(Object msg, ChannelPromise promise);

// 将当前通道写入缓冲区的数据转移到底层 Socket
ChannelOutboundInvoker flush();

// 对当前通道执行写操作并刷新缓冲区。如果携带参数 ChannelPromise，则在操作完成后进行回调
ChannelFuture writeAndFlush(Object msg, ChannelPromise promise);
ChannelFuture writeAndFlush(Object msg);

// 创建新的 ChannelPromise 实例
ChannelPromise newPromise();

// 创建待进度回调的 ChannelPromise 实例
ChannelProgressivePromise newProgressivePromise();

// 创建新的 ChannelFuture 实例，同时在创建时将其设置为已经执行完毕
ChannelFuture newSucceededFuture();

// 创建新的 ChannelFuture 实例，同时在创建时将其设置为执行失败，失败异常对象由参数指定
ChannelFuture newFailedFuture(Throwable cause);

// 返回特殊的 ChannelPromise 实例，它可以被不同的操作复用
ChannelPromise voidPromise();
}
```

12.2　ChannelInboundInvoker 接口

ChannelInboundInvoker 通道输入操作接口用于定义通道输入数据时的回调方法。通过接口定义可知。
（1）所有方法均返回当前 ChannelInboundInvoker 实例。
（2）所有方法均针对于通道数据信息，与 Inbound 含义相同。
（3）标准的观察者模式，当发生某些事件时进行回调通知。

```java
public interface ChannelInboundInvoker {

    // 通道注册到 EventLoop 时回调该方法
    ChannelInboundInvoker fireChannelRegistered();

    // 通道从所属 EventLoop 解除注册时回调该方法
    ChannelInboundInvoker fireChannelUnregistered();

    // 通道激活时回调，即通道已经连接
    ChannelInboundInvoker fireChannelActive();

    // 通道激活时回调，即通道已经关闭
```

```
    ChannelInboundInvoker fireChannelInactive();

    // 通道执行异常时回调
    ChannelInboundInvoker fireExceptionCaught(Throwable cause);

    // 通道收到用户自定义事件时回调
    ChannelInboundInvoker fireUserEventTriggered(Object event);

    // 通道收到数据时回调
    ChannelInboundInvoker fireChannelRead(Object msg);

    // 通道收到完整数据时回调
    ChannelInboundInvoker fireChannelReadComplete();

    // 通道是否可写在发生变化时回调
    ChannelInboundInvoker fireChannelWritabilityChanged();
}
```

12.3 Channel 通道接口

Channel 接口继承自 ChannelOutboundInvoker，定义了 Netty 中的网络 Socket，能够执行读、写、连接、绑定等操作。该通道为 Netty 用户提供了如下信息。

（1）当前通道状态。

（2）ChannelConfig 通道配置参数（读者对于通道配置项不必太过深究，在涉及自定义配置时，根据 Socket 和 ServerSocket 或其他底层 IO Java 原生操作类时，找到可配置项自定义即可）。

（3）通道支持的操作包括 read、write、connect、bind。

（4）处理所有与通道相关的 I/O 事件和请求。

由源码可知。

（1）每个通道拥有自己唯一的 id。

（2）通道之间通过 parent 方法形成父子关系。

（3）定义了判断当前通道状态的方法。

（4）每个通道拥有与之关联的分配内存的分配器。

（5）每个通道拥有与之关联的用于处理数据的 ChannelPipeline 通道流水线。

（6）Channel 实例所有的底层 IO 操作都将由 Unsafe 内部接口实现，同时用户不应直接调用 Unsafe 操作通道，而是应该通过 Channel 定义的方法来操作。

```
// 通道 ID 实例定义接口
public interface ChannelId extends Serializable, Comparable<ChannelId> {
    // 返回唯一通道的短名字符串
    String asShortText();

    // 返回唯一通道的长名字符串
    String asLongText();
}
```

```java
public interface Channel extends AttributeMap, ChannelOutboundInvoker, Comparable<Channel> {
    // 当前通道唯一 ID
    ChannelId id();

    // 所注册到的事件循环组（读者还记得 SingleThreadEventLoop 的 register 方法吗？）
    EventLoop eventLoop();

    // 获取父通道对象
    Channel parent();

    // 通道配置信息实例
    ChannelConfig config();

    // 判断通道是否已经打开
    boolean isOpen();

    // 判断通道是否已经注册到事件循环组
    boolean isRegistered();

    // 判断通道是否已经激活
    boolean isActive();

    // 返回描述当前通道元数据的实例
    ChannelMetadata metadata();

    // 返回当前通道绑定的本地地址信息
    SocketAddress localAddress();

    // 返回当前通道连接的远程地址信息
    SocketAddress remoteAddress();

    // 获取一个 ChannelFuture，当通道关闭时发出通知
    ChannelFuture closeFuture();

    // 判断当前通道是否可写
    boolean isWritable();

    // 获取 isWritable 方法返回 false 时，通道可写字节数
    long bytesBeforeUnwritable();

    // 获取 isWritable 方法返回 true 时，通道可写字节数
    long bytesBeforeWritable();

    // 获取实际执行 IO 底层操作的 Unsafe 实例
    Unsafe unsafe();

    // 获取当前关联的 ChannelPipeline 通道流水线实例
    ChannelPipeline pipeline();

    // 获取与当前通道关联的用于分配内存的内存分配器
```

```java
    ByteBufAllocator alloc();

    // 定义执行底层 IO 实际操作的方法，不会被 Netty 用户直接访问
    interface Unsafe {

        // 返回用于分配接收数据缓冲区的内存分配器
        RecvByteBufAllocator.Handle recvBufAllocHandle();

        // 获取当前通道绑定的本地地址，如果不存在，则返回 null
        SocketAddress localAddress();

        // 获取当前通道连接的远程地址，如果不存在，则返回 null
        SocketAddress remoteAddress();

        // 将当前通道注册到指定的 EventLoop 线程上，在完成时回调 ChannelPromise promise
        void register(EventLoop eventLoop, ChannelPromise promise);

        // 将当前通道绑定到指定的 SocketAddress 地址上，在完成时回调 ChannelPromise promise
        void bind(SocketAddress localAddress, ChannelPromise promise);

        // 连接并绑定，完成时回调 ChannelPromise promise
        void connect(SocketAddress remoteAddress, SocketAddress localAddress, ChannelPromise promise);

        // 将当前通道断开连接，完成时回调 ChannelPromise promise
        void disconnect(ChannelPromise promise);

        // 将当前通道关闭，完成时回调 ChannelPromise promise
        void close(ChannelPromise promise);

        // 将当前通道强制关闭，不会触发任何事件
        void closeForcibly();

        // 将当前通道从所属循环组中移除，完成时回调 ChannelPromise promise
        void deregister(ChannelPromise promise);

        // 对当前通道执行读取操作
        void beginRead();

        // 对当前通道执行写操作，完成时回调 ChannelPromise promise
        void write(Object msg, ChannelPromise promise);

        // 对当前通道执行写缓冲区刷新操作
        void flush();

        // 返回特殊的可共享的 ChannelPromise
        ChannelPromise voidPromise();

        // 获取当前写缓冲区对象
        ChannelOutboundBuffer outboundBuffer();
    }
}
```

12.4　ServerChannel 服务端

ServerChannel 接口是 tag 接口，即标志性接口，用于表示接收客户端连接，并且负责创建与客户端连接的 Channel 通道对象，本身并不定义任何方法和常量。源码如下。

```java
public interface ServerChannel extends Channel {
}
```

12.5　ChannelMetadata 类

ChannelMetadata 通道元数据类用于表示通道的元数据，只包含如下两个变量。

（1）hasDisconnect 表示当前通道是否可以调用 disconnect 方法断开当前连接，然后又可以再次调用 connect(SocketAddress)方法连接到对端，通常只有在 UDP 面向无连接的协议时设置为 true。

（2）defaultMaxMessagesPerRead 在 MaxMessagesRecvByteBufAllocator 接收缓冲区分配器时使用，用于设置 MaxMessagesRecvByteBufAllocator.maxMessagesPerRead()返回的当前通道在读取数据时一次读取的数量。

```java
public final class ChannelMetadata {
    private final boolean hasDisconnect;
    private final int defaultMaxMessagesPerRead;

    // 构造器：使用 1 作为 defaultMaxMessagesPerRead 的默认值
    public ChannelMetadata(boolean hasDisconnect) {
        this(hasDisconnect, 1);
    }

    // 构造器：自定义 defaultMaxMessagesPerRead
    public ChannelMetadata(boolean hasDisconnect, int defaultMaxMessagesPerRead) {
        if (defaultMaxMessagesPerRead <= 0) {
            throw new IllegalArgumentException("defaultMaxMessagesPerRead: " +
defaultMaxMessagesPerRead + " (expected > 0)");
        }
        this.hasDisconnect = hasDisconnect;
        this.defaultMaxMessagesPerRead = defaultMaxMessagesPerRead;
    }

    public boolean hasDisconnect() {
        return hasDisconnect;
    }

    public int defaultMaxMessagesPerRead() {
        return defaultMaxMessagesPerRead;
    }
}
```

12.6 ChannelOutboundBuffer 通道输出缓冲区

ChannelOutboundBuffer 类用于在 AbstractChannel 类中作为调用 write 方法写入数据时临时存放数据的写缓冲区对象。在通道调用 flush 方法时，会把该缓冲区中的数据写入底层 IO 通道中。由于该类较为复杂，笔者将其进行拆分描述。

12.6.1 核心变量与构造器

由源码可知。

（1）内部使用 FastThreadLocal<ByteBuffer[]> NIO_BUFFERS 线程本地缓存保存当前线程处理的通道，用于写入底层 Socket 的 ByteBuffer[]数组。由于在 Netty 中，一个通道唯一绑定一个事件循环线程，所以这些操作都是线程安全的。

（2）内部使用 Entry 链表表示，即 flushedEntry 链表（待写入底层 IO 的 Entry 链表）、unflushedEntry 链表（当前通道写入 ChannelOutboundBuffer 写缓冲区的数据。在挂入 flushedEntry 链表前，不可写入底层 IO），使用 tailEntry 表示当前操作的链表的尾结点。

（3）使用 totalPendingSize 来表示当前写缓冲区中总共需要写入底层 IO 的字节数（flushedEntry 链表中的消息大小加上 unflushedEntry 链表中的消息大小）。

（4）使用 unwritable 来表示当前输出缓冲区是否可写。为 0 时可写，大于 0 时不可写（在后面 Writable 可写标志位原理一节时会描述其中的缘由）。

```
public final class ChannelOutboundBuffer {
    // 线程本地缓存，用于缓存当前线程 Java 原生 ByteBuffer[]数组
    private static final FastThreadLocal<ByteBuffer[]> NIO_BUFFERS = new FastThreadLocal<ByteBuffer[]>() {
        @Override
        protected ByteBuffer[] initialValue() throws Exception {
            return new ByteBuffer[1024];
        }
    };
    private final Channel channel;            // 所属通道

    // Entry 链表轮廓：Entry(flushedEntry) --> ... Entry(unflushedEntry) --> ... Entry(tailEntry)。读者如果对此处
    // 有疑惑，可以查看 addMessage 和 addFlush 两个方法的描述，便可理解该链表的意义，即一个单链表，只不过
    // 头部可能为 flushedEntry 或 unflushedEntry
    private Entry flushedEntry;              // 已完成 flush 的链表头结点
    private Entry unflushedEntry;            // 未完成 flush 的链表头结点
    private Entry tailEntry;                 // Entry 尾结点，可能指向 flushedEntry 链表或 unflushedEntry 链表
    private int flushed;                     // flushedEntry 的结点数
    private int nioBufferCount;              // 待写入底层 IO 的 nio ByteBuffer 缓冲区数量
    private long nioBufferSize;              // 待写入底层 IO 的 nio ByteBuffer 缓冲区大小
    private boolean inFail;                  // 是否在失败处理中
    private static final AtomicLongFieldUpdater<ChannelOutboundBuffer> TOTAL_PENDING_SIZE_UPDATER;
                                             // 原子更新 totalPendingSize 变量的包装对象
    private volatile long totalPendingSize;  // 总共需要写入底层 IO 的字节数（flushedEntry 链表中的消息大小
                                             // 加上 unflushedEntry 链表中的消息大小）
```

```
    private static final AtomicIntegerFieldUpdater<ChannelOutboundBuffer> UNWRITABLE_UPDATER; // 原
子更新 unwritable 变量的包装对象
    private volatile int unwritable;    // 表示当前通道是否可写。使用最后一位,为 0 时表示当前写缓冲区可写、
为 1 时表示不可写

    private volatile Runnable fireChannelWritabilityChangedTask;  // 封装通道发生写状态改变的任务

    static {                                                      // 初始化原子更新包装对象
        AtomicIntegerFieldUpdater<ChannelOutboundBuffer> unwritableUpdater =
            PlatformDependent.newAtomicIntegerFieldUpdater(ChannelOutboundBuffer.class, "unwritable");
        if (unwritableUpdater == null) {
            unwritableUpdater = AtomicIntegerFieldUpdater.newUpdater(ChannelOutboundBuffer.class,
"unwritable");
        }
        UNWRITABLE_UPDATER = unwritableUpdater;
        AtomicLongFieldUpdater<ChannelOutboundBuffer> pendingSizeUpdater =
            PlatformDependent.newAtomicLongFieldUpdater(ChannelOutboundBuffer.class,
"totalPendingSize");
        if (pendingSizeUpdater == null) {
            pendingSizeUpdater = AtomicLongFieldUpdater.newUpdater(ChannelOutboundBuffer.class,
"totalPendingSize");
        }
        TOTAL_PENDING_SIZE_UPDATER = pendingSizeUpdater;
    }

    // 构造器初始化所属通道对象
    ChannelOutboundBuffer(AbstractChannel channel) {
        this.channel = channel;
    }
}
```

12.6.2 Entry 核心内部类

Entry 内部类用于表示放入缓冲区中的消息,用于包装消息的元数据。由源码可知。

(1) 内部定义了 Recycler RECYCLER 对象池,用于缓存分配的 Entry 对象。
(2) 使用 next 变量将 Entry 形成单向链表。
(3) 保存了当前 msg 的元数据信息:从 ByteBuf 转为 Java 原生 ByteBuffer 对象、当前写入总数量、当前写入进度。

```
static final class Entry {
    // entry 对象池
    private static final Recycler<Entry> RECYCLER = new Recycler<Entry>() {
        @Override
        protected Entry newObject(Handle handle) {
            return new Entry(handle);
        }
    };
    private final Handle<Entry> handle;            // 对象池句柄,用于回收 Entry 对象
    Entry next;                                     // 单向链表,指向后继结点
    Object msg;                                     // 包装的消息对象
```

```java
    ByteBuffer[] bufs;              // 代表消息对象转换的 ByteBuf 的缓冲区数组
    ByteBuffer buf;                 // 代表消息对象转换的 ByteBuf 缓冲区对象
    ChannelPromise promise;         // 当消息写入底层 IO 时回调的 promise
    long progress;                  // 当前消息写入底层 IO 的进度
    long total;                     // 当前消息总共需要写入底层 IO 的数量
    int pendingSize;                // 还未写入的大小
    int count = -1;                 // 当前写入数量
    boolean cancelled;              // 是否已经被取消

    // 私有构造器,用于给对象池创建 Entry 对象使用
    private Entry(Handle<Entry> handle) {
        this.handle = handle;
    }

    // 静态方法,从对象池中获取 Entry 对象,并将 msg 以及元数据作为 Entry 对象的实例变量
    static Entry newInstance(Object msg, int size, long total, ChannelPromise promise) {
        Entry entry = RECYCLER.get();
        entry.msg = msg;
        entry.pendingSize = size;
        entry.total = total;
        entry.promise = promise;
        return entry;
    }

    // 取消当前 Entry 所包装的 msg
    int cancel() {
        if (!cancelled) {
            cancelled = true;
            int pSize = pendingSize;
            // 释放消息并将 msg 替换为空缓冲区
            ReferenceCountUtil.safeRelease(msg);
            msg = Unpooled.EMPTY_BUFFER;

            pendingSize = 0;
            total = 0;
            progress = 0;
            bufs = null;
            buf = null;
            return pSize;
        }
        return 0;
    }

    // 重置实例变量,放入对象池,等待复用
    void recycle() {
        next = null;
        bufs = null;
        buf = null;
        msg = null;
        promise = null;
        progress = 0;
        total = 0;
        pendingSize = 0;
```

```
        count = -1;
        cancelled = false;
        handle.recycle(this);
    }

    // 回收当前 entry，然后获取链表中的下一个 Entry 结点
    Entry recycleAndGetNext() {
        Entry next = this.next;
        recycle();
        return next;
    }
}
```

12.6.3　addMessage 核心方法

addMessage 方法用于将 msg 消息添加到写缓冲区中。流程如下。

（1）通过 Object msg、int size、ChannelPromise promise 创建 Entry 对象。

（2）如果尾结点为空，表明当前添加的 entry 为写缓冲区中的第一个 entry，则将当前 entry 作为 tailEntry，同时将刷新到底层 IO 中的 flushedEntry 置空。

（3）如果尾结点不为空，则当前 entry 不是写缓冲区中的第一个 entry，将其插入 entry 链表的末尾。

（4）如果未刷新的 entry 为空，则将当前 entry 作为未刷新 entry 的头结点。

（5）将消息添加到未刷新的数组后，增加挂起未写入底层 IO 通道的字节数。

```java
public void addMessage(Object msg, int size, ChannelPromise promise) {
    Entry entry = Entry.newInstance(msg, size, total(msg), promise);
    if (tailEntry == null) {          // 尾结点为空，表明当前添加的 entry 为写缓冲区中的第一个 entry，将当前 entry 作为 tailEntry 即可，同时将刷新到底层 IO 中的 flushedEntry 置空
        flushedEntry = null;
        tailEntry = entry;
    } else {                          // 当前 entry 不是写缓冲区中的第一个 entry，将其插入到 entry 链表的末尾
        Entry tail = tailEntry;
        tail.next = entry;
        tailEntry = entry;
    }
    if (unflushedEntry == null) {     // 未刷新的 entry 为空，则将当前 entry 作为未刷新 entry 的头结点
        unflushedEntry = entry;
    }

    // 将消息添加到未刷新的数组后，增加挂起未写入底层 IO 通道的字节。false 表示在增加未写入字节数后，
    // 缓冲区中数据大于高水位线，立即在当前线程通知通道处理器
    incrementPendingOutboundBytes(size, false);
}

// 计算写入消息对象需要写入的数量
private static long total(Object msg) {
    if (msg instanceof ByteBuf) {     // 直接取缓冲区中可读字节数
        return ((ByteBuf) msg).readableBytes();
    }
    if (msg instanceof FileRegion) {  // 直接取需要写入底层 IO 通道的文件数量
        return ((FileRegion) msg).count();
    }
```

```java
        if (msg instanceof ByteBufHolder) {  // 取包装的 ByteBuf 的可读字节数
            return ((ByteBufHolder) msg).content().readableBytes();
        }
        // 其他类型返回-1
        return -1;
    }

    // 消息被添加到 entry 链表时，增加写缓冲区的未写入字节数
    private void incrementPendingOutboundBytes(long size, boolean invokeLater) {
        if (size == 0) {
            return;
        }
        // 原子更新 totalPendingSize 变量
        long newWriteBufferSize = TOTAL_PENDING_SIZE_UPDATER.addAndGet(this, size);
        if (newWriteBufferSize >= channel.config().getWriteBufferHighWaterMark()) {  // 如果写缓冲区中需要写
        // 入的数据量超过通道配置的写缓冲区高水位线，则设置当前写缓冲区设置为不可写状态。invokeLater 用于在触发
        // fireChannelWritabilityChanged 通知通道处理器时使用
            setUnwritable(invokeLater);
        }
    }

    // 写缓冲区中放入的数据过多时，则设置为不可写状态，并将当前通道写状态发生变化通知通道处理器
    private void setUnwritable(boolean invokeLater) {
        for (;;) {                                   // CAS 将写缓冲区的状态的最后一位设置为1，表示不可写状态
            final int oldValue = unwritable;
            final int newValue = oldValue | 1;
            if (UNWRITABLE_UPDATER.compareAndSet(this, oldValue, newValue)) {
                if (oldValue == 0 && newValue != 0) {  // 设置成功，且之前值为 0，通知通道处理器（不为 0 表示
                // 已经由其他线程完成通知）
                    fireChannelWritabilityChanged(invokeLater);
                }
                break;
            }
        }
    }

    // 通知通道处理器，当前通道的写状态发生变化。invokeLater 表示是否在当前线程立即调用通道处理器的
    // ChannelWritabilityChanged 方法，否则将其封装为 fireChannelWritabilityChangedTask，放入事件循环队列中，
    // 等待事件循环线程执行
    private void fireChannelWritabilityChanged(boolean invokeLater) {
        final ChannelPipeline pipeline = channel.pipeline();
        if (invokeLater) {                           // 将回调操作放入事件循环队列，等待下一次执行
            Runnable task = fireChannelWritabilityChangedTask;
            if (task == null) {
                fireChannelWritabilityChangedTask = task = new Runnable() {
                    @Override
                    public void run() {
                        pipeline.fireChannelWritabilityChanged();
                    }
                };
            }
            channel.eventLoop().execute(task);
        } else {                                     // 在当前线程立即通知通道处理器
```

```
            pipeline.fireChannelWritabilityChanged();
    }
}
```

12.6.4 addFlush 核心方法

addFlush 方法用于标记当前已经添加的消息处于刷新状态，通常在完成实际写入底层 IO 时，会先调用该方法将写入的消息标记为刷新状态，然后在后续进行写入处理。流程如下。

（1）如果未刷新的 entry 不为空，则进行刷新操作，否则直接返回。

（2）如果当前处于刷新状态的 flushedEntry 为空，即头结点为空，将当前 entry 作为刷新 flushedEntry 的头结点。

（3）循环处理 unflushedEntry 链表中未处于刷新状态的 Entry 对象，并增加已经刷新的消息数量。

（4）如果当前刷新的 entry 设置 promise 不能被取消失败，说明 entry 代表的消息已经失效，减少等待写入的消息字节数。

（5）所有 unflushedEntry 链表中未刷新的消息都已处理完毕，置空头结点。

```
public void addFlush() {
    Entry entry = unflushedEntry;
    if (entry != null) {              // 未刷新的 entry 不为空，进行刷新操作
        if (flushedEntry == null) {   // 当前处于刷新状态的 flushedEntry 为空，即头结点为空，将当前 entry 作为刷新 entry 的头结点
            flushedEntry = entry;
        }
        do {                          // 循环处理 unflushedEntry 链表中未处于刷新状态的 Entry 对象
            flushed ++;               // 增加已经刷新的消息数量
            if (!entry.promise.setUncancellable()) { // 如果当前刷新的 entry，设置 promise 不能被取消失败，说明 entry 代表的消息已经失效。这里为什么要执行这个操作？因为消息一旦处于刷新状态，在写入底层 IO 之前，不能被取消
                // 消息被取消，减少等待写入的消息字节数
                int pending = entry.cancel();
                decrementPendingOutboundBytes(pending, false, true);
            }
            entry = entry.next;       // 处理链表中的下一个结点
        } while (entry != null);
        // 所有未刷新的消息都已处理完毕，置空头结点
        unflushedEntry = null;
    }
}
```

12.6.5 current 核心方法

current 方法用于获取当前要写入的消息，如果之前没有刷新任何内容，则返回 null。前面分析了 addFlush 方法，该方法用于将 unflushedEntry 链表的结点转移到 flushedEntry 结点上。在转移期间，将每个需要写入的消息设置为不可取消状态。通过 current 方法的源码可知：当前需要写入底层 IO 的消息为 flushedEntry 链表的头结点。源码如下。

```
public Object current() {
    Entry entry = flushedEntry;
```

```
        if (entry == null) {
            return null;
        }
        return entry.msg;
}
```

12.6.6　progress 核心方法

progress 方法用于通过传入的 amount 数量，回调当前 flushedEntry 结点，即当前待写入消息的 promise 方法，通知当前写入进度。注意这里只针对 ChannelProgressivePromise 类型的 promise。源码如下。

```
public void progress(long amount) {
    Entry e = flushedEntry;
    assert e != null;
    ChannelPromise p = e.promise;
    if (p instanceof ChannelProgressivePromise) {
        long progress = e.progress + amount;
        e.progress = progress;
        ((ChannelProgressivePromise) p).tryProgress(progress, e.total);
    }
}
```

12.6.7　remove 核心方法

remove 方法用于删除当前需要写入的消息，同时将 ChannelPromise 标记为成功。如果在调用此方法时不存在任何待写入的消息（即 flushedEntry 链表为空），则返回 false。源码如下。

```
public boolean remove() {
    Entry e = flushedEntry;
    if (e == null) {        // 待写入 IO 通道的消息为空
        clearNioBuffers();
        return false;
    }
    Object msg = e.msg;
    ChannelPromise promise = e.promise;
    int size = e.pendingSize;
    removeEntry(e);         // 从 flushedEntry 链表中移除 entry
    if (!e.cancelled) {     // 消息还没有被取消，减少消息计数，同时标识 promise 完成，并减少待写入字节数
        ReferenceCountUtil.safeRelease(msg);
        safeSuccess(promise);
        decrementPendingOutboundBytes(size, false, true);
    }
    // 回收 entry
    e.recycle();
    return true;
}

// 清空当前线程的本地 NIO 缓冲区
private void clearNioBuffers() {
    int count = nioBufferCount;
    if (count > 0) {
        nioBufferCount = 0;
```

```
            Arrays.fill(NIO_BUFFERS.get(), 0, count, null);
    }
}

// 从 flushedEntry 链表中移除当前 entry，并减少一个待写入消息计数（-- flushed）
private void removeEntry(Entry e) {
    if (-- flushed == 0) {                    // 当前消息为最后一个待写入消息，置空 flushedEntry
        flushedEntry = null;
        if (e == tailEntry) {
            tailEntry = null;
            unflushedEntry = null;
        }
    } else {
        flushedEntry = e.next;
    }
}
```

12.6.8 removeBytes 核心方法

removeBytes 方法用于删除完全写入底层 IO 通道的 entry，该方法假设这个缓冲区中的所有消息都是 ByteBuf 类型。流程如下。

（1）循环处理所有 flushedEntry 链表消息。
（2）获取 flushedEntry 的头结点。
（3）如果当前消息类型不是 bytebuf 类型，则结束循环。
（4）如果当前消息中的内容可以全部写入，则从 flushedEntry 链表中移除当前消息。
（5）如果当前消息不能完全写入底层 IO，则设置当前消息的可读下标为实际写入数量（readerIndex + (int) writtenBytes）并结束循环。
（6）清空 NIO 缓冲区。

```
public void removeBytes(long writtenBytes) {
    for (;;) {                             // 循环处理所有消息（从 flushedEntry 头结点开始，找到待写入消息，
                                           // 将消息可读字节数与 writtenBytes 进行比较，找到最后一个可读消息正好大于或等于 writtenBytes 的 Entry）
        Object msg = current();            // 获取 flushedEntry 的头结点
        if (!(msg instanceof ByteBuf)) {   // 消息不为 bytebuf 类型，则结束循环
            assert writtenBytes == 0;      // writtenBytes 必须为 0，否则消息类型应该为 ByteBuf
            break;
        }
        final ByteBuf buf = (ByteBuf) msg;
        final int readerIndex = buf.readerIndex();
        final int readableBytes = buf.writerIndex() - readerIndex;
        if (readableBytes <= writtenBytes) {    // 当前消息可以全部写入，从 flushedEntry 链表中移除当前消息
            if (writtenBytes != 0) {            // 通知 promise 当前进度
                progress(readableBytes);
                writtenBytes -= readableBytes;
            }
            remove();
        } else { // readableBytes > writtenBytes 表明不能把当前消息完全写入底层 IO
            if (writtenBytes != 0) {            // 通知 promise 当前进度
                buf.readerIndex(readerIndex + (int) writtenBytes);  // 设置当前可读下标为实际写入数量
                progress(writtenBytes);
```

```
            }
            break;              // 当前 entry 是最后一个不完全写入的消息
        }
    }
    clearNioBuffers();          // 清空 NIO 缓冲区
}
```

12.6.9　nioBuffers 核心方法

如果当前待写入的消息（flushedEntry 链表）仅由 ByteBuf 组成，则返回一个堆外 NIO 缓冲区数组。nioBufferCount()方法和 nioBufferSize 方法将分别返回数组中 NIO 缓冲区的数量和 NIO 缓冲区的可读字节总数。流程如下。

（1）声明 nioBufferSize 变量，记录当前 ByteBuffer[]数组中数据字节数。

（2）声明 nioBufferCount 变量，记录当前 ByteBuffer[]数组中有效的 ByteBuffer 缓冲区个数。

（3）从 ThreadLocal 中获取当前线程的 ByteBuffer 数组 ByteBuffer[] nioBuffers。

（4）获取当前待写入 IO 的 flushedEntry 链表的头结点。

（5）循环处理 flushedEntry 链表中的结点，将其转为 ByteBuffer 并放入 ByteBuffer[]数组，这里处理对象必须是 ByteBuf 类型。

（6）如果 nioBufferSizelk 加 readableBytes（待写入 IO 通道的字节数）大于整型最大值，此时发生符号溢出，停止转换 ByteBuffer。

（7）否则增加总写入字节数，即 nioBufferSize 加 readableBytes。

（8）获取当前存放 msg 消息的 ByteBuffer 数量（通常为 1）。

（9）当前消息只需要一个 ByteBuffer，创建一个新的 ByteBuffer 并将其放入 nioBuffers 数组。

（10）如果当前消息需要多个 ByteBuffer，则分配所需数量的 ByteBuffer 缓冲区数组，并将其复制到 ByteBuffer[] nioBuffers 中。

```
public ByteBuffer[] nioBuffers() {
    long nioBufferSize = 0;         // 记录当前 ByteBuffer[]数组中数据字节数
    int nioBufferCount = 0;         // 记录当前 ByteBuffer[]数组中有效的 ByteBuffer 缓冲区个数
    // 从 ThreadLocal 中获取当前线程的 ByteBuffer 数组
    final InternalThreadLocalMap threadLocalMap = InternalThreadLocalMap.get();
    ByteBuffer[] nioBuffers = NIO_BUFFERS.get(threadLocalMap);
    Entry entry = flushedEntry;     // 当前待写入 IO 的 flushedEntry 链表的头结点
    while (isFlushedEntry(entry) && entry.msg instanceof ByteBuf) { // 循环处理 flushedEntry 链表中的结点,
    将其转为 ByteBuffer 并放入 ByteBuffer[]数组,这里处理对象必须是 ByteBuf 类型
        if (!entry.cancelled) {     // 消息有效,没有被取消
            ByteBuf buf = (ByteBuf) entry.msg;
            final int readerIndex = buf.readerIndex();
            final int readableBytes = buf.writerIndex() - readerIndex;   // 消息可读字节数
            if (readableBytes > 0) {                                      // 存在可读字节
                if (Integer.MAX_VALUE - readableBytes < nioBufferSize) { // nioBufferSize + readableBytes
    大于整型最大值,发生符号溢出,停止转换 ByteBuffer。这是因为在 bsd/osx 操作系统上,不允许写入比 Integer
    最大值更大的字节。当 writev(...)系统调用时,如果写入数量超过整型最大值,将返回 EINVAL（无效参数）值,
    这会引发 IOException。在 Linux 上,取决于体系结构和内核,但为了安全起见,在这里也做了限制
                    break;
                }
```

```java
                    nioBufferSize += readableBytes;     // 增加总写入字节数
                    int count = entry.count;
                    if (count == -1) {                  // 初始化存放当前 msg 消息的 ByteBuffer 数量（通常为 1）
                        entry.count = count = buf.nioBufferCount();
                    }
                    int neededSpace = nioBufferCount + count;
                    if (neededSpace > nioBuffers.length) { // 如果当前需要的 ByteBuffer 数量大于 nioBuffers 数组
// 长度，则对该数组进行扩容，并将新数组放入 ThreadLocal
                        nioBuffers = expandNioBufferArray(nioBuffers, neededSpace, nioBufferCount);
                        NIO_BUFFERS.set(threadLocalMap, nioBuffers);
                    }
                    if (count == 1) {       // 当前消息只需要一个 ByteBuffer，创建一个新的 ByteBuffer，并将其放入
// nioBuffers 数组中
                        ByteBuffer nioBuf = entry.buf;
                        if (nioBuf == null) {
                            entry.buf = nioBuf = buf.internalNioBuffer(readerIndex, readableBytes);
                        }
                        nioBuffers[nioBufferCount ++] = nioBuf;
                    } else {                // 当前消息需要多个 ByteBuffer，分配所需数量的 ByteBuffer 缓冲区数组，并将
// 其复制到 ByteBuffer[] nioBuffers 中
                        ByteBuffer[] nioBufs = entry.bufs;
                        if (nioBufs == null) {
                            entry.bufs = nioBufs = buf.nioBuffers();
                        }
                        nioBufferCount = fillBufferArray(nioBufs, nioBuffers, nioBufferCount);
                    }
                }
            }
            entry = entry.next;                          // 继续处理下一个消息
        }
        this.nioBufferCount = nioBufferCount;            // 保存当前写入底层 IO 的缓冲区个数
        this.nioBufferSize = nioBufferSize;              // 保存当前写入底层 IO 的缓冲区的总字节数
        return nioBuffers;
    }
}

// 当前 entry 是不是可以写入底层 IO 的消息，这里直接判断是否为空，且不等于 unflushedEntry 头结点，因为在
// 将消息转为 flushed 状态时，会把它们挪到 flushedEntry 链表上，此时 tail 结点将指向最后一个 flushedEntry，如
// 果将新的 Entry 添加到链表上时，会将其链接当前 tail 结点上，同时将其设置为头结点 unflushedEntry，所以这里
// 判断 e != unflushedEntry
private boolean isFlushedEntry(Entry e) {
    return e != null && e != unflushedEntry;
}

// 将 nioBufs 数组内容复制到 nioBuffers 数组中，nioBufferCount 表示当前 nioBuffers 的操作下标，即从该下标开
// 始复制
private static int fillBufferArray(ByteBuffer[] nioBufs, ByteBuffer[] nioBuffers, int nioBufferCount) {
    for (ByteBuffer nioBuf: nioBufs) {
        if (nioBuf == null) {
            break;
        }
        nioBuffers[nioBufferCount ++] = nioBuf;
    }
    return nioBufferCount;
```

```java
}

// 对 array 数组进行扩容，参数 neededSpace 表示需要的数组长度，size 表示当前 array 中有效的索引下标
private static ByteBuffer[] expandNioBufferArray(ByteBuffer[] array, int neededSpace, int size) {
    int newCapacity = array.length;
    do {    // 不断进行二倍扩容，直到满足需要的数组长度
        newCapacity <<= 1;
        if (newCapacity < 0) {
            throw new IllegalStateException();
        }

    } while (neededSpace > newCapacity);
    // 扩容完成，创建新的数组，把旧数组中的有效位复制到新数组中
    ByteBuffer[] newArray = new ByteBuffer[newCapacity];
    System.arraycopy(array, 0, newArray, 0, size);
    return newArray;
}
```

12.6.10 Writable 可写标志位原理

Writable 系列方法用于获取和修改当前写缓冲区的状态。由源码可知。

（1）isWritable()用于判断当前写缓冲区是否可写，只有 unwritable 为 0 时才可写。

（2）getUserDefinedWritability、setUserDefinedWritability 与 setUserDefinedWritability 分别用于用户对 unwritable 变量的其他位进行设置，从而表示写缓冲区是否可写。

（3）setWritable 与 setUnwritable 方法用于 Netty 自身设置通道缓冲区是否可写，即使用 unwritable 的最后一位标识是否可写：1（不可写），0（可写）。

```java
// 判断当前写缓冲区是否可写
public boolean isWritable() {
    return unwritable == 0;
}

// 判断指定索引 index 处，用户定义的可写性标志位是否为 1，通过与运算判断 index 位是否置 1。例如，unwritable
// 为 000，index 为 3，此时 writabilityMask(index)返回 100，000 & 100 == 0
public boolean getUserDefinedWritability(int index) {
    return (unwritable & writabilityMask(index)) == 0;
}


// 设置 unwritable 变量的 index 位为 writable
public void setUserDefinedWritability(int index, boolean writable) {
    if (writable) {         // 设置可写
        setUserDefinedWritability(index);
    } else {                // 设置不可写
        clearUserDefinedWritability(index);
    }
}

// 设置下标 index 位为 0，表示可写。例如，unwritable 为 000，index 为 3，此时 writabilityMask(index)返回 011，
// 000 & 011 = 0
private void setUserDefinedWritability(int index) {
```

```java
        final int mask = ~writabilityMask(index);
        for (;;) { // CAS 更新 unwritable 的值
            final int oldValue = unwritable;
            final int newValue = oldValue & mask;
            if (UNWRITABLE_UPDATER.compareAndSet(this, oldValue, newValue)) {
                if (oldValue != 0 && newValue == 0) {
                    fireChannelWritabilityChanged(true);
                }
                break;
            }
        }
    }

    // 设置下标 index 位为 1, 表示可写。例如，unwritable 为 000, index 为 3, 此时 writabilityMask(index)返回 100,
    000 | 100 = 100
    private void clearUserDefinedWritability(int index) {
        final int mask = writabilityMask(index);
        for (;;) {
            final int oldValue = unwritable;
            final int newValue = oldValue | mask;
            if (UNWRITABLE_UPDATER.compareAndSet(this, oldValue, newValue)) {
                if (oldValue == 0 && newValue != 0) {
                    fireChannelWritabilityChanged(true);
                }
                break;
            }
        }
    }

    // 获取当前 index 处的掩码，即将 1 左移到 index 处，此时其他位为 0，只有 index 处为 1
    private static int writabilityMask(int index) {
        if (index < 1 || index > 31) {
            throw new IllegalArgumentException("index: " + index + " (expected: 1~31)");
        }
        return 1 << index;
    }

    // 该方法用于设置 unwritable 变量为可写状态，invokeLater 用于在 ChannelWritabilityChanged 中传递
    private void setWritable(boolean invokeLater) {
        for (;;) {
            final int oldValue = unwritable;
            final int newValue = oldValue & ~1; // 只设置最后一位为 0
            if (UNWRITABLE_UPDATER.compareAndSet(this, oldValue, newValue)) {
                if (oldValue != 0 && newValue == 0) {
                    fireChannelWritabilityChanged(invokeLater);
                }
                break;
            }
        }
    }

    // 该方法用于设置 unwritable 变量为不可写状态，invokeLater 用于在 ChannelWritabilityChanged 中传递
    private void setUnwritable(boolean invokeLater) {
```

```
        for (;;) {
            final int oldValue = unwritable;
            final int newValue = oldValue | 1; // 只设置最后一位为 1
            if (UNWRITABLE_UPDATER.compareAndSet(this, oldValue, newValue)) {
                if (oldValue == 0 && newValue != 0) {
                    fireChannelWritabilityChanged(invokeLater);
                }
                break;
            }
        }
    }
```

12.6.11 close 核心方法

close 方法用于关闭当前写缓冲区,参数 cause 表示导致关闭的异常对象。流程如下。

(1) 如果 inFail 为 true, 表明当前已经开始处理 close 操作,则将任务添加到事件循环队列中,等待下一次事件循环时执行。因为对于该方法而言,有可能判断通道状态不满足方法执行条件,所以需要再次尝试执行。

(2) 判断通道是否仍旧处于打开状态,该方法必须在通道关闭后调用。

(3) 判断写缓冲区的 flushedEntry 链表是否为空,该方法必须在所有待写入数据都写入底层 IO 后调用。

(4) 释放 unflushedEntry 链表上的数据,因为该链表上的数据还未转移到 flushedEntry, 所以可以删除。

(5) 清空写入底层 IO 的数组。

```
void close(final ClosedChannelException cause) {
    if (inFail) { // 当前已经开始处理 close 操作,将任务添加到事件循环队列中,等待下一次事件循环时执行。因
为对于该方法而言,有可能判断通道状态不满足方法执行条件,所以需要再次尝试执行
        channel.eventLoop().execute(new Runnable() {
            @Override
            public void run() {
                close(cause);
            }
        });
        return;
    }
    inFail = true;
    if (channel.isOpen()) {     // 该方法必须在通道关闭后调用
        throw new IllegalStateException("close() must be invoked after the channel is closed.");
    }
    if (!isEmpty()) {           // 该方法必须在所有待写入数据都写入底层 IO 后调用
        throw new IllegalStateException("close() must be invoked after all flushed writes are handled.");
    }
    // 释放 unflushedEntry 链表上的数据,因为该链表上的数据还未转移到 flushedEntry, 所以可以删除
    try {
        Entry e = unflushedEntry;
        while (e != null) { // 循环处理链表上的 entry, 减少 pendingSize, 即当前等待变为 flushedEntry 的字节数
            int size = e.pendingSize;
            TOTAL_PENDING_SIZE_UPDATER.addAndGet(this, -size);
```

```
            if (!e.cancelled) {                    // 释放 msg 对象并设置 promise 失败
                ReferenceCountUtil.safeRelease(e.msg);
                safeFail(e.promise, cause);
            }
            e = e.recycleAndGetNext();            // 回收当前 entry 并获取链表中下一个待处理 entry
        }
    } finally {
        inFail = false;
    }
    clearNioBuffers();                             // 清空用于写入底层 IO 的数组
}
```

12.6.12 bytesBeforeUnwritable 核心方法

bytesBeforeUnwritable 方法在 isWritable()返回 false 之前获取可以写入的字节数。这里的计算为写缓冲区的高水位线减去 unflushedEntry 链表上的消息大小（addMessage 中介绍过超过高水位线时，设置写缓冲区不可写）。源码如下。

```
public long bytesBeforeUnwritable() {
    long bytes = channel.config().getWriteBufferHighWaterMark() - totalPendingSize;
    if (bytes > 0) {
        return isWritable() ? bytes : 0;         // 如果通道可写，则返回字节数，否则返回 0
    }
    return 0;
}
```

12.6.13 bytesBeforeWritable 核心方法

bytesBeforeWritable 方法在 isWritable()返回 true 之前获取可以写入的字节数。这里的计算为 unflushedEntry 链表上的消息大小减去写缓冲区的高水位线。源码如下。

```
public long bytesBeforeWritable() {
    long bytes = totalPendingSize - channel.config().getWriteBufferLowWaterMark();
    if (bytes > 0) {
        return isWritable() ? 0 : bytes;
    }
    return 0;
}
```

12.7 RecvByteBufAllocator 接收缓冲区分配器

RecvByteBufAllocator 接口定义了一个新的接收缓冲区，该缓冲区的容量足以存放读取所有输入数据，但又不会浪费空间。通道的 Unsafe 接口会使用该接口分配输入缓冲区接收底层 Socket 传递过来的数据。在 DefaultChannelConfig 默认通道配置类中会使用 AdaptiveRecvByteBufAllocator 自适应的接收缓冲区分配器，而它的子类 DefaultSocketChannelConfig 与 DefaultServerSocketChannelConfig 如果不进行修改，也将保持默认配置。代码如下。

```
public class DefaultChannelConfig implements ChannelConfig {
```

```
public DefaultChannelConfig(Channel channel) {
    this(channel, new AdaptiveRecvByteBufAllocator());
}
```

本节首先分析 RecvByteBufAllocator 接口，再分析它们的实现类。

12.7.1 RecvByteBufAllocator 接口

由源码可知。

（1）RecvByteBufAllocator 接口定义了 newHandle 方法，用于创建 Handle 接口实例。

（2）Handle 接口定义了所有处理操作。

```
public interface RecvByteBufAllocator {

    // 创建新的操作 Handle 对象，Handle 对象提供实际操作
    Handle newHandle();

    // 实际操作接口定义
    interface Handle {

        // 通过指定内存分配器，创建新的接收缓冲区
        ByteBuf allocate(ByteBufAllocator alloc);

        // 类似 allocate(ByteBufAllocator)方法，除了它不分配任何东西，只是返回需要分配的容量
        int guess();

        // 重置所有计数器，并设置读取下一个数据时应该读取多少消息/字节
        void reset(ChannelConfig config);

        // 增加当前读取循环中已读取的消息数
        void incMessagesRead(int numMessages);

        // 保存前一次读取操作中已读取的字节数，可以用来增加已读取的字节数
        void lastBytesRead(int bytes);

        // 返回前一次读取操作已读取的字节数
        int lastBytesRead();

        // 设置读取操作尝试读取的字节数
        void attemptedBytesRead(int bytes);

        // 获取读取操作尝试读取的字节数
        int attemptedBytesRead();

        // 判断当前是否应该继续读取数据
        boolean continueReading();

        // 读操作完成时回调
        void readComplete();
    }
}
```

12.7.2　MaxMessagesRecvByteBufAllocator 接口

MaxMessagesRecvByteBufAllocator 即最大接收消息接收缓冲区分配器接口。它继承自 RecvByteBufAllocator 接口，定义了当事件循环尝试读取操作时，限制读取操作的次数的方法。源码如下。

```
public interface MaxMessagesRecvByteBufAllocator extends RecvByteBufAllocator {
    // 获取在每次读取循环中要读取的最大消息数
    int maxMessagesPerRead();

    // 设置在每次读取循环中要读取的最大消息数
    MaxMessagesRecvByteBufAllocator maxMessagesPerRead(int maxMessagesPerRead);
}
```

12.7.3　DefaultMaxMessagesRecvByteBufAllocator 类

DefaultMaxMessagesRecvByteBufAllocator 即默认最大消息接收缓冲区分配器类。它提供了 MaxMessagesRecvByteBufAllocator 的默认实现，定义了变量 maxMessagesPerRead 用于保存每次读取的最大消息数。如果子类不定义 maxMessagesPerRead，则该值默认为 1。MaxMessageHandle 内部抽象类提供了 Handle 接口的默认实现，定义了计数变量来保存当前读取的数量，该内部抽象类的核心方法为 continueReading()。源码如下。

```
public abstract class DefaultMaxMessagesRecvByteBufAllocator implements
MaxMessagesRecvByteBufAllocator {
    private volatile int maxMessagesPerRead; // 每次读取的最大消息数

    // 指定 maxMessagesPerRead 为 1
    public DefaultMaxMessagesRecvByteBufAllocator() {
        this(1);
    }

    public DefaultMaxMessagesRecvByteBufAllocator(int maxMessagesPerRead) {
        maxMessagesPerRead(maxMessagesPerRead);
    }

    @Override
    public int maxMessagesPerRead() {
        return maxMessagesPerRead;
    }

    // 参数校验后保存 maxMessagesPerRead
    public MaxMessagesRecvByteBufAllocator maxMessagesPerRead(int maxMessagesPerRead) {
        if (maxMessagesPerRead <= 0) {
            throw new IllegalArgumentException("maxMessagesPerRead: " + maxMessagesPerRead +
" (expected: > 0)");
        }
        this.maxMessagesPerRead = maxMessagesPerRead;
        return this;
    }

    // 重点为 continueReading()方法，该方法在后面描述 NioChannel 时控制是否继续从底层 IO 通道中读取数据
```

```java
public abstract class MaxMessageHandle implements Handle {
    private ChannelConfig config;              // 通道配置对象
    private int maxMessagePerRead;             // 每次读取的最大消息数
    private int totalMessages;                 // 总读取的消息数
    private int totalBytesRead;                // 总读取的字节数
    private int attemptedBytesRead;            // 尝试读取的字节数
    private int lastBytesRead;                 // 前一次读到的字节数

    // 重置变量
    public void reset(ChannelConfig config) {
        this.config = config;
        maxMessagePerRead = maxMessagesPerRead();
        totalMessages = totalBytesRead = 0;
    }

    // 分配保存输入数据的缓冲区
    public ByteBuf allocate(ByteBufAllocator alloc) {
        return alloc.ioBuffer(guess());
    }

    // 增加已经读取的消息数
    public final void incMessagesRead(int amt) {
        totalMessages += amt;
    }

    // 保存最后一次的字节数
    public final void lastBytesRead(int bytes) {
        lastBytesRead = bytes;
        totalBytesRead += bytes;              // 增加总读取的字节数
        if (totalBytesRead < 0) {
            totalBytesRead = Integer.MAX_VALUE;
        }
    }

    public final int lastBytesRead() {
        return lastBytesRead;
    }

    // 满足以下条件时，控制事件循环线程操作 AbstractNioUnsafe 继续从底层 IO 读取数据
    // 1. 通道配置自动读取数据
    // 2. 当前尝试读取的字节数量为上一次读取的字节数量（即通道中仍可能有数据可读，因为上一次读取
    //    将 allocate(ByteBufAllocator alloc) 分配的缓冲区写满了）
    // 3. 总读取到的消息数小于设置读取的最大消息数
    // 4. 总读取的字节数没有超出整型的最大值（底层 IO 通常最多只能写入整型最大值的字节数，如果超过
    //    该值可能会发生 IO 异常）
    public boolean continueReading() {
        return config.isAutoRead() &&
            attemptedBytesRead == lastBytesRead &&
            totalMessages < maxMessagePerRead &&
            totalBytesRead < Integer.MAX_VALUE;
    }

    @Override
```

```java
    public void readComplete() {
    }

    @Override
    public int attemptedBytesRead() {
        return attemptedBytesRead;
    }

    @Override
    public void attemptedBytesRead(int bytes) {
        attemptedBytesRead = bytes;
    }

    protected final int totalBytesRead() {
        return totalBytesRead;
    }
}
```

12.7.4　AdaptiveRecvByteBufAllocator 分配器

AdaptiveRecvByteBufAllocator 自适应接收缓冲区分配器继承自 DefaultMaxMessagesRecvByteBufAllocator 抽象类，也是默认的接收缓冲区分配器。该分配器通过接收数据反馈自动增加或减少预测的缓冲区大小，使分配的接收缓冲区能存放数据，同时不会浪费太多内存。如果之前读取的数据完全填满了已分配的缓冲区，它会逐渐增加读缓冲区大小，如果之前读取的数据不能连续两次填满所分配的缓冲区，则会逐渐减少读缓冲区大小，否则它会一直返回相同的缓冲区大小。由源码可知。

（1）使用 int[] SIZE_TABLE 来保存尺寸表，在静态代码块中对尺寸表初始化，其结构为：缓冲区小于 512 时，以 16 递增；缓冲区大于 512 时，以 2 的倍数递增。界限为[16，0x 40000000]。

（2）默认接收缓冲区最小值 64 所处的 SIZE_TABLE 下标为 3。

（3）默认接收缓冲区最大值 65536 所处的 SIZE_TABLE 下标为 38。

（4）默认接收缓冲区初始值 1024 所处的 SIZE_TABLE 下标为 32。

（5）getSizeTableIndex(final int size)方法根据给定 size，返回该 size 所处 SIZE_TABLE 的索引下标，其中使用二分查找法来找到传入 size 对应的下标。

（6）HandleImpl 继承自 MaxMessageHandle 类，重写了 guess 方法，该方法将保存的 nextReceiveBufferSize 变量作为当前分配缓冲区的大小值。重写了 readComplete 方法，该方法会在读取完缓冲区的数据时调用 record(int actualReadBytes)方法，根据实际读取到的数量来调整 nextReceiveBufferSize 变量。

（7）通过 record(int actualReadBytes)方法的描述，调整的算法如下。

① 如果连续两次读取到的实际数量小于缓冲区大小时，则对其以 INDEX_DECREMENT 的增量减少容量，以当前容量的 size 下标计算 Math.max(index - INDEX_DECREMENT, minIndex)（注意，这里得到的 size 下标 minIndex 为设置的最小值界限，默认 64 SIZE_TABLE 下标为 3），其中 INDEX_DECREMENT 增量为 1。

② 如果当前读取的实际数量大于缓冲区大小时，则对其以 INDEX_INCREMENT 的增量增加容量 Math.min(index + INDEX_INCREMENT, maxIndex)（注意，这里得到的 size 下标 maxIndex 为设置的最大值界限，默认 65536 SIZE_TABLE 下标为 38），其中 INDEX_INCREMENT 增量为 4。

```java
public class AdaptiveRecvByteBufAllocator extends DefaultMaxMessagesRecvByteBufAllocator {
    static final int DEFAULT_MINIMUM = 64;         // 默认接收缓冲区最小值
    static final int DEFAULT_INITIAL = 1024;       // 默认接收缓冲区初始值
    static final int DEFAULT_MAXIMUM = 65536;      // 默认接收缓冲区最大值

    private static final int INDEX_INCREMENT = 4; // 增加缓冲区大小时 SIZE_TABLE 索引增量
    private static final int INDEX_DECREMENT = 1; // 减少缓冲区大小时 SIZE_TABLE 索引增量

    private static final int[] SIZE_TABLE;         // 缓冲区尺寸表

    static {                                       // 初始化尺寸表
        List<Integer> sizeTable = new ArrayList<Integer>();
        for (int i = 16; i < 512; i += 16) {       // 缓冲区小于 512 时,以 16 递增:16, 32, 48 ... 496
            sizeTable.add(i);
        }
        for (int i = 512; i > 0; i <<= 1) {        // 缓冲区大于 512 时,以 2 的倍数递增:512、1024、2048 ...
// 0x40000000(如果再左移 1 位,则变为 0x80000000,此时超出整型最大值 0x7fffffff,发生符号溢出,退出循环)
            sizeTable.add(i);
        }
        // 将计算好的尺寸放入 SIZE_TABLE,从 0 下标开始放入
        SIZE_TABLE = new int[sizeTable.size()];
        for (int i = 0; i < SIZE_TABLE.length; i ++) {
            SIZE_TABLE[i] = sizeTable.get(i);
        }
    }

    // 根据给定 size,返回该 size 所处 SIZE_TABLE 的索引下标
    private static int getSizeTableIndex(final int size) {
        for (int low = 0, high = SIZE_TABLE.length - 1;;) {    // 采用二分查找
            if (high < low) {                                  // 最小值 16
                return low;
            }
            if (high == low) {                                 // 最大值 0x40000000
                return high;
            }
            int mid = low + high >>> 1;
            int a = SIZE_TABLE[mid];
            int b = SIZE_TABLE[mid + 1];
            if (size > b) {
                low = mid + 1;
            } else if (size < a) {
                high = mid - 1;
            } else if (size == a) {
                return mid;
            } else {
                return mid + 1;
            }
        }
    }

    // 实现 MaxMessageHandle 的 guess 方法,将用于分配当前使用的缓冲区大小,readComplete 方法在当前
    // 读取完毕时,记录当前读取的总字节数,用于计算下次分配读缓冲区时的缓冲区的大小
```

```java
private final class HandleImpl extends MaxMessageHandle {
    private final int minIndex;                    // 最小缓冲区大小在 SizeTable 中的索引下标
    private final int maxIndex;                    // 最大缓冲区大小在 SizeTable 中的索引下标
    private int index;                             // 当前缓冲区大小在 SizeTable 中的索引下标
    private int nextReceiveBufferSize;             // 下一次分配缓冲区时的大小
    private boolean decreaseNow;                   // 是否需要立即减少缓冲区大小

    // 保存最小、最大下标，同时根据初始值设置当前缓冲区大小在 SizeTable 中的索引下标
    public HandleImpl(int minIndex, int maxIndex, int initial) {
        this.minIndex = minIndex;
        this.maxIndex = maxIndex;
        index = getSizeTableIndex(initial);
        nextReceiveBufferSize = SIZE_TABLE[index];  // 初始化下次分配大小为当前传入的初始值
    }

    // 父类依靠该值分配读缓冲区大小
    public int guess() {
        return nextReceiveBufferSize;
    }

    // 根据实际读取的字节数，调整缓冲区大小
    private void record(int actualReadBytes) {
        if (actualReadBytes <= SIZE_TABLE[Math.max(0, index - INDEX_DECREMENT - 1)]) { // 实际读
// 取字节数小于或等于设置的增量值，INDEX_DECREMENT 默认为 1，此时 SIZE_TABLE[Math.max(0, index)]，
// 即小于或等于当前索引下标值的大小，默认为 1024
            if (decreaseNow) { // 立即缩小内存，下次分配时为当前 index 减 1 的索引下标处的大小，最小
// 不会低于传入的 minIndex 处的大小，默认为 64B
                index = Math.max(index - INDEX_DECREMENT, minIndex);
                nextReceiveBufferSize = SIZE_TABLE[index];
                decreaseNow = false;
            } else {           // 下一次分配时减少（这里可以看出：当第一次读取小于设置值时，不会立即减
// 少，第二次仍小于时才进行缩容）
                decreaseNow = true;
            }
        } else if (actualReadBytes >= nextReceiveBufferSize) {      // 实际读取的数量大于读缓冲区的字
// 节数（注意，在后面会看到当 continueReading 方法为 true 时，会一直使用该缓冲区存放底层 IO 数据，直到
// continueReading()返回 false，此时 actualReadBytes 等于该值。而要做的就是：优化 buffer 的大小，尽量一次读
// 取完底层 IO 的数据，而不是分多次构建缓冲区读取）
            index = Math.min(index + INDEX_INCREMENT, maxIndex); // 根据增量来增加缓冲区大小，
// INDEX_INCREMENT 增量默认为 4，最大不会超过 maxIndex 下标的值：默认为 65536
            nextReceiveBufferSize = SIZE_TABLE[index];
            decreaseNow = false;
        }
    }

    // 当前读取完成时记录当前读取的总字节数
    public void readComplete() {
        record(totalBytesRead());
    }
}

private final int minIndex;                    // 最小缓冲区大小在 SizeTable 中的索引下标
private final int maxIndex;                    // 最大缓冲区大小在 SizeTable 中的索引下标
```

```java
    private final int initial;              // 初始缓冲区大小

    // 使用默认值构建
    public AdaptiveRecvByteBufAllocator() {
        this(DEFAULT_MINIMUM, DEFAULT_INITIAL, DEFAULT_MAXIMUM);
    }

    // 指定缓冲区容量最小值、初始值、最大值构建
    public AdaptiveRecvByteBufAllocator(int minimum, int initial, int maximum) {
        if (minimum <= 0) {
            throw new IllegalArgumentException("minimum: " + minimum);
        }
        if (initial < minimum) {
            throw new IllegalArgumentException("initial: " + initial);
        }
        if (maximum < initial) {
            throw new IllegalArgumentException("maximum: " + maximum);
        }
        // 设置最小缓冲区大小在 SizeTable 中的索引下标，默认为 3
        int minIndex = getSizeTableIndex(minimum);
        if (SIZE_TABLE[minIndex] < minimum) {
            this.minIndex = minIndex + 1;
        } else {
            this.minIndex = minIndex;
        }
        // 设置最大缓冲区大小在 SizeTable 中的索引下标，默认为 38
        int maxIndex = getSizeTableIndex(maximum);
        if (SIZE_TABLE[maxIndex] > maximum) {
            this.maxIndex = maxIndex - 1;
        } else {
            this.maxIndex = maxIndex;
        }
        // 缓冲区初始大小，注意这里不是下标，默认为 1024
        this.initial = initial;
    }

    // 使用内部 HandleImpl 完成实际操作
    public Handle newHandle() {
        return new HandleImpl(minIndex, maxIndex, initial);
    }
}
```

12.7.5 FixedRecvByteBufAllocator 分配器

FixedRecvByteBufAllocator 固定接收缓冲区分配器继承自 DefaultMaxMessagesRecvByteBufAllocator，实现了 newHandle 方法，还实现了 MaxMessageHandle 中 guess 方法，用于定义使用固定大小接收缓冲区的分配器。源码如下。

```java
public class FixedRecvByteBufAllocator extends DefaultMaxMessagesRecvByteBufAllocator {
    private final int bufferSize;          // 固定接收缓冲区大小
    private final class HandleImpl extends MaxMessageHandle {
        private final int bufferSize;
        public HandleImpl(int bufferSize) {
```

```
            this.bufferSize = bufferSize;
        }
        @Override
        public int guess() {         // MaxMessageHandle 在创建读缓冲区时，将使用该固定值分配
            return bufferSize;
        }
    }

    // 设置固定读缓冲区值
    public FixedRecvByteBufAllocator(int bufferSize) {
        if (bufferSize <= 0) {
            throw new IllegalArgumentException(
                "bufferSize must greater than 0: " + bufferSize);
        }
        this.bufferSize = bufferSize;
    }

    // 创建 HandleImpl 处理读缓冲区的操作
    public Handle newHandle() {
        return new HandleImpl(bufferSize);
    }
}
```

12.8　AbstractChannel 类

AbstractChannel 通道模板类提供了对于 Channel 接口的基本实现，提供了 Netty 中通道的架构实现。注意，该类的核心实现其实是内部的 AbstractUnsafe 类，该类实现了 Unsafe 接口，并实现了具体的通道动作。

12.8.1　核心变量与构造器

由源码可知。

（1）ClosedChannelException 异常类定义了导致通道关闭的异常信息，在模板类中通过构造器定义导致异常的方法名，当相关方法发生异常时会抛出这些定义的常量异常对象。

（2）由于底层 IO 操作与 JVM 运行时环境相关，所以将其中的操作定义在 Unsafe 接口中。

（3）DefaultChannelPipeline 用于对通道进行流水线处理，即责任链模式。

```
public abstract class AbstractChannel extends DefaultAttributeMap implements Channel {

    // 定义调用相关方法异常对象，后缀方法名例如"flush0()"发生异常时使用
    private static final ClosedChannelException FLUSH0_CLOSED_CHANNEL_EXCEPTION =
ThrowableUtil.unknownStackTrace(
        new ClosedChannelException(), AbstractUnsafe.class, "flush0()");
    private static final ClosedChannelException ENSURE_OPEN_CLOSED_CHANNEL_EXCEPTION =
ThrowableUtil.unknownStackTrace(
        new ClosedChannelException(), AbstractUnsafe.class, "ensureOpen(...)");
    private static final ClosedChannelException CLOSE_CLOSED_CHANNEL_EXCEPTION =
ThrowableUtil.unknownStackTrace(
        new ClosedChannelException(), AbstractUnsafe.class, "close(...)");
```

```java
    private static final ClosedChannelException WRITE_CLOSED_CHANNEL_EXCEPTION =
ThrowableUtil.unknownStackTrace(
        new ClosedChannelException(), AbstractUnsafe.class, "write(...)");
    private static final NotYetConnectedException FLUSH0_NOT_YET_CONNECTED_EXCEPTION =
ThrowableUtil.unknownStackTrace(
        new NotYetConnectedException(), AbstractUnsafe.class, "flush0()");

    private final Channel parent;                                  // 父通道引用
    private final ChannelId id;                                    // 当前通道唯一 id
    private final Unsafe unsafe;                                   // 实际操作 IO 底层对象引用
    private final DefaultChannelPipeline pipeline;                 // 处理通道相关操作的流水线对象
    private final VoidChannelPromise unsafeVoidPromise = new VoidChannelPromise(this, false);
    private final CloseFuture closeFuture = new CloseFuture(this); // 当通道关闭时回调对象引用

    private volatile SocketAddress localAddress;                   // 绑定的本地地址
    private volatile SocketAddress remoteAddress;                  // 连接的远程地址
    private volatile EventLoop eventLoop;                          // 所属事件循环组
    private volatile boolean registered;                           // 当前通道是否已经注册到事件循环组中
    private boolean strValActive;
    private String strVal;                                         // 缓存代表该通道的字符串信息

    // 构造器：指定所属父通道，并且在其中创建 unsafe 和处理流水线对象
    protected AbstractChannel(Channel parent) {
        this.parent = parent;
        id = newId();                                              // 生成新的通道 id
        unsafe = newUnsafe();
        pipeline = newChannelPipeline();
    }
    protected abstract AbstractUnsafe newUnsafe();                 // 由子类完成创建

    // 直接生成默认处理流水线对象
    protected DefaultChannelPipeline newChannelPipeline() {
        return new DefaultChannelPipeline(this);
    }

    // 生成通道 ID 实例（其中通过 machineId、processId、sequence、timestamp、random 生成唯一通道 ID）
    protected ChannelId newId() {
        return DefaultChannelId.newInstance();
    }

    // 构造器：指定父通道与通道 id 对象
    protected AbstractChannel(Channel parent, ChannelId id) {
        this.parent = parent;
        this.id = id;
        unsafe = newUnsafe();
        pipeline = newChannelPipeline();
    }
}
```

12.8.2 实现方法原理

由于该类为模板类实现，大部分方法会在该抽象类进行简单实现。由源码可知。

（1）可写方法将由 ChannelOutboundBuffer 输出缓冲区的方法完成（毕竟数据优先写入缓冲区）。

（2）地址相关信息，由 unsafe 类完成（底层与运行时相关，使用 Unsafe 类解耦）。

（3）实际通道的绑定、连接、关闭、刷新缓冲区等操作，由 DefaultChannelPipeline 类完成（不同操作均是对通道进行处理，可以采用责任链模式来将通道操作与具体实现解耦）。

读者只需要知道哪些方法由哪些类完成即可，笔者会在后面详细讲解这些核心类，现在先逐个击破。源码如下：

```java
/* 可写方法由 ChannelOutboundBuffer 输出缓冲区的方法完成 */
public boolean isWritable() {
    ChannelOutboundBuffer buf = unsafe.outboundBuffer();
    return buf != null && buf.isWritable();
}

@Override
public long bytesBeforeUnwritable() {
    ChannelOutboundBuffer buf = unsafe.outboundBuffer();
    return buf != null ? buf.bytesBeforeUnwritable() : 0;
}

@Override
public long bytesBeforeWritable() {
    ChannelOutboundBuffer buf = unsafe.outboundBuffer();
    return buf != null ? buf.bytesBeforeWritable() : Long.MAX_VALUE;
}

/* 地址相关信息，由 unsafe 类完成 */
public SocketAddress localAddress() {
    SocketAddress localAddress = this.localAddress;
    if (localAddress == null) {
        try {
            this.localAddress = localAddress = unsafe().localAddress();
        } catch (Throwable t) {
            return null;
        }
    }
    return localAddress;
}

@Override
public SocketAddress remoteAddress() {
    SocketAddress remoteAddress = this.remoteAddress;
    if (remoteAddress == null) {
        try {
            this.remoteAddress = remoteAddress = unsafe().remoteAddress();
        } catch (Throwable t) {
            return null;
        }
    }
    return remoteAddress;
}
```

```java
/* 实际通道的绑定、连接、关闭、刷新缓冲区等操作，由 DefaultChannelPipeline 类完成 */

@Override
public ChannelFuture bind(SocketAddress localAddress) {
    return pipeline.bind(localAddress);
}

@Override
public ChannelFuture connect(SocketAddress remoteAddress) {
    return pipeline.connect(remoteAddress);
}

@Override
public ChannelFuture connect(SocketAddress remoteAddress, SocketAddress localAddress) {
    return pipeline.connect(remoteAddress, localAddress);
}

@Override
public ChannelFuture disconnect() {
    return pipeline.disconnect();
}

@Override
public ChannelFuture close() {
    return pipeline.close();
}

@Override
public ChannelFuture deregister() {
    return pipeline.deregister();
}

@Override
public Channel flush() {
    pipeline.flush();
    return this;
}

@Override
public ChannelFuture bind(SocketAddress localAddress, ChannelPromise promise) {
    return pipeline.bind(localAddress, promise);
}

@Override
public ChannelFuture connect(SocketAddress remoteAddress, ChannelPromise promise) {
    return pipeline.connect(remoteAddress, promise);
}

@Override
public ChannelFuture connect(SocketAddress remoteAddress, SocketAddress localAddress, ChannelPromise promise) {
    return pipeline.connect(remoteAddress, localAddress, promise);
}
```

```java
@Override
public ChannelFuture disconnect(ChannelPromise promise) {
    return pipeline.disconnect(promise);
}

@Override
public ChannelFuture close(ChannelPromise promise) {
    return pipeline.close(promise);
}

@Override
public ChannelFuture deregister(ChannelPromise promise) {
    return pipeline.deregister(promise);
}

@Override
public Channel read() {
    pipeline.read();
    return this;
}

@Override
public ChannelFuture write(Object msg) {
    return pipeline.write(msg);
}

@Override
public ChannelFuture write(Object msg, ChannelPromise promise) {
    return pipeline.write(msg, promise);
}

@Override
public ChannelFuture writeAndFlush(Object msg) {
    return pipeline.writeAndFlush(msg);
}

@Override
public ChannelFuture writeAndFlush(Object msg, ChannelPromise promise) {
    return pipeline.writeAndFlush(msg, promise);
}

@Override
public ChannelPromise newPromise() {
    return pipeline.newPromise();
}

@Override
public ChannelProgressivePromise newProgressivePromise() {
    return pipeline.newProgressivePromise();
}

@Override
public ChannelFuture newSucceededFuture() {
```

```
        return pipeline.newSucceededFuture();
}

@Override
public ChannelFuture newFailedFuture(Throwable cause) {
        return pipeline.newFailedFuture(cause);
}
```

12.8.3　AbstractUnsafe 核心内部类

AbstractUnsafe 抽象类用于对 Unsafe 接口进行架构实现，即模板方法模式，实现子类的共用方法，减少子类的重复代码。同时，由于 IO 操作与运行时相关，所以用该抽象类将通道的具体操作与 IO 具体实现分离解耦。读者可以通过源码得出结论：一切判断状态操作在 AbstractUnsafe 中完成，实际 IO 操作由子类实现。

12.8.4　核心变量与构造器

```
protected abstract class AbstractUnsafe implements Unsafe {
        private volatile ChannelOutboundBuffer outboundBuffer = new
ChannelOutboundBuffer(AbstractChannel.this);        // 子类 Unsafe 实例化时，便初始化了输出缓冲区对象
        private RecvByteBufAllocator.Handle recvHandle;  // 分配接收缓冲区内存的分配器
        private boolean inFlush0;                         // 标志位标识是否正在刷新中
        private boolean neverRegistered = true;           // 标志位标识当前通道是否从来没有被注册到事件循环中
}
```

12.8.5　recvBufAllocHandle 核心方法

recvBufAllocHandle 方法用于获取分配输入缓冲区内存分配器。调用 Channel 的 Config 方法，从配置项中获取该分配器。源码如下。

```
public RecvByteBufAllocator.Handle recvBufAllocHandle() {
        if (recvHandle == null) {
                recvHandle = config().getRecvByteBufAllocator().newHandle();
        }
        return recvHandle;
}
```

12.8.6　localAddress 与 remoteAddress 核心方法

localAddress 与 remoteAddress 两个方法分别用于获取本地地址和远程连接地址。最终实现由子类完成。源码如下。

```
@Override
public final SocketAddress localAddress() {
        return localAddress0();
}

@Override
public final SocketAddress remoteAddress() {
        return remoteAddress0();
}
```

```
protected abstract SocketAddress localAddress0();           // 子类实现
protected abstract SocketAddress remoteAddress0();          // 子类实现
```

12.8.7　register 核心方法

register 方法用于将当前通道注册到指定的 EventLoop eventLoop 中，成功后回调 ChannelPromise promise。流程如下。

（1）如果通道已经被注册，则设置 ChannelPromise 为失败完成。

（2）如果事件循环与当前通道不兼容，不能将其进行绑定，则设置 ChannelPromise 为失败完成。

（3）如果当前 register 的线程为传入的事件循环所属线程，则直接调用 register0 完成注册。

（4）否则封装 Runnable 任务放入 eventLoop 的队列，由事件循环线程进行注册，但最终都是调用 register0 方法。

```java
public final void register(EventLoop eventLoop, final ChannelPromise promise) {
    if (eventLoop == null) {
        throw new NullPointerException("eventLoop");
    }
    if (isRegistered()) {              // 通道已经被注册，设置 ChannelPromise 为失败完成
        promise.setFailure(new IllegalStateException("registered to an event loop already"));
        return;
    }
    if (!isCompatible(eventLoop)) {    // 事件循环与当前通道不兼容，不能将其进行绑定，设置 ChannelPromise 为失败完成
        promise.setFailure(
            new IllegalStateException("incompatible event loop type: " + eventLoop.getClass().getName()));
        return;
    }
    AbstractChannel.this.eventLoop = eventLoop;
    if (eventLoop.inEventLoop()) {     // 如果当前调用 register 的线程为传入的事件循环所属线程，则直接调用 register0 完成注册
        register0(promise);
    } else {                            // 否则封装 Runnable 任务放入 eventLoop 的队列，由事件循环线程来进行注册，但最终都要调用 register0 方法
        try {
            eventLoop.execute(new Runnable() {
                @Override
                public void run() {
                    register0(promise);
                }
            });
        } catch (Throwable t) {
            logger.warn(
                "Force-closing a channel whose registration task was not accepted by an event loop: {}",
                AbstractChannel.this, t);
            closeForcibly();
            closeFuture.setClosed();
            safeSetFailure(promise, t);
        }
    }
}
```

12.8.8　register0 核心方法

register0 方法用于完成实际的绑定操作。流程如下。

（1）如果当前通道已被关闭，即 ChannelPromise 被取消，则直接返回，不进行注册。
（2）调用子类实现的 doRegister 方法完成子类自身的注册逻辑。
（3）设置 registered 为 true，标识注册成功。
（4）判断当前通道是不是第一次注册。如果是，则回调通道处理器的 handlerAdded 方法。
（5）将 promise 设置为完成状态，表示当前已完成注册。
（6）如果通道处于活动状态，即 isActive() 为 true，则判断当前通道是否为第一次注册。如果是第一次注册，则回调通道处理器的 ChannelActive 回调方法。否则，判断是否配置了自动读取通道数据选项，调用 beginRead 从通道中读取数据。

```
private void register0(ChannelPromise promise) {
    try {
        if (!promise.setUncancellable() || !ensureOpen(promise)) {  // 确保当前通道是否有效，没有关闭
            return;
        }
        boolean firstRegistration = neverRegistered;       // 标识是否为第一次注册
        doRegister();
        neverRegistered = false;
        registered = true;                                  // 标识注册成功
        pipeline.invokeHandlerAddedIfNeeded();              // 判断当前通道是不是第一次注册，如果是，则回调通
道处理器的 handlerAdded 方法
        safeSetSuccess(promise);                            // 将 promise 设置为完成状态，表示当前已完成注册
        pipeline.fireChannelRegistered();                   // 触发通道处理器当前通道注册成功回调方法
        if (isActive()) {                                   // 只有通道处于活动状态，即连接状态时操作
            if (firstRegistration) {        // 如果是第一次注册，则回调通道处理器的 ChannelActive 回调方法
                pipeline.fireChannelActive();
            } else if (config().isAutoRead()) {             // 如果不是第一次注册，且配置了自动读取操作，则调
用 beginRead 从通道中读取数据
                beginRead();
            }
        }
    } catch (Throwable t) {                                 // 发生任何异常，都会导致通道关闭
        closeForcibly();
        closeFuture.setClosed();
        safeSetFailure(promise, t);
    }
}

// 安全设置 ChannelPromise 成功执行，主要是判断 promise 是否为 VoidChannelPromise，然后调用 trySuccess
方法尝试将其设置为完成状态，如果设置失败，则打印警告日志
protected final void safeSetSuccess(ChannelPromise promise) {
    if (!(promise instanceof VoidChannelPromise) && !promise.trySuccess()) {
        logger.warn("Failed to mark a promise as success because it is done already: {}", promise);
    }
}

// 子类完成回调，实现自身注册逻辑
```

```
protected void doRegister() throws Exception {
    // NOOP
}
```

12.8.9 beginRead 核心方法

beginRead 方法用于从通道中读取数据。流程如下。

（1）如果通道不属于活动状态，即已经连接且可用，则直接返回。

（2）调用子类完成实际读取操作。

（3）如果在读取期间发生了异常，则封装 Runnable 任务触发通道处理器 ExceptionCaught 方法，然后关闭通道。

```
public final void beginRead() {
    if (!isActive()) {                          // 通道不处于活动状态，即已经连接且可用，直接返回
        return;
    }
    try {
        doBeginRead();
    } catch (final Exception e) {
        invokeLater(new Runnable() {
            @Override
            public void run() {                 // 触发通道处理器回调
                pipeline.fireExceptionCaught(e);
            }
        });
        close(voidPromise());
    }
}

// 将任务放入事件循环线程的队列中，在执行完前面的任务后，执行该 task 任务
private void invokeLater(Runnable task) {
    try {
        eventLoop().execute(task);
    } catch (RejectedExecutionException e) {
        logger.warn("Can't invoke task later as EventLoop rejected it", e);
    }
}

// 子类具体实现读取操作
protected abstract void doBeginRead() throws Exception;
```

12.8.10 bind 核心方法

bind 方法用于将当前通道绑定到 SocketAddress localAddress 地址上，同时在完成后回调 ChannelPromise promise。流程如下。

（1）确保当前操作通道有效，即打开状态。

（2）如果通道配置了广播选项（用于 UDP，了解即可），且不是通配符地址（*），并运行在 UNIX 系统上，则警告用户；如果 Socket 绑定在非通配符地址上，则非 root 用户无法在 UNIX 系统上接收广播包。

（3）调用子类实现的 doBind(localAddress)方法完成实际绑定操作。

（4）如果子类在绑定后，激活了通道，即顺带与对端进行连接，则向事件循环线程队列中添加一个 Runnable 任务，稍后调用通道处理器的 ChannelActive 方法。

（5）设置 promise 成功完成。

```java
public final void bind(final SocketAddress localAddress, final ChannelPromise promise) {
    if (!promise.setUncancellable() || !ensureOpen(promise)) {      // 确保通道有效
        return;
    }
    if (Boolean.TRUE.equals(config().getOption(ChannelOption.SO_BROADCAST)) &&
        localAddress instanceof InetSocketAddress &&
        !((InetSocketAddress) localAddress).getAddress().isAnyLocalAddress() &&
        !PlatformDependent.isWindows() && !PlatformDependent.isRoot()) { // 警告用户，如果 Socket 绑定在
                                                                          // 非通配符地址上，则非 root 用户无法在 UNIX 系统上接收广播包
        logger.warn(
            "A non-root user can't receive a broadcast packet if the socket " +
            "is not bound to a wildcard address; binding to a non-wildcard " +
            "address (" + localAddress + ") anyway as requested.");
    }
    boolean wasActive = isActive();
    try {
        doBind(localAddress);                                   // 完成实际绑定操作
    } catch (Throwable t) {                                     // 发生异常关闭通道
        safeSetFailure(promise, t);
        closeIfClosed();
        return;
    }
    if (!wasActive && isActive()) {   // 如果子类在绑定后激活了通道，即与对端进行连接，则向事件循环线程队
                                      // 列中添加一个 Runnable 任务，稍后调用通道处理器的 ChannelActive 方法
        invokeLater(new Runnable() {
            @Override
            public void run() {
                pipeline.fireChannelActive();
            }
        });
    }
    safeSetSuccess(promise);                                    // 设置 promise 成功完成
}

protected abstract void doBind(SocketAddress localAddress) throws Exception;   // 子类实现绑定操作
```

12.8.11 disconnect 核心方法

disconnect 方法用于从连接的对端中断开当前通道的连接，同时在完成后回调 ChannelPromise promise。流程如下。

（1）如果 promise 已经设置取消执行，则直接返回。

（2）调用 doDisconnect 方法，由子类完成断开连接操作。

（3）子类在断开连接后，将通道状态修改为失效状态，并向事件循环线程队列中添加一个 Runnable 任务，稍后调用通道处理器的 ChannelInactive 方法。

(4) 设置 promise 成功完成。

(5) 如果此时需要关闭通道,则执行关闭操作。

```java
public final void disconnect(final ChannelPromise promise) {
    if (!promise.setUncancellable()) {              // promise 已经设置取消执行
        return;
    }
    boolean wasActive = isActive();
    try {
        doDisconnect();                              // 调用子类断开连接
    } catch (Throwable t) {
        safeSetFailure(promise, t);
        closeIfClosed();
        return;
    }
    if (wasActive && !isActive()) {     // 子类在断开连接后,将通道状态修改为失效状态,并向事件循环线程队列
                                        // 中添加一个 Runnable 任务,稍后调用通道处理器的 ChannelInactive 方法
        invokeLater(new Runnable() {
            @Override
            public void run() {
                pipeline.fireChannelInactive();
            }
        });
    }
    safeSetSuccess(promise);                         // 设置 promise 成功完成
    closeIfClosed();                                 // 如果需要关闭通道,则执行关闭操作
}

// 如果通道仍处于打开状态,则返回,否则调用 close 方法完成关闭
protected final void closeIfClosed() {
    if (isOpen()) {
        return;
    }
    close(voidPromise());
}

protected abstract void doDisconnect() throws Exception;     // 子类回调
```

12.8.12 close 核心方法

close 方法用于关闭当前通道,关闭完成后,回调 ChannelPromise promise。流程如下。

(1) 实际关闭方法由 close(final ChannelPromise promise, final Throwable cause, final ClosedChannelException closeCause, final boolean notify)完成。

(2) 在 close(final ChannelPromise promise)方法中传入。CLOSE_CLOSED_CHANNEL_EXCEPTION 表示在 close(..)方法中关闭通道。

(3) 具体关闭流程如下。

① 如果 promise 已经被取消,表明当前关闭操作不需要进行,则直接返回。

② 如果输出缓冲区不存在,判断 promise 不是 VoidChannelPromise 实例,则向 closeFuture 中添加监听器,closeFuture 设置完成时,将 promise 设置成功完成并返回。

③ 如果存在缓冲区，则判断是否已经关闭；如果已经关闭，则直接返回。
④ 置空成员变量缓冲区引用，避免在操作中放入数据。
⑤ 调用 prepareToClose 方法，返回执行关闭任务的执行器。
⑥ 如果执行器不为空，则向执行器中添加任务，由执行器完成实际关闭操作，实际关闭方法为 close0。
⑦ 如果执行器为空，则在当前线程中执行关闭和回调通道处理器 ChannelInactiveAndDeregister 回调方法。

```java
public final void close(final ChannelPromise promise) {
    close(promise,CLOSE_CLOSED_CHANNEL_EXCEPTION,CLOSE_CLOSED_CHANNEL_EXCEPTION,false);
}

private void close(final ChannelPromise promise, final Throwable cause,
                   final ClosedChannelException closeCause, final boolean notify) {
    if (!promise.setUncancellable()) { // 如果 promise 已经被取消，表明当前关闭操作不需要进行，则直接返回
        return;
    }
    final ChannelOutboundBuffer outboundBuffer = this.outboundBuffer;
    if (outboundBuffer == null) {                            // 不存在输出缓冲区
        if (!(promise instanceof VoidChannelPromise)) {      // promise 不是 VoidChannelPromise 实例，向
//closeFuture 中添加监听器；closeFuture 被设置完成时，将 promise 设置成功完成（如果不是 VoidChannelPromise
//实例，这意味着之前调用了 close()，所以只需要注册一个监听器并返回）
            closeFuture.addListener(new ChannelFutureListener() {
                @Override
                public void operationComplete(ChannelFuture future) throws Exception {
                    promise.setSuccess();
                }
            });
        }
        return;
    }
    // 存在缓冲区，判断是否已经关闭。如果已经关闭，则直接返回
    if (closeFuture.isDone()) {
        safeSetSuccess(promise);
        return;
    }

    final boolean wasActive = isActive();
    this.outboundBuffer = null;                  // 将成员变量缓冲区置空，避免在操作中被放入数据
    Executor closeExecutor = prepareToClose();   // 子类实现返回执行关闭任务的执行器
    if (closeExecutor != null) {                 // 执行器不为空，向执行器中添加任务，由执行器完成实际关闭操作
        closeExecutor.execute(new Runnable() {
            @Override
            public void run() {
                try {
                    // 执行最终关闭操作
                    doClose0(promise);
                } finally {
                    // 调用 invokeLater，使通道管理的回调方法 ChannelInactiveAndDeregister 再次在
//EventLoop 事件循环线程中执行（每个通道均属于一个事件循环，这样可以保证线程安全，在多线程间不需要上
//锁，所有事件均由所属事件循环线程处理）
                    invokeLater(new Runnable() {
                        @Override
```

```java
                    public void run() {
                        // 丢弃所有写缓冲区数据并关闭
                        outboundBuffer.failFlushed(cause, notify);
                        outboundBuffer.close(closeCause);
                        fireChannelInactiveAndDeregister(wasActive); // 触发通道处理器回调方法
                    }
                });
            }
        }
    });
} else {           // 关闭执行器，不能异步关闭，在当前线程中执行关闭动作
    try {
        doClose0(promise);
    } finally {    // 关闭写缓冲区
        outboundBuffer.failFlushed(cause, notify);
        outboundBuffer.close(closeCause);
    }
    if (inFlush0) {  // 目前正在刷新缓冲区的话，向当前事件循环队列中添加一个回调 handler 任务
        invokeLater(new Runnable() {
            @Override
            public void run() {
                fireChannelInactiveAndDeregister(wasActive);
            }
        });
    } else {         // 否则直接回调通道处理器的回调方法 ChannelInactiveAndDeregister
        fireChannelInactiveAndDeregister(wasActive);
    }
}
}

// 实际处理关闭操作方法
private void doClose0(ChannelPromise promise) {
    try {
        doClose();
        closeFuture.setClosed();      // 回调 closeFuture 方法，表示通道已经关闭
        safeSetSuccess(promise);      // 设置 promise 对象完成执行
    } catch (Throwable t) {           // 子类在关闭时发生异常，设置通道关闭，同时把 promise 设置为失败完成
        closeFuture.setClosed();
        safeSetFailure(promise, t);
    }
}

protected abstract void doClose() throws Exception;    // 子类实现完成关闭操作

// 子类实现，返回执行关闭任务的执行器
protected Executor prepareToClose() {
    return null;
}
```

12.8.13 closeForcibly 核心方法

closeForcibly 方法与 close(final ChannelPromise promise)方法不同，该方法进行一系列缓冲区以及事件循环判断，直接调用子类实现的 doClose 方法完成实际关闭操作。源码如下。

```java
public final void closeForcibly() {
    try {
        doClose();
    } catch (Exception e) {     // 子类关闭失败，打印警告日志
        logger.warn("Failed to close a channel.", e);
    }
}
```

12.8.14　deregister 核心方法

deregister 方法用于将通道从所属事件循环中解除注册。流程如下。

（1）如果 promise 被取消，表明当前解除操作被取消，则直接返回。
（2）如果通道未注册到事件循环，则直接设置 promise 成功并返回。
（3）向事件循环队列中添加一个 Runnable 任务，在下一次执行时，完成解除注册操作。
（4）实际解除注册的动作由子类重写 doDeregister 方法实现。

```java
public final void deregister(final ChannelPromise promise) {
    deregister(promise, false);
}

private void deregister(final ChannelPromise promise, final boolean fireChannelInactive) {
    if (!promise.setUncancellable()) {      // promise 被取消，表明当前解除操作被取消，直接返回
        return;
    }
    if (!registered) {                       // 通道未注册到事件循环，直接设置成功并返回
        safeSetSuccess(promise);
        return;
    }
    invokeLater(new Runnable() {
        @Override
        public void run() {
            try {
                doDeregister();
            } catch (Throwable t) {
                logger.warn("Unexpected exception occurred while deregistering a channel.", t);
            } finally { // 解除注册后，如果设置触发通道失效事件，则回调通道处理器的 ChannelInactive 方法
                if (fireChannelInactive) {
                    pipeline.fireChannelInactive();
                }
                if (registered) {             // 如果之前注册成功，则回调事件通道处理器 ChannelUnregistered
                    registered = false;
                    pipeline.fireChannelUnregistered();
                }
                safeSetSuccess(promise); // 设置解除注册成功
            }
        }
    });
}

// 子类实现
protected void doDeregister() throws Exception {
```

```
        // NOOP
}
```

12.8.15　write 核心方法

write 方法用于写入 Object msg 数据，写入成功后回调 ChannelPromise。流程如下。

（1）由于通道写出数据，必须写入缓冲区中；如果不存在写缓冲区对象，则设置完成失败，同时释放 msg 消息。
（2）调用 filterOutboundMessage 方法，获取过滤后的消息对象。
（3）计算消息大小。
（4）将消息写入写缓冲区中。

```
public final void write(Object msg, ChannelPromise promise) {
    ChannelOutboundBuffer outboundBuffer = this.outboundBuffer;
    if (outboundBuffer == null) {                        // 不存在写缓冲区，设置完成失败，同时释放 msg 消息
        safeSetFailure(promise, WRITE_CLOSED_CHANNEL_EXCEPTION);
        ReferenceCountUtil.release(msg);
        return;
    }
    int size;
    try {
        msg = filterOutboundMessage(msg);                // 获取过滤后的消息对象
        size = pipeline.estimatorHandle().size(msg);     // 计算消息大小
        if (size < 0) {
            size = 0;
        }
    } catch (Throwable t) {
        safeSetFailure(promise, t);
        ReferenceCountUtil.release(msg);
        return;
    }
    outboundBuffer.addMessage(msg, size, promise);       // 将消息写入写缓冲区
}

// 子类复写用于过滤输出消息
protected Object filterOutboundMessage(Object msg) throws Exception {
    return msg;
}
```

12.8.16　flush 核心方法

flush 方法用于将通道写缓冲区中的数据刷出到底层 Socket 中。流程如下。

（1）如果不存在写缓冲区，则直接返回。
（2）标记写缓冲区中的数据已经刷出。
（3）如果当前已经刷新了，则直接返回。
（4）如果缓冲区数据不需要刷新，则直接返回。
（5）设置 inFlush0 标志位为 true，标记当前正在刷新。
（6）如果通道处于非活动状态，则将所有挂起的写请求标记为失败。

（7）如果通道处于正常活动状态，则调用 doWrite(outboundBuffer)方法，由子类完成实际刷新操作。

```java
public final void flush() {
    ChannelOutboundBuffer outboundBuffer = this.outboundBuffer;
    if (outboundBuffer == null) {      // 不存在写缓冲区，直接返回
        return;
    }
    outboundBuffer.addFlush();         // 标记写缓冲区中的数据已经被刷出
    flush0();                          // 完成实际刷新缓冲区数据操作
}

protected void flush0() {
    if (inFlush0) {                    // 当前已经刷新了，直接返回
        return;
    }
    final ChannelOutboundBuffer outboundBuffer = this.outboundBuffer;
    if (outboundBuffer == null || outboundBuffer.isEmpty()) { // 不存在缓冲区或不需要刷新数据，直接返回
        return;
    }
    inFlush0 = true;                   // 标记当前正在刷新
    if (!isActive()) {                 // 如果通道处于非活动状态，则将所有挂起的写请求标记为失败
        try {
            if (isOpen()) {            // 通道属于打开但不是激活状态（即没有连接），则抛出 NOT_YET_CONNECTED_EXCEPTION
                outboundBuffer.failFlushed(FLUSH0_NOT_YET_CONNECTED_EXCEPTION, true);
            } else {                   // 通道状态已经关闭，则抛出 CLOSED_CHANNEL_EXCEPTION
                outboundBuffer.failFlushed(FLUSH0_CLOSED_CHANNEL_EXCEPTION, false);
            }
        } finally {
            inFlush0 = false;          // 退出刷新状态
        }
        return;
    }
    try {
        doWrite(outboundBuffer);       // 子类完成实际刷新操作
    } catch (Throwable t) {
        if (t instanceof IOException && config().isAutoClose()) {
            close(voidPromise(), t, FLUSH0_CLOSED_CHANNEL_EXCEPTION, false);
        } else {
            outboundBuffer.failFlushed(t, true);
        }
    } finally {
        inFlush0 = false;              // 退出刷新状态
    }
}

protected abstract void doWrite(ChannelOutboundBuffer in) throws Exception;   // 子类完成实际写缓冲区刷新操作
```

12.8.17　CloseFuture 核心内部类

CloseFuture 类用于返回 CloseFuture，外部线程可以获取到 Future 对象。通道关闭时，会回调 CloseFuture 的方法。可以通过该类的实例获取关闭的状态。由源码可知，不允许直接调用完成操作，

而是通过 setClosed 方法调用父类的 trySuccess 方法。源码如下。

```java
static final class CloseFuture extends DefaultChannelPromise {

    CloseFuture(AbstractChannel ch) {
        super(ch);              // 保存所属通道对象
    }

    @Override
    public ChannelPromise setSuccess() {
        throw new IllegalStateException();
    }

    @Override
    public ChannelPromise setFailure(Throwable cause) {
        throw new IllegalStateException();
    }

    @Override
    public boolean trySuccess() {
        throw new IllegalStateException();
    }

    @Override
    public boolean tryFailure(Throwable cause) {
        throw new IllegalStateException();
    }

    boolean setClosed() {       // 唯一可以调用的方法，由子类完成
        return super.trySuccess();
    }
}
```

12.9 AbstractServerChannel 类

AbstractServerChannel 模板类用于定义服务端通道的架构。由源码可知。

（1）服务端通道只需要实现 accept 接收客户端请求方法，其余如 connect(SocketAddress, ChannelPromise)、disconnect(ChannelPromise)、write(Object, ChannelPromise)、flush() 等相似的方法都将抛出 UnsupportedOperationException。因为服务端并不与客户端直接沟通，而是接收客户端连接，创建客户端 Channel（JDK 的 Socket 包装）。

（2）内部类 DefaultServerUnsafe 对父类 AbstractUnsafe 的连接操作进行实现，因为不存在服务端进行连接，所以这里直接设置 UnsupportedOperationException 异常失败 ChannelPromise promise。

```java
public abstract class AbstractServerChannel extends AbstractChannel implements ServerChannel {
    private static final ChannelMetadata METADATA = new ChannelMetadata(false, 16); // 每次读取的最大客户端数量为 16

    // 构造器初始化父类
    protected AbstractServerChannel() {
```

```java
        super(null);
    }

    @Override
    public ChannelMetadata metadata() {
        return METADATA;
    }

    @Override
    public SocketAddress remoteAddress() {
        return null;
    }

    @Override
    protected SocketAddress remoteAddress0() {
        return null;
    }

    @Override
    protected void doDisconnect() throws Exception {
        throw new UnsupportedOperationException();
    }

    @Override
    protected AbstractUnsafe newUnsafe() {
        return new DefaultServerUnsafe();
    }

    @Override
    protected void doWrite(ChannelOutboundBuffer in) throws Exception {
        throw new UnsupportedOperationException();
    }

    @Override
    protected final Object filterOutboundMessage(Object msg) {
        throw new UnsupportedOperationException();
    }

    // 对父类 Unsafe 的连接操作进行实现，因为不存在服务端进行连接，所以这里直接设置
    // UnsupportedOperationException 异常，ChannelPromise promise 失败
    private final class DefaultServerUnsafe extends AbstractUnsafe {
        @Override
        public void connect(SocketAddress remoteAddress, SocketAddress localAddress, ChannelPromise promise) {
            safeSetFailure(promise, new UnsupportedOperationException());
        }
    }
}
```

12.10　LocalChannel 类

LocalChannel 代表了在内存中虚拟的本地通道，不涉及任何网络 IO 相关流程。仅在两个对端通道

中的读写缓冲区中进行内存操作，同时操作 LocalServerChannel 服务端与 LocalChannel 客户端均在同一个事件循环线程中，通过状态机转换服务端与客户端的处理流程，较为简单，所以笔者将其放在第一个通道实现类中进行讲解。读者在学习时可以类比一下 JDK 中的 ByteArrayInputStream 和 ByteArrayOutStream，只不过这里引入了一些状态校验。

12.10.1 核心变量与构造器

通过构造器和变量定义可知。

（1）State 枚举类定义了通道的四个状态。

（2）Queue inboundBuffer 定义了输入缓冲区队列。

（3）构造器中传递了 LocalChannel peer 对象作为连接的对端通道。

```java
public class LocalChannel extends AbstractChannel {
    private enum State { OPEN, BOUND, CONNECTED, CLOSED } // 通道状态枚举
    private static final AtomicReferenceFieldUpdater<LocalChannel, Future>
FINISH_READ_FUTURE_UPDATER;                                // 更新 finishReadFuture 对象原子更新对象
    private static final ChannelMetadata METADATA = new ChannelMetadata(false);   // 通道元数据，传入参
数为 false，表明当前没有 Disconnect 方法可以断开连接
    private static final int MAX_READER_STACK_DEPTH = 8;    // 最大读数据调用栈深度

    // 不同调用方法产生异常对象
    private static final ClosedChannelException DO_WRITE_CLOSED_CHANNEL_EXCEPTION =
ThrowableUtil.unknownStackTrace(
        new ClosedChannelException(), LocalChannel.class, "doWrite(...)");
    private static final ClosedChannelException DO_CLOSE_CLOSED_CHANNEL_EXCEPTION =
ThrowableUtil.unknownStackTrace(
        new ClosedChannelException(), LocalChannel.class, "doClose()");

    private final ChannelConfig config = new DefaultChannelConfig(this);        // 通道配置对象
    private final Queue<Object> inboundBuffer = PlatformDependent.newSpscQueue(); // 输入缓冲区队列
    private final Runnable readTask = new Runnable() {
        @Override
        public void run() {
            ChannelPipeline pipeline = pipeline();
            for (;;) {
                Object m = inboundBuffer.poll();
                if (m == null) {
                    break;
                }
                pipeline.fireChannelRead(m);
            }
            pipeline.fireChannelReadComplete();
        }
    };                                                      // 读取通道数据任务包装对象
    private final Runnable shutdownHook = new Runnable() {
        @Override
        public void run() {
            unsafe().close(unsafe().voidPromise());
        }
```

```java
    };                                                  // 关闭通道包装对象

    private volatile State state;                       // 通道状态
    private volatile LocalChannel peer;                 // 连接当前通道的对端通道
    private volatile LocalAddress localAddress;         // 本地地址
    private volatile LocalAddress remoteAddress;        // 连接对端地址
    private volatile ChannelPromise connectPromise;     // 完成连接时回调 promise
    private volatile boolean readInProgress;            // 当前是否读取通道数据
    private volatile boolean registerInProgress;        // 当前是否注册通道
    private volatile boolean writeInProgress;           // 当前是否向通道写入数据
    private volatile Future<?> finishReadFuture;        // 当完成通道数据读取时回调

    static {                                            // 初始化 finishReadFuture 原子更新对象
        AtomicReferenceFieldUpdater<LocalChannel, Future> finishReadFutureUpdater =
            PlatformDependent.newAtomicReferenceFieldUpdater(LocalChannel.class, "finishReadFuture");
        if (finishReadFutureUpdater == null) {
            finishReadFutureUpdater =
                AtomicReferenceFieldUpdater.newUpdater(LocalChannel.class,Future.class,"finishReadFuture");
        }
        FINISH_READ_FUTURE_UPDATER = finishReadFutureUpdater;
    }

    // 构造器初始化父通道为 null
    public LocalChannel() {
        super(null);
    }

    // 构造器初始化父通道为 parent，同时设置需要连接的本地通道 peer
    LocalChannel(LocalServerChannel parent, LocalChannel peer) {
        super(parent);
        this.peer = peer;
        localAddress = parent.localAddress();           // 获取父通道的本地地址
        remoteAddress = peer.localAddress();            // 获取连接的 peer 通道的远程地址
    }
}
```

12.10.2 doRegister 核心方法

doRegister 方法实现了 AbstractChannel 的模板方法，用于处理实际通道注册操作。流程如下。

（1）如果父通道和连接的对端通道为空，则向事件循环中添加 shutdownHook 任务关闭通道。

（2）如果父通道和连接的对端通道不为空，则修改状态为连接状态，在对端通道所属事件循环组中添加任务，用于触发 pipeline 中通道处理器的 ChannelActive 回调方法，表明通道已经激活。

```java
protected void doRegister() throws Exception {
    if (peer != null && parent() != null) {             // 只有父通道和连接的对端通道不为空才进行注册
        final LocalChannel peer = this.peer;
        registerInProgress = true;                      // 标识当前正处于注册流程中
        state = State.CONNECTED;                        // 修改状态为连接状态
        peer.remoteAddress = parent() == null ? null : parent().localAddress();
        peer.state = State.CONNECTED;
```

```
            peer.eventLoop().execute(new Runnable() {        // 在对端通道所属事件循环组中添加任务，用于触发
pipeline 中通道处理器的 ChannelActive 回调方法，表明通道已经激活
                @Override
                public void run() {
                    registerInProgress = false;
                    ChannelPromise promise = peer.connectPromise;
                    if (promise != null && promise.trySuccess()) {        // 设置对端的 promise 为完成状态
                        peer.pipeline().fireChannelActive();
                    }
                }
            });
        }
        ((SingleThreadEventExecutor) eventLoop()).addShutdownHook(shutdownHook);
}
```

12.10.3 doBind 核心方法

doBind 方法实现了 AbstractChannel 的模板方法，用于处理实际通道绑定地址操作。流程如下。
（1）如果之前已经绑定过地址，不允许重复绑定，则抛出异常。
（2）如果需要绑定的地址不是 LocalAddress 本地地址实例，则抛出异常。
（3）如果绑定地址为通配符地址，则将 channel 包装到 LocalAddress 对象中。
（4）将地址与通道放入映射集中。
（5）修改状态为绑定状态。

```
protected void doBind(SocketAddress localAddress) throws Exception {
    this.localAddress = LocalChannelRegistry.register(this, this.localAddress,
                                                      localAddress);        // 处理实际注册操作，同时返回本地地址
    state = State.BOUND;                                                    // 修改状态为绑定状态
}

final class LocalChannelRegistry {

    private static final ConcurrentMap<LocalAddress, Channel> boundChannels =
PlatformDependent.newConcurrentHashMap();                                   // 本地地址与通道的映射集

    static LocalAddress register(Channel channel, LocalAddress oldLocalAddress, SocketAddress localAddress) {
        if (oldLocalAddress != null) {                                      // 之前已经绑定过地址，不允许重复绑定
            throw new ChannelException("already bound");
        }
        if (!(localAddress instanceof LocalAddress)) {                      // 需要绑定的地址不是 LocalAddress 本地地址实
例，则抛出异常
            throw new ChannelException("unsupported address type: " +
StringUtil.simpleClassName(localAddress));
        }
        LocalAddress addr = (LocalAddress) localAddress;
        if (LocalAddress.ANY.equals(addr)) {    // 地址为通配符地址，将 channel 包装到 LocalAddress 对象中
            addr = new LocalAddress(channel);
        }
        Channel boundChannel = boundChannels.putIfAbsent(addr, channel);    // 将地址与通道放入映射集中
        if (boundChannel != null) {
```

```
            throw new ChannelException("address already in use by: " + boundChannel);
        }
        return addr;
    }

    static Channel get(SocketAddress localAddress) {
        return boundChannels.get(localAddress);
    }

    // 移除时，直接删除 map 映射即可
    static void unregister(LocalAddress localAddress) {
        boundChannels.remove(localAddress);
    }

    private LocalChannelRegistry() {
    }
}
```

12.10.4　doDisconnect 核心方法

doDisconnect 方法实现了 AbstractChannel 的模板方法，用于处理实际通道绑定断开连接操作。由于本地通道不支持断开连接，所以这里直接使用关闭操作代替断开连接。流程如下。

（1）如果当前通道状态不属于关闭状态，则通道可能正处于读写状态，需要进一步执行清理操作。首先，将其从 LocalChannelRegistry 的映射集中移除，接着修改状态为关闭，回调 connectPromise，设置状态为失败。最后，如果当前通道正处于写流程中，且对端对象不为空，则处理任何挂起的读操作。

（2）如果连接的对端通道不为空且处于活动状态，则进一步判断当前线程是否为对端事件循环所属线程，同时当前没有处于注册流程中，则执行对端通道关闭操作。否则，只能将关闭操作放入对端的事件循环线程中执行。

```
protected void doDisconnect() throws Exception {
    doClose();
}

@Override
protected void doClose() throws Exception {
    final LocalChannel peer = this.peer;
    if (state != State.CLOSED) {              // 状态不属于关闭状态
        if (localAddress != null) {           // 之前已经注册过，则将其从映射集中移除
            if (parent() == null) {
                LocalChannelRegistry.unregister(localAddress);
            }
            localAddress = null;
        }
        state = State.CLOSED;                 // 修改状态为关闭状态
        ChannelPromise promise = connectPromise;
        if (promise != null) {                // 回调 connectPromise，设置为失败状态，因为当前通道已经关闭
            promise.tryFailure(DO_CLOSE_CLOSED_CHANNEL_EXCEPTION);
            connectPromise = null;
        }
        if (writeInProgress && peer != null) {  // 当前通道正处于写流程中且对端对象不为空，则处理任何挂起
```

的读操作
```java
                finishPeerRead(peer);
            }
        }
        if (peer != null && peer.isActive()) {        // 连接的对端通道不为空，且处于活动状态
            if (peer.eventLoop().inEventLoop() && !registerInProgress) {    // 如果当前线程为对端事件循环所属
线程，同时当前没有处于注册流程中，则执行对端通道关闭操作
                doPeerClose(peer, peer.writeInProgress);
            } else {                                  // 否则只能将关闭任务提交给对端通道所属事件循环执行
                final boolean peerWriteInProgress = peer.writeInProgress;
                try {
                    peer.eventLoop().execute(new Runnable() {
                        @Override
                        public void run() {
                            doPeerClose(peer, peerWriteInProgress);
                        }
                    });
                } catch (RuntimeException e) {       // 发生任何异常，则释放输入缓冲队列
                    releaseInboundBuffers();
                    throw e;
                }
            }
            this.peer = null;
        }
    }

// 处理对端连接通道的关闭操作
private void doPeerClose(LocalChannel peer, boolean peerWriteInProgress) {
    if (peerWriteInProgress) {        // 如果对端通道正处于写数据流程中，则完成对端可能存在的读数据操作
        finishPeerRead0(this);
    }
    peer.unsafe().close(peer.unsafe().voidPromise());    // 最后，调用对端的 unsafe 对象关闭对端通道，使用
voidPromise 对象表示不需要进行 promise 通知
}


// 完成对端可能存在的读操作
private void finishPeerRead0(LocalChannel peer) {
    Future<?> peerFinishReadFuture = peer.finishReadFuture;
    if (peerFinishReadFuture != null) {        // 对端通道的读任务存在
        if (!peerFinishReadFuture.isDone()) {  // 读任务没有完成，将完成读操作的任务放到对端通道的事件循环中执行
            runFinishPeerReadTask(peer);
            return;
        } else {
            // 否则将 peerFinishReadFuture 重置为 null，在当前线程中完成读取
            FINISH_READ_FUTURE_UPDATER.compareAndSet(peer, peerFinishReadFuture, null);
        }
    }
    ChannelPipeline peerPipeline = peer.pipeline();
    if (peer.readInProgress) {            // 对端通道处于读数据流程中
        peer.readInProgress = false;    // 将对端通道状态修改为 false，因为此时需要在当前线程中手动完成读取
        for (;;) {        // 从对端的接收队列中获取到数据，然后触发对端通道处理器的 ChannelRead 回调方法，完成
```

```
对读缓冲区的数据处理
            Object received = peer.inboundBuffer.poll();
            if (received == null) {
                break;
            }
            peerPipeline.fireChannelRead(received);
        }
        peerPipeline.fireChannelReadComplete();        // 所有数据均读取完成后，回调通道处理器的
ChannelReadComplete 回调方法，表示所有数据都读取完毕
    }
}

// 对端通道的读任务存在，且未执行完毕
private void runFinishPeerReadTask(final LocalChannel peer) {
    final Runnable finishPeerReadTask = new Runnable() {
        @Override
        public void run() {
            finishPeerRead0(peer);
        }
    }; // 封装完成对端读取任务
    try {
        if (peer.writeInProgress) {                    // 对端正在写数据，提交到事件循环队列中，并保存到
finishReadFuture 变量中，等待写操作完成后回调
            peer.finishReadFuture = peer.eventLoop().submit(finishPeerReadTask);
        } else {                                       // 否则直接放入事件循环队列即可，等待该 Runnable 执行
            peer.eventLoop().execute(finishPeerReadTask);
        }
    } catch (RuntimeException e) {
        peer.releaseInboundBuffers();
        throw e;
    }
}
```

12.10.5　doBeginRead 核心方法

doBeginRead 方法实现了 AbstractChannel 的模板方法，用于处理实际通道读取数据操作。流程如下。

（1）如果当前通道已经处于读流程中，则直接返回。

（2）如果接收缓冲区为空，则设置 readInProgress 为 true 并返回，表明当前处于读数据中，但是缓冲区中没有数据。

（3）判断当前读取数据的调用栈帧是否在最大栈深限制以下，循环读取并触发通道处理器回调方法。

（4）如果当前读取数据的调用栈帧达到最大栈深限制，则只能向事件循环队列中添加读任务，在下一次执行时继续处理读操作。

```
protected void doBeginRead() throws Exception {
    if (readInProgress) {                              // 如果已经处于读流程中，则直接返回
        return;
    }
    ChannelPipeline pipeline = pipeline();
    Queue<Object> inboundBuffer = this.inboundBuffer;
    if (inboundBuffer.isEmpty()) {    // 接收缓冲区为空，则设置 readInProgress 为 true，表明当前处于读数据中，
```

但是缓冲区中没有数据
        readInProgress = **true**;
        return;
    }
    final InternalThreadLocalMap threadLocals = InternalThreadLocalMap.get();
    final Integer stackDepth = threadLocals.localChannelReaderStackDepth();
    if (stackDepth < MAX_READER_STACK_DEPTH) {   // 当前读取数据的调用栈帧是否在最大栈深限制以下
（注意思考这里为什么要这样设计？如果在ChannelRead和ChannelReadComplete中继续触发读操作会发生什么？）
        threadLocals.setLocalChannelReaderStackDepth(stackDepth + 1); // 调用次数加 1
        try {
            for (;;) {   // 循环读取并触发通道处理器回调方法
                Object received = inboundBuffer.poll();
                if (received == **null**) {
                    break;
                }
                pipeline.fireChannelRead(received);
            }
            pipeline.fireChannelReadComplete();
        } **finally** {   // 读取完成后还原调用栈的栈深
            threadLocals.setLocalChannelReaderStackDepth(stackDepth);
        }
    } **else** { // 达到最大调用栈深，只能向事件循环队列中添加读任务，在下一次执行时继续处理读操作
        try {
            eventLoop().execute(readTask);
        } **catch** (RuntimeException e) {
            releaseInboundBuffers();   // 发生异常，释放读缓冲区队列
            throw e;
        }
    }
}
```

## 12.10.6　doWrite 核心方法

doWrite 方法实现了 AbstractChannel 的模板方法，用于处理实际通道写入数据操作。流程如下。

（1）判断当前通道状态是否为连接状态。
（2）设置 writeInProgress 为 true，标识当前正处于写数据处理流程中。
（3）通过循环将缓冲区的数据写入对端通道的输入缓冲区中。
（4）写入完成后，尝试在当前线程中完成对端数据的读取操作。

```
protected void doWrite(ChannelOutboundBuffer in) **throws** Exception {
    switch (state) { // 当前状态是否为连接状态
        case OPEN:
        case BOUND:
            throw new NotYetConnectedException();
        case CLOSED:
            throw DO_WRITE_CLOSED_CHANNEL_EXCEPTION;
        case CONNECTED:
            break;
    }
    final LocalChannel peer = **this**.peer;
    writeInProgress = **true**;   // 标识当前正处于写数据处理流程中
```
```

```
    try {
        for (;;) {                                  // 循环将写缓冲区的数据写入对端通道的输入缓冲区中
            Object msg = in.current();
            if (msg == null) {
                break;
            }
            try {
                if (peer.state == State.CONNECTED) {    // 当前通道已经连接，写入对端缓冲区即可
                    peer.inboundBuffer.add(ReferenceCountUtil.retain(msg));
                    in.remove();
                } else {                                // 否则在移除写缓冲区数据时，携带异常对象
                    in.remove(DO_WRITE_CLOSED_CHANNEL_EXCEPTION);
                }
            } catch (Throwable cause) {
                in.remove(cause);
            }
        }
    } finally {
        writeInProgress = false;
    }
    finishPeerRead(peer);                          // 写入完成，尝试在当前线程中完成对端数据的读取操作
}
```

12.10.7　LocalUnsafe 核心内部类

LocalUnsafe 类继承自 AbstractUnsafe，重写了其中的 connect 方法。流程如下。

（1）如果连接操作已经取消，则返回。
（2）如果当前通道已经连接，则不允许重复连接。
（3）如果连接 promise 已经存在，表明当前连接任务处于执行状态，则抛出异常。
（4）将 promise 设置为 connectPromise 变量，标识当前正在执行连接。
（5）如果当前通道未绑定地址，则生成 LocalAddress 对象（不涉及任何相关协议，仅一个字符标识而已）。
（6）调用 doBind 方法，完成地址绑定。其实就是把通道和 LocalAddress 建立映射。
（7）从 LocalChannelRegistry 中获取到对端的通道对象，并判断对端通道类型只能为 LocalServerChannel，因为必须要满足 C/S 架构。
（8）调用 serverChannel.serve(LocalChannel.this)，生成对端通道对象与当前通道的连接。

上述流程当中，虽然 LocalChannel 是虚拟的内存之间的连接，但是它们也应该遵守标准的 C/S 架构，即客户端和服务端，客户端请求与服务端连接。源码如下。

```
private class LocalUnsafe extends AbstractUnsafe {

    @Override
    public void connect(final SocketAddress remoteAddress,
                        SocketAddress localAddress, final ChannelPromise promise) {
        if (!promise.setUncancellable() || !ensureOpen(promise)) {   // 连接操作已经被取消
            return;
        }
```

```
        if (state == State.CONNECTED) {          // 已经连接，不允许重复连接
            Exception cause = new AlreadyConnectedException();
            safeSetFailure(promise, cause);
            pipeline().fireExceptionCaught(cause);
            return;
        }
        if (connectPromise != null) {             // 连接 promise 已经存在，表明当前连接任务处于执行状态
            throw new ConnectionPendingException();
        }
        connectPromise = promise;                 // 将 promise 设置为连接 promise，标识当前正在执行连接
        if (state != State.BOUND) {               // 如果未绑定地址，则生成一个 LocalAddress 对象
            if (localAddress == null) {
                localAddress = new LocalAddress(LocalChannel.this);
            }
        }
        if (localAddress != null) {
            try {
                doBind(localAddress);             // 完成地址绑定，其实就是把通道和 LocalAddress 建立映射
            } catch (Throwable t) {
                safeSetFailure(promise, t);
                close(voidPromise());
                return;
            }
        }
        Channel boundChannel = LocalChannelRegistry.get(remoteAddress); // 获取对端的通道对象
        if (!(boundChannel instanceof LocalServerChannel)) { // 类型只能为 LocalServerChannel，因为必须满足 C/S 架构
            Exception cause = new ConnectException("connection refused: " + remoteAddress);
            safeSetFailure(promise, cause);
            close(voidPromise());
            return;
        }
        LocalServerChannel serverChannel = (LocalServerChannel) boundChannel;
        peer = serverChannel.serve(LocalChannel.this);
    }
}
```

12.11　LocalServerChannel 类

LocalChannel 实现了客户端的操作，而需要能够让客户端连接的服务端存在，这样客户端就可以连接到服务端，而服务端创建 LocalChannel 代表客户端，然后双端进行通信。LocalServerChannel 类便是实现了内存本地通道的实现，读者可以通过 LocalChannel 和 LocalServerChannel 的源码看到，这里其实就是个状态机，根据 inboundBuffer 状态转换执行。源码如下。

```
public class LocalServerChannel extends AbstractServerChannel {

    private final ChannelConfig config = new DefaultChannelConfig(this);   // 本地服务端通道配置项

    private final Queue<Object> inboundBuffer = new ArrayDeque<Object>(); // 服务端输入缓冲区（LocalChannel
```

对象）

```java
    private final Runnable shutdownHook = new Runnable() {
        @Override
        public void run() {
            unsafe().close(unsafe().voidPromise());
        }
    };                                                  // 封装关闭服务端通道钩子操作

    private volatile int state;                         // 服务端通道状态：0 为打开，1 为激活，2 为关闭
    private volatile LocalAddress localAddress;         // 本地地址
    private volatile boolean acceptInProgress;          // 是否处于接收客户端连接中

    // 只支持 SingleThreadEventLoop 单线程事件循环对象
    protected boolean isCompatible(EventLoop loop) {
        return loop instanceof SingleThreadEventLoop;
    }

    // 实现父类注册模板方法。直接向所属事件循环添加关闭事件循环时的钩子函数
    protected void doRegister() throws Exception {
        ((SingleThreadEventExecutor) eventLoop()).addShutdownHook(shutdownHook);
    }

    // 实现父类绑定方法。向 LocalChannelRegistry 的 map 集合中放入本地地址与当前通道的映射，并将状态修改为激活状态
    protected void doBind(SocketAddress localAddress) throws Exception {
        this.localAddress = LocalChannelRegistry.register(this, this.localAddress, localAddress);
        state = 1;
    }

    // 实现父类关闭模板方法。将状态修改为关闭，同时将 LocalChannelRegistry 映射集合删除
    protected void doClose() throws Exception {
        if (state <= 1) {
            if (localAddress != null) {
                LocalChannelRegistry.unregister(localAddress);
                localAddress = null;
            }
            state = 2;
        }
    }

    // 实现父类取消注册模板方法。去掉注册到所属事件循环的关闭钩子函数
    protected void doDeregister() throws Exception {
        ((SingleThreadEventExecutor) eventLoop()).removeShutdownHook(shutdownHook);
    }

    // 实现父类读取输入数据模板方法
    protected void doBeginRead() throws Exception {
        if (acceptInProgress) {                         // 当前已经处于读状态
            return;
        }
        Queue<Object> inboundBuffer = this.inboundBuffer;
```

```
        if (inboundBuffer.isEmpty()) {          // 输入缓冲区为空，由于当前还没有通道放入缓冲区（因为是内
存中虚拟的通道，在 LocalChannel 中直接将 LocalChannel 对象放入该通道），设置 acceptInProgress 并返回
            acceptInProgress = true;
            return;
        }
        ChannelPipeline pipeline = pipeline();
        for (;;) {                               // 否则循环处理所有 LocalChannel 对象
            Object m = inboundBuffer.poll();
            if (m == null) {
                break;
            }
            pipeline.fireChannelRead(m);         // 触发通道处理器的 ChannelRead 回调方法，处理接收连接
        }
        pipeline.fireChannelReadComplete();   // 处理完所有连接后，触发通道处理器的 ChannelReadComplete
回调方法，表示当前已经处理完所有连接数据
    }

    // 服务于客户端连接方法。将传递过来的 LocalChannel 包装为 LocalChannel，并将该通道放入 inboundBuffer 中
    LocalChannel serve(final LocalChannel peer) {
        final LocalChannel child = new LocalChannel(this, peer);
        if (eventLoop().inEventLoop()) {
            serve0(child);
        } else {
            eventLoop().execute(new Runnable() {
                @Override
                public void run() {
                    serve0(child);
                }
            });
        }
        return child;
    }

    // 将客户端连接放入 inboundBuffer 中，并根据 acceptInProgress 状态，选择是否立即处理该通道。如果之
前已经调用过 doBeginRead，但是当时没有数据，则会设置 acceptInProgress，这时会在通道放入后立即处理该
通道
    private void serve0(final LocalChannel child) {
        inboundBuffer.add(child);
        if (acceptInProgress) {                 // 处理缓冲区中的通道对象
            acceptInProgress = false;
            ChannelPipeline pipeline = pipeline();
            for (;;) {
                Object m = inboundBuffer.poll();
                if (m == null) {
                    break;
                }
                pipeline.fireChannelRead(m);
            }
            pipeline.fireChannelReadComplete();
        }
    }
}
```

12.12 AbstractOioChannel 类

AbstractOioChannel 类也相对简单，用于支持 Old-Blocking-IO（也称之为 OIO，即阻塞式 IO。一个事件循环一个通道，第 1 篇曾介绍过）通道的操作。源码较为简单，笔者这里一并讲解。由源码可知。

（1）定义了 OIO 的常量 SO_TIMEOUT 超时时间为 1s。

（2）DefaultOioUnsafe 继承自 AbstractUnsafe，自定义了 connect 连接方法，该方法中进行了状态判断与通道处理器的回调，但是实际连接方法将由子类完成。

（3）实际通道读取方法，将由子类实现 doRead 方法完成读取。

```java
public abstract class AbstractOioChannel extends AbstractChannel {
    protected static final int SO_TIMEOUT = 1000;        // 超时时间
    boolean readPending;                                  // 读操作是否执行

    private final Runnable readTask = new Runnable() {
        @Override
        public void run() {
            doRead();
        }
    };                                                    // 读任务包装对象

    private final Runnable clearReadPendingRunnable = new Runnable() {
        @Override
        public void run() {
            readPending = false;
        }
    };                                                    // 修改 readPending 变量包装对象

    // 构造器：初始化父通道对象
    protected AbstractOioChannel(Channel parent) {
        super(parent);
    }

    // 获取执行底层 IO 操作的 Unsafe 实现对象
    protected AbstractUnsafe newUnsafe() {
        return new DefaultOioUnsafe();
    }

    // 实现 AbstractUnsafe，自定义 connect 连接方法
    private final class DefaultOioUnsafe extends AbstractUnsafe {
        @Override
        public void connect(
                final SocketAddress remoteAddress,
                final SocketAddress localAddress, final ChannelPromise promise) {
            if (!promise.setUncancellable() || !ensureOpen(promise)) {    // 连接操作已经取消,且通道未处于关闭状态（设置了 CLOSED 标志位）
                return;
            }
```

```java
            try {
                boolean wasActive = isActive();
                doConnect(remoteAddress, localAddress);       // 子类完成实际连接操作
                safeSetSuccess(promise);                      // 把 promise 设置为成功状态
                if (!wasActive && isActive()) { // 如果在子类执行过连接后,激活了通道,则触发通道处理器的
ChannelActive 事件
                    pipeline().fireChannelActive();
                }
            } catch (Throwable t) {
                safeSetFailure(promise, annotateConnectException(t, remoteAddress));
                closeIfClosed();
            }
        }
    }

    // 与 OIO 通道绑定的事件循环组,只能是 ThreadPerChannelEventLoop 对象
    protected boolean isCompatible(EventLoop loop) {
        return loop instanceof ThreadPerChannelEventLoop;
    }

    // 由子类实现实际连接操作
    protected abstract void doConnect(
        SocketAddress remoteAddress, SocketAddress localAddress) throws Exception;

    // 从通道中读取数据
    protected void doBeginRead() throws Exception {
        if (readPending) {                                    // 读任务仍在等待处理中
            return;
        }
        readPending = true;
        eventLoop().execute(readTask);                        // 将读任务放入所属事件循环组中执行
    }

    // 由子类实现实际读取操作
    protected abstract void doRead();

    // 清除 readPending 标志位
    protected final void clearReadPending() {
        if (isRegistered()) {                                 // 通道已经注册
            EventLoop eventLoop = eventLoop();
            if (eventLoop.inEventLoop()) {                    // 当前线程为事件循环线程,直接修改
                readPending = false;
            } else {                        // 否则将其包装为 Runnable 任务,交给所属事件循环线程处理
                eventLoop.execute(clearReadPendingRunnable);
            }
        } else { // 在通道初始化期间,通道还没有注册到事件循环 EventLoop,则直接修改 readPending
            readPending = false;
        }
    }
}
```

12.13 AbstractOioByteChannel 类

AbstractOioByteChannel 抽象类定义了 OIO 的 Socket 操作，即客户端操作，在 AbstractOioChannel 类中，子类需要实现的方法为 doConnect 和 doRead。通过 AbstractOioByteChannel 的源码可知。

（1）在 closeOnRead 方法中，ALLOW_HALF_CLOSURE 半关闭状态的定义为允许通道仅关闭输入端，而可以保持当前通道的单向关联。读操作发生异常时回调。

（2）实现了 doRead 读数据方法和 doWrite 写数据方法。doRead 方法循环缓冲区 socket 中的数据，将其读入读缓冲区中，然后回调通道处理器回调方法完成数据处理。doWrite 方法中循环处理 ChannelOutboundBuffer 输出缓冲区中的数据，将数据调用子类实现的 doWriteBytes 方法写入底层 Socket 中。

（3）handleReadException 方法用于处理在执行读操作中发生异常时的处理，该方法会释放缓冲区，并触发相应的通道处理器回调方法。

```
public abstract class AbstractOioByteChannel extends AbstractOioChannel {
    private static final ChannelMetadata METADATA = new ChannelMetadata(false); // 定义元数据常量，传入参数为 false，表明没有实现 disconnect 方法，然后又可以再次允许调用 connect 方法，即断开连接后再次重连，这只能在 UDP 上支持
    private static final String EXPECTED_TYPES =
        " (expected: " + StringUtil.simpleClassName(ByteBuf.class) + ", " +
        StringUtil.simpleClassName(FileRegion.class) + ')';         // 异常类型字符串

    // 使用父通道初始化父类
    protected AbstractOioByteChannel(Channel parent) {
        super(parent);
    }

    // 判断当前通道的输入端是否已关闭
    protected abstract boolean isInputShutdown();

    // 关闭当前通道的输入端
    protected abstract ChannelFuture shutdownInput();

    // 读操作发生异常时，关闭当前通道
    private void closeOnRead(ChannelPipeline pipeline) {
        if (isOpen()) {    // 通道处于打开状态，从配置项中判断是否支持 ALLOW_HALF_CLOSURE，即半关闭状态，如果支持，则仅关闭输入端，并触发通道处理器的 UserEventTriggered 方法
            if (Boolean.TRUE.equals(config().getOption(ChannelOption.ALLOW_HALF_CLOSURE))) {
                shutdownInput();
                pipeline.fireUserEventTriggered(ChannelInputShutdownEvent.INSTANCE);
            } else {                        // 不允许半关闭状态，直接关闭整个通道
                unsafe().close(unsafe().voidPromise());
            }
        }
    }

    // 处理执行读操作时发生的异常
```

```java
    private void handleReadException(ChannelPipeline pipeline, ByteBuf byteBuf, Throwable cause, boolean close,RecvByteBufAllocator.Handle allocHandle) {
        if (byteBuf != null) {    // 存在输入缓冲区且有数据需要处理，触发通道处理器的ChannelRead回调方法，并释放缓冲区
            if (byteBuf.isReadable()) {
                readPending = false;
                pipeline.fireChannelRead(byteBuf);
            } else {
                byteBuf.release();
            }
        }
        allocHandle.readComplete();              // 通知接收缓冲区当前已经读取完数据
        pipeline.fireChannelReadComplete();      // 触发通道处理器的 ChannelReadComplete 回调方法
        pipeline.fireExceptionCaught(cause);     // 触发通道处理器的 ExceptionCaught 回调方法
        if (close || cause instanceof IOException) { // 如果需要关闭通道或异常类型为 IO 异常，则关闭通道
            closeOnRead(pipeline);
        }
    }

    // 处理读数据操作（循环从缓冲区中读取数据，然后回调通道处理器的回调方法完成数据处理）
    protected void doRead() {
        final ChannelConfig config = config();
        if (isInputShutdown() || !readPending) {  // 输入通道已关闭或正在处理当前读操作，直接返回
            return;
        }
        readPending = false;
        final ChannelPipeline pipeline = pipeline();
        final ByteBufAllocator allocator = config.getAllocator();          // 分配读缓冲区对象的内存分配器
        final RecvByteBufAllocator.Handle allocHandle = unsafe().recvBufAllocHandle(); // 读缓冲区分配器
        allocHandle.reset(config);               // 根据配置对象重置内部属性
        ByteBuf byteBuf = null;
        boolean close = false;
        boolean readData = false;
        try {
            byteBuf = allocHandle.allocate(allocator);              // 分配读缓冲区
            do {
                allocHandle.lastBytesRead(doReadBytes(byteBuf));    // 子类完成实际读取操作,并将读到的字节数记录在 RecvByteBufAllocator.Handle 中
                if (allocHandle.lastBytesRead() <= 0) {
                    if (!byteBuf.isReadable()) {    // 没有从 socket 中读出数据，释放缓冲区
                        byteBuf.release();
                        byteBuf = null;
                        close = allocHandle.lastBytesRead() < 0; // 读到通道传输数据的最后一个字节，在读取完成后关闭通道
                    }
                    break;
                } else {
                    readData = true;
                }
                final int available = available();
                if (available <= 0) {                // 通道可读取字节数为 0
                    break;
                }
```

```java
                if (!byteBuf.isWritable()) {                              // 缓冲区不可写
                    final int capacity = byteBuf.capacity();
                    final int maxCapacity = byteBuf.maxCapacity();
                    if (capacity == maxCapacity) { // 缓冲区达到最大写入容量，触发通道的 ChannelRead 方法处理读缓冲区中的数据
                        allocHandle.incMessagesRead(1);          // 增加触发通道的 ChannelRead 方法计数
                        readPending = false;
                        pipeline.fireChannelRead(byteBuf);
                        byteBuf = allocHandle.allocate(allocator); // 重新分配新的读缓冲区，存放下一次读取数据
                    } else {                                // 没有达到缓冲区支持的最大容量，对缓冲区扩容
                        final int writerIndex = byteBuf.writerIndex();
                        if (writerIndex + available > maxCapacity) {
                            byteBuf.capacity(maxCapacity);
                        } else {
                            byteBuf.ensureWritable(available);
                        }
                    }
                }
            } while (allocHandle.continueReading()); // 接收缓冲区处理器判定可以继续读取数据时，继续读取
            if (byteBuf != null) { // 读缓冲区存在（注意，上述循环不一定会调用 pipeline.fireChannelRead(byteBuf)，因为可能读缓冲区没有写满数据，所以这里需要判断并触发通道读取事件）
                if (byteBuf.isReadable()) {               // 通道可读，触发通道处理器的 ChannelRead 方法处理读缓冲区中的数据
                    readPending = false;
                    pipeline.fireChannelRead(byteBuf);
                } else {                                  // 否则释放读缓冲区
                    byteBuf.release();
                }
                byteBuf = null;
            }
            if (readData) {            // 从 socket 中读取到了数据，读缓冲区中的数据已经读完，触发通道处理器的 ChannelReadComplete 方法
                allocHandle.readComplete();
                pipeline.fireChannelReadComplete();
            }
            if (close) {                              // 如果在通道中读取到了最后一个字节，则关闭该通道
                closeOnRead(pipeline);
            }
        } catch (Throwable t) {
            handleReadException(pipeline, byteBuf, t, close, allocHandle);
        } finally {
            if (readPending || config.isAutoRead() || !readData && isActive()) {  // 没有读到通道数据，则可能意味着存在 SocketTimeout，并且没有实际读取数据，应该再次执行 read()
                read();
            }
        }
    }

    // 处理写数据操作
    protected void doWrite(ChannelOutboundBuffer in) throws Exception {
        for (;;) {                                 // 循环处理输出缓冲区中的数据
```

```java
            Object msg = in.current();
            if (msg == null) {                              // 输出缓冲区中没有数据可写，直接返回
                break;
            }
            if (msg instanceof ByteBuf) {                    // 当前消息类型为缓冲区类型，将数据写入底层 Socket 中
                ByteBuf buf = (ByteBuf) msg;
                int readableBytes = buf.readableBytes();
                while (readableBytes > 0) {
                    doWriteBytes(buf);
                    int newReadableBytes = buf.readableBytes();
                    in.progress(readableBytes - newReadableBytes); // 回调 ChannelProgressivePromise，
通知当前处理进度
                    readableBytes = newReadableBytes;
                }
                in.remove();                                 // 从输出缓冲区中移除当前消息
            } else if (msg instanceof FileRegion) {          // 当前消息类型为零复制的文件域对象，调用
doWriteFileRegion 子类方法，完成写入，并通知 ChannelProgressivePromise 当前处理进度
                FileRegion region = (FileRegion) msg;
                long transferred = region.transferred();
                doWriteFileRegion(region);
                in.progress(region.transferred() - transferred);
                in.remove();
            } else {
                in.remove(new UnsupportedOperationException(
                    "unsupported message type: " + StringUtil.simpleClassName(msg)));
            }
        }
    }

    // 执行消息过滤操作，这里其实就是对写入的 msg 对象进行参数判断，只能为 ByteBuf 或 FileRegion 类型
    protected final Object filterOutboundMessage(Object msg) throws Exception {
        if (msg instanceof ByteBuf || msg instanceof FileRegion) {
            return msg;
        }

        throw new UnsupportedOperationException(
            "unsupported message type: " + StringUtil.simpleClassName(msg) + EXPECTED_TYPES);
    }

    // 返回底层 Socket 可读取的字节数
    protected abstract int available();

    // 从底层 Socket 中读取数据放入 ByteBuf 缓冲区
    protected abstract int doReadBytes(ByteBuf buf) throws Exception;

    // 将数据写入底层 Socket
    protected abstract void doWriteBytes(ByteBuf buf) throws Exception;

    // 将保存在 FileRegion 对象中的数据写入底层 Socket。FileRegion 对象用于包装零复制的文件数据传输，即
调用 FileChannel.transferTo 方法完成数据传输
    protected abstract void doWriteFileRegion(FileRegion region) throws Exception;
}
```

12.14 OioByteStreamChannel 类

AbstractOioByteChannel 类用于定义 Socket 的操作，但这个 Socket 并没有指明具体实现类型，从字面上来看，OioByteStreamChannel 就是定义基于字节流的通道，而字节流又是 TCP 协议的特性，即面向流。所以该类的作用就很明显了：定义 TCP 协议，并且使用 Java 的字节流对象操作通道数据，同时还是 OIO 阻塞式的操作。由源码可知。

（1）OutputStream CLOSED_OUT 表示已经关闭的输出流常量对象。

（2）InputStream CLOSED_IN 表示已经关闭的输入流常量对象。

（3）根据 activate(InputStream is, OutputStream os)方法可知，通道激活状态指的是将输入和输出流赋值。

（4）写入 FileRegion 代表的文件对象时，由于需要使用 JDK 的零复制，并使用 WritableByteChannel 实例，这时可以通过调用通道的 transferTo 方法完成文件数据到 socket 的传输。

```java
public abstract class OioByteStreamChannel extends AbstractOioByteChannel {
    private static final InputStream CLOSED_IN = new InputStream() {
        @Override
        public int read() {
            return -1;
        }
    }; // 表示已经关闭的输入流常量对象：读取数据永远返回-1

    private static final OutputStream CLOSED_OUT = new OutputStream() {
        @Override
        public void write(int b) throws IOException {
            throw new ClosedChannelException();
        }
    }; // 表示已经关闭的输出流常量对象，不允许写入数据

    private InputStream is;                        // 当前通道输入流对象
    private OutputStream os;                       // 当前通道输出流对象
    private WritableByteChannel outChannel;        // 将要写入数据的通道对象，用于支持零复制的 FileRegion 对象
    (JDK 的通道零复制输出技术，可以参考特别篇的原理)

    // 构造器：初始化父通道
    protected OioByteStreamChannel(Channel parent) {
        super(parent);
    }

    // 使用指定输入、输出流对象激活当前通道，使用 isActive 方法会返回 true 。这里其实就是判空，然后赋值即可
    protected final void activate(InputStream is, OutputStream os) {
        if (this.is != null) {
            throw new IllegalStateException("input was set already");
        }
        if (this.os != null) {
            throw new IllegalStateException("output was set already");
        }
```

```java
        if (is == null) {
            throw new NullPointerException("is");
        }
        if (os == null) {
            throw new NullPointerException("os");
        }
        this.is = is;
        this.os = os;
    }

    // 存在输入、输出流，且不为关闭常量对象，则当前通道处于激活状态
    public boolean isActive() {
        InputStream is = this.is;
        if (is == null || is == CLOSED_IN) {
            return false;
        }

        OutputStream os = this.os;
        return !(os == null || os == CLOSED_OUT);
    }

    // 通道可读字节数，直接读取输入流的方法
    protected int available() {
        try {
            return is.available();
        } catch (IOException ignored) {
            return 0;
        }
    }

    // 实现父类定义的模板方法。将输入流中的数据读入缓冲区
    protected int doReadBytes(ByteBuf buf) throws Exception {
        // 使用接收缓冲区分配器处理器猜测当前可以读取的字节数（后面会专门接触到该处理器的相关内容，这里了解即可）
        final RecvByteBufAllocator.Handle allocHandle = unsafe().recvBufAllocHandle();
        allocHandle.attemptedBytesRead(Math.max(1, Math.min(available(), buf.maxWritableBytes())));
        return buf.writeBytes(is, allocHandle.attemptedBytesRead());    // 将通道中的数据写入缓冲区
    }

    // 实现父类定义的模板方法。用于将缓冲区中的数据写入输出流
    protected void doWriteBytes(ByteBuf buf) throws Exception {
        OutputStream os = this.os;
        if (os == null) {
            throw new NotYetConnectedException();
        }
        buf.readBytes(os, buf.readableBytes());
    }

    // 实现父类定义的模板方法。将FileRegion中的数据写入输出流
    protected void doWriteFileRegion(FileRegion region) throws Exception {
        OutputStream os = this.os;
        if (os == null) {
            throw new NotYetConnectedException();
        }
```

```java
        if (outChannel == null) {              // 根据输出流创建可写的通道对象
            outChannel = Channels.newChannel(os);
        }
        long written = 0;                       // 指向文件开始写入的 position，初始从文件第一个 byte 开始写入
        for (;;) {
            long localWritten = region.transferTo(outChannel, written);  // 将 region 中代表文件的数据写入 outChannel 通道
            if (localWritten == -1) {           // 达到文件末尾
                checkEOF(region);
                return;
            }
            written += localWritten;            // 增加写入量
            if (written >= region.count()) {    // 达到 region 指定的写入大小，则直接返回
                return;
            }
        }
    }

    // 判断写入数量是否超过设置的 region 数量
    private static void checkEOF(FileRegion region) throws IOException {
        if (region.transferred() < region.count()) {
            throw new EOFException("Expected to be able to write " + region.count() + " bytes, " +
                    "but only wrote " + region.transferred());
        }
    }

    // 实现父类定义的模板方法，将 FileRegion 中的数据写入输出流
    protected void doClose() throws Exception {
        InputStream is = this.is;
        OutputStream os = this.os;
        this.is = CLOSED_IN;
        this.os = CLOSED_OUT;

        try {
            if (is != null) {
                is.close();
            }
        } finally {
            if (os != null) {
                os.close();
            }
        }
    }
}
```

12.15　OioSocketChannel 类

OioSocketChannel 类继承自 OioByteStreamChannel，指定通过 SocketChannel 完成通道的具体操作。由源码可知。

（1）DuplexChannel 接口继承自 Channel，提供了双端（Duplex）的数据操作，即可以单独关闭某一端（输入端或输出端）。

（2）OioSocketChannel 实现类使用原生的 JDK 的 socket 对象完成数据输入和输出。

（3）DuplexChannel 接口定义的关闭 socket 的输入和输出流方法，会调用原生 JDK 的对应方法完成关闭。

（4）执行连接时，OioSocketChannel 通道会判断是否绑定地址。如果没有绑定，则先进行绑定，然后调用 socket 的 connect 方法完成连接。连接完成后，获取 socket 的输入输出流激活通道对象。

在具体的算法实现中，一切又回到了 JDK 的网络编程基础。源码如下：

```java
public interface DuplexChannel extends Channel {
    // 输入端是否已经关闭
    boolean isInputShutdown();

    // 异步关闭输入端
    ChannelFuture shutdownInput();

    // 异步关闭输入端，完成后回调 promise
    ChannelFuture shutdownInput(ChannelPromise promise);

    // 输出端是否已经关闭
    boolean isOutputShutdown();

    // 异步关闭输出端
    ChannelFuture shutdownOutput();

    // 异步关闭输出端，完成后回调 promise
    ChannelFuture shutdownOutput(ChannelPromise promise);

    // 输入端和输出端是否已经被关闭
    boolean isShutdown();

    // 关闭当前通道的输入和输出端
    ChannelFuture shutdown();

    // 关闭当前通道的输入和输出端，完成后回调 promise
    ChannelFuture shutdown(ChannelPromise promise);
}
// 该接口表示 TCP/IP 四层协议的 Socket 的通道
public interface SocketChannel extends DuplexChannel {
}

public class OioSocketChannel extends OioByteStreamChannel implements SocketChannel {
    private final Socket socket;                    // JDK 底层的 Socket 客户端套接字对象
    private final OioSocketChannelConfig config;    // 通道配置对象，配置 Socket 对象相关参数

    // 构造器：初始化 JDK 底层的 Socket 客户端套接字对象
    public OioSocketChannel() {
        this(new Socket());                         // 直接创建空 Socket 对象
    }
```

```java
// 构造器：指定 socket 进行创建
public OioSocketChannel(Socket socket) {
    this(null, socket);
}

// 构造器：指定父通道对象和 socket 进行创建
public OioSocketChannel(Channel parent, Socket socket) {
    super(parent);
    this.socket = socket;
    config = new DefaultOioSocketChannelConfig(this, socket);    // 创建默认配置对象
    boolean success = false;
    try {
        if (socket.isConnected()) {              // 已经连接 socket，获取输入和输出流对通道进行激活
            activate(socket.getInputStream(), socket.getOutputStream());
        }
        socket.setSoTimeout(SO_TIMEOUT); // 设置 socket 读数据超时时间
        success = true;
    } catch (Exception e) {
        throw new ChannelException("failed to initialize a socket", e);
    } finally {
        if (!success) {                          // 激活通道失败，关闭套接字
            try {
                socket.close();
            } catch (IOException e) {
                logger.warn("Failed to close a socket.", e);
            }
        }
    }
}

/* 省略掉普通 get set 方法，读者可自行打开源码查看 */

// 重写父类实现的模板方法。在实际读取前进行 socket 状态校验
protected int doReadBytes(ByteBuf buf) throws Exception {
    if (socket.isClosed()) {
        return -1;
    }
    try {
        return super.doReadBytes(buf);
    } catch (SocketTimeoutException ignored) {
        return 0;
    }
}

// 实现 DuplexChannel 的方法，关闭当前通道的输出端
public ChannelFuture shutdownOutput(final ChannelPromise promise) {
    EventLoop loop = eventLoop();
    if (loop.inEventLoop()) {      // 当前执行线程属于通道所属 EventLoop 循环组执行线程，直接执行关闭输出流操作
        shutdownOutput0(promise);
    } else {                       // 否则封装 Runnable，由事件循环线程执行
        loop.execute(new Runnable() {
```

```java
            @Override
            public void run() {
                shutdownOutput0(promise);
            }
        });
    }
    return promise;
}

// 完成实际关闭输出端操作
private void shutdownOutput0(ChannelPromise promise) {
    try {
        socket.shutdownOutput();
        promise.setSuccess();
    } catch (Throwable t) {
        promise.setFailure(t);
    }
}

// 实现 DuplexChannel 的方法，用于关闭当前通道的输入端
public ChannelFuture shutdownInput(final ChannelPromise promise) {
    EventLoop loop = eventLoop();
    if (loop.inEventLoop()) {      // 当前执行线程属于通道所属 EventLoop 循环组执行线程, 直接执行关闭输入流操作
        shutdownInput0(promise);
    } else {                        // 否则封装 Runnable 由事件循环线程执行
        loop.execute(new Runnable() {
            @Override
            public void run() {
                shutdownInput0(promise);
            }
        });
    }
    return promise;
}

// 调用原生 Socket 的方法，完成实际关闭输入端操作
private void shutdownInput0(ChannelPromise promise) {
    try {
        socket.shutdownInput();
        promise.setSuccess();
    } catch (Throwable t) {
        promise.setFailure(t);
    }
}

// 实现 DuplexChannel 的方法，关闭当前通道的输入端和输出端
public ChannelFuture shutdown(final ChannelPromise promise) {
    EventLoop loop = eventLoop();
    if (loop.inEventLoop()) {
        shutdown0(promise);
    } else {
        loop.execute(new Runnable() {
            @Override
```

```java
            public void run() {
                shutdown0(promise);
            }
        });
    }
    return promise;
}

// 调用原生 Socket 的方法，关闭当前通道的输入端和输出端
private void shutdown0(ChannelPromise promise) {
    Throwable cause = null;
    try {
        socket.shutdownOutput();
    } catch (Throwable t) {
        cause = t;
    }
    try {
        socket.shutdownInput();
    } catch (Throwable t) {
        if (cause == null) {
            promise.setFailure(t);
        } else {
            promise.setFailure(cause);
        }
        return;
    }
    if (cause == null) {
        promise.setSuccess();
    } else {
        promise.setFailure(cause);
    }
}

// 绑定通道，直接调用原生 socket 方法绑定
protected void doBind(SocketAddress localAddress) throws Exception {
    socket.bind(localAddress);
}

// 通道连接，直接调用原生 socket 方法绑定
protected void doConnect(SocketAddress remoteAddress,
                         SocketAddress localAddress) throws Exception {
    if (localAddress != null) {    // 通道还未绑定地址，先绑定本地地址信息
        socket.bind(localAddress);
    }
    boolean success = false;
    try {                          // 连接到远程地址。连接成功后，获取 socket 的输入和输出流激活当前通道
        socket.connect(remoteAddress, config().getConnectTimeoutMillis());
        activate(socket.getInputStream(), socket.getOutputStream());
        success = true;
    } catch (SocketTimeoutException e) {
        ConnectTimeoutException cause = new ConnectTimeoutException("connection timed out: " + remoteAddress);
        cause.setStackTrace(e.getStackTrace());
        throw cause;
```

```
        } finally {
            if (!success) {
                doClose();
            }
        }
    }

    // 断开连接，直接执行关闭动作
    protected void doDisconnect() throws Exception {
        doClose();
    }

    // 关闭操作，直接调用原生 Scoket 的 close 方法
    protected void doClose() throws Exception {
        socket.close();
    }
}
```

12.16 AbstractOioMessageChannel 原理

AbstractOioByteChannel 类定义了 Socket 客户端的操作。现在要介绍的 AbstractOioMessageChannel 类定义了 Socket 服务端的方法，即定义了 ServerSocket 的操作，所以直接把读取的数据当作 Socket 对象即可。由源码可知，实现父类的 doRead 方法，该方法需要由子类实现的 doReadMessages 读取连接的客户端对象，放入定义的 List readBuf socket 队列。源码如下。

```
public abstract class AbstractOioMessageChannel extends AbstractOioChannel {
    private final List<Object> readBuf = new ArrayList<Object>();   // 读缓冲区列表，真实对象为 acccept 方法获取到的 socket 对象

    // 构造器初始化父通道对象
    protected AbstractOioMessageChannel(Channel parent) {
        super(parent);
    }

    // 实现父类定义的模板方法。从 Socket 中读取数据
    protected void doRead() {
        if (!readPending) {              // 已经存在读任务，则直接返回
            return;
        }
        readPending = false;
        final ChannelConfig config = config();
        final ChannelPipeline pipeline = pipeline();
        final RecvByteBufAllocator.Handle allocHandle = unsafe().recvBufAllocHandle();// 获取接收缓冲区处理器
        allocHandle.reset(config);
        boolean closed = false;
        Throwable exception = null;
        try {
            do {
                // 调用子类完成实际读取操作
                int localRead = doReadMessages(readBuf);
```

```java
            if (localRead == 0) {                // 所有 socket 读取完成，停止读取
                break;
            }
            if (localRead < 0) {                 // 子类判断应该关闭当前通道，则执行关闭操作
                closed = true;
                break;
            }
            allocHandle.incMessagesRead(localRead);  // 增加接收缓冲区处理器读取的通道计数
        } while (allocHandle.continueReading());     // 接收缓冲区处理器判断能够继续读取，继续从
socket 中读取数据
    } catch (Throwable t) {
        exception = t;
    }
    boolean readData = false;
    int size = readBuf.size();
    if (size > 0) {                              // 从 ServerSocket 的 accpet 队列中读取到连接的客户端对象
        readData = true;
        for (int i = 0; i < size; i++) {         // 遍历每一个 socket，调用通道处理器 ChannelRead 处理已经连接的
socket 对象
            readPending = false;
            pipeline.fireChannelRead(readBuf.get(i));
        }
        readBuf.clear();
        allocHandle.readComplete();
        pipeline.fireChannelReadComplete();
    }
    if (exception != null) {  // 若发生异常，则关闭通道，并调用通道处理器 ExceptionCaught 方法处理异常
        if (exception instanceof IOException) {
            closed = true;
        }
        pipeline.fireExceptionCaught(exception);
    }
    if (closed) {                    // 关闭通道
        if (isOpen()) {
            unsafe().close(unsafe().voidPromise());
        }
    } else if (readPending || config.isAutoRead() || !readData && isActive()) { // 没有读到数据，需要执行
read 操作，避免缓冲区中存在未处理的数据
        read();
    }
}

// 子类实现实际读取操作，即调用 accept 方法把获取到连接的 socket 放入 List
protected abstract int doReadMessages(List<Object> msgs) throws Exception;
}
```

12.17 OioServerSocketChannel 类

OioServerSocketChannel 类定义了使用 JDK 原生 ServerSocket 的通道完整实现。由源码可知。
（1）OioServerSocketChannel 实现了 ServerSocketChannel 接口，该接口为标识接口，表示 TCP 协

议的 ServerSocket 实现。

（2）OioServerSocketChannel 内部使用 ServerSocket 完成实际通道接收客户端连接操作，即 ServerSocket.accept。

（3）OioServerSocketChannel 中的 doWrite、doConnect、doDisconnect 的方法实现为抛出不支持操作异常。因为对于 ServerSocket 来说，它只负责接收客户端对象，然后创建与对端相连接的 Socket 对象，自身并不与客户端 Socket 联系。

```java
// 标识接口，表示接收来自客户端的连接，并返回连接的 socket 对象
public interface ServerChannel extends Channel {
}

// 标识接口，表示基于 TCP/IP 协议连接的通道
public interface ServerSocketChannel extends ServerChannel {
}

public class OioServerSocketChannel extends AbstractOioMessageChannel
    implements ServerSocketChannel {
    private static final ChannelMetadata METADATA = new ChannelMetadata(false, 1); // 通道元数据信息，定义
不能在 disconnect 断开连接后再次调用 connect 方法进行连接，参数 1 表示 MaxMessagesRecvByteBufAllocator
每次最多读取的数量为 1

    // 创建 ServerSocket 对象
    private static ServerSocket newServerSocket() {
        try {
            return new ServerSocket();
        } catch (IOException e) {
            throw new ChannelException("failed to create a server socket", e);
        }
    }
    final ServerSocket socket;                      // 原生 JDK 服务端套接字对象
    private final OioServerSocketChannelConfig config; // 通道配置对象，用于设置 ServerSocket 对象

    // 构造器：创建服务端套接字 ServerSocket
    public OioServerSocketChannel() {
        this(newServerSocket());
    }

    // 构造器：使用创建好的 ServerSocket 对象完成初始化
    public OioServerSocketChannel(ServerSocket socket) {
        super(null);
        if (socket == null) {
            throw new NullPointerException("socket");
        }
        boolean success = false;
        try {                                       // 设置读取超时时间为父类定义的常量 1s
            socket.setSoTimeout(SO_TIMEOUT);
            success = true;
        } catch (IOException e) {
            throw new ChannelException(
                "Failed to set the server socket timeout.", e);
        } finally {
```

```java
            if (!success) {                                       // 设置超时时间失败，则关闭套接字
                try {
                    socket.close();
                } catch (IOException e) {
                    if (logger.isWarnEnabled()) {
                        logger.warn(
                            "Failed to close a partially initialized socket.", e);
                    }
                }
            }
        }
        this.socket = socket;
        config = new DefaultOioServerSocketChannelConfig(this, socket); // 初始化配置对象
    }

    /* 省略 set 和 get 方法 */

    // 绑定操作，使用原生 ServerSocket
    protected void doBind(SocketAddress localAddress) throws Exception {
        socket.bind(localAddress, config.getBacklog());
    }

    // 关闭操作，使用原生 ServerSocket
    protected void doClose() throws Exception {
        socket.close();
    }

    // 完成实际读取待连接的 socket 方法
    protected int doReadMessages(List<Object> buf) throws Exception {
        if (socket.isClosed()) {                                  // 服务端通道已经关闭
            return -1;
        }
        try {
            Socket s = socket.accept();                           // 接收客户端连接
            try {
                buf.add(new OioSocketChannel(this, s));           // 接收成功，将 JDK 原生 Socket 封装为 OioSocketChannel 对象，放入 Socket 列表中
                return 1;
            } catch (Throwable t) {                               // 发生任何异常，关闭客户端通道
                logger.warn("Failed to create a new channel from an accepted socket.", t);
                try {
                    s.close();
                } catch (Throwable t2) {
                    logger.warn("Failed to close a socket.", t2);
                }
            }
        } catch (SocketTimeoutException e) {
            // Expected
        }
        return 0;
    }

    /* 自身并不与客户端关联，所以相关方法均抛出 UnsupportedOperationException 不支持操作异常 */
```

```java
    protected void doWrite(ChannelOutboundBuffer in) throws Exception {
        throw new UnsupportedOperationException();
    }
    protected Object filterOutboundMessage(Object msg) throws Exception {
        throw new UnsupportedOperationException();
    }
    protected void doConnect(
            SocketAddress remoteAddress, SocketAddress localAddress) throws Exception {
        throw new UnsupportedOperationException();
    }
    protected SocketAddress remoteAddress0() {
        return null;
    }
    protected void doDisconnect() throws Exception {
        throw new UnsupportedOperationException();
    }
    final void clearReadPending0() {
        super.clearReadPending();
    }
}
```

12.18 AbstractNioChannel NIO 模板类

AbstractNioChannel NIO 类继承自 AbstractChannel 类，定义了基于 JDK NIO 操作的通道方法，即使用 Selector 选择器进行 IO 多路复用的方法。由于该类的实现代码较多，笔者将其分为几个部分进行描述。读者在阅读时，需要对基础的 NIO 编程有所了解。

12.18.1 核心变量与构造器

由源码可知。

（1）AbstractNioChannel 实际操作的 Java 原生对象为 SelectableChannel 接口的实例，在 Java 的 Socket 编程中，该接口的实例可以被 Selector 选择。

（2）在构造器中将传入的 SelectableChannel 通道配置为非阻塞式通道，此时可以将其注册到 Selector 中进行选择。

```java
public abstract class AbstractNioChannel extends AbstractChannel {
    private static final ClosedChannelException DO_CLOSE_CLOSED_CHANNEL_EXCEPTION =
ThrowableUtil.unknownStackTrace(
            new ClosedChannelException(), AbstractNioChannel.class, "doClose()"); // 由于执行关闭操作时失败，使用异常常量对象
    private final SelectableChannel ch;   // 可选择的操作通道对象,通常为ServerSocketChannel/SocketChannel 对象
    protected final int readInterestOp;   // 当前通道感兴趣事件集（读、写等事件参考 SelectionKey）
    volatile SelectionKey selectionKey;   // 当前与 Selector 交互，代表当前 Channel 的选择键
    boolean readPending;                  // 当前是否正在处理读操作
    private final Runnable clearReadPendingRunnable = new Runnable() {
        @Override
```

```
            public void run() {
                clearReadPending0();
            }
        };                                              // 封装清除读取标志位任务对象
        private ChannelPromise connectPromise;          // 通道连接时回调的 promise
        private ScheduledFuture<?> connectTimeoutFuture; // 判断连接超时的 Future 对象
        private SocketAddress requestedRemoteAddress;   // 当前需要连接的远程 socket 地址

        // 构造器初始化当前 NIO 通道。parent 表示父通道,ch 表示当前包装的 JDK 原生 SocketChannel 或
        // ServerSocketChannel 实际用于 IO 操作的通道,readInterestOp 表示当前通道感兴趣的事件集:读、写、连接、
        // accept 等
        protected AbstractNioChannel(Channel parent, SelectableChannel ch, int readInterestOp) {
            super(parent);
            this.ch = ch;
            this.readInterestOp = readInterestOp;
            try {
                ch.configureBlocking(false); // 配置当前通道为非阻塞式 IO,Selector 可以选择 SelectableChannel
                                             // 通道对象
            } catch (IOException e) {                   // 发生任何异常时,关闭当前通道,并抛出异常
                try {
                    ch.close();
                } catch (IOException e2) {
                    if (logger.isWarnEnabled()) {
                        logger.warn(
                            "Failed to close a partially initialized socket.", e2);
                    }
                }

                throw new ChannelException("Failed to enter non-blocking mode.", e);
            }
        }
```

12.18.2 doRegister 核心方法

doRegister 方法用于实现 AbstractNioChannel 中的通道注册模板方法。流程如下。

(1) 调用原生 Java SelectableChannel 的 register 方法,将通道注册到选择器中。

(2) 如果发生取消键异常,说明当前通道已经被注册且被取消,根据 selected 标识符选择抛出异常或尝试执行 selectNow 方法,完成一次选择,强制 Selector 清除该通道所代表的选择键。

```
        protected void doRegister() throws Exception {
            boolean selected = false;
            for (;;) {
                try { // 调用原生 Java SelectableChannel 的 register 方法,将通道注册到选择器中,并指定 attr 为当
                      // 前对象。此时可以从 SelectionKey 中取出当前通道的绑定对象,这里的 0 表示只对当前对象通道进行注册,但是
                      // 没有感兴趣事件集,因为只是注册,后面还需要进行初始化,完毕后才设置正确的感兴趣事件集:读、写、连接、
                      // 接收,这时才开始处于运行状态
                    selectionKey = javaChannel().register(eventLoop().selector, 0, this);
                    return;
                } catch (CancelledKeyException e) { // 发生取消键异常,说明当前通道已经被注册且被取消,根据
                                                    // selected 标识符选择抛出异常,或尝试执行 selectNow 方法,完成一次选择,强制 Selector 将该通道所代表的选
```

择键清除，这样在下一次循环时可以注册成功
```
            if (!selected) {
                eventLoop().selectNow();
                selected = true;
            } else {      // 对选择器进行强制选择，此时本该清理取消的选择键，但是却清理失败了。读者想想这
是为什么？应该能想到那一定是 JDK 的问题
                throw e;
            }
        }
    }
}
```

12.18.3　doDeregister 核心方法

doDeregister 方法用于实现 AbstractNioChannel 中的取消通道注册模板方法。这里直接将当前代表通道的选择键从所属事件循环中取消。源码如下。

```
protected void doDeregister() throws Exception {
    eventLoop().cancel(selectionKey());
}
```

12.18.4　doBeginRead 核心方法

doBeginRead 方法用于实现 AbstractNioChannel 中的读取通道数据模板方法。流程如下。

（1）判断当前代表通道的选择键是否有效，即没有被取消。

（2）设置 readPending 标志位，表明当前处于开始读取流程中。

（3）如果当前选择键没有注册读感兴趣事件集，将读感兴趣事件集加入选择键中，该通道发生读事件时，事件循环会得到通知，并处理该通道（详细参考第 1 篇）。

```
protected void doBeginRead() throws Exception {
    final SelectionKey selectionKey = this.selectionKey;
    if (!selectionKey.isValid()) {                    // 选择键有效
        return;
    }
    readPending = true;
    final int interestOps = selectionKey.interestOps();
    if ((interestOps & readInterestOp) == 0) {        // 注册读感兴趣事件集
        selectionKey.interestOps(interestOps | readInterestOp);
    }
}
```

12.18.5　newDirectBuffer 核心方法

newDirectBuffer 方法用于提供给子类将指定的 ByteBuf 转为堆外缓冲区，用于加速数据访问。流程如下。

（1）如果需要数据转移的缓冲区没有可用数据，则直接释放该缓冲区，并返回空缓冲区对象。

（2）如果分配器为池化堆外缓冲区分配器，则调用分配器分配新的堆外缓冲区，并将 buf 中的数据复制到分配的堆外缓冲区对象中，然后释放数据转移的缓冲区，并返回该堆外缓冲区对象。

（3）否则使用 ThreadLocal 缓存的直接缓冲区对象，如果获取成功，则进行数据转移并释放缓冲区，并返回该堆外缓冲区对象。

（4）如果 ThreadLocal 缓存的直接缓冲区对象不存在，由于使用非池化的堆外缓冲区的分配和释放消耗较大，所以不进行数据转移，返回当前 buf。

```
protected final ByteBuf newDirectBuffer(ByteBuf buf) {
    final int readableBytes = buf.readableBytes();
    if (readableBytes == 0) {        // 操作缓冲区没有可用数据，则直接释放该缓冲区，并返回空缓冲区对象
        ReferenceCountUtil.safeRelease(buf);
        return Unpooled.EMPTY_BUFFER;
    }
    final ByteBufAllocator alloc = alloc();
    if (alloc.isDirectBufferPooled()) { // 分配器为池化堆外缓冲区分配器，调用分配器分配新的堆外缓冲区，并将
buf 中的数据复制到分配的堆外缓冲区对象中
        ByteBuf directBuf = alloc.directBuffer(readableBytes);
        directBuf.writeBytes(buf, buf.readerIndex(), readableBytes);
        ReferenceCountUtil.safeRelease(buf);        // 释放数据已经转移的缓冲区
        return directBuf;
    }
    // 否则使用 ThreadLocal 缓存的直接缓冲区对象，如果获取成功，则进行数据转移并释放缓冲区
    final ByteBuf directBuf = ByteBufUtil.threadLocalDirectBuffer();
    if (directBuf != null) {
        directBuf.writeBytes(buf, buf.readerIndex(), readableBytes);
        ReferenceCountUtil.safeRelease(buf);
        return directBuf;
    }
    // 由于使用非池化的堆外缓冲区的分配和释放消耗较大，所以不进行数据转移，返回当前 buf
    return buf;
}

// 该方法同 newDirectBuffer(ByteBuf buf)，不过在释放时将传递指定的引用计数 ReferenceCounted 对象
protected final ByteBuf newDirectBuffer(ReferenceCounted holder, ByteBuf buf) {
    final int readableBytes = buf.readableBytes();
    if (readableBytes == 0) {
        ReferenceCountUtil.safeRelease(holder);
        return Unpooled.EMPTY_BUFFER;
    }

    final ByteBufAllocator alloc = alloc();
    if (alloc.isDirectBufferPooled()) {
        ByteBuf directBuf = alloc.directBuffer(readableBytes);
        directBuf.writeBytes(buf, buf.readerIndex(), readableBytes);
        ReferenceCountUtil.safeRelease(holder);   // 传递 ReferenceCounted 对象而不是 buf
        return directBuf;
    }

    final ByteBuf directBuf = ByteBufUtil.threadLocalDirectBuffer();
    if (directBuf != null) {
        directBuf.writeBytes(buf, buf.readerIndex(), readableBytes);
        ReferenceCountUtil.safeRelease(holder);   // 传递 ReferenceCounted 对象而不是 buf
        return directBuf;
    }
```

```
        // 引用计数与进行转移的 buf 属于不同对象，由于没有进行数据转移，此时调用 retain 方法，增加一个引用
        // 计数，然后释放 holder 指定的缓冲区
        if (holder != buf) {
            buf.retain();
            ReferenceCountUtil.safeRelease(holder);
        }
        return buf;
    }

    // 工具类对象用于获取线程本地缓存的堆外缓冲区对象
    private static final int THREAD_LOCAL_BUFFER_SIZE =
    SystemPropertyUtil.getInt("io.Netty.threadLocalDirectBufferSize", 64 * 1024);  // 默认缓存大小为 64KB

    public static ByteBuf threadLocalDirectBuffer() {
        if (THREAD_LOCAL_BUFFER_SIZE <= 0) {                                       // 不用缓存，则直接返回 null
            return null;
        }
        // 根据运行时是否存在 Unsafe 创建不同对象
        if (PlatformDependent.hasUnsafe()) {
            return ThreadLocalUnsafeDirectByteBuf.newInstance();
        } else {
            return ThreadLocalDirectByteBuf.newInstance();
        }
    }
}
```

12.18.6　doClose 核心方法

doClose 方法用于实现 AbstractNioChannel 中的关闭通道数据模板方法。流程如下。

（1）如果存在连接回调 connectPromise，则回调 promise 设置通道关闭异常对象。

（2）如果连接超时判断 connectTimeoutFuture 存在，则取消该 future 的执行。

```
protected void doClose() throws Exception {
    ChannelPromise promise = connectPromise;
    if (promise != null) {  // 连接回调 promise 存在，回调 promise 设置通道关闭异常对象
        promise.tryFailure(DO_CLOSE_CLOSED_CHANNEL_EXCEPTION);
        connectPromise = null;
    }
    ScheduledFuture<?> future = connectTimeoutFuture;
    if (future != null) {   // 连接超时判断 Future 存在，则取消该 future 的执行
        future.cancel(false);
        connectTimeoutFuture = null;
    }
}
```

12.18.7　AbstractNioUnsafe 核心内部类

AbstractNioUnsafe 类继承自 AbstractUnsafe 类，定义了 NioChannel 底层 IO 通道行为。由源码可知。

（1）NioUnsafe 接口继承自底层接口 Unsafe，定义了对 SelectableChannel 实例的操作行为。

（2）AbstractNioUnsafe 重写了父类的 connect 方法实现 SocketChannel 客户端的连接方法，在该方法中将调用子类实现的模板函数 doConnect 来完成实际连接。完成连接时，回调 promise，同时回调流

水线 ChannelPipeline 的对应方法。判断超时的实现是在事件循环的调度线程中放入一个延迟调度 Runnable 对象，延迟调度时间为超时时间。

（3）AbstractNioUnsafe 重写了父类的 finishConnect 方法，该方法将在连接建立后，回调执行，在该方法中调用子类需要实现的模板方法 doFinishConnect 完成连接，连接失败或成功都需要回调 promise 进行通知。

（4）AbstractNioUnsafe 重写了父类的 flush0 方法，该方法将实现将缓冲区中的数据写入底层 IO 通道。注意通道对 Selector 注册了写操作时，即 isFlushPending 判断为 true，不允许进行刷新，因为在写入完成后，会调用 forceFlush 方法，该方法直接调用父类 AbstractUnsafe 的 flush0 方法，在该方法中将写缓冲区的数据调用 doWrite 方法写入底层通道中。

```java
public interface NioUnsafe extends Unsafe {
    // 返回当前可选择通道
    SelectableChannel ch();

    // 完成 SelectableChannel 通道连接
    void finishConnect();

    // 从 SelectableChannel 通道读取信息
    void read();

    // 将写缓冲区中的数据强制刷新到 SelectableChannel 中
    void forceFlush();
}

protected abstract class AbstractNioUnsafe extends AbstractUnsafe implements NioUnsafe {
    // 异常表示当前通道的 SelectionKey 选择键的读感兴趣事件集，该通道在有读事件发生时，不会在所属事件
    // 循环中被选择执行
    protected final void removeReadOp() {
        SelectionKey key = selectionKey();
        if (!key.isValid()) {                          // 选择键无效则返回
            return;
        }
        int interestOps = key.interestOps();
        if ((interestOps & readInterestOp) != 0) {     // 去掉选择键的读感兴趣事件
            key.interestOps(interestOps & ~readInterestOp);
        }
    }

    public final SelectableChannel ch() {
        return javaChannel();
    }

    // 重写父类连接模板方法
    public final void connect(final SocketAddress remoteAddress, final SocketAddress localAddress, final ChannelPromise promise) {
        if (!promise.setUncancellable() || !ensureOpen(promise)) { // 连接操作被取消，则直接返回
            return;
        }
        try {
```

```
            if (connectPromise != null) {                              // 连接 promise 不为空，说明连接已经开始
                throw new IllegalStateException("connection attempt already made");
            }
            boolean wasActive = isActive();                            // 通道是否处于激活状态
            if (doConnect(remoteAddress, localAddress)) {              // 由子类完成实际连接（注意这里为 NIO 连
接，所以不会导致当前线程阻塞。在描述子类时，将看到该方法的定义，读者需要了解对于可选择通道的连接操
作，如果该通道配置了非阻塞 IO，连接操作执行时会返回 false，这是 JDK 的基础知识），如果连接成功，则调
用 fulfillConnectPromise 方法完成 ConnectPromise 的回调
                fulfillConnectPromise(promise, wasActive);
            } else {       // 连接失败，由于是 NIO，所以不会阻塞。如果设置了连接超时事件，则构建 Runnable
对象，并放入事件循环线程中调度执行；到达超时时间后，设置超时异常，并关闭通道
                connectPromise = promise;
                requestedRemoteAddress = remoteAddress;
                int connectTimeoutMillis = config().getConnectTimeoutMillis();
                if (connectTimeoutMillis > 0) { // 设置连接超时时间
                    connectTimeoutFuture = eventLoop().schedule(new Runnable() {
                        @Override
                        public void run() {   // 在超时时间过后，connectPromise 没有设置成功，此时代表了
连接超时，设置超时异常，并关闭通道
                            ChannelPromise connectPromise = AbstractNioChannel.this.connectPromise;
                            ConnectTimeoutException cause =
                                new ConnectTimeoutException("connection timed out: " + remoteAddress);
                            if (connectPromise != null && connectPromise.tryFailure(cause)) {
                                close(voidPromise());
                            }
                        }
                    }, connectTimeoutMillis, TimeUnit.MILLISECONDS);
                }
                // 对 promise 添加监听器，当 promise 完成时回调，用于在代表异步执行的任务 ChannelFuture
被取消时，将通道关闭
                promise.addListener(new ChannelFutureListener() {
                    @Override
                    public void operationComplete(ChannelFuture future) throws Exception {
                        if (future.isCancelled()) {
                            if (connectTimeoutFuture != null) {
                                connectTimeoutFuture.cancel(false);
                            }
                            connectPromise = null;
                            close(voidPromise());
                        }
                    }
                });
            }
        } catch (Throwable t) {
            promise.tryFailure(annotateConnectException(t, remoteAddress));
            closeIfClosed();
        }
    }
```

// 连接成功时回调，当通道在连接成功后被激活，回调流水线的 ChannelActive 回调方法。如果 promise 失
败，则表示用户取消了连接操作。如果 trySuccess 时返回 false，则需要调用 close(voidPromise())关闭通道，这
时会回调流水线的 ChannelInActive 回调方法
```
    private void fulfillConnectPromise(ChannelPromise promise, boolean wasActive) {
```

```java
    if (promise == null) {
        return;
    }
    boolean promiseSet = promise.trySuccess();
    if (!wasActive && isActive()) {           // 通道激活
        pipeline().fireChannelActive();
    }
    if (!promiseSet) {                         // 用户取消了连接
        close(voidPromise());
    }
}

// 发生异常时回调，参数 cause 指定异常对象
private void fulfillConnectPromise(ChannelPromise promise, Throwable cause) {
    if (promise == null) {                    // promise 为空，表示已经进行回调，直接返回
        return;
    }
    // 设置 promise 连接失败
    promise.tryFailure(cause);
    closeIfClosed();    // 如果当前 JDK 原生通道已经关闭，则调用 close 方法将 Netty 的 NIOChannel 关闭
}

// 重写父类的完成连接模板方法。只有在连接没有取消或超时且成功连接时，由事件循环线程调用该方法
public final void finishConnect() {
    assert eventLoop().inEventLoop();         // 调用线程必须是当前通道绑定的事件循环线程
    try {
        boolean wasActive = isActive();
        doFinishConnect();                    // 调用子类完成连接模板函数
        fulfillConnectPromise(connectPromise, wasActive); // 完成连接回调 promise
    } catch (Throwable t) {                   // 连接失败，则同样需要通过异常对象回调 promise
        fulfillConnectPromise(connectPromise, annotateConnectException(t, requestedRemoteAddress));
    } finally {
        // 如果设置了连接超时 Future，在连接完成时，取消该超时判断 Future，因为连接已完成
        if (connectTimeoutFuture != null) {
            connectTimeoutFuture.cancel(false);
        }
        connectPromise = null;
    }
}

// 重写父类的刷新写缓冲区模板方法。该方法只有在没有挂起的刷新操作时，才执行父类的 flush0 完成刷新。
// 如果有一个挂起的刷新操作，事件循环线程将在处理该刷新操作后调用 forceFlush()，因此现在不需要调用它
protected final void flush0() {
    // 存在未处理的刷新操作
    if (isFlushPending()) {
        return;
    }
    super.flush0();    // 完成实际刷新
}

// 强制刷新写缓冲区，将其中的数据写入底层 IO 通道，这里直接调用父类完成刷新
public final void forceFlush() {
    super.flush0();
```

```java
}
// 判断当前是否有挂起的刷新操作，这里仅判断选择键是否有效，且当前注册了写感兴趣事件集
private boolean isFlushPending() {
    SelectionKey selectionKey = selectionKey();
    return selectionKey.isValid() && (selectionKey.interestOps() & SelectionKey.OP_WRITE) != 0;
}
}
```

12.19 AbstractNioByteChannel NIO 客户端模板类

AbstractNioByteChannel NIO 类继承自 AbstractNioChannel 通道，实现了客户端 SocketChannel 的操作。该类实现较多，笔者将其拆分并逐个描述。

12.19.1 核心变量与构造器

由源码可知。

（1）Runnable flushTask 用于包装刷新任务。

（2）在构造器中指定父类的 readInterestOp 为 OP_READ，因为这里为客户端 SocketChannel 的模板类，所以读感兴趣事件便是 READ。如果是 ServerSocketChannel 呢？读者很快将看到为 ACCEPT 事件。

```java
public abstract class AbstractNioByteChannel extends AbstractNioChannel {
    private static final ChannelMetadata METADATA = new ChannelMetadata(false, 16); // 通道元数据对象，
// 16 指定每次读取的最大消息数（由 MaxMessagesRecvByteBufAllocator 类使用）
    private static final String EXPECTED_TYPES =
        " (expected: " + StringUtil.simpleClassName(ByteBuf.class) + ", " +
        StringUtil.simpleClassName(FileRegion.class) + ')';         // 异常类型字符串常量
    private Runnable flushTask;

// 构造器，初始化父类，指定父通道和 SocketChannel 通道对象，同时指定父类的 readInterestOp 为 OP_READ
    protected AbstractNioByteChannel(Channel parent, SelectableChannel ch) {
        super(parent, ch, SelectionKey.OP_READ);
    }
}
```

12.19.2 doWrite 核心方法

doWrite 方法重写了父类的输出缓冲区，将数据写入 IO 通道，在前面涉及过该方法在 flush 中的调用。流程如下。

（1）所有 msg 信息循环写入缓冲区。

（2）如果当前写入类型为普通 ByteBuf 类型，则调用子类实现的 doWriteBytes 方法完成写入。该方法返回 0 时，表示底层 IO 不能执行写入操作，停止继续写入其他缓冲区数据。

（3）如果当前写入类型为 FileRegion 零复制文件类型，则调用子类实现的 doWriteFileRegion 方法完成写入。该方法返回 0 时，表示底层 IO 不能执行写入操作，则停止继续写入其他缓冲区数据。

（4）当缓冲区中存在未写入数据时，调用 incompleteWrite(setOpWrite) 方法，该方法根据 setOpWrite

值决定以下策略。

① 如果 setOpWrite 为 true，则重新向代表当前通道的 SelectionKey 设置写感兴趣事件，底层通道再次可写时，会通知 Selector，这时事件循环线程会获取该通道并尝试继续写入。

② 如果 setOpWrite 为 false，则重新封装一个 flush 刷新写缓冲区的任务，并放入事件循环队列的末尾，等其他任务执行完毕后再尝试刷新。

```java
protected void doWrite(ChannelOutboundBuffer in) throws Exception {
    int writeSpinCount = -1;           // 写自旋计数
    boolean setOpWrite = false;
    for (;;) {                          // 循环写入缓冲区中的 msg 信息
        Object msg = in.current();
        if (msg == null) {              // 已经写完缓冲区的数据，表示取消在当前通道的选择键中写感兴趣事件
            clearOpWrite();             // 取消选择键写事件
            return;
        }
        if (msg instanceof ByteBuf) {   // 写入普通 ByteBuf 缓冲区数据
            ByteBuf buf = (ByteBuf) msg;
            int readableBytes = buf.readableBytes();
            if (readableBytes == 0) {   // 缓冲区中没有可写入数据，则继续循环
                in.remove();
                continue;
            }
            boolean done = false;
            long flushedAmount = 0;
            if (writeSpinCount == -1) { // 初始化写自旋计数
                writeSpinCount = config().getWriteSpinCount();
            }
            for (int i = writeSpinCount - 1; i >= 0; i--) { // 自旋写入（重试）缓冲区中所有数据
                int localFlushedAmount = doWriteBytes(buf);
                if (localFlushedAmount == 0) {  // 实际写入数量为 0，表示当前底层 IO 通道不可写，重新设置代表当前通道的选择键的写感兴趣事件即可
                    setOpWrite = true;
                    break;
                }
                flushedAmount += localFlushedAmount;
                if (!buf.isReadable()) {        // 已经写完缓冲区中的数据
                    done = true;
                    break;
                }
            }
            in.progress(flushedAmount); // 回调 ChannelProgressivePromise 当前写入进度
            if (done) {                 // ByteBuf 缓冲区中所有数据写入完毕，将其从写缓冲区中移除
                in.remove();
            } else {                    // 写入未完成，表示通道不可写，则退出当前循环，结束写入操作
                break;
            }
        } else if (msg instanceof FileRegion) {  // 写入代表文件的零复制对象
            FileRegion region = (FileRegion) msg;
            boolean done = region.transferred() >= region.count();
            if (!done) {                // 文件仍有数据还未写入
                long flushedAmount = 0;
```

```java
                if (writeSpinCount == -1) {                          // 初始化重试次数
                    writeSpinCount = config().getWriteSpinCount();
                }
                for (int i = writeSpinCount - 1; i >= 0; i--) {       // 循环写入
                    long localFlushedAmount = doWriteFileRegion(region);  // 子类完成真实写入
                    if (localFlushedAmount == 0) {                    // 写入数量为 0，结束循环
                        setOpWrite = true;
                        break;
                    }
                    flushedAmount += localFlushedAmount;
                    if (region.transferred() >= region.count()) {     // 写入完成
                        done = true;
                        break;
                    }
                }
                in.progress(flushedAmount);                           // 回调 ChannelProgressivePromise 当前写入进度
            }
            if (done) {                                               // 文件传输完毕，重写缓冲区移除该对象
                in.remove();
            } else {              // 否则停止循环，因为底层 IO 不可写，所以调用 incompleteWrite 方法
                break;
            }
        } else {              // 不应该出现第三种类型的 msg，所以到达这里一定是出现了 Netty 的 Bug
            throw new Error();
        }
    }
    incompleteWrite(setOpWrite);                                      // 写缓冲区中的数据未完全写入时调用
}

// 子类实现真实写入
protected abstract int doWriteBytes(ByteBuf buf) throws Exception;
protected abstract long doWriteFileRegion(FileRegion region) throws Exception;

// 写缓冲区中的数据尚未完全写入时调用
protected final void incompleteWrite(boolean setOpWrite) {
    if (setOpWrite) {                                 // 重新向代表当前通道的 SelectionKey 设置写感兴趣事件
        setOpWrite();
    } else {
        // 否则重新封装一个 flush 刷新写缓冲区的任务，并放入事件循环队列的末尾，等其他任务执行完毕后再尝试刷新
        Runnable flushTask = this.flushTask;
        if (flushTask == null) {
            flushTask = this.flushTask = new Runnable() {
                @Override
                public void run() {
                    flush();
                }
            };
        }
        eventLoop().execute(flushTask);
    }
}
```

12.19.3 filterOutboundMessage 核心方法

filterOutboundMessage 方法实现父类对放入写缓冲区 ChannelOutboundBuffer 中的 msg 进行包装。这里其实只将 ByteBuf 类型数据转移到堆外内存中。而对于 FileRegion 将不进行包装。源码如下。

```java
protected final Object filterOutboundMessage(Object msg) {
    if (msg instanceof ByteBuf) { // 将 ByteBuf 的数据写入堆外内存中,如果已经是堆外内存 ByteBuf,则直接返回
        ByteBuf buf = (ByteBuf) msg;
        if (buf.isDirect()) {
            return msg;
        }
        return newDirectBuffer(buf);
    }
    if (msg instanceof FileRegion) {
        return msg;
    }
    // 只能写入 ByteBuf 或 FileRegion 类型
    throw new UnsupportedOperationException(
            "unsupported message type: " + StringUtil.simpleClassName(msg) + EXPECTED_TYPES);
}
```

12.19.4 NioByteUnsafe 核心内部类

NioByteUnsafe 类继承自 AbstractNioUnsafe 类,完成对客户端字节信息写入的底层 IO 支持。由源码可知。

(1) 该类实现了 read 模板方法,完成了对底层 IO 数据的读取。

(2) 在读取过程中,根据 RecvByteBufAllocator.Handle 的策略读取底层 IO 通道的数据,子类需要实现 doReadBytes(byteBuf) 完成实际读取工作。读取完毕后,回调通道处理器的 ChannelRead 完成对读数据处理。

(3) 读取完成后,对 readPending 和 config.isAutoRead() 进行判断,决定是否需要去除代表当前通道的 SelectionKey 中的读感兴趣事件。

```java
protected class NioByteUnsafe extends AbstractNioUnsafe {
    // 没有读取到数据时(allocHandle.lastBytesRead() < 0),关闭当前通道
    private void closeOnRead(ChannelPipeline pipeline) {
        if (isOpen()) {            // 底层 IO 通道处于打开状态,将其设置为半关闭状态
            if (Boolean.TRUE.equals(config().getOption(ChannelOption.ALLOW_HALF_CLOSURE))) {
                shutdownInput();   // 关闭输入缓冲区
                SelectionKey key = selectionKey();
                key.interestOps(key.interestOps() & ~readInterestOp); // 取消读感兴趣事件,此时 Selector 不会选中该通道的读事件
                pipeline.fireUserEventTriggered(ChannelInputShutdownEvent.INSTANCE); // 回调通道处理器方法
            } else {               // 底层 IO 通道处于关闭状态,关闭当前 NIOChannel 对象
                close(voidPromise());
            }
```

```java
        }
    }
    // 在处理读数据发生异常时回调
    private void handleReadException(ChannelPipeline pipeline, ByteBuf byteBuf, Throwable cause, boolean close,
                RecvByteBufAllocator.Handle allocHandle) {
        if (byteBuf != null) { // 存在未处理数据,由回调通道处理器的 ChannelRead 方法处理数据,否则释放缓冲区
            if (byteBuf.isReadable()) {
                readPending = false;
                pipeline.fireChannelRead(byteBuf);
            } else {
                byteBuf.release();
            }
        }
        // 回调通道处理器的 ChannelReadComplete、ExceptionCaught,处理所有数据与异常
        allocHandle.readComplete();
        pipeline.fireChannelReadComplete();
        pipeline.fireExceptionCaught(cause);
        if (close || cause instanceof IOException) { // 异常类型为通道 IO 异常或判定需要关闭通道,则关闭该通道
            closeOnRead(pipeline);
        }
    }

    // 实现读取通道数据模板方法
    public final void read() {
        final ChannelConfig config = config();
        final ChannelPipeline pipeline = pipeline();
        final ByteBufAllocator allocator = config.getAllocator();
        final RecvByteBufAllocator.Handle allocHandle = recvBufAllocHandle();
        allocHandle.reset(config);                          // 根据配置项初始化接收缓冲区分配处理器
        ByteBuf byteBuf = null;
        boolean close = false;
        try {
            do {                                            // 循环读取通道数据
                byteBuf = allocHandle.allocate(allocator);  // 分配读缓冲区对象
                allocHandle.lastBytesRead(doReadBytes(byteBuf)); // 调用子类实现的 doReadBytes 方法,将
通道数据读入读缓冲区中,并将读取的字节数记录在 RecvByteBufAllocator.Handle 中
                if (allocHandle.lastBytesRead() <= 0) {     // 没有读取到任何数据,释放读缓冲区
                    byteBuf.release();
                    byteBuf = null;
                    close = allocHandle.lastBytesRead() < 0; // 如果之前没有读取到任何数据,则关闭该通道
                    break;
                }
                allocHandle.incMessagesRead(1);
                readPending = false;
                pipeline.fireChannelRead(byteBuf); // 触发通道处理器的 ChannelRead 方法,处理读取到的数据
                byteBuf = null;
            } while (allocHandle.continueReading()); // 接收缓冲区处理器决定,可以继续读取时,则继续读取
底层 IO 通道的数据,通常根据设置的读缓冲区大小进行限制
            allocHandle.readComplete();
            pipeline.fireChannelReadComplete();     // 读取完成,回调通道处理器的 ChannelReadComplete 方
法,表示当前底层 IO 中的数据已经全部读取完毕
            if (close) {                            // 关闭当前通道
```

```
                closeOnRead(pipeline);
            }
        } catch (Throwable t) {
            handleReadException(pipeline, byteBuf, t, close, allocHandle);
        } finally {   // 完成读取数据，如果没有配置自动读取，则移除代表当前通道的 SelectionKey 的读感兴趣
事件，此时 Selector 不会选中该通道的读事件。判断 readPending 主要可能有如下原因
            // 用户在 channelRead(...)或 channelReadComplete(...)方法中调用 Channel.read()或
ChannelHandlerContext.read()方法
            if (!readPending && !config.isAutoRead()) {
                removeReadOp();
            }
        }
    }
}
```

12.20 NioSocketChannel NIO TCP 客户端实现类

NioSocketChannel 类继承自 AbstractNioByteChannel，实现了对于 SocketChannel 对象实现的 TCP 客户端。该类完成了所有父类定义的模板方法实现，由于类代码较多，笔者将其拆分讲解。注意，NioSocketChannel 外部类实现了 Netty Channel 方法，对于底层 IO 的实现还是在内部类 NioSocketChannelUnsafe 中。

12.20.1 核心变量与构造器

由源码可知。

（1）在构造器中调用 newSocket 创建新的 SocketChannel 对象，该对象是 Java 原生网络编程的底层 IO 通道对象。该方法通过 SelectorProvider 的 openSocketChannel 方法创建实例，而不是直接使用 SocketChannel.open()。因为在 SocketChannel.open 方法中会使用同步锁保证线程安全，这样会降低性能，所以使用 provider.openSocketChannel()完成创建。

（2）在构造器中创建 NioSocketChannelConfig，用于对通道进行配置的对象。

```
public class NioSocketChannel extends AbstractNioByteChannel implements
io.Netty.channel.socket.SocketChannel {
    private static final SelectorProvider DEFAULT_SELECTOR_PROVIDER = SelectorProvider.provider();
    // 用于生成 Selector 对象的 provider

    // 通过指定的 SelectorProvider 生成新的 SocketChannel 实例
    private static SocketChannel newSocket(SelectorProvider provider) {
        try {
            return provider.openSocketChannel();
        } catch (IOException e) {
            throw new ChannelException("Failed to open a socket.", e);
        }
    }
    private final SocketChannelConfig config; // socketchannel 配置对象

    // 使用默认选择器创建 NioSocketChannel 实例
    public NioSocketChannel() {
```

```
        this(DEFAULT_SELECTOR_PROVIDER);
    }

    // 使用指定 SelectorProvider 创建 NioSocketChannel 实例
    public NioSocketChannel(SelectorProvider provider) {
        this(newSocket(provider));
    }

    // 使用指定 SocketChannel 创建 NioSocketChannel 实例
    public NioSocketChannel(SocketChannel socket) {
        this(null, socket);
    }

    // 使用指定父通道与 SocketChannel 创建 NioSocketChannel 实例
    public NioSocketChannel(Channel parent, SocketChannel socket) {
        super(parent, socket);
        config = new NioSocketChannelConfig(this, socket.socket());
    }
}
```

12.20.2　isActive 核心方法

isActive 方法用于判断当前通道是否处于激活状态。这里直接调用 java 原生通道的 isOpen 和 isConnected 判断是否激活，即是否打开，且已经处于连接状态。源码如下。

```
public boolean isActive() {
    SocketChannel ch = javaChannel();
    return ch.isOpen() && ch.isConnected();
}
```

12.20.3　shutdownInput 核心方法

shutdownInput 方法用于关闭 Socket 的输入端，关闭完成时回调 ChannelPromise。流程如下。
（1）获取进行异步关闭的 Executor。
（2）如果 Executor 不为空，则在执行器中执行实际关闭操作。
（3）如果 Executor 为空，则根据当前线程是否为通道所属事件循环线程，决定是直接执行关闭操作，还是把关闭操作放入事件循环线程中执行。
（4）在实际关闭操作 shutdownInput0 中，在该方法中将直接调用原生 java 通道的 shutdownInput 方法，该方法会把 Socket 读取到的数据静默丢弃。

```
public ChannelFuture shutdownInput(final ChannelPromise promise) {
    Executor closeExecutor = ((NioSocketChannelUnsafe) unsafe()).prepareToClose();
    if (closeExecutor != null) {           // 异步执行器不为空，则封装关闭操作执行
        closeExecutor.execute(new Runnable() {
            @Override
            public void run() {
                shutdownInput0(promise);
            }
        });
    } else {                                // 执行器为空，根据当前线程是否属于事件循环线程，选择是否直接执行
```

```
            EventLoop loop = eventLoop();
            if (loop.inEventLoop()) {
                shutdownInput0(promise);
            } else {                    // 不属于事件循环执行线程，封装关闭操作放入执行线程中执行
                loop.execute(new Runnable() {
                    @Override
                    public void run() {
                        shutdownInput0(promise);
                    }
                });
            }
        }
        return promise;
    }

    // 实际关闭操作，直接调用 SocketChannel 的 shutdownInput 方法完成关闭，并回调 promise 设置成功
    private void shutdownInput0(final ChannelPromise promise) {
        try {
            javaChannel().socket().shutdownInput();
            promise.setSuccess();
        } catch (Throwable t) {
            promise.setFailure(t);
        }
    }
```

12.20.4　shutdownOutput 核心方法

shutdownOutput 方法用于关闭 Socket 的输出端，关闭完成时回调 ChannelPromise。流程如下。

（1）获取进行异步关闭的 Executor。

（2）如果 Executor 不为空，则把实际关闭操作放入执行器中执行。

（3）如果 Executor 为空，则根据当前线程是否为通道所属事件循环线程，决定是直接执行关闭操作，还是将关闭操作放入事件循环线程中执行。

（4）实际关闭操作 shutdownInput0 时，在该方法中，直接调用原生 java 通道的 shutdownInput 方法，该方法会把 Socket 的输出端关闭，对于 TCP 协议而言，即发送 FIN 关闭报文，进入四次挥手的过程。

```
public ChannelFuture shutdownOutput(final ChannelPromise promise) {
    Executor closeExecutor = ((NioSocketChannelUnsafe) unsafe()).prepareToClose();
    if (closeExecutor != null) {        // 执行器不为空
        closeExecutor.execute(new Runnable() {
            @Override
            public void run() {
                shutdownOutput0(promise);
            }
        });
    } else {                    // 执行器为空，根据当前线程是否属于事件循环执行线程，选择是否直接执行关闭操作
        EventLoop loop = eventLoop();
        if (loop.inEventLoop()) {
            shutdownOutput0(promise);
        } else {
            loop.execute(new Runnable() {
                @Override
```

```java
            public void run() {
                shutdownOutput0(promise);
            }
        });
    }
}
return promise;
}

// 实际关闭操作，这里直接调用 SocketChannel 的 shutdownOutput 方法完成关闭，并回调 promise 设置成功
private void shutdownOutput0(final ChannelPromise promise) {
    try {
        javaChannel().socket().shutdownOutput();
        promise.setSuccess();
    } catch (Throwable t) {
        promise.setFailure(t);
    }
}
```

12.20.5　shutdown 核心方法

shutdown 方法用于关闭 Socket 的输入端和输出端，关闭完成时，回调 ChannelPromise。流程如下。

（1）获取进行异步关闭的 Executor。

（2）如果 Executor 不为空，则把实际关闭操作放入该执行器中执行。

（3）如果 Executor 为空，则根据当前线程是否为通道所属事件循环线程，决定是直接执行关闭操作，还是将关闭操作放入事件循环线程中执行。

（4）实际关闭操作 shutdownInput0 时，在该方法中，直接调用原生 java 通道的 shutdownInput 和 shutdownOutput 方法，执行成功后，根据是否发生异常，回调 promise 的 setSuccess() 或 setFailure(cause) 方法。

```java
public ChannelFuture shutdown(final ChannelPromise promise) {
    Executor closeExecutor = ((NioSocketChannelUnsafe) unsafe()).prepareToClose();
    if (closeExecutor != null) { // 执行器不为空
        closeExecutor.execute(new Runnable() {
            @Override
            public void run() {
                shutdown0(promise);
            }
        });
    } else {                    // 执行器为空，根据当前线程是否属于事件循环执行线程，选择是否直接执行关闭操作
        EventLoop loop = eventLoop();
        if (loop.inEventLoop()) {
            shutdown0(promise);
        } else {
            loop.execute(new Runnable() {
                @Override
                public void run() {
                    shutdown0(promise);
                }
            });
        }
    }
```

```java
    }
    return promise;
}

// 实际关闭操作,直接调用 SocketChannel 的 shutdownOutput 方法和 shutdownInput 方法完成关闭,并回调
// promise,设置成功
private void shutdown0(final ChannelPromise promise) {
    Socket socket = javaChannel().socket();
    Throwable cause = null;
    try {
        socket.shutdownOutput();    // 关闭输出端,进入 FIN 状态
    } catch (Throwable t) {
        cause = t;
    }
    try {
        socket.shutdownInput();     // 关闭输入端,停止接收对端数据
    } catch (Throwable t) {
        if (cause == null) {
            promise.setFailure(t);
        } else {
            logger.debug("Exception suppressed because a previous exception occurred.", t);
            promise.setFailure(cause);
        }
        return;
    }
    // 根据异常信息回调 promise
    if (cause == null) {
        promise.setSuccess();
    } else {
        promise.setFailure(cause);
    }
}
```

12.20.6　doBind 核心方法

doBind 方法完成对当前 SocketChannel 的绑定操作,直接调用 SocketChannel 的 bind 方法完成绑定。源码如下。

```java
protected void doBind(SocketAddress localAddress) throws Exception {
    javaChannel().socket().bind(localAddress);
}
```

12.20.7　doConnect 核心方法

doConnect 方法完成对当前 SocketChannel 的连接操作。流程如下。

（1）如果通道还未绑定地址,则先调用 bind 方法完成地址绑定。

（2）调用 SocketChannel 的 connect 方法完成连接。如果连接失败,则说明当前异步阻塞通道当前状态为正在连接中,因为当前是非阻塞通道所以立即返回 false,注册 OP_CONNECT 感兴趣事件集,Selector 选中该通道时,由所属事件循环线程来执行进一步操作。

（3）如果连接失败,则关闭当前通道。

```java
protected boolean doConnect(SocketAddress remoteAddress, SocketAddress localAddress) throws Exception
{
    if (localAddress != null) {
        javaChannel().socket().bind(localAddress);
    }

    boolean success = false;
    try {
        boolean connected = javaChannel().connect(remoteAddress);
        if (!connected) {
            selectionKey().interestOps(SelectionKey.OP_CONNECT);
        }
        success = true;
        return connected;
    } finally { // 如果连接失败，则关闭当前通道
        if (!success) {
            doClose();
        }
    }
}
```

12.20.8　doClose 核心方法

doClose 方法用于完成关闭 SocketChannel。这里先调用父类的 doClose 方法对 connectPromise 和 connectTimeoutFuture 进行回调，然后调用 Java 原生 SocketChannel 的 close 方法关闭 socket。源码如下。

```java
protected void doClose() throws Exception {
    super.doClose();
    javaChannel().close();
}
```

12.20.9　doReadBytes 核心方法

doReadBytes 方法用于把 SocketChannel 读取的数据放入指定的 bytebuf 缓冲区中。这里直接调用 byteBuf.writeBytes 读取通道数据，读取的数量由 RecvByteBufAllocator.Handle 的 attemptedBytesRead() 决定。源码如下。

```java
protected int doReadBytes(ByteBuf byteBuf) throws Exception {
    final RecvByteBufAllocator.Handle allocHandle = unsafe().recvBufAllocHandle();
    allocHandle.attemptedBytesRead(byteBuf.writableBytes());        // bytebuf 缓冲区中剩余的写入空间
    return byteBuf.writeBytes(javaChannel(), allocHandle.attemptedBytesRead());
}
```

12.20.10　doWriteBytes 核心方法

doWriteBytes 方法用于把 SocketChannel 的数据写入指定的 bytebuf 缓冲区中。这里直接调用 byteBuf.readBytes 完成对通道数据的写入，写入数量为当前 bytebuf 中的可读数据数量。源码如下。

```java
protected int doWriteBytes(ByteBuf buf) throws Exception {
    final int expectedWrittenBytes = buf.readableBytes();
    return buf.readBytes(javaChannel(), expectedWrittenBytes); // 将缓冲区中的可读数据写入 socketchannel
}
```

12.20.11 doWriteFileRegion 核心方法

doWriteFileRegion 方法完成对当前 SocketChannel 的写入指定零复制文件对象 FileRegion 的数据操作。这里直接调用 FileRegion 的 transferTo 方法完成写入。源码如下。

```
protected long doWriteFileRegion(FileRegion region) throws Exception {
    final long position = region.transferred();          // 当前文件数据的写入点
    return region.transferTo(javaChannel(), position);
}
```

12.20.12 doWrite 核心方法

doWrite 方法用于把 SocketChannel 的数据写入 ChannelOutboundBuffer 输出缓冲区。流程如下。

（1）循环写入写缓冲区中的数据。

（2）如果写缓冲区中的数据已经完全写入 SocketChannel 中，则清除 SelectionKey 的写感兴趣事件，然后结束循环。

（3）获取需要写出的 java 原生 ByteBuffer 缓冲区数组。

（4）如果数组为 0，则表明没有 ByteBuffer 数据需要写入，那么调用父类 super.doWrite(in)完成 FileRegion 的写入。

（5）如果 ByteBuffer 只有一个（数组长度为1），则进行普通写入。此时可能由于底层 IO 写缓冲区满，无法写入，则注册写感兴趣事件集。由 Selector 选中后继续写入。否则在到达重试次数后，由于写缓冲区虽然可用，但是网络较慢导致很长时间无法写入整个 ByteBuffer 中的数据，则将其放入事件循环线程的末尾待下次执行时继续写入。

（6）如果 ByteBuffer 数量大于 1 个，则聚合写入。整体流程同上所述，只是调用 socketchannel 方法不一样而已。

（7）写入完成后，释放 ChannelOutboundBuffer 中完全写入的缓冲区，并更新部分写入的缓冲区索引。

（8）如果没有全部写完写缓冲区数据，则将写入操作放入下一次操作进行写入。

```
protected void doWrite(ChannelOutboundBuffer in) throws Exception {
    for (;;) {                    // 循环写入写缓冲区中的数据
        int size = in.size();
        if (size == 0) {          // 写缓冲区中的数据已经完全写入 SocketChannel 中，则清除 SelectionKey 的写感兴趣事件，然后结束循环
            clearOpWrite();
            break;
        }
        long writtenBytes = 0;
        boolean done = false;
        boolean setOpWrite = false;
        // 获取需要写出的 java 原生 ByteBuffer 缓冲区数组（因为 SocketChannel 只支持对 ByteBuffer 进行操作，所以需要把 Netty 的 ByteBuf 对象转为 ByteBuffer）
        ByteBuffer[] nioBuffers = in.nioBuffers();
        int nioBufferCnt = in.nioBufferCount();
        long expectedWrittenBytes = in.nioBufferSize();
```

```java
            SocketChannel ch = javaChannel();
            switch (nioBufferCnt) {
                case 0:
                    // 除了 ByteBuffers，还有其他可以写入的内容，所以可以退回到正常写入，此时调用父类方法
完成对其他数据类型的写入
                    super.doWrite(in);
                    return;
                case 1:
                    // 只有一个 ByteBuffer，执行普通写入
                    ByteBuffer nioBuffer = nioBuffers[0];
                    for (int i = config().getWriteSpinCount() - 1; i >= 0; i --) {  // 写入 ByteBuffer 中的数据，如果在
到达尝试次数后仍未写完，或 SocketChannel 由于底层写缓冲区满无法写入返回 0，则退出当前循环执行
incompleteWrite 方法，在稍后重试写入
                        final int localWrittenBytes = ch.write(nioBuffer);
                        if (localWrittenBytes == 0) {  // 底层 IO 写缓冲区满，无法写入，则注册写感兴趣事件，
等待写缓冲区可用时，由 Selector 选中后，继续写入
                            setOpWrite = true;
                            break;
                        }
                        expectedWrittenBytes -= localWrittenBytes;
                        writtenBytes += localWrittenBytes;
                        if (expectedWrittenBytes == 0) {              // 当前 ByteBuffer 中数据已经写完
                            done = true;
                            break;
                        }
                    }
                    break;
                default:               // 2 个及以上缓冲区写入，则需要聚合写入，调用方法为 ch.write(nioBuffers,
0, nioBufferCnt)，流程同上
                    for (int i = config().getWriteSpinCount() - 1; i >= 0; i --) {
                        final long localWrittenBytes = ch.write(nioBuffers, 0, nioBufferCnt);
                        if (localWrittenBytes == 0) {
                            setOpWrite = true;
                            break;
                        }
                        expectedWrittenBytes -= localWrittenBytes;
                        writtenBytes += localWrittenBytes;
                        if (expectedWrittenBytes == 0) {
                            done = true;
                            break;
                        }
                    }
                    break;
            }
            // 完全释放写入的缓冲区，并更新部分写入的缓冲区索引
            in.removeBytes(writtenBytes);
            if (!done) {              // 没有全部写完写缓冲区数据，将写入操作放入下一次操作进行写入
                incompleteWrite(setOpWrite);
                break;
            }
        }
    }
}
```

12.20.13　NioSocketChannelUnsafe 核心内部类

NioSocketChannelUnsafe 内部类继承自 NioByteUnsafe，重写了 prepareToClose 方法，该方法将返回用于关闭 Socket 使用的异步执行器对象。

如果 SocketChannel 当前处于打开状态，且设置了 TCP 的 SO_LINGER 选项，该选项设置延迟关闭的时间，等待套接字发送缓冲区中的数据发送完成。没有设置该选项时，在调用close()后，将立即发送 FIN，随后 Socket 会立即进行一些清理工作并返回。

如果设置了 SO_LINGER 选项，并且等待时间为正值，则在清理之前会等待一段时间。在发送完 FIN 包后，Socket 会进入 FIN_WAIT_1 状态。如果没有设置延迟关闭（即设置 SO_LINGER 选项），在底层内核调用 tcp_send_fin()发送 FIN 后，会立即调用 sock_orphan()将 sock 结构从当前进程上下文中分离。分离后，用户层进程不会再接收到套接字的读写事件，也不知道套接字发送缓冲区中的数据是否被对端接收。

如果设置了 SO_LINGER 选项，并且等待时间为大于 0 的值，会等待套接字的状态从 FIN_WAIT_1 迁移到 FIN_WAIT_2 状态。套接字进入 FIN_WAIT_2 状态是在发送的 FIN 包被确认后，而 FIN 包肯定是在发送缓冲区中的最后一个字节，所以 FIN 包的确认就表明了发送缓冲区中的数据已经被全部接收。源码如下。

```
private final class NioSocketChannelUnsafe extends NioByteUnsafe {
    @Override
    protected Executor prepareToClose() {
        try {
            if (javaChannel().isOpen() && config().getSoLinger() > 0) {
                doDeregister();                                    // 将 SelectionKey 取消，即将通道从 Selector 中解除注册
                return GlobalEventExecutor.INSTANCE;  // 返回全局事件执行器对象，将关闭操作放入其中执行，此时不会在 SO_LINGER 期间对 Netty 的事件循环线程造成影响（阻塞或异常）
            }
        } catch (Throwable ignore) {
        }
        return null;
    }
}
```

12.21　AbstractNioMessageChannel NIO 服务端模板类原理

AbstractNioMessageChannel 类继承自 AbstractNioChannel，实现了服务端（注意，由于本书只涉及 TCP 的 ServerSocketChannel，所以该类统称为服务端实现。实际上，对于 UDP 来说，没有服务端之分，操作的基本数据为 IP 数据报，读者在实际浏览非 TCP 协议的实现时，请注意里面 Message 的含义为一组 byte 数据的组合）

NIO 通道的模板定义。AbstractNioByteChannel 类定义了客户端的 NIO 通道，代码涉及连接、读、写等操作。对于服务端而言，仅仅是接收客户端连接生成客户端 SocketChannel，然后双方进行通信。所以服务端代码较为简单，这里笔者直接将源码去掉简单的 get/set 部分统一讲解。由源码可知。

（1）在构造器中通过传入父通道、底层可选择通道、读感兴趣事件初始化父类。

（2）引入 inputShutdown 变量标识当前通道是否已经关闭输入端，在 doBeginRead 读取通道数据前验证输入端没有被关闭，然后调用父类的 doBeginRead，由父类模板算法读取。

（3）NioMessageUnsafe 继承自 AbstractNioUnsafe，实现了 read 和 doWrite 方法完成对 Message 消息的读写。注意，由于这里只研究 TCP，对于 TCP 而言，存在 Server 端，而 Server 端不参与 Socket 的通信，所以不调用 doWrite 方法。这里了解即可，了解 read 方法的实现过程即可。

① NioMessageUnsafe 定义了 List readBuf 作为读缓冲区，将从底层 IO 接收的 SocketChannel 放入其中。

② 循环读取底层 IO 数据。

③ 调用子类实现的 doReadMessages(readBuf)方法完成对 SocketChannel 的创建。

④ 如果读取完毕，则结束当前循环。

⑤ 调用流水线中的通道处理器链的 ChannelRead 方法，处理每一个接收到的 SocketChannel，然后 List readBuf 缓冲区已处理的 SocketChannel。

⑥ 调用 ChannelReadComplete 表明当前读缓冲区中的数据全部处理完毕。

⑦ 如果子类接收客户端连接时发生异常，则判断子类实例是否为 ServerChannel。即使抛出 IOException，ServerChannel 也不应该被关闭，因为它经常可以继续接收传入的连接。如果为其他类型的实例，则关闭通道。

⑧ 调用流水线中的通道处理器链的 ExceptionCaught 方法，完成对异常的处理。

⑨ 判断需要关闭当前通道，调用 close 方法关闭通道。

⑩ 最后判断 readPending 与 config.isAutoRead，决定是否移除读感兴趣事件集。

```
public abstract class AbstractNioMessageChannel extends AbstractNioChannel {
    boolean inputShutdown; // 标识当前通道是否已经关闭输入端

    // 构造器初始化：父通道对象、底层实际 IO 通道、读感兴趣事件集
    protected AbstractNioMessageChannel(Channel parent, SelectableChannel ch, int readInterestOp) {
        super(parent, ch, readInterestOp);
    }

    // 判断输入端没有关闭后，调用父类开始读取客户端连接（由于这里研究的是服务端，读的必定是客户端的连接）
    protected void doBeginRead() throws Exception {
        if (inputShutdown) {
            return;
        }
        super.doBeginRead();
    }

    // 实现读写 Message 的模板方法
    private final class NioMessageUnsafe extends AbstractNioUnsafe {
        private final List<Object> readBuf = new ArrayList<Object>();        // 用于保存读取的 Message 缓冲区（这里自然为 SocketChannel）

        // 读取通道 Message 数据模板方法实现
        public void read() {
```

```java
            final ChannelConfig config = config();
            final ChannelPipeline pipeline = pipeline();
            final RecvByteBufAllocator.Handle allocHandle = unsafe().recvBufAllocHandle(); // 接收缓冲区处理器

            allocHandle.reset(config);
            boolean closed = false;
            Throwable exception = null;
            try {
                try {
                    do {                                                    // 循环读取底层 IO 数据
                        int localRead = doReadMessages(readBuf);            // 子类实现实际读取操作
                        if (localRead == 0) {                               // 读取完毕
                            break;
                        }
                        if (localRead < 0) { // 返回-1,说明此时底层 IO 发生异常,关闭该通道(对于 TCP 而言,不会返回-1。注意后面 NioServerSocketChannel 对于 doReadMessages 方法的描述)
                            closed = true;
                            break;
                        }
                        allocHandle.incMessagesRead(localRead);             // 增加已读取数量
                    } while (allocHandle.continueReading());                // 接收缓冲区处理器,根据设置的接收个数判定是否继续读取
                } catch (Throwable t) {
                    exception = t;
                }
                // 调用流水线中的通道处理器链的 ChannelRead 方法,处理每一个接收到的 Message 数据(对 TCP 而言为 SocketChannel),处理完成后清空读缓冲区,调用 ChannelReadComplete 表明当前读缓冲区中的数据已全部处理完毕
                int size = readBuf.size();
                for (int i = 0; i < size; i ++) {
                    readPending = false;
                    pipeline.fireChannelRead(readBuf.get(i));
                }
                readBuf.clear();
                allocHandle.readComplete();
                pipeline.fireChannelReadComplete();
                if (exception != null) {        // 子类接收客户端连接时发生异常
                    if (exception instanceof IOException && !(exception instanceof PortUnreachableException)) {
                        closed = !(AbstractNioMessageChannel.this instanceof ServerChannel); // 不关闭 TCP 的 ServerChannel,即使在 IOException 的情况下,因为它经常可以继续接收传入的连接(例如,可能由于打开的文件太多,抛出异常)
                    }
                    // 调用流水线中的通道处理器链的 ExceptionCaught 方法,完成对异常的处理
                    pipeline.fireExceptionCaught(exception);
                }
                if (closed) {                   // 子类返回-1,由子类决定是否需要关闭当前通道
                    inputShutdown = true;
                    if (isOpen()) {             // 当前通道处于打开状态,需要关闭
                        close(voidPromise());
                    }
                }
            } finally {          // 判断是否需要继续自动读取数据,如果继续自动读取,则不移除代表通道的
```

SelectionKey 的读感兴趣集，在下次通道可读时将被 Selector 选中，由事件循环线程处理。readPending 为 false 时，由用户所在通道处理器回调方法 channelRead(...)/hannelReadComplete(...)中调用 Channel.read() 或 ChannelHandlerContext.read()方法

```java
            if (!readPending && !config.isAutoRead()) {
                removeReadOp();
            }
        }
    }
}

// 将写缓冲区中的数据写入底层 IO，对于 ServerChannel 而言，不会调用该方法。这里了解即可
protected void doWrite(ChannelOutboundBuffer in) throws Exception {
    final SelectionKey key = selectionKey();
    final int interestOps = key.interestOps();
    for (;;) {                                    // 循环写入写缓冲区中的数据
        Object msg = in.current();
        if (msg == null) {
            // 写入完毕，将写感兴趣事件集清空，因为已经完成写入，不需要再对底层 IO 缓冲区执行写操作
            if ((interestOps & SelectionKey.OP_WRITE) != 0) {
                key.interestOps(interestOps & ~SelectionKey.OP_WRITE);
            }
            break;
        }
        try {
            boolean done = false;
            // 调用子类处理当前 Message 的发送，在重试次数达到后，仍未将 Message 所有内容发送完毕，则注册写感兴趣事件集，等待底层 IO 可写时继续写入
            for (int i = config().getWriteSpinCount() - 1; i >= 0; i--) {
                if (doWriteMessage(msg, in)) {
                    done = true;
                    break;
                }
            }
            if (done) {     // 完全写入当前 Messsage，则从写缓冲区中移除
                in.remove();
            } else {
                if ((interestOps & SelectionKey.OP_WRITE) == 0) {
                    key.interestOps(interestOps | SelectionKey.OP_WRITE);
                }
                break;
            }
        } catch (IOException e) {
            if (continueOnWriteError()) {
                in.remove(e);
            } else {
                throw e;
            }
        }
    }
}

// 子类完成实现实际读操作
protected abstract int doReadMessages(List<Object> buf) throws Exception;
```

```java
// 子类完成实现实际写操作
protected abstract boolean doWriteMessage(Object msg, ChannelOutboundBuffer in) throws Exception;
}
```

12.22 NioServerSocketChannel NIO TCP 服务端实现类原理

NioServerSocketChannel 类继承自 AbstractNioMessageChannel，同时实现了 ServerSocketChannel 接口，完整实现了基于 Java 原生 ServerSocketChannel 实现的服务端类。由于大部分工作都在父类中完成，同时由于 ServerSocketChannel 只负责接收客户端连接创建 NioSocketChannel 对象，并不参与实际 IO 写操作，所以该类较为简单，笔者去除了 get/set 方法，进行统一分析。由源码可如。

（1）该类将通过 SelectorProvider 创建支持 IO 连接的 ServerSocketChannel 实例。

（2）该类在构造器中初始化父类的读感兴趣事件集为 SelectionKey.OP_ACCEPT，因为 ServerSocketChannel 只用于接收连接而不是读取数据。

（3）该类所有 IO 操作均通过原生 Java 的 ServerSocketChannel 的对应方法完成处理。

（4）该类只实现 doReadMessages 方法完成对客户端的连接读取，其他方法实现将抛出 UnsupportedOperationException 异常。

```java
public class NioServerSocketChannel extends AbstractNioMessageChannel
    implements io.Netty.channel.socket.ServerSocketChannel {
    private static final ChannelMetadata METADATA = new ChannelMetadata(false, 16); // 通道元数据，指定当前最大读取通道的个数为 16
    private static final SelectorProvider DEFAULT_SELECTOR_PROVIDER = SelectorProvider.provider();
// 默认通道选择器提供器，用于创建 ServerSocketChannel 实例（使用该类完成创建，而不是
// ServerSocketChannel.open()，原因同 NioSocketChannel）

    // 创建新的 ServerSocketChannel 实例
    private static ServerSocketChannel newSocket(SelectorProvider provider) {
        try {
            return provider.openServerSocketChannel();
        } catch (IOException e) {
            throw new ChannelException(
                "Failed to open a server socket.", e);
        }
    }
    private final ServerSocketChannelConfig config;

    // 构造器：使用默认的 SelectorProvider 创建新的 ServerSocketChannel，完成对 NioServerSocketChannel 初始化
    public NioServerSocketChannel() {
        this(newSocket(DEFAULT_SELECTOR_PROVIDER));
    }
    public NioServerSocketChannel(SelectorProvider provider) {
        this(newSocket(provider));
    }

    // 构造器：由于 ServerSocketChannel 没有父通道，因为没有类作为其父通道创建它，所以这里置空。同时
```

注意，这里的读感兴趣不同于 NioSocketChannel 的 OP_READ，因为这里需要感兴趣的是对客户端的接收，而不是数据读写

```java
public NioServerSocketChannel(ServerSocketChannel channel) {
    super(null, channel, SelectionKey.OP_ACCEPT);
    config = new NioServerSocketChannelConfig(this, javaChannel().socket());
}

// 通道是否激活条件：ServerSocketChannel 已经绑定地址
public boolean isActive() {
    return javaChannel().socket().isBound();
}

// 通道绑定，使用 config 配置对象中设置的 BackLog（TCP 三次握手成功后的 accept 队列长度）
protected void doBind(SocketAddress localAddress) throws Exception {
    javaChannel().socket().bind(localAddress, config.getBacklog());
}

// 通道关闭
protected void doClose() throws Exception {
    javaChannel().close();
}

// 通道读取 SocketChannel
protected int doReadMessages(List<Object> buf) throws Exception {
    SocketChannel ch = javaChannel().accept();    // 调用原生 Java 的 accept 方法对完成 TCP 三次握手的通道接收
    try {
        if (ch != null) {
            buf.add(new NioSocketChannel(this, ch));   // 将其包装为 Netty 的 NioSocketChannel 对象，注意这里把 Parent 设置为当前 NioServerSocketChannel，因为 NioServerSocketChannel 该通道创建了它
            return 1;
        }
    } catch (Throwable t) {  // 发生异常，关闭 SocketChannel
        logger.warn("Failed to create a new channel from an accepted socket.", t);
        try {
            ch.close();
        } catch (Throwable t2) {
            logger.warn("Failed to close a socket.", t2);
        }
    }

    return 0;
}

/* ServerSocketChannel 不需要实现的方法 */
@Override
protected boolean doConnect(
    SocketAddress remoteAddress, SocketAddress localAddress) throws Exception {
    throw new UnsupportedOperationException();
}

@Override
protected void doFinishConnect() throws Exception {
```

```java
        throw new UnsupportedOperationException();
    }

    @Override
    protected SocketAddress remoteAddress0() {
        return null;
    }

    @Override
    protected void doDisconnect() throws Exception {
        throw new UnsupportedOperationException();
    }

    @Override
    protected boolean doWriteMessage(Object msg, ChannelOutboundBuffer in) throws Exception {
        throw new UnsupportedOperationException();
    }

    @Override
    protected final Object filterOutboundMessage(Object msg) throws Exception {
        throw new UnsupportedOperationException();
    }
}
```

第 13 章 Netty 通道流水线与通道处理器原理

13.1 ChannelPipeline 接口

ChannelPipeline 通道处理器流水线接口表示一组 ChannelHandler 通道处理器的集合。通道发生读、写、异常等操作时，需要用通道处理器处理这些事件。设计时，需要详细考虑使用哪种设计模式将通道实际处理代码与事件处理解耦，同时还需要对这些处理器进行不同定义。很明显，使用责任链模式最合适，可以根据实际业务对 ChannelHandler 进行自定义，将其组成一个责任链一起处理这些数据。例如有责任链 A、B、C，则可以对最原始的数据在 A 和 B 中处理。例如将 TCP 报文解析为 HTTP 报文，然后将业务写在 C 处理器，这样 C 就可以安心编写业务即可，不需要参与对报文的解析。

在 Channel 通道实例的代码中，每一个通道都将在创建时关联唯一一个 ChannelPipeline 接口实例。图 13-1 是一个 IO 事件在 ChannelPipeline 中的处理流程，I/O 事件由 ChannelInboundHandler（数据输入事件处理器）或 ChannelOutboundHandler（数据输出事件处理器）处理并通过调用 ChannelHandlerContext 类中定义的事件传播方法，将处理结果传递到处理器链中下一个通道处理器。例如 ChannelHandlerContext.fireChannelRead(Object)方法和 ChannelHandlerContext.write(Object)方法。

inbound 输入事件由 Inbound Handler 输入处理器按照自底向上的方向处理，如图 13-1 左侧所示。Inbound Handler 处理器通常处理图 13-1 底部的 Netty 事件循环组 I/O 线程生成的输入数据，输入数据通常通过底层实际 IO Socket 的输入操作从远程对端读取得到。例如，SocketChannel.read(ByteBuffer)方法将 Socket 中的输入数据读取到 ByteBuffer 缓冲区中，如果输入事件到达责任链顶部的 Inbound Handler 处理器仍未被处理，则会将其静默丢弃，或在需要应用程序打开日志时将其记录下来。

outbound 输出事件由 Outbound Handler 输出处理器按照自顶向下的方向处理，如图 13-1 右侧所示。Outbound Handler 处理器通常生成或转换输出到底层 Socket IO 的报文，例如 HTTP 和其他自定义协议。如果 outbound 输出事件传输到了底部 Outbound Handler 处理器，则由与 Channel 对象关联的 Netty 事件循环 I/O 线程处理，该 I/O 线程通常执行实际的输出操作，例如，调用 SocketChannel.write(ByteBuffer)将 ByteBuffer 中的数据写入底层 SocketChannel 中传输给对端。

例如，可以定义如下责任链。在下面的例子中，名称以 Inbound 开头的类意味着它是一个 Inbound Handler 输入处理器，名称以 Outbound 开头的类意味着它是一个 Outbound Handler 输出处理器。在给定的示例配置中，当读取数据事件发生时，处理顺序为 1、2、3、4、5。当数据写出事件发生时，处理顺序为 5、4、3、2、1。在这个顺序下，ChannelPipeline 流水线会按照以下规则，来跳过某些不必要的通道处理器来减少调用栈深度。

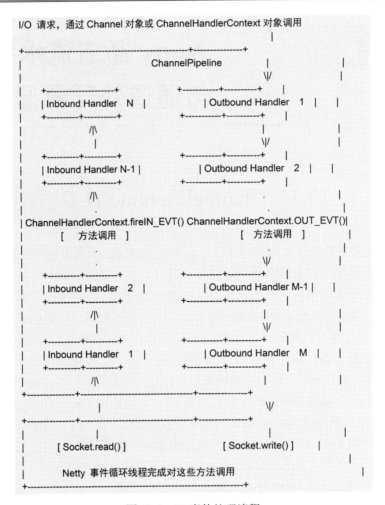

图 13-1 IO 事件处理流程

（1）通道处理器 3 和 4 没有实现 ChannelInboundHandler 接口，因此读取数据事件的实际执行顺序为 1、2、5。

（2）通道处理器 1 和 2 没有实现 ChannelOutboundHandler 接口，因此写入数据事件的实际执行顺序为 5、4、3。

（3）通道处理器 5 同时实现 ChannelInboundHandler 和 ChannelOutboundHandler 接口，因此会同时处理输入和输出事件。

（4）最终读取数据事件和写入数据事件分别为 1、2、5 和 5、4、3。

```
ChannelPipeline p = ...;
p.addLast("1", new InboundHandlerA());
p.addLast("2", new InboundHandlerB());
p.addLast("3", new OutboundHandlerA());
p.addLast("4", new OutboundHandlerB());
p.addLast("5", new InboundOutboundHandlerX());
```

另外，读者可以在图 13-1 看到：处理程序必须调用 ChannelHandlerContext 类中定义的事件传播方法，将事件转发到它的下一个处理器。这些方法包括。

（1）数据输入事件传播方法。
- ☑ ChannelHandlerContext.fireChannelRegistered()通道注册。
- ☑ ChannelHandlerContext.fireChannelActive()通道激活。
- ☑ ChannelHandlerContext.fireChannelRead(Object)接收到通道数据。
- ☑ ChannelHandlerContext.fireChannelReadComplete()当前读缓冲区中的数据读取完成。
- ☑ ChannelHandlerContext.fireExceptionCaught(Throwable)通道发生异常。
- ☑ ChannelHandlerContext.fireUserEventTriggered(Object)用户自定义事件。
- ☑ ChannelHandlerContext.fireChannelWritabilityChanged()通道写入能力改变。
- ☑ ChannelHandlerContext.fireChannelInactive()通道变为非激活状态。
- ☑ ChannelHandlerContext.fireChannelUnregistered()通道解除注册。

（2）数据输出事件传播方法。
- ☑ ChannelHandlerContext.bind(SocketAddress, ChannelPromise)通道地址绑定。
- ☑ ChannelHandlerContext.connect(SocketAddress, SocketAddress, ChannelPromise)通道连接。
- ☑ ChannelHandlerContext.write(Object, ChannelPromise)通道数据写出写缓冲区。
- ☑ ChannelHandlerContext.flush()将通道写缓冲区数据写到底层实际IO。
- ☑ ChannelHandlerContext.read()通道读取数据。
- ☑ ChannelHandlerContext.disconnect(ChannelPromise)通道断开连接。
- ☑ ChannelHandlerContext.close(ChannelPromise)通道关闭。
- ☑ ChannelHandlerContext.deregister(ChannelPromise)通道解除与事件循环注册。

ChannelPipeline接口定义了对通道处理器操作的方法，由源码可知。

（1）它继承自ChannelInboundInvoker与ChannelOutboundInvoker接口，用于触发输入和输出事件的能力。

（2）由于责任链需要一个管理器，而该管理器就是ChannelPipeline接口的实例，所以在该接口中定义了对管理的ChannelHandler进行增、删、改、查的功能。

```java
public interface ChannelPipeline extends ChannelInboundInvoker, ChannelOutboundInvoker,
Iterable<Entry<String, ChannelHandler>> {

    // 将 ChannelHandler 处理器添加到通道责任链的头部
    // 参数 name 指定通道处理器的名字
    ChannelPipeline addFirst(String name, ChannelHandler handler);

    // 将 ChannelHandler 处理器添加到通道责任链的头部
    // 参数 name 指定通道处理器的名字
    // 参数 group 指定当前通道处理器应该放入指定的事件执行器组中执行
    ChannelPipeline addFirst(EventExecutorGroup group, String name, ChannelHandler handler);

    // 将 ChannelHandler 处理器添加到通道责任链的尾部
    // 参数 name 指定通道处理器的名字
    ChannelPipeline addLast(String name, ChannelHandler handler);

    // 将 ChannelHandler 处理器添加到通道责任链的尾部
```

```
// 参数 name 指定通道处理器的名字
// 参数 group 指定当前通道处理器应该放入指定的事件执行器组中执行
ChannelPipeline addLast(EventExecutorGroup group, String name, ChannelHandler handler);

// 将 ChannelHandler 处理器添加到通道责任链的指定通道的前面
// 参数 baseName 指定需要添加到前面的通道处理器名字
// 参数 name 指定通道处理器的名字
ChannelPipeline addBefore(String baseName, String name, ChannelHandler handler);

// 将 ChannelHandler 处理器添加到通道责任链的指定通道的前面
// 参数 baseName 指定需要添加到前面的通道处理器名字
// 参数 name 指定通道处理器的名字
// 参数 group 指定当前通道处理器应该放入指定的事件执行器组中执行
ChannelPipeline addBefore(EventExecutorGroup group, String baseName, String name, ChannelHandler handler);

// 将 ChannelHandler 处理器添加到通道责任链的指定通道的后面
// 参数 baseName 指定需要添加到后面的通道处理器名字
// 参数 name 指定通道处理器的名字
ChannelPipeline addAfter(String baseName, String name, ChannelHandler handler);

// 将 ChannelHandler 处理器添加到通道责任链的指定通道的后面
// 参数 baseName 指定需要添加到后面的通道处理器名字
// 参数 name 指定通道处理器的名字
// 参数 group 指定当前通道处理器应该放入指定的事件执行器组中执行
ChannelPipeline addAfter(EventExecutorGroup group, String baseName, String name, ChannelHandler handler);

// 批量插入通道处理器，除了可变参数，作用同上
ChannelPipeline addFirst(ChannelHandler... handlers);
ChannelPipeline addFirst(EventExecutorGroup group, ChannelHandler... handlers);
ChannelPipeline addLast(ChannelHandler... handlers);
ChannelPipeline addLast(EventExecutorGroup group, ChannelHandler... handlers);
ChannelPipeline remove(ChannelHandler handler);

// 移除指定名字的通道处理器
ChannelHandler remove(String name);

// 移除指定类型的通道处理器
<T extends ChannelHandler> T remove(Class<T> handlerType);

// 移除责任链头部的通道处理器
ChannelHandler removeFirst();

// 移除责任链尾部的通道处理器
ChannelHandler removeLast();

// 将指定 oldHandler 通道处理器替换为 newHandler
// 并设置新的名字为 newName
ChannelPipeline replace(ChannelHandler oldHandler, String newName, ChannelHandler newHandler);

// 将指定名字为 oldName 通道处理器替换为 newHandler
// 并设置新的名字为 newName
```

```java
ChannelHandler replace(String oldName, String newName, ChannelHandler newHandler);

// 将指定类型为 oldHandlerType 通道处理器替换为 newHandler
// 并设置新的名字为 newName
<T extends ChannelHandler> T replace(Class<T> oldHandlerType, String newName,
                ChannelHandler newHandler);

// 获取责任链头部的通道处理器
ChannelHandler first();

// 获取代表责任链头部通道处理器的 ChannelHandlerContext 上下文对象
ChannelHandlerContext firstContext();

// 获取责任链尾部的通道处理器
ChannelHandler last();

// 获取代表责任链尾部通道处理器的 ChannelHandlerContext 上下文对象
ChannelHandlerContext lastContext();

// 获取指定名字为 name 的通道处理器
ChannelHandler get(String name);

// 获取指定类型为 handlerType 的通道处理器
<T extends ChannelHandler> T get(Class<T> handlerType);

// 获取代表 ChannelHandler 通道处理器的 ChannelHandlerContext 上下文对象
ChannelHandlerContext context(ChannelHandler handler);

// 获取代表名为 name 通道处理器的 ChannelHandlerContext 上下文对象
ChannelHandlerContext context(String name);

// 获取类型为 handlerType 通道处理器的 ChannelHandlerContext 上下文对象
ChannelHandlerContext context(Class<? extends ChannelHandler> handlerType);

// 返回与流水线绑定的通道对象
Channel channel();

// 获取所有通道名
List<String> names();

// 将所有通道和名字转为 map 集合
Map<String, ChannelHandler> toMap();
}
```

13.2　ChannelHandlerContext 接口

ChannelHandlerContext 通道处理器上下文接口用于包装通道处理器，使通道处理器可以在 ChannelPipeline 中形成责任链模式，同时也可以将处理事件传递到责任链的下游。包括 ChannelHandlerContext-A（Channel - A）、ChannelHandlerContext - B（Channel - B）、ChannelHandlerContext - C（Channel - C），

包装通道管理器提供责任链的相关功能。由源码可知。

（1）该接口继承自ChannelInboundInvoker和ChannelOutboundInvoker，将具备触发数据输入和数据输出的能力，将事件传递到责任链的后继处理器。

（2）该接口实例将用于包装ChannelHandler，提供ChannelHandler的元数据：名字、关联的通道等。

```java
public interface ChannelHandlerContext extends AttributeMap, ChannelInboundInvoker, ChannelOutboundInvoker {
    // 获取当前所属通道对象
    Channel channel();

    // 获取当前可以执行事件的执行器
    EventExecutor executor();

    // 当前通道处理器的名字
    String name();

    // 包装的通道管理器
    ChannelHandler handler();

    // 判断是否已经从所属 ChannelPipeline 流水线中移除了当前通道处理器的上下文
    boolean isRemoved();

    // 获取所属流水线对象
    ChannelPipeline pipeline();

    // 获取用于支持内存分配的内存分配器
    ByteBufAllocator alloc();
}
```

13.3　ChannelHandler 接口

ChannelHandler 通道处理器接口定义了通道处理器回调事件方法，读者可以类比观察者模式中的被观察者定义观察者列表、观察者实现观察者接口接收对应事件处理。这里便是定义了可以处理特定通道事件的方法。由源码可知。

（1）由于通道处理器分为数据输入和输出处理器，所以相关回调方法会在子接口 ChannelInboundHandler 和 ChannelOutboundHandler 中定义，本接口只定义两者共同的点：通道添加和移除回调方法。

（2）内部 Sharable 注解为标志性注解，通过该注解标识的通道处理器可以在不同流水线中共享对象，即多线程安全。

```java
public interface ChannelHandler {

    // 通道在添加通道处理器时回调，参数 ctx 表明添加的处理器上下文对象
    void handlerAdded(ChannelHandlerContext ctx) throws Exception;

    // 通道在移除通道处理器时回调，参数 ctx 表明移除的处理器上下文对象
    void handlerRemoved(ChannelHandlerContext ctx) throws Exception;
```

```
    // 标识该注解的 ChannelHandler 通道处理器对象的同一个实例可以被多次添加到一个或多个流水线中,即多
线程共享。
    // 如果没有指定此注释,则必须在每次将其添加到流水线时,创建一个新的通道处理器对象
    @Inherited
    @Documented
    @Target(ElementType.TYPE)
    @Retention(RetentionPolicy.RUNTIME)
    @interface Sharable {
        // no value
    }
}
```

13.4　ChannelInboundHandler 接口

ChannelInboundHandler 数据输入通道处理器接口继承自 ChannelHandler,定义了用于回调处理通道输入事件的方法。由源码可知。

所有处理方法均与通道的输入事件相关,子类只需要实现对应方法便可完成对数据的处理,同时入参(即 Input Parameters)ChannelHandlerContext 表示当前通道处理器的上下文,即责任链的对象,可以通过它将事件传递到后继通道处理器当中。源码如下。

```
public interface ChannelInboundHandler extends ChannelHandler {

    // 通道被注册到事件循环时回调
    void channelRegistered(ChannelHandlerContext ctx) throws Exception;

    // 通道从事件循环解除注册时回调
    void channelUnregistered(ChannelHandlerContext ctx) throws Exception;

    // 通道被激活时回调
    void channelActive(ChannelHandlerContext ctx) throws Exception;

    // 通道失活时回调
    void channelInactive(ChannelHandlerContext ctx) throws Exception;

    // 通道读取到数据时回调,msg 表示读取到的数据
    void channelRead(ChannelHandlerContext ctx, Object msg) throws Exception;

    // 当前读缓冲区中的所有数据都处理完成时回调
    void channelReadComplete(ChannelHandlerContext ctx) throws Exception;

    // 触发用户自定义事件时回调,evt 表示事件对象
    void userEventTriggered(ChannelHandlerContext ctx, Object evt) throws Exception;

    // 通道可写状态发生改变时回调
    void channelWritabilityChanged(ChannelHandlerContext ctx) throws Exception;

    // 通道发生异常时回调
    void exceptionCaught(ChannelHandlerContext ctx, Throwable cause) throws Exception;
}
```

13.5 ChannelOutboundHandler 接口

ChannelOutboundHandler 数据输出通道处理器接口继承自 ChannelHandler，定义了用于回调处理通道输出事件的方法。

所有处理方法均与通道的输出事件相关，子类只需要实现对应方法便可完成对数据的处理，同时入参（即 Input Parameters）ChannelHandlerContext 表示当前通道处理器的上下文，即责任链的对象，可以通过它将事件传递到后继通道处理器当中。源码如下。

```java
public interface ChannelOutboundHandler extends ChannelHandler {

    // 通道绑定时回调
    void bind(ChannelHandlerContext ctx, SocketAddress localAddress, ChannelPromise promise) throws Exception;

    // 通道连接时回调
    void connect(
        ChannelHandlerContext ctx, SocketAddress remoteAddress,
        SocketAddress localAddress, ChannelPromise promise) throws Exception;

    // 通道断开连接时回调
    void disconnect(ChannelHandlerContext ctx, ChannelPromise promise) throws Exception;

    // 通道关闭时回调
    void close(ChannelHandlerContext ctx, ChannelPromise promise) throws Exception;

    // 通道从事件循环中解除注册时回调
    void deregister(ChannelHandlerContext ctx, ChannelPromise promise) throws Exception;

    // 拦截 ChannelHandlerContext.read 方法
    void read(ChannelHandlerContext ctx) throws Exception;

    // 通道写出数据时回调
    void write(ChannelHandlerContext ctx, Object msg, ChannelPromise promise) throws Exception;

    // 通道将写缓冲区数据写入底层 IO 时回调
    void flush(ChannelHandlerContext ctx) throws Exception;
}
```

13.6 AbstractChannelHandlerContext 模板类

AbstractChannelHandlerContext 通道处理器上下文抽象类用于定义 ChannelHandlerContext 接口的模板类。该类实现较为复杂，所以笔者将其拆分描述。注意了解该类如何形成责任链，以及如何包装通道处理器元数据信息。

13.6.1 核心变量与构造器

由源码可知。

(1) AbstractChannelHandlerContext 继承自 DefaultAttributeMap，实现了对 attr 属性 key 和 value 的存储。读者这里不了解也没关系，本书由于篇幅有限，省略了工具类相关的内容，读者只需要知道通过继承 DefaultAttributeMap 类，便可以在上下文对象中存储 key-value 即可。

(2) 通道上下文的状态由 handlerState 表示，并且使用 AtomicIntegerFieldUpdater 对其进行原子性更新，状态值有如下四个。

- ☑ ADD_PENDING：ChannelHandler.handlerAdded(ChannelHandlerContext)方法即将被调用。
- ☑ ADD_COMPLETE：ChannelHandler.handlerAdded(ChannelHandlerContext)方法已经完成调用。
- ☑ REMOVE_COMPLETE：ChannelHandler.handlerRemoved(ChannelHandlerContext)方法已经完成调用。
- ☑ INIT：通道上下文对象刚创建时的状态。

(3) 使用 inbound 和 outbound 标志位，表示通道上下文处理的数据事件类型：输入数据事件、输出数据事件。

```java
abstract class AbstractChannelHandlerContext extends DefaultAttributeMap
    implements ChannelHandlerContext, ResourceLeakHint {
    volatile AbstractChannelHandlerContext next;   // 通道上下文后继结点
    volatile AbstractChannelHandlerContext prev;   // 通道上下文前驱结点
    private static final AtomicIntegerFieldUpdater<AbstractChannelHandlerContext>
HANDLER_STATE_UPDATER;                              // 原子更新通道上下文状态的更新对象

    static {                                        // 初始化 HANDLER_STATE_UPDATER
        AtomicIntegerFieldUpdater<AbstractChannelHandlerContext> handlerStateUpdater =
PlatformDependent.newAtomicIntegerFieldUpdater(AbstractChannelHandlerContext.class, "handlerState");
        if (handlerStateUpdater == null) {
            handlerStateUpdater = AtomicIntegerFieldUpdater
                .newUpdater(AbstractChannelHandlerContext.class, "handlerState");
        }
        HANDLER_STATE_UPDATER = handlerStateUpdater;
    }

    // 当前上下文添加到流水线的任务已经放入事件循环线程的任务队列中，即将调用 ChannelHandler.
handlerAdded(ChannelHandlerContext) 方法
    private static final int ADD_PENDING = 1;

    // ChannelHandler.handlerAdded(ChannelHandlerContext)方法已经调用完成，此时通道上下文已经处于流
水线责任链中
    private static final int ADD_COMPLETE = 2;

    // ChannelHandler.handlerRemoved(ChannelHandlerContext)方法已经调用完成，从流水线责任链中移除了
通道上下文
    private static final int REMOVE_COMPLETE = 3;

    // 通道上下文刚创建状态
    private static final int INIT = 0;
```

```
    private final boolean inbound;                    // 通道处理器是否为输入数据处理器
    private final boolean outbound;                   // 通道处理器是否为输出数据处理器
    private final DefaultChannelPipeline pipeline;    // 所关联的流水线对象
    private final String name;                        // 通道处理器名
    private final boolean ordered;                    // 执行当前通道处理器回调的事件执行器为 null 或为
OrderedEventExecutor 实例
    final EventExecutor executor;                     // 处理当前通道事件的执行器，如果没有，则为 null
    private ChannelFuture succeededFuture;            // 状态设置为成功的 future，用于支持 newSucceededFuture 方法

    // 用于包装事件到事件执行器队列的处理程序的延迟实例化任务
    private Runnable invokeChannelReadCompleteTask;   // 包装触发 ChannelReadComplete 回调方法任务对象
    private Runnable invokeReadTask;                  // 包装触发 read 回调方法任务对象
    private Runnable invokeChannelWritableStateChangedTask; // 包装触发 ChannelWritableStateChanged
回调方法任务对象
    private Runnable invokeFlushTask;                 // 包装触发 flush 回调方法任务对象
    private volatile int handlerState = INIT;         // 通道状态，默认为初始化状态

    // 构造器，使用流水线、事件执行器、名字、是否为输入数据通道处理器、是否为输出处理器，初始化成员变量
    AbstractChannelHandlerContext(DefaultChannelPipeline pipeline, EventExecutor executor, String name,
                                   boolean inbound, boolean outbound) {
        this.name = ObjectUtil.checkNotNull(name, "name");
        this.pipeline = pipeline;
        this.executor = executor;
        this.inbound = inbound;
        this.outbound = outbound;
        ordered = executor == null || executor instanceof OrderedEventExecutor; // 初始化 ordered 变量
    }
}
```

13.6.2 fireChannelRegistered 核心方法

fireChannelRegistered 方法用于通道在注册到事件循环线程中时调用，用于回调责任链中的通道处理器的 ChannelRegistered 方法。流程如下。

（1）调用 findContextInbound 方法，获取当前上下文中后继结点的第一个输入数据处理器。

（2）调用静态方法 invokeChannelRegistered，在判断当前线程与通道处理器需要执行的事件执行器线程后，调用 invokeChannelRegistered 完成实际方法回调。

（3）在 invokeChannelRegistered 方法中，判断可以调用该通道处理器后，调用通道处理器的 channelRegistered 回调方法，传入了当前通道上下文对象（此时，在通道处理器中就可以调用 ctx.fireChannelRegistered 方法将事件传递到责任链的下一个结点）。

```
public ChannelHandlerContext fireChannelRegistered() {
    invokeChannelRegistered(findContextInbound());
    return this;
}

// 获取当前上下文中第一个输入数据处理器
private AbstractChannelHandlerContext findContextInbound() {
    AbstractChannelHandlerContext ctx = this;
    do {                                      // 从当前上下文开始查找，直到找到下一个输入数据通道处理器
```

```java
        ctx = ctx.next;
    } while (!ctx.inbound);
    return ctx;
}

// 静态方法，判断当前线程是否为设置执行通道回调事件的事件执行器的所属线程。如果不是，则封装实际调用
// 方法由所属事件执行线程执行，否则直接调用回调方法通知通道处理器
static void invokeChannelRegistered(final AbstractChannelHandlerContext next) {
    EventExecutor executor = next.executor();
    if (executor.inEventLoop()) {    // 当前线程是事件执行器对象的所属线程，则直接调用
        next.invokeChannelRegistered();
    } else {
        executor.execute(new Runnable() {
            @Override
            public void run() {
                next.invokeChannelRegistered();
            }
        });
    }
}

// 实际回调操作，判断可以调用该通道处理器后，调用通道处理器的 channelRegistered 回调方法，传入当前通道
// 上下文对象
private void invokeChannelRegistered() {
    if (invokeHandler()) {
        try {
            ((ChannelInboundHandler) handler()).channelRegistered(this);
        } catch (Throwable t) {
            notifyHandlerException(t);
        }
    } else {                             // 通道状态不符合调用条件，则调用责任链的下一个结点完成调用
        fireChannelRegistered();
    }
}

// 尽最大努力判断 ChannelHandler.handlerAdded(ChannelHandlerContext)方法是否已经被调用。如果没有，则
// 返回 false，如果已经被调用，则返回 true。如果这个方法返回 false，不会调用 ChannelHandler 回调方法，而只
// 是转发事件。这是必须要判断的条件，因为 DefaultChannelPipeline 流水线对象可能已经把 ChannelHandler 放在
// 链表中，但还没有调用 ChannelHandler.handlerAdded(ChannelHandlerContext)方法。该方法判断了当前通道上
// 下文状态必须为：ADD_COMPLETE 或 ordered 为 false 且当前已经放入事件执行器队列中
private boolean invokeHandler() {
    // 把 volatile 变量存储到当前栈帧的局部变量表中，避免后面判断时产生更多的 volatile 读操作，影响性能
    int handlerState = this.handlerState;
    return handlerState == ADD_COMPLETE || (!ordered && handlerState == ADD_PENDING);
}

// 处理通道事件发生异常时，判断异常是否由用户定义的 exceptionCaught 方法抛出。如果是，则打印日志，并
// 返回，否则调用通道处理器的 exceptionCaught 回调方法
private void notifyHandlerException(Throwable cause) {
    if (inExceptionCaught(cause)) {
        if (logger.isWarnEnabled()) {
            logger.warn(
                "An exception was thrown by a user handler " +
```

```
                "while handling an exceptionCaught event", cause);
        }
        return;
    }
    invokeExceptionCaught(cause);
}

// 遍历异常堆栈信息，判断当前异常是否在 exceptionCaught 方法中抛出。如果是，则返回 true
private static boolean inExceptionCaught(Throwable cause) {
    do {
        StackTraceElement[] trace = cause.getStackTrace();
        if (trace != null) {
            for (StackTraceElement t : trace) {
                if (t == null) {
                    break;
                }
                if ("exceptionCaught".equals(t.getMethodName())) {
                    return true;
                }
            }
        }
        cause = cause.getCause();
    } while (cause != null);
    return false;
}
```

13.6.3　fireChannelUnregistered 核心方法

fireChannelUnregistered 方法用于通道从事件循环线程中解除注册时调用，用于回调责任链中通道处理器的 ChannelUnregistered 方法。处理流程同 fireChannelRegistered 方法。源码如下。

```
static void invokeChannelUnregistered(final AbstractChannelHandlerContext next) {
    EventExecutor executor = next.executor();
    if (executor.inEventLoop()) {      // 当前线程是事件执行器对象的所属线程
        next.invokeChannelUnregistered();
    } else {
        executor.execute(new Runnable() {
            @Override
            public void run() {
                next.invokeChannelUnregistered();
            }
        });
    }
}

private void invokeChannelUnregistered() {
    if (invokeHandler()) {             // 判断通道上下文状态后，进行回调
        try {
            ((ChannelInboundHandler) handler()).channelUnregistered(this);
        } catch (Throwable t) {
```

```
            notifyHandlerException(t);
        }
    } else {
        fireChannelUnregistered();
    }
}
```

13.6.4　fireChannelActive 核心方法

fireChannelActive 方法用于通道转变为激活状态时调用，用于回调责任链中的通道处理器的 ChannelActive 方法。处理流程同 fireChannelRegistered 方法。源码如下。

```
public ChannelHandlerContext fireChannelActive() {
    final AbstractChannelHandlerContext next = findContextInbound();
    invokeChannelActive(next);
    return this;
}

static void invokeChannelActive(final AbstractChannelHandlerContext next) {
    EventExecutor executor = next.executor();
    if (executor.inEventLoop()) {          // 当前线程是事件执行器对象的所属线程
        next.invokeChannelActive();
    } else {
        executor.execute(new Runnable() {
            @Override
            public void run() {
                next.invokeChannelActive();
            }
        });
    }
}

private void invokeChannelActive() {
    if (invokeHandler()) {                 // 判断通道上下文状态后，进行回调
        try {
            ((ChannelInboundHandler) handler()).channelActive(this);
        } catch (Throwable t) {
            notifyHandlerException(t);
        }
    } else {                               // 传递事件到责任链的下一个结点
        fireChannelActive();
    }
}
```

13.6.5　invokeChannelInactive 核心方法

invokeChannelInactive 方法在通道转换为失活状态时调用，用于回调责任链中的通道处理器的 ChannelInactive 方法。处理流程同 fireChannelRegistered 方法。源码如下。

```
static void invokeChannelInactive(final AbstractChannelHandlerContext next) {
    EventExecutor executor = next.executor();
    if (executor.inEventLoop()) {          // 当前线程是事件执行器对象的所属线程
```

```
                next.invokeChannelInactive();
            } else {
                executor.execute(new Runnable() {
                    @Override
                    public void run() {
                        next.invokeChannelInactive();
                    }
                });
            }
        }

        private void invokeChannelInactive() {
            if (invokeHandler()) {           // 判断通道上下文状态后，进行回调
                try {
                    ((ChannelInboundHandler) handler()).channelInactive(this);
                } catch (Throwable t) {
                    notifyHandlerException(t);
                }
            } else {                         // 传递事件到责任链的下一个结点
                fireChannelInactive();
            }
        }
```

13.6.6　fireExceptionCaught 核心方法

fireExceptionCaught 方法用于通道处理发生异常时调用，用于回调责任链中的通道处理器的 ExceptionCaught 方法。处理流程同 fireChannelRegistered 方法。源码如下。

```
public ChannelHandlerContext fireExceptionCaught(final Throwable cause) {
    invokeExceptionCaught(next, cause);
    return this;
}

static void invokeExceptionCaught(final AbstractChannelHandlerContext next, final Throwable cause) {
    ObjectUtil.checkNotNull(cause, "cause");
    EventExecutor executor = next.executor();
    if (executor.inEventLoop()) {          // 当前线程是事件执行器对象的所属线程
        next.invokeExceptionCaught(cause);
    } else {
        try {
            executor.execute(new Runnable() {
                @Override
                public void run() {
                    next.invokeExceptionCaught(cause);
                }
            });
        } catch (Throwable t) {
            if (logger.isWarnEnabled()) {
                logger.warn("Failed to submit an exceptionCaught() event.", t);
                logger.warn("The exceptionCaught() event that was failed to submit was:", cause);
            }
        }
    }
}
```

```java
private void invokeExceptionCaught(final Throwable cause) {
    if (invokeHandler()) {                    // 判断通道上下文状态后，进行回调
        try {
            handler().exceptionCaught(this, cause);
        } catch (Throwable error) {           // 处理过程中发生了异常，则打印日志
            if (logger.isDebugEnabled()) {
                logger.debug(
                    "An exception {}" +
                    "was thrown by a user handler's exceptionCaught() " +
                    "method while handling the following exception:",
                    ThrowableUtil.stackTraceToString(error), cause);
            } else if (logger.isWarnEnabled()) {
                logger.warn(
                    "An exception '{}' [enable DEBUG level for full stacktrace] " +
                    "was thrown by a user handler's exceptionCaught() " +
                    "method while handling the following exception:", error, cause);
            }
        }
    } else {                                  // 传递事件到责任链的下一个结点
        fireExceptionCaught(cause);
    }
}
```

13.6.7 fireUserEventTriggered 核心方法

fireUserEventTriggered 方法在通道产生用户自定义事件时调用，用于回调责任链中的通道处理器的 UserEventTriggered 方法。处理流程同 fireChannelRegistered 方法。源码如下。

```java
@Override
public ChannelHandlerContext fireUserEventTriggered(final Object event) {
    invokeUserEventTriggered(findContextInbound(), event);
    return this;
}

static void invokeUserEventTriggered(final AbstractChannelHandlerContext next, final Object event) {
    ObjectUtil.checkNotNull(event, "event");
    EventExecutor executor = next.executor();
    if (executor.inEventLoop()) {             // 当前线程是事件执行器对象的所属线程
        next.invokeUserEventTriggered(event);
    } else {
        executor.execute(new Runnable() {
            @Override
            public void run() {
                next.invokeUserEventTriggered(event);
            }
        });
    }
}

private void invokeUserEventTriggered(Object event) {
    if (invokeHandler()) {                    // 判断通道上下文状态后，进行回调
        try {
```

```
            ((ChannelInboundHandler) handler()).userEventTriggered(this, event);
        } catch (Throwable t) {
            notifyHandlerException(t);
        }
    } else {                          // 传递事件到责任链的下一个结点
        fireUserEventTriggered(event);
    }
}
```

13.6.8　fireChannelRead 核心方法

fireChannelRead 方法用于通道产生读取事件调用，用于回调责任链中的通道处理器的 ChannelRead 方法。处理流程同 fireChannelRegistered 方法。源码如下。

```
public ChannelHandlerContext fireChannelRead(final Object msg) {
    invokeChannelRead(findContextInbound(), msg);
    return this;
}

static void invokeChannelRead(final AbstractChannelHandlerContext next, Object msg) {
    final Object m = next.pipeline.touch(ObjectUtil.checkNotNull(msg, "msg"), next);
    EventExecutor executor = next.executor();
    if (executor.inEventLoop()) {     // 当前线程是事件执行器对象的所属线程
        next.invokeChannelRead(m);
    } else {
        executor.execute(new Runnable() {
            @Override
            public void run() {
                next.invokeChannelRead(m);
            }
        });
    }
}

private void invokeChannelRead(Object msg) {
    if (invokeHandler()) {            // 判断通道上下文状态后，进行回调
        try {
            ((ChannelInboundHandler) handler()).channelRead(this, msg);
        } catch (Throwable t) {
            notifyHandlerException(t);
        }
    } else {                          // 传递事件到责任链的下一个结点
        fireChannelRead(msg);
    }
}
```

13.6.9　fireChannelReadComplete 核心方法

fireChannelReadComplete 方法用于通道读取，且在处理完输入缓冲区中的数据时调用，用于回调责任链中的通道处理器的 ChannelReadComplete 方法。处理流程同 fireChannelRegistered 方法。源码如下。

```
public ChannelHandlerContext fireChannelReadComplete() {
```

```
        invokeChannelReadComplete(findContextInbound());
        return this;
}

static void invokeChannelReadComplete(final AbstractChannelHandlerContext next) {
    EventExecutor executor = next.executor();
    if (executor.inEventLoop()) {      // 当前线程是事件执行器对象的所属线程
        next.invokeChannelReadComplete();
    } else {
        Runnable task = next.invokeChannelReadCompleteTask;
        if (task == null) {
            next.invokeChannelReadCompleteTask = task = new Runnable() {
                @Override
                public void run() {
                    next.invokeChannelReadComplete();
                }
            };
        }
        executor.execute(task);
    }
}

private void invokeChannelReadComplete() {
    if (invokeHandler()) {              // 判断通道上下文状态后,进行回调
        try {
            ((ChannelInboundHandler) handler()).channelReadComplete(this);
        } catch (Throwable t) {
            notifyHandlerException(t);
        }
    } else {                            // 传递事件到责任链的下一个结点
        fireChannelReadComplete();
    }
}
```

13.6.10　fireChannelWritabilityChanged 核心方法

fireChannelWritabilityChanged 方法用于通道可写状态发生变化时调用,用于回调责任链中的通道处理器的 ChannelWritabilityChanged 方法。处理流程同 fireChannelRegistered 方法。源码如下。

```
public ChannelHandlerContext fireChannelWritabilityChanged() {
    invokeChannelWritabilityChanged(findContextInbound());
    return this;
}

static void invokeChannelWritabilityChanged(final AbstractChannelHandlerContext next) {
    EventExecutor executor = next.executor();
    if (executor.inEventLoop()) {      // 当前线程是事件执行器对象的所属线程
        next.invokeChannelWritabilityChanged();
    } else {
        Runnable task = next.invokeChannelWritableStateChangedTask;
        if (task == null) {
            next.invokeChannelWritableStateChangedTask = task = new Runnable() {
                @Override
```

```java
            public void run() {
                next.invokeChannelWritabilityChanged();
            }
        };
        executor.execute(task);
    }
}

private void invokeChannelWritabilityChanged() {
    if (invokeHandler()) {                          // 判断通道上下文状态后，进行回调
        try {
            ((ChannelInboundHandler) handler()).channelWritabilityChanged(this);
        } catch (Throwable t) {
            notifyHandlerException(t);
        }
    } else {                                        // 传递事件到责任链的下一个结点
        fireChannelWritabilityChanged();
    }
}
```

13.6.11　bind 核心方法

bind 方法用于通道进行绑定地址时调用，用于回调责任链中的通道处理器的 bind 方法。处理流程同 fireChannelRegistered 方法。源码如下。

```java
public ChannelFuture bind(final SocketAddress localAddress, final ChannelPromise promise) {
    if (localAddress == null) {                     // 绑定的本地地址不能为空
        throw new NullPointerException("localAddress");
    }
    if (!validatePromise(promise, false)) {         // 判断完成时进行回调的 promise 的状态：不能为空、不能为完成状态、不能为 voidpromise 实例、所属通道为当前通道对象等
        return promise;
    }
    // 从当前上下文对象往前一直找到第一个处理输出通道数据事件的通道处理器上下文
    final AbstractChannelHandlerContext next = findContextOutbound();
    EventExecutor executor = next.executor();
    if (executor.inEventLoop()) {                   // 当前线程是事件执行器对象的所属线程
        next.invokeBind(localAddress, promise);
    } else {
        safeExecute(executor, new Runnable() {
            @Override
            public void run() {
                next.invokeBind(localAddress, promise);
            }
        }, promise, null);
    }
    return promise;
}

private void invokeBind(SocketAddress localAddress, ChannelPromise promise) {
    if (invokeHandler()) {                          // 判断通道上下文状态后，进行回调
        try {
```

```java
                ((ChannelOutboundHandler) handler()).bind(this, localAddress, promise);
            } catch (Throwable t) {
                notifyOutboundHandlerException(t, promise);
            }
        } else {                                        // 传递事件到责任链的下一个结点
            bind(localAddress, promise);
        }
    }

    // 从后往前找到处理输出数据的通道处理器
    private AbstractChannelHandlerContext findContextOutbound() {
        AbstractChannelHandlerContext ctx = this;
        do {
            ctx = ctx.prev;
        } while (!ctx.outbound);
        return ctx;
    }
}
```

13.6.12　connect 核心方法

connect 方法在通道进行连接时调用，用于回调责任链中的通道处理器的 connect 方法。处理流程同 bind 方法。源码如下。

```java
public ChannelFuture connect(SocketAddress remoteAddress, ChannelPromise promise) {
    return connect(remoteAddress, null, promise);
}

@Override
public ChannelFuture connect(
        final SocketAddress remoteAddress, final SocketAddress localAddress, final ChannelPromise promise)
{
    if (remoteAddress == null) {                        // 连接的远程地址不能为空
        throw new NullPointerException("remoteAddress");
    }
    if (!validatePromise(promise, false)) {
        return promise;
    }

    final AbstractChannelHandlerContext next = findContextOutbound();// 获取责任链中下一个处理输出数据事件的通道处理器
    EventExecutor executor = next.executor();
    if (executor.inEventLoop()) {                       // 当前线程是事件执行器对象的所属线程
        next.invokeConnect(remoteAddress, localAddress, promise);
    } else {
        safeExecute(executor, new Runnable() {
            @Override
            public void run() {
                next.invokeConnect(remoteAddress, localAddress, promise);
            }
        }, promise, null);
    }
    return promise;
}
```

```java
private void invokeConnect(SocketAddress remoteAddress, SocketAddress localAddress, ChannelPromise promise) {
    if (invokeHandler()) {                              // 判断通道上下文状态后，进行回调
        try {
            ((ChannelOutboundHandler) handler()).connect(this, remoteAddress, localAddress, promise);
        } catch (Throwable t) {
            notifyOutboundHandlerException(t, promise);
        }
    } else {                                            // 传递事件到责任链的下一个结点
        connect(remoteAddress, localAddress, promise);
    }
}
```

13.6.13　disconnect 核心方法

disconnect 方法在通道断开连接时调用，用于回调责任链中的通道处理器的 disconnect 方法。处理流程同 bind 方法。源码如下。

```java
public ChannelFuture disconnect(final ChannelPromise promise) {
    if (!validatePromise(promise, false)) {             // promise 已被取消，则直接返回，不进行回调
        return promise;
    }
    final AbstractChannelHandlerContext next = findContextOutbound();// 获取责任链中下一个处理输出数据事件的通道处理器
    EventExecutor executor = next.executor();
    if (executor.inEventLoop()) {                       // 当前线程是事件执行器对象的所属线程
        if (!channel().metadata().hasDisconnect()) {    // 如果通道不支持调用 disconnect（注意，之前在介绍通道对象时传入 ChannelMetadata 对象中的参数），则直接触发 close 回调
            next.invokeClose(promise);
        } else {                                        // 否则触发 Disconnect 回调
            next.invokeDisconnect(promise);
        }
    } else {
        safeExecute(executor, new Runnable() {
            @Override
            public void run() {
                if (!channel().metadata().hasDisconnect()) {
                    next.invokeClose(promise);
                } else {
                    next.invokeDisconnect(promise);
                }
            }
        }, promise, null);
    }
    return promise;
}

private void invokeDisconnect(ChannelPromise promise) {
    if (invokeHandler()) {                              // 判断通道上下文状态后，进行回调
        try {
            ((ChannelOutboundHandler) handler()).disconnect(this, promise);
        } catch (Throwable t) {
```

```
                notifyOutboundHandlerException(t, promise);
            }
        } else {                                       // 传递事件到责任链的下一个结点
            disconnect(promise);
        }
    }
}
```

13.6.14 close 核心方法

close 方法在通道关闭时调用，用于回调责任链中的通道处理器的 close 方法。处理流程同 bind 方法。源码如下。

```
public ChannelFuture close(final ChannelPromise promise) {
    if (!validatePromise(promise, false)) {        // promise 被取消
        return promise;
    }
    final AbstractChannelHandlerContext next = findContextOutbound();// 获取责任链中下一个处理输出数据事件的通道处理器
    EventExecutor executor = next.executor();
    if (executor.inEventLoop()) {                  // 当前线程是事件执行器对象的所属线程
        next.invokeClose(promise);
    } else {
        safeExecute(executor, new Runnable() {
            @Override
            public void run() {
                next.invokeClose(promise);
            }
        }, promise, null);
    }
    return promise;
}

private void invokeClose(ChannelPromise promise) {
    if (invokeHandler()) {                         // 判断通道上下文状态后，进行回调
        try {
            ((ChannelOutboundHandler) handler()).close(this, promise);
        } catch (Throwable t) {
            notifyOutboundHandlerException(t, promise);
        }
    } else {
        close(promise);
    }
}
```

13.6.15 deregister 核心方法

deregister 方法在通道解除注册时调用，用于回调责任链中的通道处理器的 close 方法。处理流程同 bind 方法。源码如下。

```
public ChannelFuture deregister(final ChannelPromise promise) {
    if (!validatePromise(promise, false)) {
```

```
            return promise;
        }
        final AbstractChannelHandlerContext next = findContextOutbound(); // 获取责任链中下一个处理输出数据
事件的通道处理器
        EventExecutor executor = next.executor();
        if (executor.inEventLoop()) {     // 当前线程是事件执行器对象的所属线程
            next.invokeDeregister(promise);
        } else {
            safeExecute(executor, new Runnable() {
                @Override
                public void run() {
                    next.invokeDeregister(promise);
                }
            }, promise, null);
        }
        return promise;
    }

    private void invokeDeregister(ChannelPromise promise) {
        if (invokeHandler()) {            // 判断通道上下文状态后，进行回调
            try {
                ((ChannelOutboundHandler) handler()).deregister(this, promise);
            } catch (Throwable t) {
                notifyOutboundHandlerException(t, promise);
            }
        } else {
            deregister(promise);
        }
    }
```

13.6.16 read 核心方法

read 方法用于通道配置自动读取时调用（对于 NIO 通道来说，在处理完当前通道事件时，是否自动向代表通道对象的 SelectionKey 中注册读事件集），用于回调责任链中的通道处理器的 read 方法。处理流程同 bind 方法。源码如下。

```
    public ChannelHandlerContext read() {
        final AbstractChannelHandlerContext next = findContextOutbound(); // 获取责任链中下一个处理输出数据
事件的通道处理器
        EventExecutor executor = next.executor();
        if (executor.inEventLoop()) {     // 当前线程是事件执行器对象的所属线程
            next.invokeRead();
        } else {        // 否则封装任务对象，放入实例变量中（懒加载即 Load on Demand，也叫延迟加载），待下次
使用时不需要再创建新的对象
            Runnable task = next.invokeReadTask;
            if (task == null) {
                next.invokeReadTask = task = new Runnable() {
                    @Override
                    public void run() {
                        next.invokeRead();
                    }
```

```
            };
        }
        executor.execute(task);
    }
    return this;
}

private void invokeRead() {
    if (invokeHandler()) {              // 判断通道上下文状态后，进行回调
        try {
            ((ChannelOutboundHandler) handler()).read(this);
        } catch (Throwable t) {
            notifyHandlerException(t);
        }
    } else {
        read();
    }
}
```

13.6.17 write 核心方法

write 方法在通道写入数据事件时调用，用于回调责任链中的通道处理器的 read 方法。处理流程同 bind 方法。源码如下。

```
public ChannelFuture write(Object msg) {
    return write(msg, newPromise());
}

@Override
public ChannelFuture write(final Object msg, final ChannelPromise promise) {
    if (msg == null) {                          // 写出数据不能为空
        throw new NullPointerException("msg");
    }
    try {
        if (!validatePromise(promise, true)) {  // promise 被取消，即写操作被取消，释放内存
            ReferenceCountUtil.release(msg);
            return promise;
        }
    } catch (RuntimeException e) {
        ReferenceCountUtil.release(msg);
        throw e;
    }
    write(msg, false, promise);                 // 完成实际写入操作，这里传入 false，表明不需要刷新写缓冲区
    return promise;
}

// 完成实际写入操作
private void write(Object msg, boolean flush, ChannelPromise promise) {
    AbstractChannelHandlerContext next = findContextOutbound();
    final Object m = pipeline.touch(msg, next);  // 记录当前使用 msg 的堆栈
    EventExecutor executor = next.executor();
    if (executor.inEventLoop()) {                // 当前线程是事件执行器对象的所属线程
```

```java
            if (flush) {                        // 如果执行刷新操作，则调用 WriteAndFlush 方法
                next.invokeWriteAndFlush(m, promise);
            } else {                            // 否则调用 write 方法
                next.invokeWrite(m, promise);
            }
        } else {                                // 否则根据是否刷新创建执行任务实例，并放入执行器线程的队列中执行
            AbstractWriteTask task;
            if (flush) {
                task = WriteAndFlushTask.newInstance(next, m, promise);
            } else {
                task = WriteTask.newInstance(next, m, promise);
            }
            safeExecute(executor, task, promise, m);
        }
    }
}

// 安全执行，其实就是捕捉放入事件执行器时发生了异常，设置 promise 完成失败，同时释放 msg 内存
private static void safeExecute(EventExecutor executor, Runnable runnable, ChannelPromise promise, Object msg) {
    try {
        executor.execute(runnable);
    } catch (Throwable cause) {
        try {
            promise.setFailure(cause);
        } finally {
            if (msg != null) {
                ReferenceCountUtil.release(msg);
            }
        }
    }
}

// 回调 write 方法
private void invokeWrite(Object msg, ChannelPromise promise) {
    if (invokeHandler()) {                      // 判断通道上下文状态后，进行回调
        invokeWrite0(msg, promise);
    } else {                                    // 传递事件到下一个处理器
        write(msg, promise);
    }
}

// 回调处理器 write 方法，发生异常时通知责任链中输出通道处理器
private void invokeWrite0(Object msg, ChannelPromise promise) {
    try {
        ((ChannelOutboundHandler) handler()).write(this, msg, promise);
    } catch (Throwable t) {
        notifyOutboundHandlerException(t, promise);
    }
}

// 发生异常时通道输出通道异常处理器，这里实际上只是将 promise 设置为完成失败而已
private static void notifyOutboundHandlerException(Throwable cause, ChannelPromise promise) {
    if (!promise.tryFailure(cause) && !(promise instanceof VoidChannelPromise)) {
```

```
            if (logger.isWarnEnabled()) {
                logger.warn("Failed to fail the promise because it's done already: {}", promise, cause);
            }
        }
    }
}
```

13.6.18 flush 核心方法

flush 方法在通道刷新写缓冲区事件时调用，用于回调责任链中的通道处理器的 flush 方法。处理流程同 bind 方法。源码如下。

```
public ChannelHandlerContext flush() {
    final AbstractChannelHandlerContext next = findContextOutbound();
    EventExecutor executor = next.executor();
    if (executor.inEventLoop()) {      // 当前线程是事件执行器对象的所属线程
        next.invokeFlush();
    } else {                            // 封装刷新任务并放置到实例变量中
        Runnable task = next.invokeFlushTask;
        if (task == null) {
            next.invokeFlushTask = task = new Runnable() {
                @Override
                public void run() {
                    next.invokeFlush();
                }
            };
        }
        safeExecute(executor, task, channel().voidPromise(), null);
    }
    return this;
}

private void invokeFlush() {
    if (invokeHandler()) {              // 判断通道上下文状态后，进行回调
        invokeFlush0();
    } else {                            // 传递事件到下一个处理器
        flush();
    }
}

// 完成实际刷新，发生异常时回调通道处理器的 ExceptionCaught 方法
private void invokeFlush0() {
    try {
        ((ChannelOutboundHandler) handler()).flush(this);
    } catch (Throwable t) {
        notifyHandlerException(t);
    }
}
```

13.6.19 writeAndFlush 核心方法

writeAndFlush 方法在通道写入数据并刷新写缓冲区事件时调用，用于回调责任链中的通道处理器的 WriteAndFlush 方法。处理流程同 flush 方法。源码如下。

```java
public ChannelFuture writeAndFlush(Object msg, ChannelPromise promise) {
    if (msg == null) {                                  // 写入消息不为空
        throw new NullPointerException("msg");
    }
    if (!validatePromise(promise, true)) {              // 操作被取消
        ReferenceCountUtil.release(msg);
        return promise;
    }
    write(msg, true, promise);                          // 完成实际写入，调用上述的 write 方法，参数设置为 true，表明
写入并刷新，回调 writeAndFlush 方法
    return promise;
}

private void invokeWriteAndFlush(Object msg, ChannelPromise promise) {
    if (invokeHandler()) {                              // 判断通道上下文状态后，进行回调
        invokeWrite0(msg, promise);
        invokeFlush0();
    } else {                                            // 传递事件到下一个处理器
        writeAndFlush(msg, promise);
    }
}

// 不带 promise 写入，使用默认的 promise 对象
public ChannelFuture writeAndFlush(Object msg) {
    return writeAndFlush(msg, newPromise());
}
```

13.6.20　AbstractWriteTask 核心内部类

AbstractWriteTask 抽象类用于定义封装写和刷新任务的模板内部类。由源码可知。

（1）ESTIMATE_TASK_SIZE_ON_SUBMIT 用于控制对写入大小的预估，并使用 WRITE_TASK_OVERHEAD 与写缓冲区，预估写入数量之和得到最终写入数量。

（2）在事件循环线程实际执行的 run 方法中，回调输出通道处理器的 write 方法完成实际写入底层 IO 通道。

```java
abstract static class AbstractWriteTask implements Runnable {
    private static final boolean ESTIMATE_TASK_SIZE_ON_SUBMIT =
        SystemPropertyUtil.getBoolean("io.Netty.transport.estimateSizeOnSubmit", true); // 是否预估写入的数量

    // 假设当前执行环境为 64 位的 JVM，则 16B 的对象头（未开启指针压缩），3 个引用字段（一个 8B）和一
个 int 字段，加上对齐字节数，所以写任务的大小为 48B
    private static final int WRITE_TASK_OVERHEAD =
        SystemPropertyUtil.getInt("io.Netty.transport.writeTaskSizeOverhead", 48);
    private final Recycler.Handle<AbstractWriteTask> handle;    // 封装写任务的对象池
    private AbstractChannelHandlerContext ctx;                  // 所属上下文对象
    private Object msg;                                         // 实际写入消息对象
    private ChannelPromise promise;                             // 完成时回调 promise
    private int size;                                           // 当前写入大小

    // 私有构造器：初始化对象池
    private AbstractWriteTask(Recycler.Handle<? extends AbstractWriteTask> handle) {
        this.handle = (Recycler.Handle<AbstractWriteTask>) handle;
```

```java
    }
    // 由子类完成调用初始化 task 的成员变量
    protected static void init(AbstractWriteTask task, AbstractChannelHandlerContext ctx,
                               Object msg, ChannelPromise promise) {
        task.ctx = ctx;
        task.msg = msg;
        task.promise = promise;
        if (ESTIMATE_TASK_SIZE_ON_SUBMIT) {                    // 开启预估任务大小
            ChannelOutboundBuffer buffer = ctx.channel().unsafe().outboundBuffer();
            if (buffer != null) {        // 判断写缓冲区是否为空，如果通道已关闭，可能被设置为空
                task.size = ctx.pipeline.estimatorHandle().size(msg) + WRITE_TASK_OVERHEAD; // 任务大小为流水线的评估器大小加对象头部大小
                buffer.incrementPendingOutboundBytes(task.size);    // 增加写出数据字节数
            } else {
                task.size = 0;
            }
        } else {                                                // 不采用预估大小，设置为 0
            task.size = 0;
        }
    }

    // 事件循环线程执行体
    public final void run() {
        try {
            ChannelOutboundBuffer buffer = ctx.channel().unsafe().outboundBuffer();
            if (ESTIMATE_TASK_SIZE_ON_SUBMIT && buffer != null) { // 开启大小预测，且写缓冲区不为空，减少一个预估的写入数据量
                buffer.decrementPendingOutboundBytes(size);
            }
            write(ctx, msg, promise);    // 调用 write 函数，回调 Write 方法完成实际写入过程
        } finally {
            // 由于当前对象在写入完成后将放入对象池，因此清理引用，设置为 null，垃圾回收就可以直接收集它们
            ctx = null;
            msg = null;
            promise = null;
            handle.recycle(this);
        }
    }

    // 回调输出通道处理器的 write 方法，完成实际数据写入底层 IO 中
    protected void write(AbstractChannelHandlerContext ctx, Object msg, ChannelPromise promise) {
        ctx.invokeWrite(msg, promise);
    }
}
```

13.6.21 WriteTask 核心内部类

WriteTask 类继承自 AbstractWriteTask 类，定义了 WriteTask 对象池，同时提供了创建该实例的方法，所有操作均由父类完成。源码如下。

```java
static final class WriteTask extends AbstractWriteTask implements
```

```
SingleThreadEventLoop.NonWakeupRunnable {

    private static final Recycler<WriteTask> RECYCLER = new Recycler<WriteTask>() {
        @Override
        protected WriteTask newObject(Handle<WriteTask> handle) {
            return new WriteTask(handle);
        }
    }; // 对象池

    // 创建 WriteTask 实例，用于服务 Recycler 的 newObject，所以设置为私有构造器
    private static WriteTask newInstance(
            AbstractChannelHandlerContext ctx, Object msg, ChannelPromise promise) {
        WriteTask task = RECYCLER.get();
        init(task, ctx, msg, promise);
        return task;
    }

    private WriteTask(Recycler.Handle<WriteTask> handle) {
        super(handle);
    }
}
```

13.6.22　WriteAndFlushTask 核心内部类

WriteAndFlushTask 类继承自 AbstractWriteTask 类，封装了写出数据到写缓冲区，同时刷新缓冲区的操作，并定义了 WriteAndFlushTask 对象池，同时提供了创建该实例的方法，重写了父类的 write 方法。在调用父类的 write 方法写入缓冲区后，调用 flush 回调函数，将缓冲区中的数据写入底层 IO 通道。其他操作均由父类完成。源码如下。

```
static final class WriteAndFlushTask extends AbstractWriteTask {

    private static final Recycler<WriteAndFlushTask> RECYCLER = new Recycler<WriteAndFlushTask>() {
        @Override
        protected WriteAndFlushTask newObject(Handle<WriteAndFlushTask> handle) {
            return new WriteAndFlushTask(handle);
        }
    }; // 对象池

    // 创建 WriteAndFlushTask 实例，用于服务 Recycler 的 newObject，所以设置为私有构造器
    private static WriteAndFlushTask newInstance(
            AbstractChannelHandlerContext ctx, Object msg, ChannelPromise promise) {
        WriteAndFlushTask task = RECYCLER.get();
        init(task, ctx, msg, promise);
        return task;
    }

    private WriteAndFlushTask(Recycler.Handle<WriteAndFlushTask> handle) {
        super(handle);
    }

    // 首先调用父类方法将数据写入输出缓冲区，然后调用 flush 方法，完成数据写出到底层 IO 通道中
```

```java
public void write(AbstractChannelHandlerContext ctx, Object msg, ChannelPromise promise) {
    super.write(ctx, msg, promise);
    ctx.invokeFlush();
}
}
```

13.7　ChannelHandlerAdapter 抽象类

ChannelHandlerAdapter 通道处理器抽象类适配实现了 ChannelHandler，对定义的方法进行了默认实现，同时定义了 isSharable 方法，判断当前通道是否可以在流水线中共享。源码如下。

```java
public abstract class ChannelHandlerAdapter implements ChannelHandler {
    // 通道处理器是否已经被添加到流水线中
    boolean added;

    // 判断当前通道处理器是否可以在不同流水线共享，使用线程本地缓存的类信息标识加速访问
    public boolean isSharable() {
        Class<?> clazz = getClass();
        Map<Class<?>, Boolean> cache = InternalThreadLocalMap.get().handlerSharableCache(); // 获取当前线程对于标识 Sharable 注解的类标识缓存
        Boolean sharable = cache.get(clazz);
        if (sharable == null) { // 不存在缓存，判断当前类是否定义了 Sharable 注解，然后初始化标识缓存
            sharable = clazz.isAnnotationPresent(Sharable.class);
            cache.put(clazz, sharable);
        }
        return sharable;
    }

    // 子类可以选择性实现，默认实现为空方法
    public void handlerAdded(ChannelHandlerContext ctx) throws Exception {
        // NOOP
    }

    public void handlerRemoved(ChannelHandlerContext ctx) throws Exception {
        // NOOP
    }

    // 子类可以选择性实现，默认实现调用 fireExceptionCaught，将异常处理传递到后继通道处理器
    public void exceptionCaught(ChannelHandlerContext ctx, Throwable cause) throws Exception {
        ctx.fireExceptionCaught(cause);
    }
}
```

13.8　DefaultChannelPipeline 类

DefaultChannelPipeline 默认通道处理器流水线实现类实现了 ChannelPipeline 接口对于通道处理器责任链的管理，由于该类的实现较为复杂，笔者这里将其进行拆分分析。读者请注意，由于该类的批

量操作功能会间接调用单个通道处理器的操作功能，所以笔者这里省略了批量操作的原理，因为它们只是简单地 for 循环调用添加而已，同时由于添加、移除操作其实就是双向链表的标准操作而已，其他都是对通道处理器进行排重和共享@Shareable 的判断，大部分逻辑均相同，所以读者可以仔细阅读 addFirst 方法，笔者会详细解释该方法，在其他操作中如果有类似的方法将省略。

13.8.1 核心变量与构造器

由源码可知。
（1）使用 AbstractChannelHandlerContext head 表示通道处理器责任链的头结点。
（2）使用 AbstractChannelHandlerContext tail 表示通道处理器责任链的尾结点。
（3）在构造器中将该双向链表形成的责任链初始化，初始状态为首尾结点互相关联。

```java
public class DefaultChannelPipeline implements ChannelPipeline {
    // 头结点与尾结点名字常量
    private static final String HEAD_NAME = generateName0(HeadContext.class);
    private static final String TAIL_NAME = generateName0(TailContext.class);
    private static String generateName0(Class<?> handlerType) {
        return StringUtil.simpleClassName(handlerType) + "#0";
    }

    // 缓存通道处理器 Class 对象与名字的映射
    private static final FastThreadLocal<Map<Class<?>, String>> nameCaches =
        new FastThreadLocal<Map<Class<?>, String>>() {
        @Override
        protected Map<Class<?>, String> initialValue() throws Exception {
            return new WeakHashMap<Class<?>, String>();
        }
    };

    final AbstractChannelHandlerContext head;       // 头结点
    final AbstractChannelHandlerContext tail;       // 尾结点
    private final Channel channel;                  // 所属通道对象
    private final ChannelFuture succeededFuture;    // 执行结果为成功的Future,用于实现newSucceededFuture()
    private final VoidChannelPromise voidPromise;   // 不能操作的空 Promise，用于实现 voidPromise()
    private final boolean touch = ResourceLeakDetector.isEnabled();  // 标识是否记录调用栈,用于内存泄漏分析
    private Map<EventExecutorGroup, EventExecutor> childExecutors;  // 用于保存事件执行器组与从事件执行
    // 去组中获取到的线程映射，因为 EventExecutorGroup 包含多个 EventExecutor，而需要获取 EventExecutor 时，会选
    // 择从 EventExecutorGroup 中获取一个 EventExecutor，而需要获取到的 EventExecutor 始终为同一个 EventExecutor
    private MessageSizeEstimator.Handle estimatorHandle;   // 用于猜测 Message 消息的大小
    private boolean firstRegistration = true;      // 标识当前通道是否为第一次注册

    // 用 callHandlerAddedForAllHandlers 方法回调 PendingHandler 链表的头结点
    // 用 callHandlerAdded0(AbstractChannelHandlerContext)方法添加这些回调结点
    // 这里只保留头结点，因为该链表不经常使用，且较小
    // 为了节省内存，折衷方案为在执行插入时，遍历到链表的末尾进行插入
    private PendingHandlerCallback pendingHandlerCallbackHead;
    private boolean registered;     // 当通道被注册到事件循环中时回调，一旦设置为不会修改，即表明该通道
    // 是否曾经被注册到事件循环中

    // 构造器：初始化成员变量
```

```java
    protected DefaultChannelPipeline(Channel channel) {
        succeededFuture = new SucceededChannelFuture(channel, null);
        voidPromise =    new VoidChannelPromise(channel, true);

        // 初始化事件处理器责任链
        tail = new TailContext(this);
        head = new HeadContext(this);
        head.next = tail;
        tail.prev = head;
    }
}
```

13.8.2　newContext 核心方法

newContext 方法用于根据提供的事件执行器组、通道名、通道处理器对象创建用于包装 ChannelHandler 的 AbstractChannelHandlerContext 通道上下文对象，这里创建的上下文实例为 DefaultChannelHandlerContext。同时，在构造器中通过 EventExecutorGroup 获取了其中一个管理的事件执行器，如果 EventExecutorGroup 为 null，则表明当前通道处理器将在回调该通道处理器方法的线程中执行，而不是指定的线程。源码如下：

```java
private AbstractChannelHandlerContext newContext(EventExecutorGroup group, String name,
ChannelHandler handler) {
    return new DefaultChannelHandlerContext(this, childExecutor(group), name, handler);
}

private EventExecutor childExecutor(EventExecutorGroup group) {
    if (group == null) {                                  // 事件执行器组为空
        return null;
    }
    Boolean pinEventExecutor = channel.config().getOption(ChannelOption.SINGLE_EVENTEXECUTOR_PER_GROUP);        // 通道处理器是否需要在指定事件执行器中执行。默认为 true, 此时将在后面将流水线事件处理线程与所属事件执行器组建立映射
    if (pinEventExecutor != null && !pinEventExecutor) { // 如果不使用单线程，则从事件执行器中随机获取一个执行器
        return group.next();
    }
    // 否则建立 EventExecutorGroup 与获取到的 EventExecutor 映射。在下一次调用该方法时，如果 EventExecutorGroup 相同，则返回同一个 EventExecutor
    Map<EventExecutorGroup, EventExecutor> childExecutors = this.childExecutors;
    if (childExecutors == null) {
        // 使用大小为 4。因为大多数场景下只使用同一个 EventExecutorGroup 与 EventExecutor
        childExecutors = this.childExecutors = new IdentityHashMap<EventExecutorGroup, EventExecutor>(4);
    }
    EventExecutor childExecutor = childExecutors.get(group);
    if (childExecutor == null) {
        childExecutor = group.next();
        childExecutors.put(group, childExecutor);
    }
    return childExecutor;
}
```

13.8.3　addFirst 核心方法

addFirst 方法实现将通道处理器添加到流水线的头部。流程如下。

（1）获取流水线对象锁，保证操作责任链时线程安全。
（2）判断通道处理器是否已经被添加过，且是否具有@Shareable 注解。
（3）如果通道处理器名字为空，则生成名字。然后进行名字排重，不允许重复添加重名的处理器。
（4）创建通道上下文对象。
（5）如果 registered 为 false，则表示该通道还没有在事件循环中注册。此时，将上下文对象添加到流水线中，并添加一个回调任务，该任务将在通道注册后调用。
（6）如果当前线程不属于通道应该执行的事件执行线程，则添加一个任务包装放入通道处理器所属事件循环中执行。
（7）否则直接调用通道处理器的 HandlerAdded 回调方法，通知通道处理器当前已经添加成功。

```java
@Override
public final ChannelPipeline addFirst(String name, ChannelHandler handler) {
    return addFirst(null, name, handler);
}

@Override
public final ChannelPipeline addFirst(EventExecutorGroup group, String name, ChannelHandler handler) {
    final AbstractChannelHandlerContext newCtx;
    synchronized (this) {                              // 获取流水线对象锁，保证操作责任链时线程安全
        checkMultiplicity(handler);                    // 判断通道处理器是否已经被添加过，且是否具有@Shareable 注解
        name = filterName(name, handler);              // 生成名字并排重，不允许重复添加重名的处理器
        newCtx = newContext(group, name, handler);     // 创建通道上下文对象
        addFirst0(newCtx);                             // 将其添加到链表头部
        if (!registered) {   // 如果 registered 是 false，则表示该通道还没有在事件循环中注册。此时，将上下文
对象添加到流水线中，并添加一个任务，该任务将在通道注册后调用
            newCtx.setAddPending();
            callHandlerCallbackLater(newCtx, true);
            return this;
        }
        EventExecutor executor = newCtx.executor();
        if (!executor.inEventLoop()) {    // 如果当前线程不属于通道应该执行的事件执行线程，则添加一个任务
包装放入通道处理器所属事件循环中执行
            newCtx.setAddPending();       // 设置当前状态为 AddPending，添加到了任务队列待处理
            executor.execute(new Runnable() {
                @Override
                public void run() {
                    callHandlerAdded0(newCtx);
                }
            });
            return this;
        }
    }
    callHandlerAdded0(newCtx);
    return this;
}
```

```java
// 标准的双向链表头结点插入
private void addFirst0(AbstractChannelHandlerContext newCtx) {
    AbstractChannelHandlerContext nextCtx = head.next;
    newCtx.prev = head;
    newCtx.next = nextCtx;
    head.next = newCtx;
    nextCtx.prev = newCtx;
}

// 判断需要添加的通道处理器是否为 ChannelHandlerAdapter 实例，如果是，则需要判断是否是共享处理器，即
// 包含@Sharable 注解；如果没有该注解，且通道处理器已经被添加到其他流水线，则抛出异常
private static void checkMultiplicity(ChannelHandler handler) {
    if (handler instanceof ChannelHandlerAdapter) {
        ChannelHandlerAdapter h = (ChannelHandlerAdapter) handler;
        if (!h.isSharable() && h.added) {
            throw new ChannelPipelineException(
                h.getClass().getName() +
                " is not a @Sharable handler, so can't be added or removed multiple times.");
        }
        h.added = true;
    }
}

// 修正并排重需要添加的通道处理器名
private String filterName(String name, ChannelHandler handler) {
    if (name == null) {                              // 根据添加的 ChannelHandler 生成名字
        return generateName(handler);
    }
    checkDuplicateName(name);                        // 排重
    return name;
}

// 如果没有传入 handler 的名字，则根据 ChannelHandler 对象生成名字
private String generateName(ChannelHandler handler) {
    // 优先从名字缓存中获取生成的名字
    Map<Class<?>, String> cache = nameCaches.get();
    Class<?> handlerType = handler.getClass();
    String name = cache.get(handlerType);
    if (name == null) {                              // 如果为空，则将生成的名字放入映射
        name = generateName0(handlerType); // StringUtil.simpleClassName(handlerType) + "#0"通过类名加#0
        cache.put(handlerType, name);
    }
    if (context0(name) != null) { // 已经存在与生成的名字相同的通道处理器，将名字后面的 0 截断，对其自增，
// 然后找到一个不重复的名字
        String baseName = name.substring(0, name.length() - 1);
        for (int i = 1;; i ++) {
            String newName = baseName + i;
            if (context0(newName) == null) {
                name = newName;
                break;
            }
        }
    }
```

```java
    }
    return name;
}

// 判断当前需要添加的通道处理器是否已经在流水线中，不允许重复添加重名的处理器
private void checkDuplicateName(String name) {
    if (context0(name) != null) {
        throw new IllegalArgumentException("Duplicate handler name: " + name);
    }
}

// 根据提供的名字，遍历链表找到与之相同名字的 AbstractChannelHandlerContext 实例。如果没有找到，则返回 null
private AbstractChannelHandlerContext context0(String name) {
    AbstractChannelHandlerContext context = head.next;
    while (context != tail) {
        if (context.name().equals(name)) {
            return context;
        }
        context = context.next;
    }
    return null;
}

// 添加全局 PendingHandlerCallback 回调任务
// 参数 added 表示是否为添加通道处理器操作
private void callHandlerCallbackLater(AbstractChannelHandlerContext ctx, boolean added) {
    assert !registered;                          // 调用该方法时，通道必须还未被注册到事件循环中
    PendingHandlerCallback task = added ? new PendingHandlerAddedTask(ctx) : new PendingHandlerRemovedTask(ctx);     // 生成添加或删除的回调任务对象
    // 将其添加到全局的 PendingHandlerCallback 单链表中
    PendingHandlerCallback pending = pendingHandlerCallbackHead;
    if (pending == null) {
        pendingHandlerCallbackHead = task;
    } else {
        while (pending.next != null) {
            pending = pending.next;
        }
        pending.next = task;
    }
}

// 通知通道处理器的 handlerAdded 方法，表示当前已经被成功添加到流水线中
private void callHandlerAdded0(final AbstractChannelHandlerContext ctx) {
    try {
        ctx.handler().handlerAdded(ctx);
        ctx.setAddComplete();                    // 更新上下文状态为添加完成
    } catch (Throwable t) {                      // 发生异常，从流水线中移除通道处理器
        boolean removed = false;
        try {
            remove0(ctx);
            try {
                ctx.handler().handlerRemoved(ctx);
```

```
            } finally {
                ctx.setRemoved();
            }
            removed = true;
        } catch (Throwable t2) {
            if (logger.isWarnEnabled()) {
                logger.warn("Failed to remove a handler: " + ctx.name(), t2);
            }
        }
        // 回调所有通道处理器的 exceptionCaught 方法，removed 用于控制传递异常的信息
        if (removed) {
            fireExceptionCaught(new ChannelPipelineException(
                ctx.handler().getClass().getName() +
                ".handlerAdded() has thrown an exception; removed.", t));
        } else {
            fireExceptionCaught(new ChannelPipelineException(
                ctx.handler().getClass().getName() +
                ".handlerAdded() has thrown an exception; also failed to remove.", t));
        }
    }
}
```

13.8.4　addLast 核心方法

addLast 方法实现将通道处理器添加到流水线的尾部。流程如下。

（1）获取流水线对象锁，保证操作责任链时线程安全。

（2）判断通道处理器是否已经被添加过，且是否具有@Shareable 注解。

（3）如果通道处理器名字为空，则生成名字。然后进行名字排重，不允许重复添加重名的处理器。

（4）创建通道上下文对象。

（5）如果 registered 为 false，则表示该通道还没有在事件循环中注册。此时将上下文对象添加到流水线中，并添加一个回调任务，该任务将在通道注册后调用。

（6）如果当前线程不属于通道应该执行的事件执行线程，则添加一个任务包装放入通道处理器所属事件循环中执行。

（7）否则直接调用通道处理器的 HandlerAdded 回调方法，通知通道处理器当前已经添加成功。

```
@Override
public final ChannelPipeline addLast(String name, ChannelHandler handler) {
    return addLast(null, name, handler);
}

@Override
public final ChannelPipeline addLast(EventExecutorGroup group, String name, ChannelHandler handler) {
    final AbstractChannelHandlerContext newCtx;
    synchronized (this) {        // 流程同 addFirst，只不过链入的结点在尾结点，而不是在头结点
        checkMultiplicity(handler);
        newCtx = newContext(group, filterName(name, handler), handler);
        addLast0(newCtx);
        if (!registered) {
            newCtx.setAddPending();
            callHandlerCallbackLater(newCtx, true);
```

```
                return this;
            }
            EventExecutor executor = newCtx.executor();
            if (!executor.inEventLoop()) {
                newCtx.setAddPending();
                executor.execute(new Runnable() {
                    @Override
                    public void run() {
                        callHandlerAdded0(newCtx);
                    }
                });
                return this;
            }
        }
        callHandlerAdded0(newCtx);
        return this;
    }

// 标准的双向链表尾结点插入
private void addLast0(AbstractChannelHandlerContext newCtx) {
    AbstractChannelHandlerContext prev = tail.prev;
    newCtx.prev = prev;
    newCtx.next = tail;
    prev.next = newCtx;
    tail.prev = newCtx;
}
```

13.8.5 addBefore 核心方法

addBefore 方法实现将通道处理器添加到流水线指定的名为 baseName 的通道处理器的前面。流程如下。

（1）获取流水线对象锁，保证操作责任链时线程安全。

（2）判断通道处理器是否已经被添加过，且是否具有@Shareable 注解。

（3）如果通道处理器名字为空，则生成名字，然后进行名字排重，不允许重复添加重名的处理器。

（4）找到需要添加在前面的通道处理器对象。如果为空，则抛出 NoSuchElementException。

（5）创建通道上下文对象，并将其添加到指定通道处理器对象的前面。

（6）如果 registered 为 false，则表示该通道还没有在事件循环中注册。添加一个回调任务，该任务将在通道注册后调用。

（7）如果当前线程不属于通道应该执行的事件执行线程，则添加一个任务包装放入通道处理器所属事件循环中执行。

（8）否则直接调用通道处理器的 HandlerAdded 回调方法，通知通道处理器当前已经添加成功。

```
@Override
public final ChannelPipeline addBefore(String baseName, String name, ChannelHandler handler) {
    return addBefore(null, baseName, name, handler);
}

@Override
public final ChannelPipeline addBefore(
```

```java
        EventExecutorGroup group, String baseName, String name, ChannelHandler handler) {
    final AbstractChannelHandlerContext newCtx;
    final AbstractChannelHandlerContext ctx;
    synchronized (this) {        // 大部分流程同 addFirst，只不过这里涉及查找需要添加在前面的通道处理器
        checkMultiplicity(handler);
        name = filterName(name, handler);
        ctx = getContextOrDie(baseName);        // 找到需要添加在前面的通道处理器，如果为空，则抛出 NoSuchElementException
        newCtx = newContext(group, name, handler);
        addBefore0(ctx, newCtx);
        // 以下流程同 addFirst
        if (!registered) {
            newCtx.setAddPending();
            callHandlerCallbackLater(newCtx, true);
            return this;
        }

        EventExecutor executor = newCtx.executor();
        if (!executor.inEventLoop()) {
            newCtx.setAddPending();
            executor.execute(new Runnable() {
                @Override
                public void run() {
                    callHandlerAdded0(newCtx);
                }
            });
            return this;
        }
    }
    callHandlerAdded0(newCtx);
    return this;
}

// 标准双向链表的中间插入操作
private static void addBefore0(AbstractChannelHandlerContext ctx, AbstractChannelHandlerContext newCtx) {
    newCtx.prev = ctx.prev;
    newCtx.next = ctx;
    ctx.prev.next = newCtx;
    ctx.prev = newCtx;
}

// 根据名字查找到具体通道处理器。如果为空，则抛出 NoSuchElementException
private AbstractChannelHandlerContext getContextOrDie(String name) {
    AbstractChannelHandlerContext ctx = (AbstractChannelHandlerContext) context(name);
    if (ctx == null) {
        throw new NoSuchElementException(name);
    } else {
        return ctx;
    }
}
```

13.8.6 addAfter 核心方法

addAfter 方法实现将通道处理器添加到流水线指定的名为 baseName 的通道处理器的后面。流程如下。

(1)获取流水线对象锁,保证操作责任链时线程安全。

(2)判断通道处理器是否已经被添加过,且是否具有@Shareable 注解。

(3)如果通道处理器名字为空,则生成名字,然后进行名字排重,不允许重复添加重名的处理器。

(4)找到需要添加在前面的通道处理器对象。如果为空,则抛出 NoSuchElementException。

(5)创建通道上下文对象,并将其添加到指定通道处理器对象的后面。

(6)如果 registered 为 false,则表示该通道还没有在事件循环中注册。添加一个回调任务,该任务将在通道注册后调用。

(7)如果当前线程不属于通道应该执行的事件执行线程,则添加一个任务包装放入通道处理器所属事件循环中执行。

(8)否则直接调用通道处理器的 HandlerAdded 回调方法,通知通道处理器当前已经添加成功。

```java
@Override
public final ChannelPipeline addAfter(String baseName, String name, ChannelHandler handler) {
    return addAfter(null, baseName, name, handler);
}

@Override
public final ChannelPipeline addAfter(
        EventExecutorGroup group, String baseName, String name, ChannelHandler handler) {
    final AbstractChannelHandlerContext newCtx;
    final AbstractChannelHandlerContext ctx;
    synchronized (this) {    // 同 addBefore,只不过在操作链表时,将添加到已找到的 ctx 处理器的后面
        checkMultiplicity(handler);
        name = filterName(name, handler);
        ctx = getContextOrDie(baseName);
        newCtx = newContext(group, name, handler);
        addAfter0(ctx, newCtx);
        if (!registered) {
            newCtx.setAddPending();
            callHandlerCallbackLater(newCtx, true);
            return this;
        }
        EventExecutor executor = newCtx.executor();
        if (!executor.inEventLoop()) {
            newCtx.setAddPending();
            executor.execute(new Runnable() {
                @Override
                public void run() {
                    callHandlerAdded0(newCtx);
                }
            });
            return this;
        }
    }
    callHandlerAdded0(newCtx);
    return this;
}

// 标准双向链表的中间插入操作
private static void addAfter0(AbstractChannelHandlerContext ctx, AbstractChannelHandlerContext newCtx) {
```

```
        newCtx.prev = ctx;
        newCtx.next = ctx.next;
        ctx.next.prev = newCtx;
        ctx.next = newCtx;
}
```

13.8.7　remove 核心方法

remove 方法实现从流水线中移除通道处理器。流程如下。

（1）调用 getContextOrDie 方法，从责任链中获取到需要移除的代表 handler 通道处理器的上下文对象。

（2）对流水线对象上锁，保证操作责任链时的并发安全。

（3）调用 remove0 方法从责任链链表中移除上下文对象。

（4）调用 callHandlerRemoved0 方法回调通道处理器的 handlerRemoved 方法。

```
@Override
public final ChannelPipeline remove(ChannelHandler handler) {
    remove(getContextOrDie(handler));
    return this;
}

@Override
public final ChannelHandler remove(String name) {
    return remove(getContextOrDie(name)).handler();
}

@Override
public final <T extends ChannelHandler> T remove(Class<T> handlerType) {
    return (T) remove(getContextOrDie(handlerType)).handler();
}

private AbstractChannelHandlerContext remove(final AbstractChannelHandlerContext ctx) {
    synchronized (this) {       // 对流水线对象上锁，保证操作责任链时的并发安全
        remove0(ctx);
        if (!registered) {       // 如果通道还未注册到事件循环中，则添加一个全局回调方法，在通道注册到事
件循环中时调用，传入 false 表明任务类型为删除任务
            callHandlerCallbackLater(ctx, false);
            return ctx;
        }
        // 当前线程不是 EventExecutor 的事件循环线程，将回调通道处理器方法的操作放在事件循环线程队列中
        EventExecutor executor = ctx.executor();
        if (!executor.inEventLoop()) {
            executor.execute(new Runnable() {
                @Override
                public void run() {
                    callHandlerRemoved0(ctx);
                }
            });
            return ctx;
        }
    }
}
```

```
        callHandlerRemoved0(ctx);
        return ctx;
}

// 标准双向链表的删除操作
private static void remove0(AbstractChannelHandlerContext ctx) {
    AbstractChannelHandlerContext prev = ctx.prev;
    AbstractChannelHandlerContext next = ctx.next;
    prev.next = next;
    next.prev = prev;
}

// 回调 AbstractChannelHandlerContext 包装的通道处理器 handlerRemoved 方法
private void callHandlerRemoved0(final AbstractChannelHandlerContext ctx) {
    try {
        try {
            ctx.handler().handlerRemoved(ctx);
        } finally {    // 设置已经完成上下文移除
            ctx.setRemoved();
        }
    } catch (Throwable t) {
        fireExceptionCaught(new ChannelPipelineException(
            ctx.handler().getClass().getName() + ".handlerRemoved() has thrown an exception.", t));
    }
}
```

13.8.8　removeFirst 核心方法

removeFirst 方法用于移除通道处理器责任链头部的处理器。首先，判断是否存在通道处理器。如果不存在，则抛出异常，否则调用上述的 remove 方法完成移除。源码如下。

```
public final ChannelHandler removeFirst() {
    if (head.next == tail) {
        throw new NoSuchElementException();
    }
    return remove(head.next).handler();
}
```

13.8.9　removeLast 核心方法

removeLast 方法用于移除通道处理器责任链尾部的处理器。首先，判断是否存在通道处理器。如果不存在，则抛出异常，否则调用上述的 remove 方法完成移除。源码如下。

```
public final ChannelHandler removeLast() {
    if (head.next == tail) {
        throw new NoSuchElementException();
    }
    return remove(tail.prev).handler();
}
```

13.8.10　replace 核心方法

replace 方法用于将 oldHandler 指定的通道处理器替换为 newHandler，同时指定新的名字为

newName。流程如下。

(1) 通过 getContextOrDie 方法获取到需要替换的通道处理器上下文对象。
(2) 对流水线对象上锁,保证操作责任链时的并发安全。
(3) 判断需要替换的共享性。
(4) 如果新名字为空,则生成新的名字,并根据旧名与新名是否相同决定是否排重。
(5) 创建新上下文。
(6) 调用 replace0 方法完成对双向链表中的对象替换操作。
(7) 回调旧上下文的 handleRemoved 方法。
(8) 回调新上下文的 handleAdded 方法。

```java
// 通过指定对象替换
public final ChannelPipeline replace(ChannelHandler oldHandler, String newName, ChannelHandler newHandler) {
    replace(getContextOrDie(oldHandler), newName, newHandler);
    return this;
}

// 通过指定名字替换
public final ChannelHandler replace(String oldName, String newName, ChannelHandler newHandler) {
    return replace(getContextOrDie(oldName), newName, newHandler);
}

// 通过指定类型 Class 替换
public final <T extends ChannelHandler> T replace(
    Class<T> oldHandlerType, String newName, ChannelHandler newHandler) {
    return (T) replace(getContextOrDie(oldHandlerType), newName, newHandler);
}

private ChannelHandler replace(
    final AbstractChannelHandlerContext ctx, String newName, ChannelHandler newHandler) {
    final AbstractChannelHandlerContext newCtx;
    synchronized (this) {               // 对流水线对象上锁,保证操作责任链时的并发安全
        checkMultiplicity(newHandler);  // 判断需要替换的共享性(同 addFirst)
        if (newName == null) {          // 新名字为空,生成新的名字(同 addFirst)
            newName = generateName(newHandler);
        } else {    // 否则判断当前添加的名字是否与旧名相同,因为旧名在添加时已经排重。如果不相同,则
                    // 调用 checkDuplicateName 方法排重
            boolean sameName = ctx.name().equals(newName);
            if (!sameName) {
                checkDuplicateName(newName);
            }
        }
        // 以下方法同 addFirst,只不过这里调用 replace0 进行链表对象替换,同时由于这里替换分别表示两个
        // 操作:将旧的上下文移除,新的上下文添加。所以这里回调通道处理器的方法为 handlerRemoved 和 HandlerAdded
        newCtx = newContext(ctx.executor, newName, newHandler);
        replace0(ctx, newCtx);
        if (!registered) {
            callHandlerCallbackLater(newCtx, true);
            callHandlerCallbackLater(ctx, false);
            return ctx.handler();
```

```java
        EventExecutor executor = ctx.executor();
        if (!executor.inEventLoop()) {
            executor.execute(new Runnable() {
                @Override
                public void run() {
                    callHandlerAdded0(newCtx);
                    callHandlerRemoved0(ctx);
                }
            });
            return ctx.handler();
        }
    }
    callHandlerAdded0(newCtx);
    callHandlerRemoved0(ctx);
    return ctx.handler();
}

// 将链表中 oldCtx 对象替换为 newCtx
private static void replace0(AbstractChannelHandlerContext oldCtx, AbstractChannelHandlerContext newCtx) {
    AbstractChannelHandlerContext prev = oldCtx.prev;  // 需要替换的前驱结点
    AbstractChannelHandlerContext next = oldCtx.next;  // 需要替换的后继结点
    // 将新上下文的前驱和后继引用关联
    newCtx.prev = prev;
    newCtx.next = next;
    // 将新上下文的前驱和后继与当前上下文关联, 此时完成双向链表连接
    prev.next = newCtx;
    next.prev = newCtx;
    // 旧上下文的前驱和后继均指向新上下文, 通过这两个引用获取替换它的对象
    oldCtx.prev = newCtx;
    oldCtx.next = newCtx;
}
```

13.8.11　destroy 核心方法

从流水线通道处理器责任链中逐个移除所有处理器，从尾结点到头结点移除，并回调通道处理的 handlerRemoved 方法。由源码可知。

（1）destroy 方法将调用 destroyUp 方法，传入头结点的后继结点，同时指定当前 inEventLoop 参数为 false，表明当前处理线程默认不再需要在指定执行器中执行的通道处理器的所属线程。

（2）在 destroyUp 方法中通过 next 引用遍历责任链中的所有处理器，并将不应该在当前线程中执行移除和回调处理器方法的通道处理器上下文放入所属事件执行器中执行（注意，这样处理是为了让在不同事件处理器中回调的处理器交给它们应该执行的线程中执行）。

（3）实际移除方法为 destroyDown，在该方法中从当前参数传入的 ctx 的方法开始通过 prev 引用向前遍历，如果遍历到的通道处理器所属事件执行器的所属线程就是当前执行线程，则先调用 remove0 将当前上下文从责任链中移除，然后调用 callHandlerRemoved0(ctx) 方法回调通道处理器的 HandlerRemoved 方法。如果遍历到的通道处理器所属事件执行器的所属线程不是当前执行线程，则封装 destroyDown 执行任务交给所属执行线程中执行。

```java
private synchronized void destroy() {
```

```java
        destroyUp(head.next, false);      // 传入头结点的后继结点，并开始移除
}

private void destroyUp(AbstractChannelHandlerContext ctx, boolean inEventLoop) {
    final Thread currentThread = Thread.currentThread();
    final AbstractChannelHandlerContext tail = this.tail;
    for (;;) {                             // 循环遍历所有处理器
        if (ctx == tail) {                 // 如果遍历到尾结点，则从尾结点开始向前遍历
            destroyDown(currentThread, tail.prev, inEventLoop);
            break;
        }
        final EventExecutor executor = ctx.executor();
        if (!inEventLoop && !executor.inEventLoop(currentThread)) { // 当前线程不属于执行器的执行线程，封
装 destroyUp 任务，并放入所属执行线程队列中执行
            final AbstractChannelHandlerContext finalCtx = ctx;
            executor.execute(new Runnable() {
                @Override
                public void run() {
                    destroyUp(finalCtx, true);
                }
            });
            break;
        }
        ctx = ctx.next;                    // 继续向后遍历
        inEventLoop = false;
    }
}

// 从当前 ctx 上下文开始，通过 prev 引用向前遍历
private void destroyDown(Thread currentThread, AbstractChannelHandlerContext ctx, boolean inEventLoop) {
    final AbstractChannelHandlerContext head = this.head;
    for (;;) {
        if (ctx == head) {                 // 到达头结点，退出循环
            break;
        }
        final EventExecutor executor = ctx.executor();
        if (inEventLoop || executor.inEventLoop(currentThread)) {  // 当前线程属于事件循环线程，直接调用
remove0 从流水线中移除通道，然后回调通道处理器 HandlerRemoved 方法
            synchronized (this) {          // 注意，上锁优先从链表中移除通道处理器，不会导致多个线程同时调用
destroyDown 处理多个通道对象
                remove0(ctx);
            }
            callHandlerRemoved0(ctx);
        } else {                           // 否则将 destroyDown 任务放到所属事件循环中执行
            final AbstractChannelHandlerContext finalCtx = ctx;
            executor.execute(new Runnable() {
                @Override
                public void run() {
                    destroyDown(Thread.currentThread(), finalCtx, true); // 传递为 true，因为当前已经确定放
入的线程必定是执行通道处理器的线程，所以不需要再判断 executor.inEventLoop(currentThread)
                }
            });
            break;
```

```
        }
        ctx = ctx.prev;                          // 向前继续遍历通道处理器
        inEventLoop = false;
    }
}
```

13.8.12 TailContext 核心内部类

TailContext 类表示一个特殊的用于处理所有通道输入数据的上下文，之前在描述流水线对象时涉及过，当所有数据走到最上层的数据输入回调处理器，将静默丢弃。这个上下文便是最后一个处理输入数据事件的尾部上下文对象。由源码可知。

（1）对于通道注册、激活、写状态改变、通道添加到流水线、通道从流水线移除、通道处理完所有读缓冲区中的数据，以上回调方法均静默处理。

（2）对于通道处理异常回调和读取数据回调，会打印异常信息同时释放异常对象。

```java
final class TailContext extends AbstractChannelHandlerContext implements ChannelInboundHandler {
    // 构造器初始化：通道名为 TAIL_NAME, inbound 为 true, 表明当前为输入数据处理器
    TailContext(DefaultChannelPipeline pipeline) {
        super(pipeline, null, TAIL_NAME, true, false);
        setAddComplete();    // 由于创建流水线时就创建了该对象，同时作为责任链末尾对象，所以这里设置为完成连接状态
    }
    /* 通道输入数据事件将静默处理：空方法 */
    public void channelRegistered(ChannelHandlerContext ctx) throws Exception { }
    public void channelUnregistered(ChannelHandlerContext ctx) throws Exception { }
    public void channelActive(ChannelHandlerContext ctx) throws Exception { }
    public void channelInactive(ChannelHandlerContext ctx) throws Exception { }
    public void channelWritabilityChanged(ChannelHandlerContext ctx) throws Exception { }
    public void handlerAdded(ChannelHandlerContext ctx) throws Exception { }
    public void handlerRemoved(ChannelHandlerContext ctx) throws Exception { }

    // 用户触发事件，将释放接收到的事件对象
    public void userEventTriggered(ChannelHandlerContext ctx, Object evt) throws Exception {
        ReferenceCountUtil.release(evt);
    }

    // 通道发生异常事件，打印异常日志，并释放异常对象
    public void exceptionCaught(ChannelHandlerContext ctx, Throwable cause) throws Exception {
        onUnhandledInboundException(cause);
    }

    // 通道读取到数据时回调，打印日志，并释放 msg 对象
    public void channelRead(ChannelHandlerContext ctx, Object msg) throws Exception {
        onUnhandledInboundMessage(msg);
    }

    // 通道处理完读缓冲区中的数据时回调，静默处理
    public void channelReadComplete(ChannelHandlerContext ctx) throws Exception { }
}
```

```
// DefaultChannelPipeline 方法，打印日志，并释放 msg 对象
protected void onUnhandledInboundMessage(Object msg) {
    try {
        logger.debug(
            "Discarded inbound message {} that reached at the tail of the pipeline. " +
            "Please check your pipeline configuration.", msg);
    } finally {
        ReferenceCountUtil.release(msg);
    }
}

// DefaultChannelPipeline 方法，用于打印异常日志，并释放异常对象（ReferenceCounted 引用）
protected void onUnhandledInboundException(Throwable cause) {
    try {
        logger.warn(
            "An exceptionCaught() event was fired, and it reached at the tail of the pipeline. " +
            "It usually means the last handler in the pipeline did not handle the exception.",
            cause);
    } finally {
        ReferenceCountUtil.release(cause);
    }
}
```

13.8.13 HeadContext 核心内部类

HeadContext 类表示一个特殊的用于处理所有通道输出数据的上下文，之前在描述流水线对象时涉及过，当所有数据到这最底层的数据输出回调处理器，将实际操作作用到底层 IO 通道对象。这个上下文便是那最后一个处理输出数据事件的头部上下文对象（写事件通道处理器从后往前调用）。由源码可知。

（1）对于通道注册、激活、写状态改变、通道添加到流水线、通道从流水线移除、通道处理完所有读缓冲区中的数据，以上回调方法均静默处理。

（2）对于通道处理异常回调和读取数据回调，会打印异常信息同时释放异常对象。

```
final class HeadContext extends AbstractChannelHandlerContext
        implements ChannelOutboundHandler, ChannelInboundHandler {
    private final Unsafe unsafe; // 实际操作底层 IO 对象

    // 构造器初始化父类，这里指定通道处理器名为HEAD_NAME，同时从通道对象中获取到用于操作底层 IO
    的 Unsafe 对象
    HeadContext(DefaultChannelPipeline pipeline) {
        super(pipeline, null, HEAD_NAME, false, true);
        unsafe = pipeline.channel().unsafe();
        setAddComplete(); // 由于在创建流水线的构造器中初始化该上下文并保存到实例变量中，所以在创建该
        对象时直接将状态修改为完成状态
    }

    /* 通道添加和移除回调静默处理 */
    public void handlerAdded(ChannelHandlerContext ctx) throws Exception {
    }
    public void handlerRemoved(ChannelHandlerContext ctx) throws Exception {
    }
```

```java
/* 调用 unsafe 完成底层通道对应操作 */
public void bind(
    ChannelHandlerContext ctx, SocketAddress localAddress, ChannelPromise promise)
    throws Exception {
    unsafe.bind(localAddress, promise);
}
public void connect(
    ChannelHandlerContext ctx,
    SocketAddress remoteAddress, SocketAddress localAddress,
    ChannelPromise promise) throws Exception {
    unsafe.connect(remoteAddress, localAddress, promise);
}
public void disconnect(ChannelHandlerContext ctx, ChannelPromise promise) throws Exception {
    unsafe.disconnect(promise);
}
public void close(ChannelHandlerContext ctx, ChannelPromise promise) throws Exception {
    unsafe.close(promise);
}
public void deregister(ChannelHandlerContext ctx, ChannelPromise promise) throws Exception {
    unsafe.deregister(promise);
}
public void read(ChannelHandlerContext ctx) {
    unsafe.beginRead();
}
public void write(ChannelHandlerContext ctx, Object msg, ChannelPromise promise) throws Exception {
    unsafe.write(msg, promise);
}
public void flush(ChannelHandlerContext ctx) throws Exception {
    unsafe.flush();
}

/* 通道输出数据回调方法，默认传递到责任链的下一个处理器处理 */
public void exceptionCaught(ChannelHandlerContext ctx, Throwable cause) throws Exception {
    ctx.fireExceptionCaught(cause);
}
public void channelRegistered(ChannelHandlerContext ctx) throws Exception {
    invokeHandlerAddedIfNeeded();
    ctx.fireChannelRegistered();
}

public void channelUnregistered(ChannelHandlerContext ctx) throws Exception {
    ctx.fireChannelUnregistered();
    // 如果通道解除在事件循环注册后未关闭，则顺序删除责任链中的所有通道处理器
    if (!channel.isOpen()) {
        destroy();
    }
}
public void channelActive(ChannelHandlerContext ctx) throws Exception {
    ctx.fireChannelActive();
    readIfIsAutoRead();
}
public void channelInactive(ChannelHandlerContext ctx) throws Exception {
    ctx.fireChannelInactive();
```

```
    }
    public void channelRead(ChannelHandlerContext ctx, Object msg) throws Exception {
        ctx.fireChannelRead(msg);
    }
    public void channelReadComplete(ChannelHandlerContext ctx) throws Exception {
        ctx.fireChannelReadComplete();
        readIfIsAutoRead();
    }
    // 如果通道配置了自动读取数据，则调用 read 方法。对于 NIO 通道，会在 SelectionKey 中添加读感兴趣事
    件集。这时 Selector 便可选中该通道，交由事件循环线程处理
    private void readIfIsAutoRead() {
        if (channel.config().isAutoRead()) {
            channel.read();
        }
    }
    public void userEventTriggered(ChannelHandlerContext ctx, Object evt) throws Exception {
        ctx.fireUserEventTriggered(evt);
    }
    public void channelWritabilityChanged(ChannelHandlerContext ctx) throws Exception {
        ctx.fireChannelWritabilityChanged();
    }
}
```

13.8.14 write 核心方法

write 方法在通道输出数据时调用，从 tail 结点开始往责任链前调用。源码如下。

```
public final ChannelFuture write(Object msg) {
    return tail.write(msg);
}
```

13.8.15 fireChannelRead 核心方法

fireChannelRead 方法将在通道输入数据时调用，这里从 head 结点处开始往责任链后调用。源码如下。

```
public final ChannelPipeline fireChannelRead(Object msg) {
    AbstractChannelHandlerContext.invokeChannelRead(head, msg);
    return this;
}
```

13.9 ChannelInboundHandlerAdapter 抽象类

ChannelInboundHandlerAdapter 数据输入通道处理器适配器类实现了所有 ChannelInboundHandler 方法，默认行为均是将事件传递给责任链中下一个通道输入数据处理器，而子类可以继承该类实现自身需要的方法。源码如下。

```
public class ChannelInboundHandlerAdapter extends ChannelHandlerAdapter implements
ChannelInboundHandler {
    public void channelRegistered(ChannelHandlerContext ctx) throws Exception {
```

```
            ctx.fireChannelRegistered();
        }
        public void channelUnregistered(ChannelHandlerContext ctx) throws Exception {
            ctx.fireChannelUnregistered();
        }
        public void channelActive(ChannelHandlerContext ctx) throws Exception {
            ctx.fireChannelActive();
        }
        public void channelInactive(ChannelHandlerContext ctx) throws Exception {
            ctx.fireChannelInactive();
        }
        public void channelRead(ChannelHandlerContext ctx, Object msg) throws Exception {
            ctx.fireChannelRead(msg);
        }
        public void channelReadComplete(ChannelHandlerContext ctx) throws Exception {
            ctx.fireChannelReadComplete();
        }
        public void userEventTriggered(ChannelHandlerContext ctx, Object evt) throws Exception {
            ctx.fireUserEventTriggered(evt);
        }
        public void channelWritabilityChanged(ChannelHandlerContext ctx) throws Exception {
            ctx.fireChannelWritabilityChanged();
        }
        public void exceptionCaught(ChannelHandlerContext ctx, Throwable cause)
            throws Exception {
            ctx.fireExceptionCaught(cause);
        }
    }
```

13.10　SimpleChannelInboundHandler 处理器类

SimpleChannelInboundHandler 简单通道输入通道处理器类继承自 ChannelInboundHandlerAdapter，子类可以通过设置 TypeParameterMatcher 类型匹配器处理特定类型的输入数据。例如，以下例子定义了一个 StringHandler，该处理器通过泛型指定了当前只接收 String 的输入数据。

```
public class StringHandler extends SimpleChannelInboundHandler<String>{
    protected void channelRead0( ChannelHandlerContext ctx,  String message)throws Exception {
        System.out.println(message);
    }
}
```

由于该类较为简单，所以笔者将其一并讲解。由源码可知。

（1）TypeParameterMatcher matcher 将在构造器中初始化，用于判断当前处理器是否接收 msg。TypeParameterMatcher 的实现为 Class.isInstance(msg)或 instanceof，读者了解即可。

（2）boolean autoRelease 如果指定为 true，且当前通道已处理了该数据，还指定了自动释放，则释放该 msg。

（3）子类只需要重写 channelRead0(ChannelHandlerContext ctx, I msg)，完成自身逻辑即可。

```
public abstract class SimpleChannelInboundHandler<I> extends ChannelInboundHandlerAdapter {
```

```java
    private final TypeParameterMatcher matcher;     // 输入对象类型匹配器
    private final boolean autoRelease;              // 标识如果当前处理器除了输入对象, 是否自动释放

    // 构造器: 指定自动释放
    protected SimpleChannelInboundHandler() {
        this(true);
    }

    // 构造器: 初始化输入数据类型匹配器
    protected SimpleChannelInboundHandler(boolean autoRelease) {
        matcher = TypeParameterMatcher.find(this, SimpleChannelInboundHandler.class, "I"); // 这里表示泛型 I 的类型字符串
        this.autoRelease = autoRelease;
    }

    // 构造器: 自定义接收类型
    protected SimpleChannelInboundHandler(Class<? extends I> inboundMessageType) {
        this(inboundMessageType, true);
    }

    // 构造器: 自定义接收类型与自动释放
    protected SimpleChannelInboundHandler(Class<? extends I> inboundMessageType, boolean autoRelease) {
        matcher = TypeParameterMatcher.get(inboundMessageType);
        this.autoRelease = autoRelease;
    }

    // 判断当前处理器是否接收 msg
    public boolean acceptInboundMessage(Object msg) throws Exception {
        return matcher.match(msg);
    }

    // 实现父类通道输入数据时回调方法
    public void channelRead(ChannelHandlerContext ctx, Object msg) throws Exception {
        boolean release = true;
        try {
            if (acceptInboundMessage(msg)) {          // 当前通道可接收 msg 类型的数据, 调用子类实现的 channelRead0 完成消息处理
                I imsg = (I) msg;
                channelRead0(ctx, imsg);
            } else { // 当前通道不接收 msg 类型的数据, 将数据传递给通道责任链中其他的处理器处理
                release = false;
                ctx.fireChannelRead(msg);
            }
        } finally {
            if (autoRelease && release) { // 当前通道处理了该数据, 同时指定了自动释放, 并释放该 msg
                ReferenceCountUtil.release(msg);
            }
        }
    }

    // 子类实现该方法完成实际 msg 数据处理器
    protected abstract void channelRead0(ChannelHandlerContext ctx, I msg) throws Exception;
}
```

13.11 ChannelInitializer 类

在 Netty 的使用中经常会见到如下代码，不难发现 ChannelInitializer 通道初始化器类会在通道初始化时，对通道处理器责任链进行初始化，把重写的 initChannel 方法回调执行，从而在其中将自定义的通道处理器添加到责任链中。源码如下。

```
ServerBootstrap b = new ServerBootstrap();
b.group(bossGroup, workerGroup)
    .channel(NioServerSocketChannel.class)
    .option(ChannelOption.SO_BACKLOG, 100)
    .handler(new LoggingHandler(LogLevel.INFO))
    .childHandler(new ChannelInitializer<SocketChannel>() {
        @Override
        public void initChannel(SocketChannel ch) throws Exception {
            ChannelPipeline p = ch.pipeline();
            if (sslCtx != null) {
                p.addLast(sslCtx.newHandler(ch.alloc()));
            }
            // p.addLast(new LoggingHandler(LogLevel.INFO));
            p.addLast(new EchoServerHandler());
        }
    });
```

本节通过前面学过的知识，研究 initChannel 方法在何时进行回调。由源码可知。

（1）该类继承自 ChannelInboundHandlerAdapter 适配器，说明自身属于一个通道数据输入事件处理器。

（2）该类加上了@Sharable 注释，表明该通道输入处理器对象可以在不同流水线中共享。

（3）该类重写了 channelRegistered 方法，该方法将在通道注册到事件循环时回调，在该方法中首先回调 initChannel，完成子类对流水线的处理，然后将自身从责任链中移除，将事件传递给责任链的下一个处理器。

（4）如果处理通道时发生异常，则打印日志，并从责任链中移除。

```
@Sharable
public abstract class ChannelInitializer<C extends Channel> extends ChannelInboundHandlerAdapter {

    // 子类实现该方法，完成通道处理器责任链的处理
    protected abstract void initChannel(C ch) throws Exception;

    // 在重写适配器的通道注册到事件循环时回调方法，在该方法中首先回调 initChannel，完成子类对流水线的
    // 处理，然后将自身从责任链中移除，将事件传递给责任链的下一个处理器
    public final void channelRegistered(ChannelHandlerContext ctx) throws Exception {
        initChannel((C) ctx.channel());
        ctx.pipeline().remove(this);
        ctx.pipeline().fireChannelRegistered();
    }

    // 发生异常时，打印日志，并从责任链中移除
```

```java
public void exceptionCaught(ChannelHandlerContext ctx, Throwable cause) throws Exception {
    logger.warn("Failed to initialize a channel. Closing: " + ctx.channel(), cause);
    try {
        ChannelPipeline pipeline = ctx.pipeline();
        if (pipeline.context(this) != null) {        // 如果自身还在责任链中，则进行移除
            pipeline.remove(this);
        }
    } finally {
        ctx.close();
    }
}
```

13.12　ChannelOutboundHandlerAdapter 抽象类

ChannelOutboundHandlerAdapter 数据输出通道处理器适配器类实现了所有 ChannelOutboundHandler 方法，默认行为均是将事件传递给责任链中下一个通道输出数据处理器，而子类可以继承该类实现自身需要的方法即可。源码如下。

```java
public class ChannelOutboundHandlerAdapter extends ChannelHandlerAdapter implements ChannelOutboundHandler {
    public void bind(ChannelHandlerContext ctx, SocketAddress localAddress,
                    ChannelPromise promise) throws Exception {
        ctx.bind(localAddress, promise);
    }
    public void connect(ChannelHandlerContext ctx, SocketAddress remoteAddress,
                       SocketAddress localAddress, ChannelPromise promise) throws Exception {
        ctx.connect(remoteAddress, localAddress, promise);
    }
    public void disconnect(ChannelHandlerContext ctx, ChannelPromise promise)
            throws Exception {
        ctx.disconnect(promise);
    }
    public void close(ChannelHandlerContext ctx, ChannelPromise promise)
            throws Exception {
        ctx.close(promise);
    }
    public void deregister(ChannelHandlerContext ctx, ChannelPromise promise) throws Exception {
        ctx.deregister(promise);
    }
    public void read(ChannelHandlerContext ctx) throws Exception {
        ctx.read();
    }
    public void write(ChannelHandlerContext ctx, Object msg, ChannelPromise promise) throws Exception {
        ctx.write(msg, promise);
    }
    public void flush(ChannelHandlerContext ctx) throws Exception {
        ctx.flush();
    }
}
```

13.13　ChannelDuplexHandler 处理器

有些时候，如果想在一个类中既实现 ChannelInboundHandler 双端通道处理器接口，又实现 ChannelOutboundHandler 接口，这样一个类既可以处理输入数据又可以处理输出数据。本来可以同时继承 ChannelInboundHandlerAdapter 和 ChannelOutboundHandlerAdapter 类，但 Java 中不允许多继承，所以引入该类来定义通道双端处理器。由源码可知。该类仅是为了打破不可多继承而实现的，实现方式同两个适配器类。源码如下。

```
// 继承自 ChannelInboundHandlerAdapter，同时与 ChannelOutboundHandlerAdapter 有相同的实现 ChannelOutboundHandler 接口
public class ChannelDuplexHandler extends ChannelInboundHandlerAdapter implements ChannelOutboundHandler {
    public void bind(ChannelHandlerContext ctx, SocketAddress localAddress,
                     ChannelPromise promise) throws Exception {
        ctx.bind(localAddress, promise);
    }
    public void connect(ChannelHandlerContext ctx, SocketAddress remoteAddress,
                        SocketAddress localAddress, ChannelPromise promise) throws Exception {
        ctx.connect(remoteAddress, localAddress, promise);
    }
    public void disconnect(ChannelHandlerContext ctx, ChannelPromise promise)
            throws Exception {
        ctx.disconnect(promise);
    }
    public void close(ChannelHandlerContext ctx, ChannelPromise promise) throws Exception {
        ctx.close(promise);
    }
    public void deregister(ChannelHandlerContext ctx, ChannelPromise promise) throws Exception {
        ctx.deregister(promise);
    }
    public void read(ChannelHandlerContext ctx) throws Exception {
        ctx.read();
    }
    public void write(ChannelHandlerContext ctx, Object msg, ChannelPromise promise) throws Exception {
        ctx.write(msg, promise);
    }
    public void flush(ChannelHandlerContext ctx) throws Exception {
        ctx.flush();
    }
}
```

13.14　通道与通道处理器小结

前面介绍了很多内容，其实也就是 Netty 的通道处理相关的框架，接下来要探讨的内容依托于这些框架提供的扩展点，所以笔者认为有必要在这里进行阶段性总结。由于常用的通道类为 NIO 通道，所

以这里以这个通道类型作为复习对象。通过上述知识可知。

（1）Netty 中使用 Channel 接口和 ServerChannel 接口分别定义客户端通道与服务端通道。

（2）ServerChannel 接口为标识性接口，本身不扩展任何方法。

（3）Netty 中使用 AbstractNioChannel 类定义 Java 原生 NIO 的模板类。

（4）AbstractNioChannel 类的子类 AbstractNioByteChannel 定义了客户端通道模板。

（5）AbstractNioChannel 类的子类 AbstractNioMessageChannel 定义了服务端通道模板。

（6）NioServerSocketChannel 类与 NioSocketChannel 类分别组成 NIO 服务端和 NIO 客户端的实现对。

（7）Netty 中将通道实现与底层 IO 交流分割，将底层 IO 通道操作放在 Unsafe 接口的实现类中，对于 NIO 来说，最终实现类为 NioSocketChannelUnsafe（客户端实现）、NioMessageUnsafe（服务端实现）。

（8）Netty 中使用责任链模式处理通道数据，即 ChannelPipeline 与 ChannelHandler 机制。

（9）在 ChannelPipeline 中将一系列 ChannelHandler 组成责任链共同完成对通道的处理。

（10）Netty 中使用 ChannelInboundInvoker 接口和 ChannelOutboundInvoker 接口，分别定义输入数据和输出数据调用方法，而 ChannelPipeline 便是继承这两个接口完成对通道数据输入和数据输出事件的回调。

（11）Netty 中使用 ChannelHandlerContext 接口包装通道处理器，在 ChannelPipeline 中管理的责任链对象便是 ChannelHandlerContext 接口实例，它封装了 ChannelHandler 的元数据，同时定义了责任链的双向链表，它也实现了 ChannelInboundInvoker 和 ChannelOutboundInvoker 接口，可以将事件传递给责任链的下一个 ChannelHandlerContext。

（12）Netty 通道处理器责任链中有两个关键的 ChannelHandlerContext 接口实现：HeadContext 和 TailContext。其中 TailContext 实现了 ChannelInboundHandler，用于在责任链末尾处理前面读事件通道处理器未处理的事件，默认将收到的数据或是静默处理、或是输出日志并释放内存。HeadContext 既实现了 ChannelOutboundHandler 接口，又实现了 ChannelInboundHandler 接口。接收到数据输入事件时，它将事件传递给责任链的下一个通道处理器处理；接收到输出事件时，调用 Unsafe 的实例完成底层 IO 操作。

通过以上信息不难看出，之后的描述必然是与通道处理器相关，通过通道处理器处理接收到的数据。由于这里描述的均为 TCP 协议，所以不难猜出，后面肯定是介绍编码器相关的内容：根据 TCP 协议实现应用层的协议，由于应用层协议较多，所以笔者仅以 HTTP 协议作为描述，同时省略掉对于 HTTP 协议进行详细解码的部分，如解析头部、换行符等，毕竟这已经不属于 Netty 的范畴，而是属于 HTTP 协议详解的范畴。

第 14 章 Netty 解码器与编码器

14.1 ByteToMessageDecoder 解码器

ByteToMessageDecoder 类是特殊的 ChannelInboundHandlerAdapter 处理通道输入数据的抽象类，定义了从 ByteBuf 缓冲区解析数据并生成另外一种类型的消息对象的模板方法。例如，可以实现一个 ByteToMessageDecoder，将从一个 ByteBuf 中读取的数据放入另外一个 ByteBuf 对象中，代码如下。通常使用该类的子类完成从通道中读取的数据，子类只需要继承 ByteToMessageDecoder 并实现 decode 自身解码逻辑即可。

```
public class SquareDecoder extends ByteToMessageDecoder {
    // 子类只需要继承 ByteToMessageDecoder 并实现 decode 解码即可
    public void decode(ChannelHandlerContext ctx, ByteBuf in, List<Object> out) throws Exception {
        out.add(in.readBytes(in.readableBytes()));
    }
}
```

该类较为复杂，笔者将其进行拆分描述。

14.1.1 核心变量与构造器

由源码可知。

（1）实际存储从通道中读取的数据对象为 ByteBuf cumulation 缓冲区，该对象可能为单个 ByteBuf，也可能为组合缓冲区 CompositeByteBuf 对象，这取决于 Cumulator cumulator 对象的实现，该对象用于将每次读取的数据进行组合，提供给编码器使用，默认为实现单个 ByteBuf 的 MERGE_CUMULATOR。

（2）在构造器中判断当前类的实现类不允许出现@Shareable 注解，因为编码器类是非线程安全的，不能在不同流水线中共享。

```
public abstract class ByteToMessageDecoder extends ChannelInboundHandlerAdapter {
    // 定义组合 cumulation 和 in 两个缓冲区，并将组合后的缓冲区返回
    public interface Cumulator {
        ByteBuf cumulate(ByteBufAllocator alloc, ByteBuf cumulation, ByteBuf in);
    }
    // 封装一个混合缓冲区对象 Cumulator，具体实现为根据组合后的缓冲区大小选择是否对组合缓冲区扩容，
    随后将 in 缓存区中的数据写入 ByteBuf cumulation 缓冲区中，然后释放 in 缓冲区
    public static final Cumulator MERGE_CUMULATOR = new Cumulator() {
        @Override
```

```java
    public ByteBuf cumulate(ByteBufAllocator alloc, ByteBuf cumulation, ByteBuf in) {
        ByteBuf buffer;
        if (cumulation.writerIndex() > cumulation.maxCapacity() - in.readableBytes()
                || cumulation.refCnt() > 1) {
            buffer = expandCumulation(alloc, cumulation, in.readableBytes());
        } else {
            buffer = cumulation;
        }
        buffer.writeBytes(in);
        in.release();
        return buffer;
    }
};

// 对ByteBuf cumulation 缓冲区进行扩容，通过ByteBufAllocator alloc 分配新的cumulation，将oldCumulation
// 中的数据写入新的 cumulation 缓冲区中，并释放旧的 cumulation 缓冲区
static ByteBuf expandCumulation(ByteBufAllocator alloc, ByteBuf cumulation, int readable) {
    ByteBuf oldCumulation = cumulation;
    cumulation = alloc.buffer(oldCumulation.readableBytes() + readable);
    cumulation.writeBytes(oldCumulation);
    oldCumulation.release();
    return cumulation;
}

// 封装一个组合缓冲区对象 Cumulator，具体实现为使用组合缓冲区对象 CompositeByteBuf 完成 in 缓冲区
// 与 cumulation 缓冲区的组合
public static final Cumulator COMPOSITE_CUMULATOR = new Cumulator() {
    @Override
    public ByteBuf cumulate(ByteBufAllocator alloc, ByteBuf cumulation, ByteBuf in) {
        ByteBuf buffer;
        if (cumulation.refCnt() > 1) { // cumulation 存在外部引用，进行缓冲区组合。对 cumulation 扩容后，
                                        // 将 in 放入新的 buffer 缓冲区中，此时不引用原有的 cumulation
            buffer = expandCumulation(alloc, cumulation, in.readableBytes());
            buffer.writeBytes(in);
            in.release();
        } else {                        // 否则判断 cumulation 是否为 CompositeByteBuf 组合缓冲区，如果
                                        // 是 CompositeByteBuf 实例，则将其强转为 CompositeByteBuf，然后将 in 缓冲区添加放入组合缓冲区对象中；如
                                        // 果不是，则创建新的组合缓冲区对象，并将 cumulation 与 in 放入其中
            CompositeByteBuf composite;
            if (cumulation instanceof CompositeByteBuf) {
                composite = (CompositeByteBuf) cumulation;
            } else {
                composite = alloc.compositeBuffer(Integer.MAX_VALUE);
                composite.addComponent(true, cumulation);
            }
            composite.addComponent(true, in);
            buffer = composite;
        }
        return buffer;
    }
};
ByteBuf cumulation;             // 组合缓冲区对象（可能为单个 ByteBuf 包含所有输入数据，也可能为
                                // CompositeByteBuf 组合缓冲区对象，其中包含多个 ByteBuf 对象共同组成输入数据）
```

```
    private Cumulator cumulator = MERGE_CUMULATOR;        // 默认为混合缓冲区对象
    private boolean singleDecode;  // 如果为true,则表明在调用channelRead(ChannelHandlerContext, Object)
方法时只有一个消息被解码
    private boolean decodeWasNull;                         // 表示在调用 channelRead(ChannelHandlerContext
ctx, Object msg)解码后的消息为空,即没有发生解码
    private boolean first;                                 // 表示是否是第一次进行解码
    private int discardAfterReads = 16;                    // 限制读缓冲区中读入的消息量,numReads 超过该值
时,尝试丢弃部分缓冲区的数据
    private int numReads;                                  // 当前从通道中读取并解码处理的消息数量,每次调用
channelRead(ChannelHandlerContext ctx, Object msg)方法时该变量加 1

    // 构造器,判断当前编码器类不能拥有@Shareable 注解,因为它是非线程安全的,不能在不同流水线中共享
    protected ByteToMessageDecoder() {
        CodecUtil.ensureNotSharable(this);
    }
}
```

14.1.2 channelRead 核心方法

channelRead 方法将在通道读取到数据时调用,参数 msg 表示从通道中读取的数据,ctx 表示当前通道的包装上下文对象。处理流程如下。

(1) 本类的名字叫作 ByteToMessageDecoder,所以只处理 byte 数据,而这些数据保存在 ByteBuf 缓冲区对象中,所以该方法只处理 msg 为 ByteBuf 类型的数据。如果是其他类型,则调用 ctx.fireChannelRead(msg) 将消息传递给责任链的下一个处理器。

(2) 创建 CodecOutputList 存放编码后的对象列表,该类继承自 AbstractList,只不过内部使用了 Recycler RECYCLER 来对该对象进行对象缓存。因为对于 Netty 来说,编码解码很常见,所以建立 CodecOutputList 的对象池可以减少频繁的 YGC,以及对象创建的消耗。

(3) 如果当前是第一次调用,则将 data 缓冲区作为 cumulation 即可,否则调用 cumulator 将当前 cumulation 缓冲区对象与 data 对象进行组合,并将组合后的对象保存为 cumulation 实例变量。

(4) 调用 callDecode(ctx, cumulation, out)方法,对 cumulation 缓冲区数据进行解码,同时将结果放入 out 列表中。

(5) 如果解码完成后存在 cumulation 缓冲区,且当前已经全部解码,即!cumulation.isReadable() (缓冲区不可读,此时表明已经对其中的可读字节进行了处理),则重置 numReads 计数,同时释放缓冲区。

(6) 如果当前 cumulation 缓冲区仍然存在未处理数据,则表明已经读取了足够多的信息,这里试图丢弃一些 cumulation 缓冲区中已经解码后的字节,减少发生内存溢出(OOM)的概率。

```
    public void channelRead(ChannelHandlerContext ctx, Object msg) throws Exception {
        if (msg instanceof ByteBuf) {
            CodecOutputList out = CodecOutputList.newInstance();     // 存放编码后的对象列表
            try {
                ByteBuf data = (ByteBuf) msg;
                first = cumulation == null;
                // 如果当前是第一次调用,则将 data 缓冲区作为 cumulation 即可,否则调用 cumulator 对当前
cumulation 缓冲区对象与 data 对象进行组合
                if (first) {
```

```java
                    cumulation = data;
                } else {
                    cumulation = cumulator.cumulate(ctx.alloc(), cumulation, data);
                }
                callDecode(ctx, cumulation, out);                    // 实现解码操作
            } catch (DecoderException e) {
                throw e;
            } catch (Throwable t) {
                throw new DecoderException(t);
            } finally {
                if (cumulation != null && !cumulation.isReadable()) { // cumulation 缓冲区存在且当前已经全部解
码，即 !cumulation.isReadable()（缓冲区不可读，此时表明已经对其中的可读字节进行了处理），则重置 numReads
计数，同时释放缓冲区
                    numReads = 0;
                    cumulation.release();
                    cumulation = null;
                } else if (++ numReads >= discardAfterReads) {       // 当前已经读取了足够多的信息，这里试图丢弃
一些 cumulation 缓冲区中已经解码后的字节，以减少发生 OOM 的几率
                    numReads = 0;
                    discardSomeReadBytes();     // 调用 ByteBuf 的 discardSomeReadBytes 方法完成丢弃
                }
                // 将列表中解码的对象传递给责任链中下一个处理输入数据事件的通道处理器
                int size = out.size();
                decodeWasNull = !out.insertSinceRecycled();          // 在回收前是否有数据放入 out 列表中，表示是
否发生了解码操作并得到解码对象
                fireChannelRead(ctx, out, size);
                out.recycle();
            }
        } else {                                                     // 将其他类型对象交给责任链的下一个处理器完成
            ctx.fireChannelRead(msg);
        }
    }

    // 将列表中解码的对象传递给责任链中下一个处理输入数据事件的通道处理器，这里通过循环处理
    static void fireChannelRead(ChannelHandlerContext ctx, CodecOutputList msgs, int numElements) {
        for (int i = 0; i < numElements; i ++) {
            ctx.fireChannelRead(msgs.getUnsafe(i)); // 这里的 getUnsafe 等同于直接操作 array[index]，不进行数
组越界判断，所以为 Unsafe
        }
    }

    // 实现解码操作
    protected void callDecode(ChannelHandlerContext ctx, ByteBuf in, List<Object> out) {
        try {
            while (in.isReadable()) {                                // in 缓冲区仍然存在数据时，循环解码
                int outSize = out.size();
                if (outSize > 0) {                                   // out 列表中存在对象，表明产生解码对象，将解
码得到的值传递给责任链的下一个通道处理器
                    fireChannelRead(ctx, out, outSize);
                    out.clear();                                     // 清空列表
                    if (ctx.isRemoved()) {                           // 如果已经从责任链移除了当前 ctx，则结束解码
                        break;
                    }
```

```java
                outSize = 0;
            }
            int oldInputLength = in.readableBytes();
            decode(ctx, in, out);        // 调用子类完成实际解码操作
            if (ctx.isRemoved()) {       // 如果已经从责任链移除了当前 ctx，则结束解码
                break;
            }
            if (outSize == out.size()) { // 解码后 out 列表长度不变，说明子类未完成解码，则进行长度判断，
// 如果解码前后缓冲区中的可读数据均不变，表明当前子类无法依靠当前数据完成解码，则停止循环；如果可读数
// 据发生变化，说明子类丢弃了部分字节，则继续循环处理
                if (oldInputLength == in.readableBytes()) {
                    break;
                } else {
                    continue;
                }
            }

            if (oldInputLength == in.readableBytes()) {  // 解码后 out 列表长度发生变化，但是 in 缓冲区的可
// 读数据不变，这表明子类不消费缓冲区中的数据，然后发生了解码。这是非法操作，因为如果不消耗数据，则不
// 会发生解码。所以抛出异常
                throw new DecoderException(
                    StringUtil.simpleClassName(getClass()) +
                    ".decode() did not read anything but decoded a message.");
            }

            if (isSingleDecode()) {       // 如果当前指明为解码一次的解码器，则结束循环
                break;
            }
        }
    } catch (DecoderException e) {
        throw e;
    } catch (Throwable cause) {
        throw new DecoderException(cause);
    }
}

// 子类完成实际解码过程，将解码结果放入 out 列表中
protected abstract void decode(ChannelHandlerContext ctx, ByteBuf in, List<Object> out) throws Exception;
```

14.1.3　handlerRemoved 核心方法

handlerRemoved 方法将在通道从流水线中移除时调用。流程如下。

（1）如果 cumulation 输入缓冲区数据不为空，则将实例变量置空，同时将其引用保存在线程栈上，然后对其中的可读数据调用 ctx.fireChannelRead(bytes)处理其中尚未解码的数据，然后将其释放。

（2）调用钩子函数 handlerRemoved0(ctx)，完成子类的通道移除时的回调逻辑。

```java
public final void handlerRemoved(ChannelHandlerContext ctx) throws Exception {
    ByteBuf buf = cumulation;
    if (buf != null) {
        // 直接把它设为 null，这样就能确保不再在这里的任何其他方法中访问它
        cumulation = null;

        int readable = buf.readableBytes();
        if (readable > 0) {
```

```
                ByteBuf bytes = buf.readBytes(readable);
                buf.release();
                ctx.fireChannelRead(bytes);
            } else {
                buf.release();
            }

            numReads = 0;
            ctx.fireChannelReadComplete();
        }
        handlerRemoved0(ctx);
    }

    // 钩子函数，子类可重写该方法完成自身逻辑
    protected void handlerRemoved0(ChannelHandlerContext ctx) throws Exception { }
```

14.1.4　channelReadComplete 核心方法

channelReadComplete 方法将在通道读取完所有读缓冲区数据时调用。处理流程如下。

（1）调用 discardSomeReadBytes()丢弃已完成解码的 ByteBuf cumulation 缓冲区中的数据，因为 channelReadComplete 方法一定是在 channelRead 方法的后面调用，已经在 channelRead 中对该缓冲区进行了解码，所以这里可以尝试丢弃一些已经解码的数据，减少内存的消耗。

（2）判断当前是否产生了解码对象。如果没有，则判断通道是否配置了 AutoRead。如果没有调用 ctx.read()，则在代表当前通道的 SelectionKey 的感兴趣事件集中添加读感兴趣事件（这里以 NIO 为例）。

（3）将 ChannelReadComplete 传递给其他通道处理器。

```
public void channelReadComplete(ChannelHandlerContext ctx) throws Exception {
    numReads = 0;
    discardSomeReadBytes();
    if (decodeWasNull) { // 没有产生解码对象
        decodeWasNull = false;
        if (!ctx.channel().config().isAutoRead()) {
            ctx.read();
        }
    }
    ctx.fireChannelReadComplete();
}
```

14.1.5　channelInactive 核心方法

channelInactive 方法在通道从活动状态变为失活状态时调用。处理流程如下。

（1）调用 channelInputClosed，完成对 cumulation 缓冲区中的数据处理。

（2）处理完成后，如果 cumulation 仍然存在，则释放该缓冲区。

（3）如果 out 列表中存在还未处理器的解码数据，则调用通道处理器 fireChannelRead 方法完成处理，处理完成后调用 ChannelReadComplete，表明当前解码列表中的数据已经完全处理。

（4）如果设置调用 ChannelInactive 方法，则进行回调。

```
public void channelInactive(ChannelHandlerContext ctx) throws Exception {
    channelInputClosed(ctx, true);
```

```java
private void channelInputClosed(ChannelHandlerContext ctx, boolean callChannelInactive) throws Exception {
    CodecOutputList out = CodecOutputList.newInstance();
    try {
        channelInputClosed(ctx, out);      // 完成对 cumulation 缓冲区中数据的处理
    } catch (DecoderException e) {
        throw e;
    } catch (Exception e) {
        throw new DecoderException(e);
    } finally {
        try {
            if (cumulation != null) {      // 处理完成后，如果 cumulation 仍然存在，则释放该缓冲区
                cumulation.release();
                cumulation = null;
            }
            // out 列表中存在还未处理器的解码数据，调用通道处理器完成处理
            int size = out.size();
            fireChannelRead(ctx, out, size);
            if (size > 0) { // 处理完成后，调用 ChannelReadComplete，表明当前解码列表中的数据已经完全处理
                ctx.fireChannelReadComplete();
            }
            if (callChannelInactive) {    // 如果设置调用 ChannelInactive 方法，则在这里回调
                ctx.fireChannelInactive();
            }
        } finally {
            // 回收存放解码对象的 out 列表
            out.recycle();
        }
    }
}

// 关闭通道的输入时调用，这可能是因为通道被更改为非活动状态，或产生了 ChannelInputShutdownEvent 事件
void channelInputClosed(ChannelHandlerContext ctx, List<Object> out) throws Exception {
    if (cumulation != null) {    // 存在 cumulation 缓冲区，先调用 callDecode 循环解码，再调用 decodeLast，
                                 // 通知子类当前处理的 in 缓冲区为最后一个缓冲区对象
        callDecode(ctx, cumulation, out);
        decodeLast(ctx, cumulation, out);
    } else {    // 不存在 cumulation 缓冲区，使用 EMPTY_BUFFER 空缓冲区，调用 decode，通知子类方法当
                // 前通道输入端已经关闭
        decodeLast(ctx, Unpooled.EMPTY_BUFFER, out);
    }
}

// 通知子类当前处理的 in 缓冲区为最后一个缓冲区对象，此时输入端已经关闭
protected void decodeLast(ChannelHandlerContext ctx, ByteBuf in, List<Object> out) throws Exception {
    if (in.isReadable()) {              // 只有在缓冲区还有字节需要解码时才调用 decode()
        decode(ctx, in, out);
    }
}
```

14.1.6　userEventTriggered 核心方法

在通道接收到用户自定义事件时调用 userEventTriggered 方法。该方法只处理输入端关闭事件，其

他事件将传递给其他处理器处理，直接调用 channelInputClosed(ctx, false)完成处理。源码如下。

```
public void userEventTriggered(ChannelHandlerContext ctx, Object evt) throws Exception {
    if (evt instanceof ChannelInputShutdownEvent) {    // 只处理输入端关闭事件
        channelInputClosed(ctx, false);    // 只有在缓冲区中还有字节需要解码时才调用 decode()。遇到
channelInactive 事件时，将调用 decodeLast 方法，该方法负责在某些情况下结束当前客户端请求，并且必须在
输入已关闭时调用
    }
    super.userEventTriggered(ctx, evt);
}
```

14.2 MessageToByteEncoder 编码器

ByteToMessageDecoder 解码器会把 ByteBuf 中的字节信息编码为消息对象，MessageToByteEncoder 抽象类之所以称为编码器，因为它是 ByteToMessageDecoder 的反向操作。将消息对象通过转为 ByteBuf 中的字节信息（对象的序列化和反序列化）。如下可以这样使用该编码器。

```
public class IntegerEncoder extends MessageToByteEncoder<Integer> {
    public void encode(ChannelHandlerContext ctx, Integer msg, ByteBuf out) throws Exception {
        out.writeInt(msg);                    // 直接将整型消息写入 out 缓冲区
    }
}
```

从上面例子来看，只需要在子类重写 encode 方法完成编码即可。由于该类较为简单，所以笔者直接统一分析。由源码可知。

（1）该类继承自 ChannelOutboundHandlerAdapter，自身就是一个处理通道输出数据的通道处理器。

（2）由于在网络 IO 中，对于数据的写入，使用堆外内存可以避免一次 JVM 堆内存到 JVM 进程堆内存的复制损耗，所以默认优先分配堆外内存。

（3）该类定义了 TypeParameterMatcher 参数匹配器，用于判断当前编码器可接收编码的 msg 对象类型。读者不必深究类型匹配器的原理，因为它采用了 type.isInstance(msg)，当然 Netty 也使用了 Javassist 类实现类型匹配器来生成使用操作符 instanceof 的匹配类，但是那并不重要。

（4）在 write(ChannelHandlerContext ctx, Object msg, ChannelPromise promise)方法中完成对子类的 encode(ChannelHandlerContext ctx, I msg, ByteBuf out)方法回调，由子类完成实际编码处理。

```
public abstract class MessageToByteEncoder<I> extends ChannelOutboundHandlerAdapter {
    private final TypeParameterMatcher matcher;    // 类型匹配器
    private final boolean preferDirect;            // 优先分配堆外缓冲区

    // 默认优先使用堆外缓冲区
    protected MessageToByteEncoder() {
        this(true);
    }

    // 传入类型匹配，表明当前编码器只编码 outboundMessageType 类型的消息
    protected MessageToByteEncoder(Class<? extends I> outboundMessageType) {
        this(outboundMessageType, true);
    }
```

```java
// 默认匹配泛型 I 指定的类型的数据
protected MessageToByteEncoder(boolean preferDirect) {
    matcher = TypeParameterMatcher.find(this, MessageToByteEncoder.class, "I"); // 通过该类获取到参数类型匹配器，这里传入"I"表示找到当前编码器泛型定义的类型
    this.preferDirect = preferDirect;
}

// 完整构造器，初始化成员变量
protected MessageToByteEncoder(Class<? extends I> outboundMessageType, boolean preferDirect) {
    matcher = TypeParameterMatcher.get(outboundMessageType);
    this.preferDirect = preferDirect;
}

// 判断当前编码器是否接收 msg 对象
public boolean acceptOutboundMessage(Object msg) throws Exception {
    return matcher.match(msg);
}

// 通道处理器写入数据时调用
public void write(ChannelHandlerContext ctx, Object msg, ChannelPromise promise) throws Exception {
    ByteBuf buf = null;
    try {
        if (acceptOutboundMessage(msg)) {                       // 当前编码器可以对 msg 对象进行编码
            I cast = (I) msg;
            buf = allocateBuffer(ctx, cast, preferDirect);      // 分配存放编码后字节的缓冲区对象
            try {
                encode(ctx, cast, buf);                         // 调用子类完成实际编码
            } finally {
                ReferenceCountUtil.release(cast);               // 编码完成后释放 msg 对象
            }
            // 编码得到的数据，调用 ctx.write 将数据交给其他处理器处理
            if (buf.isReadable()) {
                ctx.write(buf, promise);
            } else {
                buf.release();
                ctx.write(Unpooled.EMPTY_BUFFER, promise);
            }
            buf = null;
        } else {        // 当传入消息类型与当前编码器可以处理的类型不匹配时，则交给其他处理器处理
            ctx.write(msg, promise);
        }
    } catch (EncoderException e) {
        throw e;
    } catch (Throwable e) {
        throw new EncoderException(e);
    } finally {
        if (buf != null) {
            buf.release();
        }
    }
}
```

```java
// 调用分配器分配内存
protected ByteBuf allocateBuffer(ChannelHandlerContext ctx, @SuppressWarnings("unused") I msg,
                                 boolean preferDirect) throws Exception {
    if (preferDirect) {
        return ctx.alloc().ioBuffer();
    } else {
        return ctx.alloc().heapBuffer();
    }
}

// 子类实现完成实际编码，编码后的字节信息放入 ByteBuf out 中
protected abstract void encode(ChannelHandlerContext ctx, I msg, ByteBuf out) throws Exception;
}
```

14.3 MessageToMessageDecoder 解码器

ByteToMessageDecoder 是将 ByteBuf 缓冲区中的 byte 数据解码为 Message 对象，而 MessageToMessageDecoder 类则用于将 Message 对象解码为另外一个 Message 对象。例如，将 String 类型的 Message 在 decode 方法中解码成了字符串的长度整型值。

```java
public class StringToIntegerDecoder extends MessageToMessageDecoder<String> {
    public void decode(ChannelHandlerContext ctx, String message, List<Object> out) throws Exception {
        outadd(message.length());
    }
}
```

该类较为简单，不涉及字节的操作，均是对象转对象。所以和 MessageToByteEncoder 类一样，需要一个 TypeParameterMatcher 类型匹配器，对传入的 Message 对象进行校验，判断当前解码器是否适用于该对象。源码如下。

```java
public abstract class MessageToMessageDecoder<I> extends ChannelInboundHandlerAdapter {
    private final TypeParameterMatcher matcher;    // 判定当前编码器是否对传入的 Message 对象进行解码

    // 构造器同 MessageToByteEncoder
    protected MessageToMessageDecoder() {
        matcher = TypeParameterMatcher.find(this, MessageToMessageDecoder.class, "I");
    }
    protected MessageToMessageDecoder(Class<? extends I> inboundMessageType) {
        matcher = TypeParameterMatcher.get(inboundMessageType);
    }

    // 判定当前解码器是否可以对 msg 对象解码
    public boolean acceptInboundMessage(Object msg) throws Exception {
        return matcher.match(msg);
    }

    // 同 MessageToByteEncoder
    public void channelRead(ChannelHandlerContext ctx, Object msg) throws Exception {
        CodecOutputList out = CodecOutputList.newInstance();
        try {
```

```java
            if (acceptInboundMessage(msg)) {      // 校验通过后，调用子类完成实际解码
                I cast = (I) msg;
                try {
                    decode(ctx, cast, out);
                } finally {
                    ReferenceCountUtil.release(cast);
                }
            } else {
                out.add(msg);
            }
        } catch (DecoderException e) {
            throw e;
        } catch (Exception e) {
            throw new DecoderException(e);
        } finally {
            int size = out.size();
            for (int i = 0; i < size; i ++) {
                ctx.fireChannelRead(out.getUnsafe(i));
            }
            out.recycle();
        }
    }

    // 子类完成实际解码过程
    protected abstract void decode(ChannelHandlerContext ctx, I msg, List<Object> out) throws Exception;
}
```

14.4 MessageToMessageEncoder 编码器

MessageToByteDecoder 是将 Message 对象进行编码为 byte 数据，将其存入 ByteBuf 缓冲区，而 MessageToMessageEncoder 类用于将 Message 对象编码为另外一个 Message 对象。例如，将 Integer 整型类型的 Message 在 encode 方法中解码成了字符串。

```java
public class IntegerToStringEncoder extends MessageToMessageEncoder<Integer> {
    public void encode(ChannelHandlerContex} ctx, Integer message, List<Object> out) throws Exception {
        out.add(message.toString());
    }
}
```

该类较为简单，不涉及字节的操作，均是对象转对象，所以同 MessageToMessageDecoder 类一样，需要一个 TypeParameterMatcher 类型匹配器，对传入的 Message 对象进行校验，判断当前解码器是否适用于该对象。源码如下。

```java
public abstract class MessageToMessageEncoder<I> extends ChannelOutboundHandlerAdapter {
    private final TypeParameterMatcher matcher;    // 判定当前编码器是否对传入的 Message 对象进行解码

    // 构造器同 MessageToByteEncoder
    protected MessageToMessageEncoder() {
        matcher = TypeParameterMatcher.find(this, MessageToMessageEncoder.class, "I");
    }
```

```java
protected MessageToMessageEncoder(Class<? extends I> outboundMessageType) {
    matcher = TypeParameterMatcher.get(outboundMessageType);
}

// 判定当前解码器是否可以对 msg 对象编码
public boolean acceptOutboundMessage(Object msg) throws Exception {
    return matcher.match(msg);
}

// 当通道数据写入时调用
public void write(ChannelHandlerContext ctx, Object msg, ChannelPromise promise) throws Exception {
    CodecOutputList out = null;
    try {
        if (acceptOutboundMessage(msg)) {              // 当前编码器可以对 msg 对象进行编码
            out = CodecOutputList.newInstance();       // 用于存放子类实现的 encode 编码后的对象
            I cast = (I) msg;
            try {                                       // 子类完成编码
                encode(ctx, cast, out);
            } finally {
                ReferenceCountUtil.release(cast);
            }
            if (out.isEmpty()) {    // 编码没有生成对象，回收对象列表对象，同时抛出异常
                out.recycle();
                out = null;

                throw new EncoderException(
                    StringUtil.simpleClassName(this) + " must produce at least one message.");
            }
        } else {                    // 当前编码器不编码 msg 对象，将该操作传递给其他处理器处理
            ctx.write(msg, promise);
        }
    } catch (EncoderException e) {
        throw e;
    } catch (Throwable t) {
        throw new EncoderException(t);
    } finally {
        if (out != null) {          // 存在编码后的对象，将其作为 msg 交给通道处理器责任链下游处理器处理
            final int sizeMinusOne = out.size() - 1;
            if (sizeMinusOne == 0) { // 只编码了一个对象，取出该对象传递
                ctx.write(out.get(0), promise);
            } else if (sizeMinusOne > 0) { // 否则遍历列表，对每个编码对象封装 promise 传递到其他处理器
                ChannelPromise voidPromise = ctx.voidPromise();
                boolean isVoidPromise = promise == voidPromise;
                for (int i = 0; i < sizeMinusOne; i++) {
                    ChannelPromise p;
                    if (isVoidPromise) {
                        p = voidPromise;
                    } else {
                        p = ctx.newPromise();
                    }
                    ctx.write(out.getUnsafe(i), p);
                }
                ctx.write(out.getUnsafe(sizeMinusOne), promise);
```

```
            }
            out.recycle();
        }
    }
}

// 子类实现完成实际编码操作,编码后对象放在 out 列表中
protected abstract void encode(ChannelHandlerContext ctx, I msg, List<Object> out) throws Exception;
}
```

14.5 ByteToMessageCodec 双端编码器原理

ByteToMessageCodec 类继承自 ChannelDuplexHandler 类,不难看出,该类既可以作为解码器用于解码,又可以作为编码器用于编码。它属于 ByteToMessageDecoder 类与 MessageToByteEncoder 类的组合。前面介绍过这两个类,所以理解起来十分容易。由于源码较为简单,所以笔者在本节一并讲解。由源码可知。

(1) 该类通过包装 MessageToByteEncoder encoder 编码器与 ByteToMessageDecoder decoder 解码器,完成实际编码与解码操作。

(2) 编码器与解码器的编码解码方法,定义为抽象方法 encode 与 decode,由子类继承 ByteToMessageCodec 类完成实际操作。

(3) 发生通道事件时,将通道事件调用 encoder 编码器和 decoder 解码器完成处理。

总结一下,包装两个对象完成实际操作,对象的实际操作又由子类实现完成。源码如下。

```
public abstract class ByteToMessageCodec<I> extends ChannelDuplexHandler {
    private final TypeParameterMatcher outboundMsgMatcher; // 判断当前通道是否可以对传入对象进行编码
    private final MessageToByteEncoder<I> encoder;          // 编码器

    // 解码器对象,内部方法间接调用 ByteToMessageCodec 类的解码方法完成
    private final ByteToMessageDecoder decoder = new ByteToMessageDecoder() {
        @Override
        public void decode(ChannelHandlerContext ctx, ByteBuf in, List<Object> out) throws Exception {
            ByteToMessageCodec.this.decode(ctx, in, out);
        }

        @Override
        protected void decodeLast(ChannelHandlerContext ctx, ByteBuf in, List<Object> out) throws Exception {
            ByteToMessageCodec.this.decodeLast(ctx, in, out);
        }
    };

    // 构造器,同 ByteToMessageDecoder
    protected ByteToMessageCodec() {
        this(true);
    }
    protected ByteToMessageCodec(Class<? extends I> outboundMessageType) {
        this(outboundMessageType, true);
    }
```

```java
protected ByteToMessageCodec(boolean preferDirect) {
    CodecUtil.ensureNotSharable(this);
    outboundMsgMatcher = TypeParameterMatcher.find(this, ByteToMessageCodec.class, "I");
    encoder = new Encoder(preferDirect);        // 创建编码器对象
}
protected ByteToMessageCodec(Class<? extends I> outboundMessageType, boolean preferDirect) {
    CodecUtil.ensureNotSharable(this);
    outboundMsgMatcher = TypeParameterMatcher.get(outboundMessageType);
    encoder = new Encoder(preferDirect);
}

// 判断是否可以对 msg 进行编码
public boolean acceptOutboundMessage(Object msg) throws Exception {
    return outboundMsgMatcher.match(msg);
}

// 通道读入数据时调用，这里完成解码处理
public void channelRead(ChannelHandlerContext ctx, Object msg) throws Exception {
    decoder.channelRead(ctx, msg);
}

// 通道写入数据时调用，这里完成编码处理
public void write(ChannelHandlerContext ctx, Object msg, ChannelPromise promise) throws Exception {
    encoder.write(ctx, msg, promise);
}

// 其他方法同 ByteToMessageDecoder 的方法
public void channelReadComplete(ChannelHandlerContext ctx) throws Exception {
    decoder.channelReadComplete(ctx);
}
public void channelInactive(ChannelHandlerContext ctx) throws Exception {
    decoder.channelInactive(ctx);
}

// 添加通道时，回调编码器和解码器对象方法
public void handlerAdded(ChannelHandlerContext ctx) throws Exception {
    try {
        decoder.handlerAdded(ctx);
    } finally {
        encoder.handlerAdded(ctx);
    }
}

// 移除通道时，回调编码器和解码器对象方法
public void handlerRemoved(ChannelHandlerContext ctx) throws Exception {
    try {
        decoder.handlerRemoved(ctx);
    } finally {
        encoder.handlerRemoved(ctx);
    }
}

// 子类完成编码操作
protected abstract void encode(ChannelHandlerContext ctx, I msg, ByteBuf out) throws Exception;
```

```java
// 子类完成解码操作
protected abstract void decode(ChannelHandlerContext ctx,ByteBuf in,List<Object> out) throws Exception;

// 关闭通道时，通知子类当前 in 缓冲区为最后一个缓冲区，因为输入端已经关闭
protected void decodeLast(ChannelHandlerContext ctx, ByteBuf in, List<Object> out) throws Exception {
    if (in.isReadable()) {
        decode(ctx, in, out);
    }
}

// 内部类，继承自 MessageToByteEncoder，实现编码器工作，不难看出，同样调用 ByteToMessageCodec
// 的方法完成实际编码操作
private final class Encoder extends MessageToByteEncoder<I> {
    Encoder(boolean preferDirect) {
        super(preferDirect);
    }

    @Override
    public boolean acceptOutboundMessage(Object msg) throws Exception {
        return ByteToMessageCodec.this.acceptOutboundMessage(msg);
    }

    @Override
    protected void encode(ChannelHandlerContext ctx, I msg, ByteBuf out) throws Exception {
        ByteToMessageCodec.this.encode(ctx, msg, out);
    }
}
}
```

14.6 MessageToMessageCodec 双端编码器原理

　　ByteToMessageCodec 类继承自 ChannelDuplexHandler 类，该类既可以作为解码器用于解码，又可以作为编码器用于编码。与 ByteToMessageCodec 不同的是，它属于 MessageToMessageDecoder 类与 MessageToMessageEncoder 类的组合。前面已介绍过 ByteToMessageCodec 类与这两个类，所以理解起来十分容易。由于源码较为简单，所以笔者这里将其一起讲解。由源码可知。

　　（1）该类通过包装 MessageToMessageEncoder encoder 编码器与 MessageToMessageDecoder decoder 解码器完成实际编码与解码操作。

　　（2）编码器与解码器的编码解码方法，定义为抽象方法 encode 与 decode，由子类继承 MessageToMessageCodec 类完成实际操作。

　　（3）发生通道事件时，将通道事件调用 encoder 编码器和 decoder 解码器完成处理。

　　例如，将 Integer 类型的数据解码为 Long 类型，同时又将 Long 类型编码为 Integer 类型。

```java
public class NumberCodec extends MessageToMessageCodec<Integer,Long> {
    // 数据输入时解码
    public Long decode(ChannelHandlerContext ctx, Integer msg, List<Object> out) throws Exception{
        out.add(msg.longValue());
    }
```

```java
    // 数据输出时编码
    public Integer encode(ChannelHandlerContext ctx,Long msg, List<Object> out) throws Exception {
        out.add(msg.intValue());
    }
}
```

总结一下，包装两个对象完成实际操作，对象的实际操作又由子类实现完成。

```java
public abstract class MessageToMessageCodec<INBOUND_IN, OUTBOUND_IN> extends ChannelDuplexHandler { // 泛型 INBOUND_IN 表示输入对象类型，泛型 OUTBOUND_IN 表示输出对象类型

    // 编码器匿名对象，所有方法由 MessageToMessageCodec 类完成
    private final MessageToMessageEncoder<Object> encoder = new MessageToMessageEncoder<Object>() {
        public boolean acceptOutboundMessage(Object msg) throws Exception {
            return MessageToMessageCodec.this.acceptOutboundMessage(msg);
        }
        protected void encode(ChannelHandlerContext ctx, Object msg, List<Object> out) throws Exception {
            MessageToMessageCodec.this.encode(ctx, (OUTBOUND_IN) msg, out);
        }
    };

    // 解码器匿名对象，所有方法都由 MessageToMessageCodec 类完成
    private final MessageToMessageDecoder<Object> decoder = new MessageToMessageDecoder<Object>() {
        public boolean acceptInboundMessage(Object msg) throws Exception {
            return MessageToMessageCodec.this.acceptInboundMessage(msg);
        }
        protected void decode(ChannelHandlerContext ctx, Object msg, List<Object> out) throws Exception {
            MessageToMessageCodec.this.decode(ctx, (INBOUND_IN) msg, out);
        }
    };

    // 由于是对象与对象的编码解码，所以需要输入与输出编码器对象匹配对象
    private final TypeParameterMatcher inboundMsgMatcher;
    private final TypeParameterMatcher outboundMsgMatcher;

    /* 两个构造器均初始化类型匹配器 */
    protected MessageToMessageCodec() {
        inboundMsgMatcher = TypeParameterMatcher.find(this, MessageToMessageCodec.class, "INBOUND_IN");
        outboundMsgMatcher = TypeParameterMatcher.find(this, MessageToMessageCodec.class, "OUTBOUND_IN");
    }
    protected MessageToMessageCodec(
        Class<? extends INBOUND_IN> inboundMessageType, Class<? extends OUTBOUND_IN> outboundMessageType) {
        inboundMsgMatcher = TypeParameterMatcher.get(inboundMessageType);
        outboundMsgMatcher = TypeParameterMatcher.get(outboundMessageType);
    }

    /* 通道输入与输出回调时，调用解码与编码器完成操作 */
    public void channelRead(ChannelHandlerContext ctx, Object msg) throws Exception {
        decoder.channelRead(ctx, msg);
    }
```

```
    public void write(ChannelHandlerContext ctx, Object msg, ChannelPromise promise) throws Exception {
        encoder.write(ctx, msg, promise);
    }

    /* 判断编码与解码器是否可接收对象 msg */
    public boolean acceptInboundMessage(Object msg) throws Exception {
        return inboundMsgMatcher.match(msg);
    }
    public boolean acceptOutboundMessage(Object msg) throws Exception {
        return outboundMsgMatcher.match(msg);
    }

    // 子类完成实际编码操作
    protected abstract void encode(ChannelHandlerContext ctx, OUTBOUND_IN msg, List<Object> out)
        throws Exception;

    // 子类完成实际解码操作
    protected abstract void decode(ChannelHandlerContext ctx, INBOUND_IN msg, List<Object> out)
        throws Exception;
}
```

14.7 编码器小结

编码器是特殊的 ChannelOutboundHandler 或 ChannelInboundHandler 处理器，只不过是因为具备特殊的对数据进行编码和解码处理的作用。通过以上源码描述可知。

（1）ByteToMessageDecoder 解码器类和 MessageToByteEncoder 编码器类分别用于将 ByteBuf 的 byte 数据解码为 Message 对象（这里 Messgae 对象可以为任意对象类型），将 Message 对象编码为 ByteBuf 数据，即对应输入数据处理器和输出数据处理器。

（2）MessageToMessageDecoder 解码器类和 MessageToMessageEncoder 编码器类分别用于将 Message 对象解码为另一个 Message 对象、将 Message 对象编码为另一个 Message 对象。

（3）ByteToMessageCodec 类用于组合 ByteToMessageDecoder 类与 MessageToByteEncoder 类，提供给子类既能实现编码又能实现解码的便捷类。

（4）MessageToMessageCodec 类用于组合 MessageToMessageDecoder 类与 MessageToMessageEncoder 类，提供给子类既能实现编码又能实现解码的便捷类。

通过以上信息，可以推理得出，通常将 ByteToMessageDecoder 解码器作为处理数据的第一个输入通道处理器，用于将传入的 ByteBuf 转为其他对象，而将 MessageToByteEncoder 作为处理数据的最后一个输出通道处理器，用于将输出结果转为 ByteBuf 数据写入底层 IO。而 MessageToMessageDecoder 和 MessageToMessageEncoder 可以用于中间的处理器用于对象相互转换。

了解了大致编码与解码器框架后，将以 HTTP 协议的解析为例，了解编码器与解码器的搭配使用及其原理。注意，这里不会涉及 HTTP 协议的具体处理细节，那属于 HTTP 协议原理范畴。在研究之前，先来看一个使用 Netty 来作为 HTTP 服务器的例子。向通道中注册了 HttpRequestDecoder、HttpContentEncoder、HttpObjectAggregator，分别用于对 HTTP 请求解码、HTTP 响应编码、对 HTTP 的消息头与消息体进行组合，然后将 HttpHelloWorldServerHandler 业务处理器放入最后，可以在 HttpHelloWorldServerHandler

的 channelRead(ChannelHandlerContext ctx, Object msg)方法中，直接获取到被 HttpObjectAggregator 聚合后的 HttpRequest，而不是分别得到请求头与请求体。源码如下。

```java
public class HttpHelloWorldServerInitializer extends ChannelInitializer<SocketChannel> {
    @Override
    public void initChannel(SocketChannel ch) {
        ChannelPipeline p = ch.pipeline();
        p.addLast(new HttpRequestDecoder());            // HTTP 请求解码器
        p.addLast(new HttpContentEncoder());            // HTTP 响应编码器
        p.addLast(new HttpObjectAggregator(2000));      // 用于对 HTTP 的消息头与消息体进行组合的聚合器
        p.addLast(new HttpHelloWorldServerHandler());   // 业务处理器
    }
}

public class HttpHelloWorldServerHandler extends ChannelInboundHandlerAdapter {
    public void channelRead(ChannelHandlerContext ctx, Object msg) {
        if (msg instanceof HttpRequest) {
            // doSomething
        }
    }
}
```

注意，由于这里涉及了 HTTP 协议的具体细节，本书不涉及 HTTP 协议的具体内容，所以涉及 HTTP 协议的部分只会给出关键部分，而不分析 HTTP 的头部、消息体等。相信读者在理解了整体流程后，对于 HTTP、Redis 等其他协议都能简单掌握，因为无外乎获取到 ByteBuf，读取里面的字节，转为消息对象，再传递给后面的处理器使用。

14.8 HttpObjectDecoder 解码器和 HttpRequestDecoder 解码器

HttpObjectDecoder 解码器把 ByteBuf 中的 HTTP 协议数据转换为 HttpMessage 与 HttpContent 对象。HttpMessage 表示消息头部（响应头部或请求头部）、HttpContent 表示消息体。由源码可知。

（1）HTTP 解码器通过状态机解析 HTTP 协议，状态机由 currentState 控制。

（2）解析的结果包含一个 HttpMessage、多个 DefaultHttpContent、一个 DefaultLastHttpContent，此时处于该解码器后面的输入数据通道处理器将可能接收到 List out 中这三个对象实体，即 out 列表中存在多少个 Message 对象，后面的处理器则回调多少次，并获取其中的对象。

```java
public abstract class HttpObjectDecoder extends ByteToMessageDecoder {
    private HttpMessage message;                                // HTTP 头部信息
    // 实现 HTTP 协议解码过程，buffer 中存放着 HTTP 数据
    protected void decode(ChannelHandlerContext ctx, ByteBuf buffer, List<Object> out) throws Exception {
        ...
        switch (currentState) {                                 // 根据当前状态进行解析
            case SKIP_CONTROL_CHARS: {                          // 解析控制符
                ...
                currentState = State.READ_INITIAL;              // 转换状态为解析请求行
            }
            case READ_INITIAL:                                  // 解析请求行并创建 HttpMessage 对象
```

```
                try {
                    ...
                    message = createMessage(initialLine);
                    currentState = State.READ_HEADER;   // 转换状态为解析请求头
                } catch (Exception e) {
                    out.add(invalidMessage(buffer, e));
                    return;
                }
            case READ_HEADER:               // 解析请求头，并将 HttpMessage 添加到 List<Object> out 中
                ...
                out.add(message);
                ...
            case READ_FIXED_LENGTH_CONTENT: {    // 解析固定 Context-Length 的请求体
                ...
                ByteBuf content = buffer.readRetainedSlice(toRead); // 获取缓冲区中的 HTTP Content 数据
                ...
                if (chunkSize == 0) {   // 当前已经是最后的 HTTP 内容，将生成的代表 HTTP 最后消息体的
对象 DefaultLastHttpContent 放入 out 列表
                    out.add(new DefaultLastHttpContent(content, validateHeaders));
                    ...
                } else {                // 否则将表示 HTTP 部分消息体的 DefaultHttpContent 对象放入 out 列表
                    out.add(new DefaultHttpContent(content));
                }
                return;
            }
            ...
        }
        ...
    }
}
```

HttpRequestDecoder 请求解码器继承自 HttpObjectDecoder，实际解码操作均由 HttpObjectDecoder 类完成。该类实现了 createMessage 方法，在创建消息对象时，将其实例化为 DefaultHttpRequest，表示请求头对象（请求行和头部信息）。源码如下。

```
public class HttpRequestDecoder extends HttpObjectDecoder {
    @Override
    protected HttpMessage createMessage(String[] initialLine) throws Exception {
        return new DefaultHttpRequest(
            HttpVersion.valueOf(initialLine[2]),
            HttpMethod.valueOf(initialLine[0]), initialLine[1], validateHeaders);
    }
}
```

14.9 HttpObjectEncoder 编码器和 HttpContentEncoder 编码器

HttpObjectEncoder 类继承 MessageToMessageEncoder 类，将 HttpMessage（响应头部）、HttpContent（响应体）编码数据放入 ByteBuf 中。由源码可知，在 encode 方法中将响应头部与响应体编码放入 out 列表中，如果 ByteBuf buf 足够存放响应头部与响应体，则传递到通道责任链处理器的下一个处理器会得到一个 ByteBuf buf 缓冲区；如果不足以存放响应头部与响应体，则会回调两次，分别为响应头部

ByteBuf、响应体 ByteBuf。源码如下。

```java
public abstract class HttpObjectEncoder<H extends HttpMessage> extends MessageToMessageEncoder
<Object> {
    // 实现具体编码操作,msg 为消息体,编码后的数据放入 List<Object> out 列表
    protected void encode(ChannelHandlerContext ctx, Object msg, List<Object> out) throws Exception {
        ByteBuf buf = null;
        if (msg instanceof HttpMessage) {                    // 编码响应头
            ...
            H m = (H) msg;
            buf = ctx.alloc().buffer();
            encodeInitialLine(buf, m);                       // 编码响应行
            encodeHeaders(m.headers(), buf);                 // 编码响应头部
            buf.writeBytes(CRLF);                            // 写入换行符
            ...
        }
        ...
        if (msg instanceof HttpContent || msg instanceof ByteBuf || msg instanceof FileRegion) { // 编码响应体
            final long contentLength = contentLength(msg);   // 计算响应体长度
            if (state == ST_CONTENT_NON_CHUNK) {
                if (contentLength > 0) {
                    // 将响应体转换为 ByteBuf,并写入 buf
                    if (buf != null && buf.writableBytes() >= contentLength && msg instanceof HttpContent) {
                        buf.writeBytes(((HttpContent) msg).content());
                        out.add(buf);
                    } else {
                        if (buf != null) {
                            out.add(buf);
                        }
                        out.add(encodeAndRetain(msg));
                    }
                } else {                                     // 不存在 contentLength,直接将 buf 放入 out 列表
                    if (buf != null) {
                        out.add(buf);
                    } else {
                        out.add(EMPTY_BUFFER);
                    }
                }
            }
            ...
        } ...
    }
}
```

HttpContentEncoder 继承自 HttpObjectEncoder,实际编码操作均由 HttpObjectEncoder 类完成。该类实现了 acceptOutboundMessage 方法和 encodeInitialLine 方法,判断当前编码消息在父类判断类型的基础上添加了 HttpRequest 类型校验,因为编码器接收的对象肯定不能为 HttpRequest 对象。源码如下。

```java
public class HttpContentEncoder extends HttpObjectEncoder<HttpContent> {
    private static final byte[] CRLF = { CR, LF };

    // 不能接收 HttpRequest 对象,因为这是请求消息对象
```

```java
public boolean acceptOutboundMessage(Object msg) throws Exception {
    return super.acceptOutboundMessage(msg) && !(msg instanceof HttpRequest);
}

// 编码响应行信息
protected void encodeInitialLine(ByteBuf buf, HttpContent response) throws Exception {
    response.protocolVersion().encode(buf);
    buf.writeByte(SP);
    response.status().encode(buf);
    buf.writeBytes(CRLF);
}
}
```

14.10 MessageAggregator 消息对象聚合器

由 HttpObjectDecoder 可知，消息体分为多个部分时，会解析出多个 Message 对象放入 out 列表中，然后回调多次后面的通道处理器对象；如果业务处理器只需要一个对象，就可以使用消息对象聚合器将多个消息聚合为一个对象，再回调后面的处理器。

例如这个责任链 handler-A->handler-B。如果不使用聚合器，则为 handler-A(a, b, c)->handler-B(a, b, c)，handler-A 解码出多个对象，handler-B 就会需要回调很多次，这时就需要 handler - B 自己处理这分开的消息对象，而如果使用聚合器：handler-A -> MessageAggregator -> handler-B，则：handler-A (a, b, c) -> MessageAggregator(a, b, c) -> handler-B(abc)，这时 MessageAggregator 回调多次，但是 handler-B 只需要回调一次。由源码可知。

（1）O currentMessage 用于保存所有消息聚合后的对象。

（2）decode 方法调用子类完成实际聚合过程。

（3）没有到达 LastContentMessage 前，将部分内容缓冲区对象 partialContent 放入 currentMessage 组合缓冲区中。

（4）到达 LastContentMessage 后，将聚合完成的对象放入 out 列表中，传递给下一个通道处理器。

```java
// 泛型 I 表示开始消息与内容消息的类型的父类类型（例如，HttpRequest 和 HttpContent 都为 HttpObject）；泛型 S 表示开始消息类型；泛型 C 表示内容类型；泛型 O 表示聚合后的消息类型
public abstract class MessageAggregator<I, S, C extends ByteBufHolder, O extends ByteBufHolder>
    extends MessageToMessageDecoder<I> {
    private O currentMessage;                                    // 当前消息对象
    protected abstract boolean isStartMessage(I msg) throws Exception;   // 子类实现判断 msg 是否为开始消息
    protected abstract boolean isContentMessage(I msg) throws Exception; // 子类实现判断 msg 是否为内容消息
    protected abstract boolean isLastContentMessage(C msg) throws Exception; // 子类实现判断 msg 是否为最后一个内容消息
    protected abstract boolean isAggregated(I msg) throws Exception;     // 子类实现判断 msg 是否已经聚合
    protected abstract O beginAggregation(S start, ByteBuf content) throws Exception; // 子类完成实际聚合过程
    protected void aggregate(O aggregated, C content) throws Exception {}  // 钩子函数，聚合完成后回调
```

```
// 实现解码过程，I msg 表示上一个通道处理器解码的对象，List<Object> out 表示当前聚合器聚合完成后将
对象放入其中
protected void decode(final ChannelHandlerContext ctx, I msg, List<Object> out) throws Exception {
    O currentMessage = this.currentMessage;
    if (isStartMessage(msg)) {                              // msg 为开始消息
        ...
        S m = (S) msg;
        ...
        // 将 CompositeByteBuf content 与 msg 进行聚合，同时将聚合后的对象放入 currentMessage 变量
        CompositeByteBuf content = ctx.alloc().compositeBuffer(maxCumulationBufferComponents);
        if (m instanceof ByteBufHolder) {
            appendPartialContent(content, ((ByteBufHolder) m).content());
        }
        this.currentMessage = beginAggregation(m, content);
        ...
    } else if (isContentMessage(msg)) {                     // msg 为消息内容对象
        final C m = (C) msg;
        final ByteBuf partialContent = ((ByteBufHolder) msg).content();
        final boolean isLastContentMessage = isLastContentMessage(m);
        ...
        CompositeByteBuf content = (CompositeByteBuf) currentMessage.content(); // 获取当前组合缓冲区
        ...
        appendPartialContent(content, partialContent);   // 将不是最后一个内容体的 partialContent 放入
组合缓冲区中
        // 调用子类完成消息聚合 currentMessage 与 msg
        aggregate(currentMessage, m);
        ...
        if (last) {       // 当前 msg 为最后一个内容体消息，调用钩子函数后，将聚合后的消息放入 out 列表
中，传递给下一个通道处理器
            finishAggregation(currentMessage);
            out.add(currentMessage);
            this.currentMessage = null;
        }
    } else {
        throw new MessageAggregationException();
    }
}
```

14.11　HttpObjectAggregator 聚合器

HttpObjectAggregator 类继承自 MessageAggregator，完成 MessageAggregator 类定义的模板方法，这里指定泛型为 HttpObject、HttpMessage、HttpContent、FullHttpMessage，表示将 HttpMessage 与 HttpContent 聚合为 FullHttpMessage，这时业务处理器就可以直接获取 FullHttpMessage 的实例完成业务，不用关心对于 HTTP 的聚合。源码如下。

```
public class HttpObjectAggregator extends MessageAggregator<HttpObject, HttpMessage, HttpContent, FullHttpMessage> {
```

```java
    // 初始聚合时调用, 这里 HttpMessage 与 ByteBuf, 分别为 HTTP 请求头与 HTTP 内容字节缓冲区
    protected FullHttpMessage beginAggregation(HttpMessage start, ByteBuf content) throws Exception {
        ...
        AggregatedFullHttpMessage ret;
        if (start instanceof HttpRequest) {              // 聚合 HTTP 请求头
            ret = new AggregatedFullHttpRequest((HttpRequest) start, content, null);
        } else if (start instanceof HttpContent) {       // 聚合 HTTP 响应头
            ret = new AggregatedFullHttpContent((HttpContent) start, content, null);
        } else {                                          // 不支持其他类型
            throw new Error();
        }
        return ret;
    }

    // 中间聚合时使用, 聚合 LastHttpContent 与 FullHttpMessage 对象
    protected void aggregate(FullHttpMessage aggregated, HttpContent content) throws Exception {
        if (content instanceof LastHttpContent) {
            ((AggregatedFullHttpMessage) aggregated).setTrailingHeaders(((LastHttpContent) content).trailingHeaders());
                                                          // 将 content 设置为 FullHttpMessage 的最后一个缓冲区
        }
    }
}
```

14.12 TCP 粘包原理

TCP 协议是网络通信协议中十分重要的协议。与 UDP 协议不同的是, 它是一个可靠的传输协议, 并且是面向数据流的协议。面向数据流是指数据以流式的方式传输, 传输的数据都是二进制且没有数据边界。UDP 协议则是面向数据包的, 收发的数据包与数据包之间是有明显的数据边界的。例如, 传输 aaabbbccc, 如果使用 TCP 协议, 传输的是 aaabbbccc; 如果是 UDP 协议, 传输的是 aaa (一个数据包)、bbb (一个数据包)、ccc (一个数据包)。

这时就可以引入粘包的概念。一堆数据同时在一个缓冲区中, 需要从中分辨出哪几个数据是一组数据。例如, 上面的 aaa bbb ccc 均为一组, 而在 TCP 中, 由于流式传输, 无法从缓冲区中分辨出哪里为数据的边界。

所以可以从 TCP/IP 四层协议以及 Socket 基础知识中得出结论。粘包情况发生在发送缓冲区或接收缓冲区中。应用程序从缓冲区中取数据时, 就有可能多个数据同时存在缓冲区(如 aaabbbccc), 而 TCP 是流式的, 数据无边界, 这时发生粘包。继续来看在发送方和接收方发生粘包的情况以及如何解决。

1. 发送方缓冲区产生粘包

采用 TCP 协议传输数据的客户端与服务器经常是保持一个长连接的状态, 双方在连接不断开的情况下, 可以不断传输数据, 但当发送方发送的数据太小时, TCP 协议默认启用 Nagle 算法, 将这些较小的数据包进行合并发送(将数据暂时保存在 Socket 的写缓冲区中), 这个合并过程发生在发送缓冲区中, 数据在发送到对端时就已经是粘包的状态了。这种情况很容易解决, 可以使用 TCP_NODELAY 选项禁用 Nagle 算法。Nagle 算法的规则(可参考 tcp_output.c 文件里 tcp_nagle_check 函数注释)如下。

(1) 如果包长度达到 MSS, 则允许发送。

（2）如果该包中有 FIN，则允许发送。

（3）设置了 TCP_NODELAY 选项，则允许发送。

（4）未设置 TCP_CORK 选项时，如果所有发出去的小数据包（包长度小于 MSS）均被确认，则允许发送。

（5）上述条件都未满足，但发生了超时（一般为 200ms），则立即发送。

2．接收方缓冲区产生粘包

接收方采用 TCP 协议接收数据时，将从网络模型的下方将数据逐层解包后，传递至传输层，传输层的 TCP 协议处理是将其放置在 Socket 接收缓冲区，然后由应用层来主动获取（C 语言中可用 recv、read 等函数。这时就会出现一个问题，在程序中调用的读取数据函数不能及时地把缓冲区中的数据取出来，而下一个数据又到来并有一部分放入缓冲区末尾，等读取数据时就是一个粘包，即存放缓冲区的速度大于应用层获取缓冲区数据的速度。解决方案如下。

（1）解析数据包头部信息，根据长度来接收。

（2）短连接传输，建立一次连接，只传输一次数据就关闭

对于第二种方案而言，连接的频繁释放和建立的消耗是不能接收的，所以一般采用第一种解决方案。

这时可以在 aaabbbccc 中加入分隔符。例如，aaa，这时不管是发送方还是接收方发生粘包，只需要按照分隔符来拆包即可。读者一看便知，这其实就是 HTTP 协议。HTTP 协议本身不会出现粘包的原因便在于此。同样，在 Netty 中给使用方提供了一些通用的解码器用于便捷解决粘包问题。

14.13　DelimiterBasedFrameDecoder 解码器

DelimiterBasedFrameDecoder 通用分隔符解码器类用一个或多个分隔符分割接收到的 ByteBuf 中的字节数据。它对于解码使用分隔符结束的帧特别有用。例如，Delimiters 静态类下提供的 delimiter.nulDelimiter 方法或 delimiter.lineDelimiter 方法返回的分隔符 ByteBuf 数组。

```java
public final class Delimiters {
    // 获取 NUL (0x00)分隔符 ByteBuf 数组
    public static ByteBuf[] nulDelimiter() {
        return new ByteBuf[] {
            Unpooled.wrappedBuffer(new byte[] { 0 }) };
    }

    // 获取 CR ('\r')与 LF ('\n')分隔符 ByteBuf 数组
    public static ByteBuf[] lineDelimiter() {
        return new ByteBuf[] {
            Unpooled.wrappedBuffer(new byte[] { '\r', '\n' }),
            Unpooled.wrappedBuffer(new byte[] { '\n' }),
        };
    }
}
```

DelimiterBasedFrameDecoder 允许指定一个以上的分隔符。如果在缓冲区中发现多个分隔符，则选择产生最短帧的分隔符。例如，如果在缓冲区中有以下数据，假如要使用 Delimiters.lineDelimiter()方法

分割数据，由于 Delimiters.lineDelimiter()方法有两个分隔符："、"和","这里使用 ABC、DEF，而不是使用 ABC。

```
+--------------+
| ABC\nDEF\r\n |
+--------------+
```

由于该类的实现较为简单，所以笔者这里将其统一讲解。由源码可知。
（1）当分隔符使用换行符时，实际解码操作使用 LineBasedFrameDecoder 类完成。
（2）用于分割数据帧的分隔符存放在 ByteBuf[] delimiters 数组中。
（3）可以指定当前分出出的数据帧的最大长度，使用 int maxFrameLength 变量保存。
（4）可以通过 failFast 变量设置是否快速失败，即当不能使用分隔符分割太长数据帧时，是否立即抛出 TooLongFrameException。如果设置 failFast 变量为 false 时，当遇到不能分割的数据大于设置的 maxFrameLength 最大帧长度时，则不会抛出异常，而是静默丢弃缓冲区数据，等待下一次缓冲区中有数据可以分割时，再抛出异常。
（5）在分割后，可以使用 stripDelimiter 标志位决定是否需要在分割后的数据帧中保留分隔符。

```java
public class DelimiterBasedFrameDecoder extends ByteToMessageDecoder {
    private final ByteBuf[] delimiters;            // 用于分割数据的分隔符数组，每个 ByteBuf 代表一个分隔符
    private final int maxFrameLength;              // 最大帧长度
    private final boolean stripDelimiter;          // 标识解码后的帧是否应该去掉分隔符
    private final boolean failFast;                // 标识快速失败
    private boolean discardingTooLongFrame;        // 是否丢弃过长的帧
    private int tooLongFrameLength;                // 保存当前丢弃最大帧的长度
    private final LineBasedFrameDecoder lineBasedDecoder; // 仅当以\n 或\r\n 作为分隔符进行解码时，设置该对象作为行解析对象

    // 构造器，指定支持的最大帧长度 maxFrameLength、解码后的帧是否应该去掉分隔符 stripDelimiter、快速失败 failFast（如果 failFast 指定为 true，一旦解码器注意到帧的长度将超过 maxFrameLength，则抛出 TooLongFrameException，无论当前是否已经读取了整个帧。如果 failFast 指定为 false，则在读取超过 maxFrameLength 的整个帧后抛出 TooLongFrameException）、分隔符 delimiters
    public DelimiterBasedFrameDecoder(
            int maxFrameLength, boolean stripDelimiter, boolean failFast, ByteBuf... delimiters) {
        validateMaxFrameLength(maxFrameLength);              // 最大帧长度不能小于 0
        if (delimiters == null) {                            // 分隔符不能为空
            throw new NullPointerException("delimiters");
        }
        if (delimiters.length == 0) {                        // 存在分隔符
            throw new IllegalArgumentException("empty delimiters");
        }
        if (isLineBased(delimiters) && !isSubclass()) {      // 当前分隔符为行分隔符，且当前类不属于 DelimiterBasedFrameDecoder 的子类，使用 LineBasedFrameDecoder 作为解码器
            lineBasedDecoder = new LineBasedFrameDecoder(maxFrameLength, stripDelimiter, failFast);
            this.delimiters = null;
        } else {        // 否则创建 delimiters 分隔符数组，将分隔符解析为 ByteBuf 放入数组
            this.delimiters = new ByteBuf[delimiters.length];
            for (int i = 0; i < delimiters.length; i ++) {
                ByteBuf d = delimiters[i];
                validateDelimiter(d);
                this.delimiters[i] = d.slice(d.readerIndex(), d.readableBytes());
```

```java
        }
        lineBasedDecoder = null;
    }
    this.maxFrameLength = maxFrameLength;
    this.stripDelimiter = stripDelimiter;
    this.failFast = failFast;
}

// 如果分隔符为\n 或\r\n，则返回 true
private static boolean isLineBased(final ByteBuf[] delimiters) {
    if (delimiters.length != 2) {                    // 长度大于 2，则必定不是\n 或\r\n
        return false;
    }
    // 比较两个 ByteBuf 中是否为\n 或\r\n
    ByteBuf a = delimiters[0];
    ByteBuf b = delimiters[1];
    if (a.capacity() < b.capacity()) {
        a = delimiters[1];
        b = delimiters[0];
    }
    return a.capacity() == 2 && b.capacity() == 1
        && a.getByte(0) == '\r' && a.getByte(1) == '\n'
        && b.getByte(0) == '\n';
}

// 查看当前对象的类是否为 DelimiterBasedFrameDecoder 的子类
private boolean isSubclass() {
    return getClass() != DelimiterBasedFrameDecoder.class;
}

// 实现父类 ByteToMessageDecoder 定义的 decode 方法，实际调用 decode(ctx, in)完成解码，如果不为空，
则添加到保存解码结果的 out 列表中
protected final void decode(ChannelHandlerContext ctx, ByteBuf in, List<Object> out) throws Exception {
    Object decoded = decode(ctx, in);
    if (decoded != null) {
        out.add(decoded);
    }
}

// 完成实际解码操作，并返回解码后的结果
protected Object decode(ChannelHandlerContext ctx, ByteBuf buffer) throws Exception {
    if (lineBasedDecoder != null) {                  // 行解码器不为空，则直接调用行解码器完成解码
        return lineBasedDecoder.decode(ctx, buffer);
    }
    int minFrameLength = Integer.MAX_VALUE;          // 最小帧长度
    ByteBuf minDelim = null;                         // 产生最小帧的分隔符
    for (ByteBuf delim: delimiters) {                // 遍历分隔符数组，使用 indexOf 完成解码，查看解码
后帧的长度，选取产生最小帧的分隔符
        int frameLength = indexOf(buffer, delim);
        if (frameLength >= 0 && frameLength < minFrameLength) {
            minFrameLength = frameLength;
            minDelim = delim;
        }
```

```java
        }
        if (minDelim != null) {                              // 存在分割数据的分隔符
            int minDelimLength = minDelim.capacity();
            ByteBuf frame;
            if (discardingTooLongFrame) {    // 之前设置过需要丢弃太长的帧，由于当前传入的 buffer 中已经
// 存在可以分割的数据，所以还原 discardingTooLongFrame 为 false，还原 this.tooLongFrameLength 为 0。同时，
// 虽然当前 buffer 中存在可以分割的数据，但是由于之前发生过丢弃长帧，但是并没有设置 failFast，所以这里需要
// 抛出异常，当然如果之前设置过 failFast 为 true，则之前已经抛出异常通知调用方，所以这里返回 null 即可
                discardingTooLongFrame = false;
                buffer.skipBytes(minFrameLength + minDelimLength);
                int tooLongFrameLength = this.tooLongFrameLength;
                this.tooLongFrameLength = 0;
                if (!failFast) {                              // 没有使用快速失败，则抛出异常
                    fail(tooLongFrameLength);
                }
                return null;
            }
            if (minFrameLength > maxFrameLength) {  // 如果当前可以分割出的最小数据帧长度大于设置的
// 最大帧限制，则丢弃读到的帧
                buffer.skipBytes(minFrameLength + minDelimLength); // 修改 readIndex 下标为 readIndex +
// minFrameLength + minDelimLength
                fail(minFrameLength);                         // 抛出异常
                return null;
            }
            if (stripDelimiter) {                             // 分割出的帧需要去掉分隔符
                frame = buffer.readRetainedSlice(minFrameLength);
                buffer.skipBytes(minDelimLength);
            } else {                                          // 分割出的帧不需要去掉分隔符，保留分隔符
                frame = buffer.readRetainedSlice(minFrameLength + minDelimLength);
            }
            return frame;
        } else {                                              // 不存在用于分割数据的分隔符
            if (!discardingTooLongFrame) {      // 初始为 false，表明再次分割时，是否直接丢弃太长的帧
                if (buffer.readableBytes() > maxFrameLength) { // 缓冲区中的数据长度大于设置的最大帧长度
                    // 直接丢弃 buffer 缓冲区中的内容，并设置 discardingTooLongFrame 为 true；如果设置
// 快速失败，则抛出异常
                    tooLongFrameLength = buffer.readableBytes();   // 保存当前丢弃的最大帧长度
                    buffer.skipBytes(buffer.readableBytes());      // 丢弃所有缓冲区中的数据
                    discardingTooLongFrame = true;
                    if (failFast) {
                        fail(tooLongFrameLength);
                    }
                }
            } else {
                // discardingTooLongFrame 为 true，依旧丢弃当前缓冲区中的数据，因为此时没有在数据缓冲
// 区中找到可以分割的数据
                tooLongFrameLength += buffer.readableBytes();
                buffer.skipBytes(buffer.readableBytes());
            }
            return null;
        }
    }
```

```java
// 抛出 TooLongFrameException 太长帧异常
private void fail(long frameLength) {
    if (frameLength > 0) {
        throw new TooLongFrameException(
            "frame length exceeds " + maxFrameLength +
            ": " + frameLength + " - discarded");
    } else {
        throw new TooLongFrameException(
            "frame length exceeds " + maxFrameLength +
            " - discarding");
    }
}

// haystack 表示输入数据缓冲区，needle 表明分隔符，返回 haystack 中找到的第一个匹配 needle 的
// readerIndex 下标。如果没有找到，则返回-1
private static int indexOf(ByteBuf haystack, ByteBuf needle) {
    for (int i = haystack.readerIndex(); i < haystack.writerIndex(); i ++) {    // 遍历 haystack 缓冲区可读字节
                                                                                 // 与 ByteBuf 中的分割符进行比较
        int haystackIndex = i;
        int needleIndex;
        for (needleIndex = 0; needleIndex < needle.capacity(); needleIndex ++) {    // 与 needle 中的每个
                                                                                     // 字节进行比较
            if (haystack.getByte(haystackIndex) != needle.getByte(needleIndex)) {
                break;
            } else {
                haystackIndex ++;
                if (haystackIndex == haystack.writerIndex() &&
                    needleIndex != needle.capacity() - 1) {
                    return -1;
                }
            }
        }
        if (needleIndex == needle.capacity()) {                    // 找到匹配的读索引下标
            return i - haystack.readerIndex();
        }
    }
    return -1;
}

// 分隔符缓冲区不能为空，且存在可读数据
private static void validateDelimiter(ByteBuf delimiter) {
    if (delimiter == null) {
        throw new NullPointerException("delimiter");
    }
    if (!delimiter.isReadable()) {
        throw new IllegalArgumentException("empty delimiter");
    }
}

// 最大帧不能设置为小于或等于 0
private static void validateMaxFrameLength(int maxFrameLength) {
    if (maxFrameLength <= 0) {
        throw new IllegalArgumentException(
```

```
            "maxFrameLength must be a positive integer: " +
            maxFrameLength);
        }
    }
}
```

14.14 FixedLengthFrameDecoder 类

FixedLengthFrameDecoder 定长解码器类比较简单，它将接收到的 ByteBuf 按固定字节数分割。例如，如果收到了以下四个分片报文，使用 FixedLengthFrameDecoder 时，指定 3 个 byte 为一帧，分割为 ABC、DEF、GHI 三个帧。

```
+---+----+------+----+
| A | BC | DEFG | HI |
+---+----+------+----+
```

该类实现很简单，这里直接一并讲解。通过源码可知，frameLength 变量指定按多少字节为一个数据帧进行分割，通过构造器设置。源码如下。

```
public class FixedLengthFrameDecoder extends ByteToMessageDecoder {
    private final int frameLength;
    public FixedLengthFrameDecoder(int frameLength) {
        if (frameLength <= 0) {
            throw new IllegalArgumentException(
                "frameLength must be a positive integer: " + frameLength);
        }
        this.frameLength = frameLength;
    }

    // 实现父类 ByteToMessageDecoder 定义的 decode 方法，实际调用 decode(ctx, in)完成解码，如果不为空，
    则添加到保存解码结果的 out 列表中
    protected final void decode(ChannelHandlerContext ctx, ByteBuf in, List<Object> out) throws Exception {
        Object decoded = decode(ctx, in);
        if (decoded != null) {
            out.add(decoded);
        }
    }

    // 完成实际编码，ByteBuf 中的数据不足以组成一帧时返回空，下一次数据到来时由父类 ByteToMessageDecoder
    再次回调 decode 方法完成解码
    protected Object decode(
        @SuppressWarnings("UnusedParameters") ChannelHandlerContext ctx, ByteBuf in) throws
Exception {
        if (in.readableBytes() < frameLength) {
            return null;
        } else {
            return in.readRetainedSlice(frameLength); // 按长度切割返回，使用 readRetainedSlice 方法，生
成新的 ByteBuf 时将 in 中的 readerIndex 加上 frameLength
        }
    }
}
```

14.15　LengthFieldBasedFrameDecoder 解码器

LengthFieldBasedFrameDecoder 长度属性解码器类实现了基于长度属性解码的解码器，它根据消息中的长度字段的值，动态分割接收到的 ByteBuf 中的数据。它在需要解码二进制的消息时特别有用，可以在二进制消息中保存代表消息体或整个消息长度的整数报头字段，然后根据该字段对数据进行分割。

LengthFieldBasedFrameDecoder 类有如下配置参数。

（1）maxFrameLength：解码成功后的数据帧的最大长度，如果帧的长度大于这个值，则抛出 TooLongFrameException 异常。

（2）lengthFieldOffset：数据帧中 length 字段的偏移量。

（3）lengthFieldLength：数据帧中 length 字段的长度。

（4）lengthAdjustment：要添加到长度字段计算的补偿值。

（5）initialBytesToStrip：在解码成功后，去掉已解码的帧中的头部字节数。

所以它可以解码任何带有长度字段的二进制消息，这通常在特定的 c/s 架构协议中查看。下面来看一个例子，length 字段的值为 12 (0x0C)，表示 "HELLO, WORLD" 字符串的长度。在默认情况下，解码器假定长度字段表示长度字段后面的字节数。因此，可以用简单的参数组合进行解码。源码如下。

```
lengthFieldOffset   = 0      // 表明当前长度属性在数据帧的第 1 个字节处
lengthFieldLength   = 2      // 表明当前长度属性占用 2B
lengthAdjustment    = 0      // 不需要添加到长度字段的补偿值
initialBytesToStrip = 0      // 表明解码后的帧不略过头部信息

解码前（14 bytes）              解码后（14 bytes）
+--------+----------------+     +--------+----------------+
| Length | Actual Content |---->| Length | Actual Content |
| 0x000C | "HELLO, WORLD" |     | 0x000C | "HELLO, WORLD" |
+--------+----------------+     +--------+----------------+
```

再来看以下例子。可以通过指定 initialBytesToStrip 变量来剥离解码后的数据帧的长度字段。在本例中，指定 initialBytesToStrip = 2，这与 length 字段的长度相同，所以去除解码后的 "HELLO, WORLD" 数据帧的前两个长度字节。详细实现如下。

```
lengthFieldOffset   = 0      // 表明当前长度属性在数据帧的第 1 个字节处
lengthFieldLength   = 2      // 表明当前长度属性占用 2B
lengthAdjustment    = 0      // 表明解码后的帧不略过头部信息
initialBytesToStrip = 2      // 表明解码后的帧略过 2B 的头部信息（等同于去掉长度字段）
解码前（14B）                    解码后（12B）
+--------+----------------+     +----------------+
| Length | Actual Content |---->| Actual Content |
| 0x000C | "HELLO, WORLD" |     | "HELLO, WORLD" |
+--------+----------------+     +----------------+
```

继续看以下例子。在大多数情况下，用 length 字段仅表示消息正文的长度，如前例所示。然而，在某些协议中，Length 字段可以表示整个消息的长度，即 length 等于消息头加消息正文。在这种情况下，指定一个非零的 lengthAdjustment 值表示当前内容中消息正文的偏移值。例如，在该示例消息中的长度值总是比正文长度大 2，所以指定-2 作为 lengthAdjustment 变量值。详细实现如下。

```
lengthFieldOffset     = 0      // 表明当前长度属性在数据帧的第 1 个字节处
lengthFieldLength     = 2      // 表明当前长度属性占用 2B
lengthAdjustment      = -2     // 表明当前长度的偏移值（可以用 Length 加 lengthAdjustment 得到消息正文的
                                  起始位置，即帧的实际长度）
initialBytesToStrip   = 0      // 表明解码后的帧不略过头部信息
解码前（14B）                    解码后（14B）
+--------+----------------+    +--------+----------------+
| Length | Actual Content |--->| Length | Actual Content |
| 0x000E | "HELLO, WORLD" |    | 0x000E | "HELLO, WORLD" |
+--------+----------------+    +--------+----------------+
```

再来看如下例子。该示例中，输入的消息是第一个示例的简单变体，在消息前添加了额外的头部信息。这时，可以指定 lengthFieldOffset 为 2，lengthFieldLength 为 3。详细实现如下。

```
lengthFieldOffset = 2          // 表明当前长度属性在数据帧的第 3 个字节处（前 2 个字节用于表示头部信息）
lengthFieldLength = 3          // 表明当前长度属性占用 3 个字节
lengthAdjustment  = 0
initialBytesToStrip = 0
解码前（17B）                                      解码后（17B）
+----------+----------+----------------+         +----------+----------+----------------+
| Header 1 |  Length  | Actual Content |-------->| Header 1 |  Length  | Actual Content |
|  0xCAFE  | 0x00000C | "HELLO, WORLD" |         |  0xCAFE  | 0x00000C | "HELLO, WORLD" |
+----------+----------+----------------+         +----------+----------+----------------+
```

再来看如下例子。在该示例中，展示了长度字段和消息体之间有一个额外头部信息的情况，这时必须指定一个正的 lengthAdjustment 值，以便解码器将额外的头部信息大小计算到帧的长度中。详细实现如下。

```
lengthFieldOffset   = 0        // 表明当前长度属性在数据帧的第 1 个字节处
lengthFieldLength   = 3        // 表明当前长度属性占用 3 个字节
lengthAdjustment>   = 2        // 表明当前长度值后面存在 2 字节的头部信息
initialBytesToStrip = 0
解码前（17B）                                      解码后（17B）
+----------+----------+----------------+         +----------+----------+----------------+
|  Length  | Header 1 | Actual Content |-------->|  Length  | Header 1 | Actual Content |
| 0x00000C |  0xCAFE  | "HELLO, WORLD" |         | 0x00000C |  0xCAFE  | "HELLO, WORLD" |
+----------+----------+----------------+         +----------+----------+----------------+
```

再来看如下例子。该示例是上面所有例子的组合，在长度字段之前有一个前缀头信息 HDR1，在长度字段之后有一个额外的头信息 HDR2。HDR1 头信息影响 lengthFieldOffset 变量，HDR2 信息影响 lengthAdjustment 变量。还指定了非零的 initialbytesodstrip 从解码后的数据帧中去除 Length 长度字段和前缀头信息 HDR1。当然，如果不想去掉这两个信息，可以指定 initialBytesToStrip 变量值为 0。详细实现如下。

```
lengthFieldOffset    = 1       // 表明当前长度属性在数据帧的第 2 个字节处（前面包含一个字节的头部信息）
lengthFieldLength    = 2       // 表明当前长度属性占用 2B
lengthAdjustment     = 1       // 表明当前长度值后面存在 1B 的头部信息
initialBytesToStrip = 3        // 表明解析后的数据帧需要移除 3B 的头信息（Length + HDR1）
```

```
解码前 (16B)                                 解码后 (13B)
+------+--------+------+----------------+    +------+----------------+
| HDR1 | Length | HDR2 | Actual Content |---→| HDR2 | Actual Content |
| 0xCA | 0x000C | 0xFE | "HELLO, WORLD" |    | 0xFE | "HELLO, WORLD" |
+------+--------+------+----------------+    +------+----------------+
```

再来看最后一个例子。现在，对上一个例子做一点改动，length 字段表示整个消息的长度，而不是消息正文。必须将 HDR1 和 length 的长度和放入 lengthAdjustment 变量中。注意，这里不需要考虑 HDR2 的长度，因为长度字段已经包含了整个报头长度。详细实现如下。

```
lengthFieldOffset   = 1     // 表明当前长度属性在数据帧的第 2 个字节处（前面包含一个字节的头部信息）
lengthFieldLength   = 2     // 表明当前长度属性占用 2B
lengthAdjustment    = -3    // 表明当前长度的偏移值（可以用 HDR1+ length 得到消息正文的起始位置）
initialBytesToStrip = 3     // 表明解析后的数据帧需要移除 3B 的头信息（Length 加 HDR1）
```

```
解码前 (16B)                                 解码后 (13B)
+------+--------+------+----------------+    +------+----------------+
| HDR1 | Length | HDR2 | Actual Content |---→| HDR2 | Actual Content |
| 0xCA | 0x0010 | 0xFE | "HELLO, WORLD" |    | 0xFE | "HELLO, WORLD" |
+------+--------+------+----------------+    +------+----------------+
```

学习了以上例子之后，读者对于该类的变量的定义和原理已不再陌生了。该类实现也较为简单，笔者将其一并讲解。由源码可知。

（1）使用 maxFrameLength 表示解码成功后的数据帧的最大长度，如果帧的长度大于这个值，抛出 TooLongFrameException 异常。

（2）使用 lengthFieldOffset 表示数据帧中 length 字段的偏移量。

（3）使用 lengthFieldLength 表示数据帧中 length 字段的长度。

（4）使用 lengthFieldOffset 加 lengthFieldLength 定位缓冲区中 length 的字段值。

（5）使用 lengthFieldEndOffset 表示数据帧中 length 字段的长度字节的结束偏移量。

（6）使用 lengthAdjustment 表示添加到帧长度计算的补偿值。

（7）使用 initialBytesToStrip 表示在解码成功后，去掉已解码的帧中的头部字节数。

（8）使用 failFast 表示是否快速失败。

（9）解码操作流程如下。

① 如果 discardingTooLongFrame 为 true，表明当前需要丢弃太长的帧，此时调用 failIfNecessary 方法，根据 failFast 变量判断是否需要抛出异常。

② 如果 ByteBuf 中的数据还未完全接收（数据长度未达到长度偏移量），则返回 null，等待下一次数据回调时再进行解码。

③ 通过缓冲区的读下标加长度属性偏移量，得到最终的处于 ByteBuf 中的实际 length 下标的索引值，并调用 getUnadjustedFrameLength 方法获取 length 字段中描述的帧长度。

④ 对 length 字段中指定的数据帧长度进行调整，通过 ByteBuf in 中 length 字段中的值加长度调整值加 length 字段实际占用的空间，得到整个包含 length 的数据帧长度。

⑤ 如果调整后的帧长度小于长度属性的结束偏移值，说明 lengthAdjustment 设置有误，则抛出 CorruptedFrameException 异常。

⑥ 如果调整过后的帧超过设置支持的最大帧长度，则丢弃当前接收到的数据，同时调用

failIfNecessary 方法判断需要快速失败，即抛出异常。

⑦ 如果缓冲区中没有 frameLength 长度的数据，则返回 null，等待下次通道读到数据时再提取帧。

⑧ 通过 initialBytesToStrip 变量略过 initialBytesToStrip 字节数。

⑨ 从缓冲区中提取数据帧，并返回。

对于丢弃模式，读者可以这么理解，假如有如下数据需要发送（包含两个帧）：length（占用 2B）aabbcc length（占用 2B）dd，如果此时缓冲区中只有 length aa，而此时 length 解析出来为 6。但是由于此时 maxFrameLength 设置为 2，由于 length 已经超出了最大帧长度，此时要做的就是丢弃后面的 6 个字节 aabbcc。但是由于当前缓冲区中只有 aa，其他数据还没有接收到，此时就需要设置丢弃模式了，直到丢弃到 cc，此时就需要重置丢弃模式和其他变量了，因为新的帧 length（占用 2B）dd 并没有超过最大帧长度。源码如下。

```java
public class LengthFieldBasedFrameDecoder extends ByteToMessageDecoder {
    private final ByteOrder byteOrder;                    // 当前字节序，默认 ByteOrder BIG_ENDIAN = new ByteOrder("BIG_ENDIAN") 大端序
    private final int maxFrameLength;                     // 解码成功后的数据帧的最大长度，如果帧的长度大于这个值，将抛出 TooLongFrameException 异常
    private final int lengthFieldOffset;                  // 数据帧中 length 字段的偏移量
    private final int lengthFieldLength;                  // 数据帧中 length 字段的长度
    private final int lengthFieldEndOffset;               // byte数据数组中长度字节的结束偏移量（lengthFieldOffset 加 lengthFieldLength）
    private final int lengthAdjustment;                   // 要添加到帧长度计算的补偿值
    private final int initialBytesToStrip;                // 在解码成功后，去掉已解码的帧中的头部字节数
    private final boolean failFast;                       // 是否快速失败，如果为 true，当解码器注意到解码后的帧长度将超过 maxFrameLength 时，将抛出 TooLongFrameException 异常，而不管是否已经读取了整个帧。如果为 false，则在整个帧超过 maxFrameLength 被读取后抛出 TooLongFrameException 异常
    private boolean discardingTooLongFrame;               // 用于支持 failFast
    private long tooLongFrameLength;                      // 当前解码成功后的超出 maxFrameLength 的帧长度
    private long bytesToDiscard;                          // 当缓冲区中没有包含所有帧数据，但是由于超过了最大帧长度而丢弃的字节数

    // 构造器：初始化 lengthAdjustment 和 initialBytesToStrip 默认值 0
    public LengthFieldBasedFrameDecoder(
        int maxFrameLength,
        int lengthFieldOffset, int lengthFieldLength) {
        this(maxFrameLength, lengthFieldOffset, lengthFieldLength, 0, 0);
    }

    // 构造器：初始化 failFast 默认值 true
    public LengthFieldBasedFrameDecoder(
        int maxFrameLength,
        int lengthFieldOffset, int lengthFieldLength,
        int lengthAdjustment, int initialBytesToStrip) {
        this(
            maxFrameLength,
            lengthFieldOffset, lengthFieldLength, lengthAdjustment,
            initialBytesToStrip, true);
    }
```

```java
// 构造器：初始化 byteOrder 默认值 BIG_ENDIAN 大端序
public LengthFieldBasedFrameDecoder(
    int maxFrameLength, int lengthFieldOffset, int lengthFieldLength,
    int lengthAdjustment, int initialBytesToStrip, boolean failFast) {
    this(
        ByteOrder.BIG_ENDIAN, maxFrameLength, lengthFieldOffset, lengthFieldLength,
        lengthAdjustment, initialBytesToStrip, failFast);
}

// 完整构造器：参数校验后，进行实例变量赋值
public LengthFieldBasedFrameDecoder(
    ByteOrder byteOrder, int maxFrameLength, int lengthFieldOffset, int lengthFieldLength,
    int lengthAdjustment, int initialBytesToStrip, boolean failFast) {
    if (byteOrder == null) {
        throw new NullPointerException("byteOrder");
    }

    if (maxFrameLength <= 0) {
        throw new IllegalArgumentException(
            "maxFrameLength must be a positive integer: " +
            maxFrameLength);
    }

    if (lengthFieldOffset < 0) {
        throw new IllegalArgumentException(
            "lengthFieldOffset must be a non-negative integer: " +
            lengthFieldOffset);
    }

    if (initialBytesToStrip < 0) {
        throw new IllegalArgumentException(
            "initialBytesToStrip must be a non-negative integer: " +
            initialBytesToStrip);
    }

    if (lengthFieldOffset > maxFrameLength - lengthFieldLength) {
        throw new IllegalArgumentException(
            "maxFrameLength (" + maxFrameLength + ") " +
            "must be equal to or greater than " +
            "lengthFieldOffset (" + lengthFieldOffset + ") + " +
            "lengthFieldLength (" + lengthFieldLength + ").");
    }

    this.byteOrder = byteOrder;
    this.maxFrameLength = maxFrameLength;
    this.lengthFieldOffset = lengthFieldOffset;
    this.lengthFieldLength = lengthFieldLength;
    this.lengthAdjustment = lengthAdjustment;
    this.lengthFieldEndOffset = lengthFieldOffset + lengthFieldLength; // 通过 byte 数据数组中的 length 偏移量和 length 占用字节数计算出 byte 数据数组中长度字节的结束偏移量
    this.initialBytesToStrip = initialBytesToStrip;
    this.failFast = failFast;
}

// 实现父类的解码操作，当通道读取到数据时，由事件循环线程调用，解码结果放到 List<Object> out 列表中
```

```java
protected final void decode(ChannelHandlerContext ctx, ByteBuf in, List<Object> out) throws Exception {
    Object decoded = decode(ctx, in);           // 该方法完成实际解码
    if (decoded != null) {
        out.add(decoded);
    }
}

// 对 ByteBuf 中的数据进行解码
protected Object decode(ChannelHandlerContext ctx, ByteBuf in) throws Exception {
    if (discardingTooLongFrame) {               // 当前需要丢弃过长的帧
        long bytesToDiscard = this.bytesToDiscard;
        int localBytesToDiscard = (int) Math.min(bytesToDiscard, in.readableBytes());
        in.skipBytes(localBytesToDiscard);
        bytesToDiscard -= localBytesToDiscard;
        this.bytesToDiscard = bytesToDiscard;
        failIfNecessary(false);  // 当 failFast 为 false 时，探测到超出长度的数据帧产生时，不会立即抛出异常，而是等待解码完毕后再抛出异常。所以这里 firstDetectionOfTooLongFrame 设置为 false，肯定不是第一次探测到超出长度的数据帧
    }
    if (in.readableBytes() < lengthFieldEndOffset) {   // ByteBuf 中的数据还未完全接收（数据长度未达到长度偏移量）
        return null;
    }
    int actualLengthFieldOffset = in.readerIndex() + lengthFieldOffset;   // 通过缓冲区的读下标加长度属性偏移量，得到最终的处于 ByteBuf 中的实际 length 下标的索引值
    long frameLength = getUnadjustedFrameLength(in, actualLengthFieldOffset, lengthFieldLength, byteOrder);                     // 获取未调整的数据帧长度
    if (frameLength < 0) {   // 解析后的数据帧不能小于 0，用于约束子类重写 getUnadjustedFrameLength 方法时返回小于 0 的情况
        in.skipBytes(lengthFieldEndOffset);
        throw new CorruptedFrameException(
            "negative pre-adjustment length field: " + frameLength);
    }
    frameLength += lengthAdjustment + lengthFieldEndOffset;   // 调整数据帧长度：ByteBuf in 中 length 字段中的值加长度调整值加 length 字段实际占用的空间，得到整个包含 length 的数据帧长度

    if (frameLength < lengthFieldEndOffset) {   // 调整后的帧长度小于长度属性的结束偏移值，说明 lengthAdjustment 设置有误（注意，笔者写到这里时发现了一个问题：如果缓冲区中没有足够的字节，则会再次调用该方法，会重新计算 frameLength，这是没有必要的。于是笔者提了一个 PR，而且最终被采纳修复了。详见：https://github.com/Netty/Netty/issues/12171）
        in.skipBytes(lengthFieldEndOffset);
        throw new CorruptedFrameException(
            "Adjusted frame length (" + frameLength + ") is less " +
            "than lengthFieldEndOffset: " + lengthFieldEndOffset);
    }

    if (frameLength > maxFrameLength) {           // 调整过后的帧超过设置支持的最大帧长度
        long discard = frameLength - in.readableBytes();   // 用于判断当前 buf 中是否包含了整个数据帧，因为计算数据帧长度时是根据 length 字段计算的，所以有可能 buf 中并没有读取到 frameLength 长度的数据
        tooLongFrameLength = frameLength;           // 保存的当前超过 maxFrameLength 的帧长度

        if (discard < 0) {   // buf 缓冲区中包含的字节比 frameLength 多，可以丢弃 buf 中的整个数据帧
            in.skipBytes((int) frameLength);
```

```java
        } else { // buf 缓冲区中没有包含 frameLength 长度的数据,设置 discardingTooLongFrame 为 true,
进入丢弃模式,丢弃掉到目前为止收到的所有内容
            discardingTooLongFrame = true;
            bytesToDiscard = discard;
            in.skipBytes(in.readableBytes());
        }
        failIfNecessary(true);  // 通过设置判断当前是否需要快速失败,firstDetectionOfTooLongFrame 为
true,表明当前第一次探测到超过 maxFrameLength 的帧
        return null;
    }

    // 最大数据帧的长度应该在整型范围内,这里直接取低 32 位即可
    int frameLengthInt = (int) frameLength;
    if (in.readableBytes() < frameLengthInt) {   // 缓冲区中没有这么多数据,则返回 null,等待下次通道读
到数据时再提取帧
        return null;
    }
    // in 缓冲区中包含了 frameLengthInt 的数据,判断需要略过数据帧的头部字节数是否大于数据帧长度,
这属于非法状态
    if (initialBytesToStrip > frameLengthInt) {
        in.skipBytes(frameLengthInt);
        throw new CorruptedFrameException(
            "Adjusted frame length (" + frameLength + ") is less " +
            "than initialBytesToStrip: " + initialBytesToStrip);
    }
    in.skipBytes(initialBytesToStrip);  // 略过 initialBytesToStrip 字节数,通常用该字段去掉缓冲区中的长度
信息:length 字段

    // 从缓冲区中提取数据帧,并返回
    int readerIndex = in.readerIndex();
    int actualFrameLength = frameLengthInt - initialBytesToStrip;
    ByteBuf frame = extractFrame(ctx, in, readerIndex, actualFrameLength);
    in.readerIndex(readerIndex + actualFrameLength);
    return frame;
}

// 将 ByteBuf buf 缓冲区的指定区域解码为未调整的帧长度(actualLengthFieldOffset(ByteBuf 中 length 的
下标)+ lengthFieldLength(length 占用字节数)= length 的结束下标->读出 length 的值)。当前实现将指定的区
域解码为无符号的 8/16/24/32/64 位整数。子类重写此方法以解码以不同方式编码的长度字段。注意,该方法不能
修改指定缓冲区的状态 readerIndex、writerIndex,因为后续还需要继续通过这些状态读取缓冲区内容,而这里只
是探测数据帧的长度
protected long getUnadjustedFrameLength(ByteBuf buf, int offset, int length, ByteOrder order) {
    buf = buf.order(order);     // 设置字节序
    long frameLength;
    switch (length) {           // 根据 length 占用的字节数读取当前数据帧的长度信息
        case 1:                 // length 占用的字节数为 1B
            frameLength = buf.getUnsignedByte(offset);
            break;
        case 2:                 // length 占用的字节数为 2B
            frameLength = buf.getUnsignedShort(offset);
            break;
        case 3:                 // length 占用的字节数为 3B
            frameLength = buf.getUnsignedMedium(offset);
```

```java
                    break;
                case 4:                    // length 占用的字节数为 4B
                    frameLength = buf.getUnsignedInt(offset);
                    break;
                case 8:                    // length 占用的字节数为 8B
                    frameLength = buf.getLong(offset);
                    break;
                default:                   // 不支持其他长度的帧
                    throw new DecoderException(
                        "unsupported lengthFieldLength: " + lengthFieldLength + " (expected: 1, 2, 3, 4, or 8)");
            }
            return frameLength;
        }

        // 在必要时，抛出 TooLongFrameException 异常
        private void failIfNecessary(boolean firstDetectionOfTooLongFrame) {
            if (bytesToDiscard == 0) {                    // bytesToDiscard 表示丢弃字节数为 0，此时缓冲区
// 中包含了所有帧数据，不设置该值。此时，上一个超过最大帧的数据包已经丢弃，只需要根据 failFast 变量和
// firstDetectionOfTooLongFrame 变量决定是否抛出异常，并重置变量到初始状态（因为下一个数据包可能不会超
// 过帧的最大值）
                long tooLongFrameLength = this.tooLongFrameLength;
                this.tooLongFrameLength = 0;
                discardingTooLongFrame = false;
                if (!failFast ||
                    failFast && firstDetectionOfTooLongFrame) {
                    fail(tooLongFrameLength);
                }
            } else {        // 此时缓冲区没有包含所有帧数据，继续丢弃接收到的数据，并在必要时通知调用方，
// 即指定了快速失败，同时为第一次探测到数据帧太长（满足 failFast 快速失败语义，一旦发现解码后帧超过
// maxFrameLength，则立即抛出异常）
                if (failFast && firstDetectionOfTooLongFrame) {
                    fail(tooLongFrameLength);
                }
            }
        }

        // 抛出异常，frameLength 表示当前丢弃的帧长度
        private void fail(long frameLength) {
            if (frameLength > 0) {
                throw new TooLongFrameException(
                    "Adjusted frame length exceeds " + maxFrameLength +
                    ": " + frameLength + " - discarded");
            } else {
                throw new TooLongFrameException(
                    "Adjusted frame length exceeds " + maxFrameLength +
                    " - discarding");
            }
        }

    // 直接使用 retainedSlice 方法，从 ByteBuf buffer 中分割出 index 下标 - index + length 下标区间的数据
    protected ByteBuf extractFrame(ChannelHandlerContext ctx, ByteBuf buffer, int index, int length){
        return buffer.retainedSlice(index, length);
    }
}
```

14.16 LineBasedFrameDecoder 解码器

相对于 LengthFieldBasedFrameDecoder 来说，LineBasedFrameDecoder 行分隔符解码器类非常简单，它支持以 " " 将 ByteBuf 中的数据分割为数据帧。其实就是一个特殊的 DelimiterBasedFrameDecoder，在 DelimiterBasedFrameDecoder 中也看到，使用 " " 和 " " 作为分隔符完成实际操作。解码流程如下。

（1）获取到缓冲区中行尾的索引下标。

（2）获取数据帧长度。

（3）如果当前没有处于丢弃模式，则判断数据帧长度是否超过 maxLength，设置缓冲区读下标略过当前帧长度，并回调通道处理器的 fireExceptionCaught 方法处理 TooLongFrameException 异常，如果数据帧长度没有超过 maxLength，则根据 stripDelimiter 变量决定是否去掉换行符，并返回解码后的数据帧对象。

（4）如果当前缓冲区中没有存在换行符，则判断当前缓冲区中的数据已经超过设置的最大数据帧长度，此时丢弃该缓冲区的字节，并设置丢弃模式 discarding 为 true。如果指定了快速失败，则通知通道处理器处理 TooLongFrameException 异常。

（5）如果当前处于丢弃模式，则判断当前缓冲区中是否已经包含了换行符。如果包含换行符，则重置变量，并且退出丢弃模式；如果没有包含换行符，则说明当前数据仍处于上一个超过最大长度的帧，记录丢弃的字节数，不会重置变量。

```java
public class LineBasedFrameDecoder extends ByteToMessageDecoder {
    private final int maxLength;            // 支持的最大帧长度
    private final boolean failFast;         // 是否在超过 maxLength 时抛出异常
    private final boolean stripDelimiter;   // 是否将解码后的帧去除分隔符
    private boolean discarding;  // 是否进入丢弃模式,表明当前数据帧已经超过了maxLength,但是failFast为false
    private int discardedBytes;             // 丢弃的字节数

    // 构造器：指定 stripDelimiter 为 true，failFast 为 false
    public LineBasedFrameDecoder(final int maxLength) {
        this(maxLength, true, false);
    }

    // 初始化成员变量
    public LineBasedFrameDecoder(final int maxLength, final boolean stripDelimiter, final boolean failFast) {
        this.maxLength = maxLength;
        this.failFast = failFast;
        this.stripDelimiter = stripDelimiter;
    }

    // 重写父类解码方法，将解码后的数据帧放入 out 列表
    protected final void decode(ChannelHandlerContext ctx, ByteBuf in, List<Object> out) throws Exception {
        Object decoded = decode(ctx, in);
        if (decoded != null) {
            out.add(decoded);
        }
    }
}
```

```java
// 实际解码操作
protected Object decode(ChannelHandlerContext ctx, ByteBuf buffer) throws Exception {
    final int eol = findEndOfLine(buffer);     // 获取到缓冲区中行尾的索引下标
    if (!discarding) {                          // 没有处于丢弃模式
        if (eol >= 0) {                         // 存在数据帧
            final ByteBuf frame;
            final int length = eol - buffer.readerIndex();          // 数据帧长度
            final int delimLength = buffer.getByte(eol) == '\r'? 2 : 1;   // 计算分隔符长度：'\r' ? '\r\n'

            if (length > maxLength) {    // 超过最大长度，设置缓冲区读下标略过当前帧长度，并回调通道处理器的 fireExceptionCaught 方法处理 TooLongFrameException
                buffer.readerIndex(eol + delimLength);
                fail(ctx, length);
                return null;
            }
            if (stripDelimiter) {            // 不保留分隔符
                frame = buffer.readRetainedSlice(length);
                buffer.skipBytes(delimLength);
            } else {
                frame = buffer.readRetainedSlice(length + delimLength);
            }
            return frame;
        } else {                                // 在当前缓冲区数据中没有找到分割符
            final int length = buffer.readableBytes();
            if (length > maxLength) {    // 当前缓冲区中的数据已经超过设置的最大数据帧长度，此时丢弃该缓冲区的字节，并设置丢弃模式 discarding 为 true，如果指定了快速失败，则通知通道处理器处理 TooLongFrameException 异常
                discardedBytes = length;
                buffer.readerIndex(buffer.writerIndex());
                discarding = true;
                if (failFast) {
                    fail(ctx, "over " + discardedBytes);
                }
            }
            return null;
        }
    } else {                                    // 丢弃模式
        if (eol >= 0) {                         // 当前缓冲区中已经包含了换行符，重置变量，并且退出丢弃模式（因为可能下一个帧就没有超过 maxLength）
            final int length = discardedBytes + eol - buffer.readerIndex();
            final int delimLength = buffer.getByte(eol) == '\r'? 2 : 1;
            buffer.readerIndex(eol + delimLength);
            discardedBytes = 0;
            discarding = false;
            if (!failFast) {
                fail(ctx, length);
            }
        } else {                                // 当前缓冲区中还是没有换行符，说明当前数据仍处于上一个超过最大长度的帧，记录丢弃的字节数，不会重置变量
            discardedBytes += buffer.readableBytes();
            buffer.readerIndex(buffer.writerIndex());
        }
        return null;
```

```java
    }
}

// 回调其他通道处理器处理 TooLongFrameException 异常

private void fail(final ChannelHandlerContext ctx, int length) {
    fail(ctx, String.valueOf(length));
}

private void fail(final ChannelHandlerContext ctx, String length) {
    ctx.fireExceptionCaught(
        new TooLongFrameException(
            "frame length (" + length + ") exceeds the allowed maximum (" + maxLength + ')'));
}

// 返回在 ByteBuf buffer 缓冲区中找到的行尾的索引下标，如果在缓冲区中没有找到行尾，则返回-1
private static int findEndOfLine(final ByteBuf buffer) {
    int i = buffer.forEachByte(ByteProcessor.FIND_LF);
    if (i > 0 && buffer.getByte(i - 1) == '\r') {      // 查看前一个字节是不是\r
        i--;
    }
    return i;
}

// 字节处理器：在发现 LF ('\n')时终止，并返回下标
ByteProcessor FIND_LF = new IndexOfProcessor((byte) '\n');
```

14.17 ReplayingDecoder 解码器

上述讲解的解码器都是用于解析数据帧的，而本节介绍的 ReplayingDecoder 可重放解码器是开发者自定义解码器的模板类。它是特殊的 ByteToMessageDecoder 解码器。开发者在实现 decode()方法时，可以假定当前已经接收到所有必需的字节，而不是判断缓冲区中是否存在满足长度的字节。下面来看一个使用 ByteToMessageDecoder 时的例子，在该例子中，需要从 ByteBuf buf 缓冲区中获取到 4B，但可能该缓冲区中没有这么多字节的数据，那么就需要直接返回，在下一次通道中存在满足长度的数据时才执行业务。代码如下：

```java
public class IntegerHeaderFrameDecoder extends ByteToMessageDecoder {
    protected void decode( ChannelHandlerContext ctx, ByteBuf buf, List<Object> out ) throws Exception {
        if (buf.readableBytes() < 4) {           // 缓冲区中的数据长度小于4
            return;
        }
        // 标记当前 readIndex 下标
        buf.markReaderIndex();
        int length = buf.readInt();              // 读取 4B 的数据，该数据表示后面数据的长度
        if (buf.readableBytes() < length) {      // 长度不满足要求，则重置索引并返回，等待 ByteBuf buf 中包含需要的数据
            buf.resetReaderIndex();
            return;
```

```
        out.add(buf.readBytes(length));    // 读取长度为 length 的数据放入 out 列表,表明读取成功
    }
}
```

上面的操作较为烦琐,如果使用 ReplayingDecoder 作为模板类,一切就变得非常简单了,只需要假定当前缓冲区中存在需要的数据即可,即直接执行业务逻辑。代码如下。

```
public class IntegerHeaderFrameDecoder extends ReplayingDecoder<Void> {
    protected void decode(ChannelHandlerContext ctx,ByteBuf buf) throws Exception {
        out.add(buf.readBytes(buf.readInt()));
    }
}
```

现在学习一下 ReplayingDecoder 是如何实现这么强大的功能的。ReplayingDecoder 使用了特殊的 ByteBuf,当缓冲区中没有足够的数据时,该 ByteBuf 会抛出一个特定类型的错误。在上面的 IntegerHeaderFrameDecoder 例子中,只需要假设调用 buffer.readint()时缓冲区中会有 4B 或更多字节。如果缓冲区中真的有 4B,它会像期望的那样返回需要的数据。否则,将引发 Error 并返回到 ReplayingDecoder 类中进行处理。如果 ReplayingDecoder 捕获了 Error,将把缓冲区的 readerIndex 回退到初始位置(即缓冲区的开始位置),并在缓冲区接收到更多数据时再次调用 decode(..)方法。

但便捷的同时肯定也会带来一些限制。这些限制如下。

(1)一些 ByteBuf 缓冲区的方法禁止调用。

(2)如果网络很慢,并且消息格式很复杂(不像上面的例子仅读取长度字节),性能可能会更差。在这种情况下,继承该类实现的解码器可能需要一遍又一遍地解码消息的同一部分。例如,当前存在 4B,先解析 4B,然后突然发现又需要 8B,但是由于没有这么多数据,则抛出异常,继续下次回调时,先解析 4B,然后又解析 8B,以此类推,太过复杂的解析与重复工作带来的性能损耗是非常大的。

(3)所以在使用该类时必须记住,在解码单个消息时,decode(..)方法可以被多次调用。

接下来来看以下代码,在 decode(.., ByteBuf buf, List out)方法中假定缓冲区中存在了 8B 的数据,此时理应通过,但由于 decode 方法会在缓冲区中的数据不足 8B 时调用多次,队列中可能会放置多个数据,则会抛出异常。代码如下。

```
public class MyDecoder extends ReplayingDecoder<Void>    {
    private final Queue Integer values = new LinkedList<Integer>();
    public void decode(.., ByteBuf  buf, List<Object> out) throws Exception {
        // 读取 8B 的数据
        values.offer(buf.readInt());
        values.offer(buf.readInt());
        // 这里的断言可能会失败,因为 decode 方法会在缓冲区中的数据不足 8B 时多次调用,这时队列中可能
会放置多个数据,则会抛出异常
        assert values.size() == 2;
        out.add(values.poll() + values.poll());
    }
}
```

正确的做法如下,在方法的调用处,还原自上次部分解码以来可能已更改的变量的状态,即队列状态,这时不会发生任何异常,代码如下。

```
public class MyDecoder extends ReplayingDecoder<Void>    {
    private final Queue Integer values = new LinkedList<Integer>();
```

```
public void decode(.., ByteBuf buf, List<Object>  out) throws Exception {
    // 还原自上次部分解码以来可能已更改的变量的状态
    values.clear();
    // 读取 8B 的数据
    values.offer(buf.readInt());
    values.offer(buf.readInt());
    // 由于在开始时清空了队列，即还原了初始状态，这里不会发生异常
    assert values.size() == 2;
    out.add(values.poll() + values.poll());
}
```

ReplayingDecoder 类提供了使用 state 状态变量与 checkpoint 方法，供子类方便快捷且显著提高复杂解码器实现的性能。checkpoint 方法可以保存缓冲区的读下标 readIndex 的初始位置，以便 ReplayingDecoder 将缓冲区的 readerIndex 回退到调用 checkpoint 方法的最后一个位置。来看如下例子，这里定义了一个 MyDecoderState 枚举类表示解码器状态，在构造器中设置初始状态为读取长度，读取完消息的长度后，调用 checkpoint(MyDecoderState.READ_CONTENT)，将状态修改为 READ_CONTENT，在下一次调用 decode 方法时，会进入解码消息内容状态。代码如下。

```
public enum MyDecoderState {                                // 解码器状态枚举
    READ_LENGTH,
    READ_CONTENT;
}
public class IntegerHeaderFrameDecoder extends ReplayingDecoder<MyDecoderState> {
    private int length;
    public IntegerHeaderFrameDecoder() {                    // 设置初始状态为读取长度
        super(MyDecoderState.READ_LENGTH);
    }

    // 实现解码过程
    protected void decode(ChannelHandlerContext ctx, ByteBuf buf, List<Object> out) throws Exception{
        switch (state()) {                                  // 根据状态执行不同操作
            case READ_LENGTH:
                !length = buf.readInt();
                checkpoint(MyDecoderState.READ_CONTENT);    // 设置状态为 READ_CONTENT，下次调用
该方法时将进入 case READ_CONTENT 执行
            case READ_CONTENT:
                ByteBuf frame = buf.readBytes(length);
                checkpoint(MyDecoderState.READ_LENGTH);
                out.add(frame);
                break;
            default:
                throw new Error("Shouldn't reach here.");
        }
    }
}
```

该类实现较为简单，笔者将其一并描述。由源码可知。

（1）该类使用了特殊的 ByteBuf 包装实际操作的缓冲区对象 ReplayingDecoderByteBuf 类，在其中的获取数据方法中对当前包装的 ByteBuf buffer 缓冲区中可读字节数判断长度要求。如果当前缓冲区中

的数据不足以满足需要的字节数，则抛出 ReplayingDecoder.REPLAY。

（2）该类使用 S state 变量方便子类完成解码的状态机。在每次调用 decode 方法时，根据当前状态实现逻辑切换。

（3）该类使用 checkpoint 变量保存调用子类 decode 方法前的缓冲区下标，当子类抛出 ReplayingDecoder.REPLAY 异常对象，表明缓冲区数据长度不足时，还原缓冲区的 readIndex 下标。

（4）子类可继承自该类并实现 decode(ChannelHandlerContext ctx, ByteBuf in, List out)方法完成自身解码逻辑。

```java
// 特殊的 ByteBuf，由 ReplayingDecoder 类使用。注意由于该类方法太多，所以只需要关注读取数据，了解原理即可
final class ReplayingDecoderByteBuf extends ByteBuf {
    private static final Signal REPLAY = ReplayingDecoder.REPLAY; // 保存当缓冲区中不存在需要的数据长度
                                                                   的数据时，则抛出该异常
    private ByteBuf buffer;        // 保存实际数据的 ByteBuf 缓冲区对象

    // 以读取 4B 数据为例，其他方法均是如此
    public int readInt() {
        checkReadableBytes(4);
        return buffer.readInt();
    }

    // 判断 buffer 中的数据长度是否大于 readableBytes，否则抛出 REPLAY 异常
    private void checkReadableBytes(int readableBytes) {
        if (buffer.readableBytes() < readableBytes) {
            throw REPLAY;
        }
    }
}

public abstract class ReplayingDecoder<S> extends ByteToMessageDecoder {
    static final Signal REPLAY = Signal.valueOf(ReplayingDecoder.class, "REPLAY");  // 特殊异常对象，由于
                                                                                     表明当前缓冲区中没有需要的数据
    private final ReplayingDecoderByteBuf replayable = new ReplayingDecoderByteBuf(); // 特殊的缓冲区对象
    private S state;                  // 用于管理解码器状态
    private int checkpoint = -1;      // 当前还原点，即缓冲区 readIndex 的下标

    // 创建没有初始状态的 ReplayingDecoder
    protected ReplayingDecoder() {
        this(null);
    }

    // 创建包含初始状态的 ReplayingDecoder
    protected ReplayingDecoder(S initialState) {
        state = initialState;
    }

    // 不保存 state 状态，但是保存当前操作的还原点，即缓冲区的 readIndex 下标
    protected void checkpoint() {
        checkpoint = internalBuffer().readerIndex();
    }
```

```java
// 保存 state 状态以及缓冲区的 readIndex 下标
protected void checkpoint(S state) {
    checkpoint();
    state(state);
}

protected S state() {
    return state;
}

// 设置新的状态，并返回旧状态
protected S state(S newState) {
    S oldState = state;
    state = newState;
    return oldState;
}

// 通道输入端关闭时回调
final void channelInputClosed(ChannelHandlerContext ctx, List<Object> out) throws Exception {
    try {
        replayable.terminate();              // 关闭缓冲区包装对象
        if (cumulation != null) {            // 组合缓冲区对象不为空，调用子类完成缓冲区剩余字节数解码
            callDecode(ctx, internalBuffer(), out);
            decodeLast(ctx, replayable, out); // 通知当前缓冲区为最后一个解码的缓冲区对象
        } else {                             // 设置 replayable 中包装的缓冲区对象为空缓冲区对象，调用
// decodeLast，表示当前为最后一个数据帧（空帧）
            replayable.setCumulation(Unpooled.EMPTY_BUFFER);
            decodeLast(ctx, replayable, out);
        }
    } catch (Signal replay) {                // 如果异常对象不是 REPLAY 对象，则抛出 IllegalStateException 异常，因
// 为子类只能复用 REPLAY 异常对象
        replay.expect(REPLAY);
    }
}

// 重写父类 callDecode 方法，而不是 decode(ChannelHandlerContext ctx, ByteBuf in, List<Object> out)方
// 法，实现自身基于状态 state 和还原点的解码逻辑
protected void callDecode(ChannelHandlerContext ctx, ByteBuf in, List<Object> out) {
    replayable.setCumulation(in);            // 包装外部组合缓冲区 in
    try {
        while (in.isReadable()) {            // 循环处理输入缓冲区 in 中的数据
            int oldReaderIndex = checkpoint = in.readerIndex();   // 保存当前读下标作为还原点
            int outSize = out.size();
            if (outSize > 0) {               // out 列表中存在子类完成解码后的数据，回调通道处理器的
// ChannelRead 方法，处理解码后的对象
                fireChannelRead(ctx, out, outSize);
                out.clear();
                if (ctx.isRemoved()) {
                    break;
                }
                outSize = 0;
            }
```

```java
            S oldState = state;                                          // 保存当前状态
            int oldInputLength = in.readableBytes();
            try {
                decode(ctx, replayable, out);                            // 调用子类完成实际解码操作，这里传
入了 ReplayingDecoderByteBuf replayable 缓冲区
                if (ctx.isRemoved()) {
                    break;
                }
                if (outSize == out.size()) {                             // 子类未完成解码
                    if (oldInputLength == in.readableBytes() && oldState == state) {   // 如果子类未读
取缓冲区中的数据，同时状态也未改变，则抛出异常。因为子类必须实现解码操作：或者读取缓冲区中的数据，
或者修改状态
                        throw new DecoderException(
                                StringUtil.simpleClassName(getClass()) + ".decode() must consume the
inbound " + "data or change its state if it did not decode anything.");
                    } else {                                             // 否则继续循环解码
                        continue;
                    }
                }
            } catch (Signal replay) {
                replay.expect(REPLAY);                                   // 判断当前异常是否为 REPLAY 对象，如果不是，则抛出
new IllegalStateException("unexpected signal: " + signal)异常
                if (ctx.isRemoved()) {                                   // 判断是否已经从流水线中移除了当前解码器
                    break;
                }
                // 由于当前异常为 REPLAY，表示当前缓冲区中的数据不足以满足子类完成解码，所以返
回之前保存的判断点 readIndex
                int checkpoint = this.checkpoint;
                if (checkpoint >= 0) {
                    in.readerIndex(checkpoint);
                } else {                                                 // 不需要维护 readerIndex
                }
                break;
            }
            // 子类未读取缓冲区的字节数，同时没有改变状态，如同什么也没做，则抛出解码异常
            if (oldReaderIndex == in.readerIndex() && oldState == state) {
                throw new DecoderException(
                        StringUtil.simpleClassName(getClass()) + ".decode() method must consume the
inbound data " + "or change its state if it decoded something.");
            }
            if (isSingleDecode()) {                                      // 只解码一次，退出循环
                break;
            }
        }
    } catch (DecoderException e) {
        throw e;
    } catch (Throwable cause) {
        throw new DecoderException(cause);
    }
}
```

小结

通道管理层篇

本篇详细介绍了 Netty 中的通道管理原理。

Netty 中的通道处理器分为 Inbound 和 Outbound，分别用于表示输入通道处理器和输出通道处理器。使用责任链模式，将其放入 ChannelPipeline 流水线对象中管理，其中包含了两个特殊的通道处理器。

（1）HeadContext 主要完成 Outbound 处理器最终的输出行为，将写操作反馈到 Unsafe 底层的 Channel 通道中。

（2）ailContext 主要完成 Inbound 处理器的输入行为，将没有经过任何责任链的 Inbound 处理器的数据静默丢弃。

（3）Netty 的编码器将基于通道处理器进行封装处理，其中定义了诸多解决 TCP 粘包的编码器与解码器。

（4）业务通道处理器只需要放置在解码器与编码器之中完成业务处理即可，不需要考虑具体的编码与解码过程。